Genetic Counseling

Clinical Practice and Ethical Considerations

A subject collection from *Cold Spring Harbor Perspectives in Medicine*

Genetic Counseling

Clinical Practice and Ethical Considerations

A subject collection from *Cold Spring Harbor Perspectives in Medicine*

EDITED BY

Laura Hercher
Sarah Lawrence College

Barbara Biesecker
RTI International

Jehannine C. Austin
University of British Columbia

COLD SPRING HARBOR LABORATORY PRESS
Cold Spring Harbor, New York • www.cshlpress.org

Genetic Counseling: Clinical Practice and Ethical Considerations
A subject collection from *Cold Spring Harbor Perspectives in Medicine*
Articles online at www.perspectivesinmedicine.org

Executive Editor	Richard Sever
Managing Editor	Maria Smit
Senior Project Manager	Barbara Acosta
Permissions Administrator	Carol Brown
Production Editor	Kathleen Bubbeo
Production Manager/Cover Designer	Denise Weiss
Publisher	John Inglis

Front cover artwork: Image provided by Grommik/Shutterstock.com.

Library of Congress Cataloging-in-Publication Data

Names: Hercher, Laura, editor.
Title: Genetic counseling : clinical practice and ethical considerations / edited by Laura Hercher,
    Sarah Lawrence College, Barbara Biesecker, RTI International, and Jehannine C. Austin,
    University of British Columbia.
Description: Cold Spring Harbor, New York : Cold Spring Harbor Laboratory Press, [2019] |
    Series: Cold Spring Harbor perspectives in medicine | A subject collection from Cold Spring
    Harbor Perspectives in Medicine. | Includes bibliographical references and index. | Summary:
    "Genetic counseling is poised to play an increasingly important role in society as genome
    sequencing becomes a routine aspect of clinical assessment of both infants and adults. This
    volume examines the changing role of genetic counselors in the genomic age and the ways in
    which genetic counseling can be integrated into modern medicine"-- Provided by publisher.
Identifiers: LCCN 2019037657 (print) | LCCN 2019037658 (ebook) | ISBN 9781621823476
    (cloth) | ISBN 9781621823964 (paperback) | ISBN 9781621823483 (epub) | ISBN
    9781621823490 (kindle edition)
Subjects: LCSH: Genetic counseling.
Classification: LCC RB155.7 .G46235 2019 (print) | LCC RB155.7 (ebook) | DDC 362.196/042--
    dc23
LC record available at https://lccn.loc.gov/2019037657
LC ebook record available at https://lccn.loc.gov/2019037658

For a complete catalog of all Cold Spring Harbor Laboratory Press publications, visit our website at
www.cshlpress.org.

Contents

Genetic Counseling and the Central Tenets of Practice, 1
Barbara Biesecker

PRENATAL

Supporting Patient Autonomy and Informed Decision-Making in Prenatal Genetic Testing, 11
Katie Stoll and Judith Jackson

Impact of Emerging Technologies in Prenatal Genetic Counseling, 25
Blair Stevens

NEUROLOGY

Predictive Genetic Counseling for Neurodegenerative Diseases: Past, Present, and Future, 43
Jill S. Goldman

Genetic Counseling in Neurodevelopmental Disorders, 59
Alyssa Blesson and Julie S. Cohen

CANCER

Cancer Genetic Counseling—Current Practice and Future Challenges, 73
Jaclyn Schienda and Jill Stopfer

Tumor-Based Genetic Testing and Familial Cancer Risk, 97
Andrea Forman and Jilliane Sotelo

PSYCHIATRY

Evidence-Based Genetic Counseling for Psychiatric Disorders: A Road Map, 115
Jehannine C. Austin

Genetic Risk Assessment in Psychiatry, 129
Holly Landrum Peay

CARDIOLOGY

Psychological Issues in Managing Families with Inherited Cardiovascular Diseases, 141
Jodie Ingles

Contents

A Person-Centered Approach to Cardiovascular Genetic Testing, 155
Julia Platt

PEDIATRICS

Genetic Counseling and Genome Sequencing in Pediatric Rare Disease, 169
Alison M. Elliott

Bridging the Gap between Scientific Advancement and Real-World Application:
Pediatric Genetic Counseling for Common Syndromes and Single-Gene Disorders, 185
Julie A. McGlynn and Elinor Langfelder-Schwind

INFERTILITY

Genetic Counseling and Assisted Reproductive Technologies, 199
Debra Lilienthal and Michelle Cahr

ETHICAL, LEGAL, AND SOCIAL ISSUES

Discouraging Elective Genetic Testing of Minors: A Norm under Siege in a New Era
of Genomic Medicine, 213
Laura Hercher

Legal Challenges in Genetics, Including Duty to Warn and Genetic Discrimination, 225
Sonia Suter

Birds of a Feather? Genetic Counseling, Genetic Testing, and Humanism, 237
Robert Resta

Regulating Preimplantation Genetic Testing across the World: A Comparison
of International Policy and Ethical Perspectives, 249
Margaret E.C. Ginoza and Rosario Isasi

GENOMICS

Evolving Roles of Genetic Counselors in the Clinical Laboratory, 263
Megan T. Cho and Carrie Guy

Informed Consent in the Genomics Era, 275
Shannon Rego, Megan E. Grove, Mildred K. Cho, and Kelly E. Ormond

THE FUTURE OF THE PROFESSION

Genetic Counseling, Personalized Medicine, and Precision Health, 289
Erica Ramos

Index, 299

Genetic Counseling and the Central Tenets of Practice

Barbara Biesecker

RTI International, Washington, D.C. 20005, USA

Correspondence: bbiesecker@rti.org

Genetic counseling is a profession growing and evolving at an extraordinary rate. This growth is driven by an explosion in what we know, as a result of progress in science, technology, and bioinformatics, and an explosion in what we do not know, as we strive to understand the impact of genomic information on the lives of our patients and clients. Genetic counselors work in an increasing number of subspecialties and diversity of settings. But although the field has evolved, it has maintained a remarkably unchanged core of shared values and beliefs. The heart of genetic counseling practice is the therapeutic relationship, with its dual role of providing information and facilitating assimilation of that information to personalize health-related decision-making and foster successful adaptation. Genetic counseling aims to communicate cutting-edge genomic science within an empathic understanding of the client/patient's concerns and needs. In pursuit of these goals, further assessment of genetic counseling's effectiveness is needed to facilitate evidence-based practices and to scale counseling resources.

Genetic counseling is a rapidly evolving professional practice that has kept pace with emerging genetic technologies and related patient/client[1] needs. Translational genomics and personalized medicine are beginning to make their mark, and genetic counseling in various subspecialties has expanded accordingly. Regardless of the changes, what remains unwavering and central to the identity of the profession is the therapeutic relationship between the counselor and the patient/client. This approach originated with recognition of the sensitivities around genetic illness and risks and the resultant difficult decisions and uncertainties. Built upon the centrality of these concerns, a typical session would involve the counselor assessing the patient/client's genetics-related needs, exchanging and discussing relevant information, and eliciting the patient/client's relevant thoughts and feelings to advance the relationship to one of empathic understanding. Within this relationship, the patient/client can be helped to reach a goal, such as making an informed decision, assimilating effective coping strategies to facilitate adaptation to living with at an increased risk, gaining relevant personal insight into living

---

[1] "Patient/client" is used throughout as genetic counseling is provided to those who are ill and those who are healthy and may not even be at increased risk.

Cite this article as *Cold Spring Harb Perspect Med* doi: 10.1101/cshperspect.a038968

with a condition, or empowerment to enhance one's quality of life.

One way in which genetic counseling remains stubbornly difficult to define is the range of genetics-related needs that patients/clients bring to counseling, and the specialization of services provided. Reasons for counseling may vary from an elevated risk for an untreatable neurodevelopmental disorder to the routine discussion of a prenatal screening test. Although genetic testing may be on offer, genetic counseling is not exclusively linked to testing. Take, for example, the case of parents of a son with an undiagnosed complex developmental disability who was previously tested; these parents may seek genetic counseling to learn about his prognosis. The counselor would facilitate a discussion about their need for new information and contact with parents of similarly affected children. S/he would help them work through a variety of challenges, including tensions in their marriage related to the stress of coping with their son's disability and barriers to addressing their relatives' concerns about their risks for having a similarly affected child. Whereas communicating and clarifying an understanding of genetic information are common elements in genetic counseling, psychotherapeutic counseling to address resultant responses to the stress of living at risk of or with a genetic condition is, at times, even more pertinent to the job of addressing patient/client needs (Austin et al. 2014; Biesecker et al. 2017).

DELINEATING PRACTICE

Because of the variety of settings and patient/client needs and a limited evidence base, genetic counseling was previously described as a black box (Biesecker and Peters 2001). However, the National Society of Genetic Counselors (Resta et al. 2006) has successfully defined genetic counseling as the "*process of helping people understand and adapt to the medical, psychological and familial implications of genetic contributions to disease. This process integrates: Interpretation of family and medical histories to assess the chance of disease occurrence or recurrence; education about inheritance, testing, manage-* *ment, prevention, resources and research; and counseling to promote informed choices and adaptation to the risk or condition.*" This definition has been cited often and taught in most, if not all graduate programs across the globe (Resta et al. 2018). In 2007, a novel practice model for client-centered care, the reciprocal engagement model (REM), was developed by directors of North American graduate programs in genetic counseling (Veach et al. 2007). The REM has at its center the relationship between the counselor and patient. The tenets of practice proposed for the REM include:

1. Genetic information is key.

2. The relationship is integral.

3. Patient autonomy must be supported.

4. Patients are resilient.

5. Patient emotions make a difference.

To develop a fuller, more nuanced understanding of the tenets, I have extended them:

1. Information is key and must be accurate in reflecting cutting-edge genomic science.

2. The relationship is integral and should be psychotherapeutic.

3. Patient autonomy is important, but at times may be challenged in the interest of mitigating health risks.

4. Patients are resilient and capable, but some are vulnerable and need further interventions.

5. Patients' affect reflects only one of a triad of psychological responses to genetic information—cognition, affect, and behavior—all of which are important patient/client experiences to be addressed in genetic counseling.

These five tenets remain true to the field despite the ways it has adapted to technological advances in genomics. From its early days, when diagnoses and etiology of disease were hard to come by, genetic counseling came to include a diverse range of practices and specialty settings. Today, the diagnosis of genetic conditions occurs regularly, and risk information is more precise because of advances in genetic/genomic testing. As such, the roles and responsibilities of genetic counselors have advanced to include making medical recommendations (Matias et al. 2019; Subas et al. 2019; Zakas et al. 2019), working to modify health behaviors (Rutherford et al. 2014; Milliron et al. 2019), and facilitating communication and testing among at-risk relatives (Montgomery et al. 2013), while engaging in psychotherapeutic counseling (Austin et al. 2014; Biesecker et al. 2017). With the growth in genetic counseling have come expanded responsibilities and opportunities for creative use of familiar skills. Many of these broadening responsibilities continue to reflect client-centered care and the tenets of genetic counseling practice (Resta et al. 2006; Veach et al. 2007). Yet because of diversification and novel interactions with patients/clients, genetic counseling now also extends beyond patient/client-centered care. Aspects of what has historically been considered genetic counseling practice can be delivered by nongenetic counselor resources using, for example, artificial intelligence and digital platforms. These challenges to the REM and its original tenets are based on the need to be relevant across subspecialty settings and to meet growing patient/client needs, requiring genetic counseling to embrace alternative service delivery models and recognize resources to augment and even replace components of genetic counseling when patient/client needs can still be met (see Cho and Guy 2019; Ramos 2019). For example, many genetic counselors work in the laboratory setting; in 2018, ~22% worked for companies that provide genetic testing (https://www.nsgc.org/p/cm/ld/fid=68), complicating their practice with parallel roles as clinical experts and business representatives. Simultaneously, in cancer genetics and cardiovascular genetics, more accurate risk prediction is coupled with greater uncertainty. The health-threatening nature of the information and risks to close relatives generate fears, worries, and concerns that may inhibit adherence to medical recommendations and affect family relationships. As such, a psychotherapeutic approach to genetic counseling is often engaged in these

settings (see Ingles 2019; Platt 2019; Schienda and Stopfer 2019). Similarly, in the neurogenetics and neurodevelopmental subspecialties, the psychological impact of the conditions on individuals and families has long been recognized as profound (see Blesson and Cohen 2019; Goldman 2019). As such, the therapeutic relationship remains at the center of genetic counseling, even as the profession recognizes that less engagement with a patient/client may be appropriate in situations in which the information is less alarming or readily addressed.

GENETIC COUNSELING PRACTICE MODELS

To compare and contrast service delivery options, the roots of psychotherapeutic counseling are important considerations. Historically, Dr. Seymour Kessler described genetic counseling as composed of two practice models: a teaching model and a counseling model (Kessler 1997). These models have distinct goals that can be at cross purposes and must be integrated to achieve the goals of practice. The teaching model involves effective communication of information relevant to the patient/client. In this model, the counselor has more authority in the relationship and is looked to as an expert by the patient/client —as the one who knows the information sought. In processing the information, the patient/client experiences psychological responses. In the counseling model, the patient/client is the authority on her/his thoughts, feelings, and behaviors. The counselor guides her/him toward opportunities to gain personal insight, identify resources, and cope effectively with any health threat that presents. Integrating these two models to address a patient/client's needs is challenging but comprises a key professional skill in genetic counseling. Use of one model without the other is insufficient to successfully meet the patient/client's needs as defined in traditional client-centered care.

The integration of teaching and counseling practice models in genetic counseling practice is often referred to as psychoeducational (Biesecker et al. 2019). More recently the practice has been described as psychotherapeutic based on the similarity of genetic counseling to the definition of psychotherapy (Austin et al. 2014; Biesecker et al. 2017, 2019). Yet antithetical to these descriptors, evidence from some studies demonstrates that rather than integrating teaching and counseling models in practice, genetic counseling tends to favor the teaching model at the cost of the counseling model (Ellington et al. 2005; Meiser et al. 2008; Joseph and Guerra 2015; Joseph et al. 2017, 2019). Here, I describe genetic counseling as psychotherapeutic to highlight the importance of the counseling model and the significant limitations of genetic counseling when it is conceptualized and delivered as an encounter centered primarily on genetic information provision. This unilateral approach fails to attend to patients/clients' psychological responses and needs that are not remarkable in all cases, but are in many (Kessler 1997; Biesecker et al. 2017).

Despite the importance of therapeutic engagement with patients/clients, genetic counseling is more often dominated by "counselor-speak," which aims to communicate genetics concepts. As of August 2019, there is no published evidence explaining the prevalence of this phenomenon. Many opinion pieces have floated ideas about why counselors practice as such, including the rapid development of genomic science that focuses their attention on the science; their affinity for explaining how genetics "works" to their patients/clients; the complexity of addressing psychological responses that may feel at times intimidating; and the sense that they may feel less well-prepared by their graduate education to provide an integrated practice model. This pattern remains a significant concern and growing challenge for the field as patients/clients are demonstrably better served through empathic engagement. It is a particularly crucial point of consideration for the profession as alternative service delivery models are being introduced to replace components of genetic counseling.

Given the nature of genetic information and its close connection to health and wellness, the burden of disease, and the implications for relatives and future reproductive decisions, the likelihood for patients/clients to concern themselves with related thoughts, feelings, worries, and behaviors is high. Patients/clients should be offered sufficient opportunity to process information, deliberate on options, express their values and beliefs, and make informed decisions. Thus, actualizing a more balanced practice model that integrates both teaching and counseling is important in meeting the needs of many patients/clients.

STATE OF THE PROFESSION

Since Kessler introduced the two models of practice, the profession of genetic counseling has expanded rapidly, with recent growth fueled by findings from a 2016 workforce report commissioned by the National Society of Genetic Counselors (NSGC) that concluded there were an inadequate number of genetic counselors to meet growing service demands and patient/client needs (https://www.nsgc.org/page/genetic-counselor-workforce-initiatives-532, accessed July 1, 2019). In response to the identified need for additional genetic counselors, existing programs have expanded to take more students; 14 new programs have opened (bringing the total in North America to 49 as of July 2019) and at least three more are under development (https://www.gceducation.org/program-directory/, accessed July 1, 2019). The existing U.S. programs graduate about 200 genetic counselors annually, and the January 2019 membership of NSGC totaled nearly 5000 members. Genetic counselors work in increasingly diverse positions beyond the clinical setting, including, for example, as founders and CEOs of telegenetic counseling companies, research scientists, laboratory managers, and academic professors. This expansion has facilitated growth in our capacity to help patients/clients within expanded delivery models, contribute evidence to guide practice, and proffer expert interpretation of genomic data. Throughout this collection, the primary focus is on the practice of genetic counseling in direct patient care, although expanded roles are delineated. Although much of genetic counseling practice is not yet informed by an evidence base, increasingly there are emerging data by which to assess the effectiveness of genetic counseling.

Cite this article as *Cold Spring Harb Perspect Med* doi: 10.1101/cshperspect.a038968

ROLE OF GENETIC TESTING

Although genetic counseling in direct patient care and genetic testing have been intertwined since the start of the profession, the onslaught of technical discoveries has led to an era in which testing plays a far more prominent role. There are increasing ways for people to undergo testing without genetic counseling, via direct to consumer testing (e.g., 23&Me) and patient-initiated testing (https://www.invitae.com/en/patient-testing/). Although expanded testing options have greatly enhanced the likelihood that patients/clients learn more definitive information about their risk status, testing is not always an option, nor is it always a solution to the challenges that face patients/clients. Further, although genetic/genomic testing expands the resources available to genetic counseling patients/clients, increasing testing uptake is not and should not be the goal of genetic counseling.

Helping counselees to determine whether testing may be useful, or which test is indicated, often does fall to the expertise of the genetic counselor. Yet in genetic counseling, the goal is to promote informed decision-making. In prenatal testing, genetic counseling facilitates patients/clients' consideration of their personal values and beliefs and fosters deliberation of the relative benefits of undergoing testing to make a preference-based decision. In the end, genetic counseling in most settings comes down to helping patients/clients to understand what genetic vulnerability and/or risk means to them—how these issues are interpreted and valued, and how they may be managed for personal or family benefit. A stark exception is in recommending medically indicated testing that can mitigate health risk and reduce morbidity and mortality. Common examples are cancer screening recommendations for those at high heritable risk, and cardiac screening for those with an inherited cardiovascular condition (see Forman and Sotelo 2019; Platt 2019). As such, the outcomes of genetic counseling include its clinical utility in addition to personal utility in whether it helped to better understand health risks and how to cope with them when there is no follow-up screening or intervention to recommend (Koh-ler et al. 2017a,b), in addition to psychological well-being (Athens et al. 2018).

SCALING OF GENETIC COUNSELING

As the demand for genetic counseling services has rapidly outpaced its availability (https://www.nsgc.org/page/genetic-counselor-work force-initiatives-532, accessed July 1, 2019), a priority for the profession is the scalability of resources. To deal with the limited availability of services, triage of patient/client needs is a logical first step. Matching the service model to patient/client needs will be essential to effective outcomes. Yet evidence to guide triage is limited and there are constraints, primarily the challenges for patients/clients to identify the most appropriate providers within a fractionated health-care system with limited third-party coverage for genetic counseling services.

Patient/client-centered triage would prioritize genetic counseling for those with the greatest psychological need. Examples include those living with high levels of risk; handling personal experiences with illness, loss, and disability; dealing with complex inheritance of life-threatening conditions; and struggling to learn how to cope effectively with genetic risk information. Yet these notions of who may benefit most from clinical services are not evidence-based; this is research that needs to be done to ensure that those with the greatest need meet with a genetic counselor (Biesecker 2018).

Alternative Service Delivery Modes

Scaling access to genetic counseling involves increasing the number of genetic counselors, but also the introduction of alternative service delivery models. These will need to be embraced following assessment of how effective and how well-accepted they are as methods of delivering lower-risk genetic information. The shift in service delivery to include the use of interactive digital platforms to provide genetic information is already underway. Although research into alternative delivery models is ongoing via randomized controlled trials (Athens et al. 2018), evidence to inform targeted use of a variety of service delivery models across settings is limited. For now, existing evidence can be used to guide triage, but the array of service delivery models will need to be assessed for their effectiveness in meeting patient/client needs. Subspecialties are working to address the demand for services and their early efforts in the use of new technology to provide information and address the needs of their patients/clients.

Prior to 2018, 54 randomized controlled trials (RCTs) of genetic counseling over 25 years were published, according to a systematic literature review by Athens et al. (2018). More have been published since. Of the trials in the review, 31 compared an alternative service delivery model, such as telegenetic counseling, to usual care (in-person genetic counseling). The services provided via telegenetic counseling followed the same client-centered protocol as those provided in person. In each trial, the alternative delivery model was as effective, or noninferior, to usual care. One such RCT in the cancer genetics setting also demonstrated significant cost savings with telegenetic counseling (Buchanan et al. 2015). As 76% of these trials were conducted in the cancer genetics setting, use of telegenetic counseling services in this setting can accurately be discussed as an evidence-based practice equal in value to in-person counseling.

Other service delivery models aim to streamline components of genetic counseling practice. Digital platforms have been developed to communicate genetic information and its health implications without the involvement of a genetic counselor. A recent study comparing information delivery systems, including an online version that can be kept current, suggests alternative ways exist to meet some patients/clients' needs for certain types of information, particularly information that is less likely to lead to significant psychological responses (Persky et al. 2019).

Digital platforms have been studied over the past 20 years as an alternative way to return certain genetic risk information. In 2001, a randomized trial comparing the return of test results to low–moderate risk participants did so effectively without incurring psychological distress and to the patient/client's satisfaction

(Green et al. 2001). This finding supports substituting genetic counseling with well-designed educational resources with the residual option of following up with a genetic counselor with questions or concerns. Yet few studies have subsequently addressed the same question, and more evidence is needed to discern when this would be sufficient in meeting patient/client needs. More recently, platforms programmed to return less threatening risk information from genome sequencing, such as carrier status among postreproductive adults, have been found to be noninferior to in-person return (Biesecker et al. 2018; Lewis et al. 2018), further suggesting that there may not be a need for genetic counselors to return all types of genetic risk information. Although results from these trials are beginning to assess differences in patient/client needs, it remains difficult to parse patients/clients according to who is more likely to have significant psychological responses to their risk and health management needs and to ensure that they are not overlooked by an inadequate mode of service delivery.

Many of the attributes of digital platforms, such as dual-channel messaging—coupling audio to video—and use of quizzes to check for understanding, can enhance learning for patients/clients who benefit from taking in new information through multiple means (Kraft et al. 2017). Digital platforms can also extend access of services to patients/clients who may otherwise not be able to travel to a medical center. As such, they represent a potentially valuable resource for meeting the needs of the expanding number of people whose lives will be touched by genomics. For example, in a minority of prenatal settings, genetic counselors provide in-person pretest genetic counseling for noninvasive prenatal screening (NIPS) (see Stoll and Jackson 2019). Yet much of prenatal screening is being done by obstetricians who have less training in genetics and little time to explain the potential outcomes of screening. Because its preference-based nature is important to uphold, one could argue that a well-designed digital platform that includes a decision-making intervention to review personal values and beliefs could be widely used. If careful study proves it to be accessible, acceptable, and effective in teaching pregnant women and their partners about their screening options, it would help to address the workforce shortages. Genetic counselors could then focus their services on those patients who are struggling with decisions about screening and those who receive abnormal findings that require intensive discussion and deliberation to facilitate an informed decision on follow-up testing and pregnancy options.

Accordingly, Kuppermann et al. (2009, 2014) assessed a prenatal testing educational platform using an RCT design that compared outcomes to in-person genetic counseling. In the 2014 trial, both groups successfully learned key information about prenatal testing. Yet, fewer people who were randomized to the platform chose to undergo invasive prenatal testing, raising questions about whether this outcome reflected enhanced preference-based decision-making or potential internal biases in the program.

A more comprehensive service is offered by Counsyl via an interactive digital platform, Counsyl Complete. It uses video to teach pregnant women about NIPS and includes a portal for return of results (Arjunan et al. 2019). The woman's obstetrics provider must authorize her to use the site, and women have the option to request and meet with a genetic counselor. Those with inconclusive results are not able to receive their results via the platform and are directed to meet with a counselor. Over 39 months, 67,122 women received results via the platform. Of those requesting genetic counseling (4673), more were likely to have received a positive screen result, to be of advanced maternal age, to have a family history of concern, to have had a past pregnancy with a chromosomal finding, or to represent a high-risk pregnancy. Patient satisfaction was uniformly high among those with both negative and positive findings, with very few exceptions. Yet, there was no assessment of whether participants made an informed choice to undergo screening or whether they sufficiently understood the implications of their results. Further, when commercial entities develop educational platforms, they need to be scrutinized for bias in how the information is presented, given that their ultimate objective is to sell tests. This is not to say that Counsyl, or any other testing company, aims to persuade women to undergo screening they prefer not to pursue, but caution is warranted in cases in which there is clear conflict of interest. Additional evidence is needed to assess the value of such platforms to women making a value-based personal decision such as whether to undergo NIPS.

Augmenting Genetic Counseling

Beyond those services aiming to replace genetic counseling or practice components, there are other promising resources to augment service provision, such as chatbots. Clear Genetics developed the AI chatbot GIA (Genetic Information Assistant) and programmed "her" to answer common questions asked by genetic counseling patients/clients as they are selecting services. In certain circumstances, this may even be preferable. For example, potential patients/clients may prefer to ask about insurance coverage for genetic services in advance of prenatal screening, which is a discussion that can take time and often distract from the true intention of the counseling session. In a pilot study conducted with research participants enrolled in MyCode, a large exome sequencing cohort study at Geisinger Laboratories, participants who received results indicating the presence of a pathogenic variant in a hereditary cancer gene found GIA to be accessible, engaging, and informative (Nazareth et al. 2019). This iteration of GIA reminds patients/clients to tell close relatives about their own genetic test results and communicate their implications for their relatives' risks while also encouraging them to undergo testing. These important follow-up issues are typically addressed by genetic counselors, who often do not have time to remind patients. In this capacity, GIA can save counselors time and effort, especially with respect to delivery of repetitive information. Use of GIA or other AI resources can also result in patients/clients bringing more complex and nuanced questions to genetic counselors for answers, arguably a better use of the professionals' time. There are many other ways that genetic counseling adjuncts will serve to help streamline genetic counseling services, preserving in-person counseling for the most challenging cases.

The expanding ways that genomics is intersecting the lives of patients/clients suggest that all who seek understanding of their genetic risk status and health management needs may not require psychotherapeutic genetic counseling. Yet scaling of services should occur without losing the gold standard of psychotherapeutic care for those patients/clients who face the greatest health threats or complex decisions.

EVIDENCE TO GUIDE PRACTICE

These alternative and augmented service delivery models for use in the wake of expanding roles and responsibilities in genetic counseling need to be properly assessed to determine their relative success in meeting a wide variety of patient/client needs. Developing and evaluating creative solutions to meet systemic health delivery needs, such as educating other medical providers about genomics, are needed. Providers are currently ill-prepared to address their patients' queries about direct-to-consumer genetic test results, among other genetics/genomics-related questions. This could be an opportunity for genetic counselors to take the lead in developing and evaluating strategies for educating practicing physicians and nurses about ways to help interpret and use genomic information, and in particular how to help patients/clients accept the uncertainties due to the limited evidence to interpret many findings.

Individuals who have pursued the still-new profession of genetic counseling are intellectually curious, thrive on change, and have demonstrated their ability to embrace the unique nature of each new professional challenge. Because of the small size and limited number of training programs, the application process has traditionally been competitive, and high academic standards have been maintained and strengthened (Stern 2012). With each new wave of genomic discoveries, genetic counselors have risen to lead in devising creative solutions for providing care. As genomics translates into mainstream medical care over the coming years, there will be even more challenges. Yet already, genetic counselors are looking ahead, playing a key role in the development of telegenetic counseling companies for alternative service delivery and chatbots pioneered as an adjunct to practice. An important component of the profession's approach to problem solving is a commitment to patient care, and innovation is accompanied by the simultaneous collection of evidence to identify the best means of addressing patients/clients' needs going forward.

As genetic counseling advances, it will be important to the profession to uphold its central tenets in the provision of client-centered care, or to revise them based on new evidence. It is of critical importance to accrue data about how the tenets are actualized and how they relate to positive patient/client outcomes. Evidence can be used to ensure that patient/client needs are met, regardless of service delivery mode. It can also be used to preserve the limited work force of counselors for casework that demands cutting-edge complex information delivery coupled with advanced counseling skills. Evidence-based triage will be vital to keeping pace with demand and successfully addressing specific patient/client needs.

REFERENCES

*Reference is also in this subject collection.

Arjunan A, Ben-Shachar R, Kostialik J, Johansen Taber K, Lazarin GA, Denne E, Muzzey D, Haverty C. 2019. Technology-driven noninvasive prenatal screening results disclosure and management. *Telemed J E Health.* doi:10.1089/tmj.2018.0253

Athens BA, Caldwell SL, Umstead KL, Connors PD, Brenna E, Biesecker BB. 2018. A systematic review of randomized controlled trials to assess outcomes of genetic counseling. *J Genet Couns* **26:** 902–933. doi:10.1007/s10897-017-0082-y

Austin J, Semaka A, Hadjipavlou G. 2014. Conceptualizing genetic counseling as psychotherapy in the era of genomic medicine. *J Genet Couns* **23:** 903–909. doi:10.1007/s10897-014-9728-1

Biesecker BB. 2018. Genetic counselors as social and behavioral scientists in the era of precision medicine. *Am J Med Genet C* **178:** 10–14. doi:10.1002/ajmg.c.31609

Biesecker BB, Peters K. 2001. Process studies in genetic counseling: peering into the black box. *Am J Med Genetics* **106:** 191–198. doi:10.1002/ajmg.10004

Cite this article as *Cold Spring Harb Perspect Med* doi: 10.1101/cshperspect.a038968

Biesecker BB, Austin J, Caleshu C. 2017. Theories for psychotherapeutic genetic counseling: fuzzy trace theory and cognitive behavior theory. *J Genet Couns* 26: 322–330. doi:10.1007/s10897-016-0023-1

Biesecker BB, Lewis KL, Umstead KL, Johnston JJ, Turbitt E, Fishler KP, Patton JH, Miller IM, Heidlebaugh AR, Biesecker LG. 2018. Web platform vs. in-person genetic counselor for return of carrier results from exome sequencing: a randomized clinical trial. *JAMA Internal Med* 178: 338–346. doi:10.1001/jamainternmed.2017.8049

Biesecker BB, Peters K, Resta R. 2019. *Advanced genetic counseling theory and practice.* Oxford University Press, Cambridge.

* Blesson A, Cohen JS. 2019. Genetic counseling in neurodevelopmental disorders. *Cold Spring Harb Perspect Med* doi:10.1101/cshperspect.a036533

Buchanan AH, Datta SK, Skinner CS, Hollowell GP, Beresford HF, Freeland T, Rogers B, Boling J, Marcom PK, Adams MB. 2015. Randomized trial of telegenetics vs. in-person cancer genetic counseling: cost, patient satisfaction and attendance. *J Genet Counsel* 24: 961–970. doi:10.1007/s10897-015-9836-6

* Cho MT, Guy C. 2019. Evolving roles of genetic counselors in the clinical laboratory. *Cold Spring Harb Perspect Med* doi:10.1101/cshperspect.a036574

Ellington L, Roter D, Dudley WN, Baty BJ, Upchurch R, Larson S, Wylie JE, Smith KR, Botkin JR. 2005. Communication analysis of *BRCA1* genetic counseling. *J Genet Couns* 14: 377–386.

* Forman A, Sotelo J. 2019. Tumor-based genetic testing and familial cancer risk. *Cold Spring Harb Perspect Med* doi:10.1101/cshperspect.a036590

* Goldman JS. 2019. Predictive genetic counseling for neurodegenerative diseases: past, present, and future. *Cold Spring Harb Perspect Med* doi:10.1101/cshperspect.a036525

Green MJ, Biesecker BB, McInerney AM, Mauger D, Fost N. 2001. An interactive computer program can effectively educate patients about genetic testing for breast cancer susceptibility. *Am J Med Genet* 103: 16–23.

* Ingles J. 2019. Psychological issues in managing families with inherited cardiovascular diseases. *Cold Spring Harb Perspect Med* doi:10.1101/cshperspect.a036558

Joseph G, Guerra C. 2015. To worry or not to worry: breast cancer genetic counseling communication with low-income Latina immigrants. *J Community Genet* 6: 63–76. doi:10.1007/s12687-014-0202-4

Joseph G, Pasick RJ, Schillinger D, Luce J, Guerra C, Cheng JKY. 2017. Information mismatch: cancer risk counseling with diverse underserved patients. *J Genet Couns* 26: 1090–1104. doi:10.1007/s10897-017-0089-4

Joseph G, Lee R, Pasick RJ, Guerra C, Schillinger D, Rubin S. 2019. Effective communication in the era of precision medicine: a pilot intervention with low health literacy patients to improve genetic counseling communication. *Eur J Med Genet* 62: 357–367. doi:10.1016/j.ejmg.2018.12.004

Kessler S. 1997. Psychological aspects of genetic counseling. IX. Teaching and counseling. *J Genet Couns* 6: 287–295. doi:10.1023/A:1025676205440

Kohler JN, Turbitt E, Biesecker BB. 2017a. Personal utility in genomic testing: a systematic literature review. *Eur J Human Genet* 25: 662–668. doi:10.1038/ejhg.2017.10

Kohler J, Turbitt E, Lewis K, Wilfond B, Jamal L, Peay H, Biesecker L, Biesecker B. 2017b. Defining personal utility in genomics: a Delphi study. *Clin Genet* 92: 290–297. doi:10.1111/cge.12998

Kraft SA, Constantine M, Magnus D, Porter KM, Lee SS, Green M, Kass NE, Wilfond BS, Cho MK. 2017. A randomized study of multimedia informational aids for research on medical practices: implications for informed consent. *Clin Trials* 14: 94–102.

Kuppermann M, Norton ME, Gates E, Gregorich SE, Learman LA, Nakagawa S, Feldstein VA, Lewis J, Washington AE, Nease RF Jr. 2009. Computerized prenatal genetic testing decision-assisting tool: a randomized controlled trial. *Obstet Gynecol* 113: 53–63.

Kuppermann M, Pena S, Bishop JT, Nakagawa S, Gregorich SE, Sit A, Vargas J, Caughey AB, Sykes S, Pierce L, et al. 2014. Effect of enhanced information, values clarification, and removal of financial barriers on use of prenatal genetic testing: a randomized clinical trial. *J Am Med Assoc* 312: 1210–1217.

Lewis L, Umstead KL, Johnston JJ, Miller IM, Thompson LJ, Fishler KP, Biesecker LG, Biesecker BB. 2018. Outcomes of counseling after education about carrier results: a randomized controlled trial. *Am J Hum Genet* 102: 540–546. doi:10.1016/j.ajhg.2018.02.009

Matias M, Wusik K, Neilson D, Zhang X, Alexander-Valencia C, Collins K. 2019. Comparison of medical management and genetic counseling options pre- and post-whole exome sequencing for patients with positive and negative results. *J Genet Couns* 28: 182–193.

Meiser B, Irle J, Lobb E, Barlow-Stewart K. 2008. Assessment of the content and process of genetic counseling: a critical review of empirical studies. *J Genet Couns* 17: 434–451. doi:10.1007/s10897-008-9173-0

Milliron BJ, Bruneau M, Obeid E, Gross L, Bealin L, Smaltz C, Giri VN. 2019. Diet assessment among men undergoing genetic counseling and genetic testing for inherited prostate cancer: exploring a teachable moment to support diet intervention. *Prostate* 79: 778–783. doi:10.1002/pros.23783

Montgomery SV, Barsevick AM, Egleston BL, Bingler R, Ruth K, Miller SM, Malick J, Cescon TP, Daly MB. 2013. Preparing individuals to communicate genetic test results to their relatives: report of a randomized control trial. *Fam Cancer* 12: 537–546. doi:10.1007/s10689-013-9609-z

National Society of Genetic Counselors' Definition Task Force, Resta R, Biesecker BB, Bennett RL, Blum S, Hahn SE, Strecker MN, Williams JL. 2006. A new definition of genetic counseling: National Society of Genetic Counselors' Task Force report. *J Genet Couns* 15: 77–83.

Nazareth S, Simmons E, Snir M, Shohat M, Goldberg J. 2019. Use of a chatbot to offer pre-test education for expanded carrier screening. *Abstract 8836 American College of Medical Genetics Annual Conference*, Seattle, WA

Persky S, Kistler WD, Klein WMP, Ferrer RA. 2019. Internet versus virtual reality settings for genomics information provision. *Cyberpsychol Behav Soc Netw* **22:** 7–14. doi:10 .1089/cyber.2017.0453

* Platt J. 2019. A person-centered approach to cardiovascular genetic testing. *Cold Spring Harb Perspect Med* doi:10 .1101/cshperspect.a036624

* Ramos E. 2019. Genetic counseling, personalized medicine, and precision health. *Cold Spring Harb Perspect Med* doi:10.1101/cshperspect.a036699

Resta R, Biesecker B, Bennett R, Blum S, Estabrooks Hahn S, Strecker M, Williams JL. 2018. How is the NSGC definition of genetic counseling being used? *Abstract C-330 presented at The National Society of Genetic Counselors Annual Education Conference*, Atlanta, GA

Rutherford S, Zhang X, Atzinger C, Rushman J, Meyers M. 2014. Medical management adherence as an outcome of genetic counseling in a pediatric setting. *Genet Med* **16:** 157–163.

* Schienda J, Stopfer J. 2019. Cancer genetic counseling—current practice and future challenges. *Cold Spring Harb Perspect Med* doi:10.1101/cshperspect.a036541

Stern A. 2012. *Telling genes: the story of genetic counseling in America.* Johns Hopkins University Press, Baltimore, MD.

* Stoll K, Jackson J. 2019. Supporting patient autonomy and informed decision-making in prenatal genetic testing. *Cold Spring Harb Perspect Med* doi:10.1101/cshperspect .a036509

Subas T, Luiten R, Hanson-Kahn A, Wheeler M, Caleshu C. 2019. Evolving decisions: perspectives of active and athletic individuals with inherited heart disease who exercise against recommendations. *J Genet Couns* **28:** 119–129. doi:10.1007/s10897-018-0297-6

Veach PM, Bartels DM, LeRoy B. 2007. Coming full circle: a reciprocal-engagement model of genetic counseling practice. *J Genet Couns* **16:** 713–728.

Zakas AL, Leifeste C, Dudley B, Karloski E, Afonso S, Grubs RE, Shaffer JR, Durst AL, Parkinson MD, Brand R. 2019. The impact of genetic counseling on patient engagement in a specialty cancer clinic. *J Genet Couns.* doi:10.1002/ jgc4.1149

Supporting Patient Autonomy and Informed Decision-Making in Prenatal Genetic Testing

Katie Stoll[1] and Judith Jackson[2,3]

[1]Genetic Support Foundation, Olympia, Washington 98502, USA

[2]Department of Genetic Counseling, Brandeis University, Waltham, Massachusetts 02453, USA

[3]Department of Maternal Fetal Medicine, South Shore Health, South Weymouth, Massachusetts 02190, USA

Correspondence: kstoll@geneticsupport.org; jjackson@brandeis.edu

Genetic counselors have both the burden and the privilege of supporting patients who are faced with making difficult decisions. In the prenatal setting, genetic counselors are responsible for reviewing a growing list of prenatal testing options for patients with the goal of helping people to anticipate the potential consequences of their decision. Prenatal genetic counselors also support patients in making decisions about the next steps after clinical evaluation has indicated a genetic condition, birth defect, or information of uncertain clinical significance in the fetus. The information provided and choices patients face in the context of prenatal and reproductive genetics can be life-altering, and decisions often must be made within a short window of time. It is imperative that the needs and preferences of each patient are considered and that individuals are empowered to make active decisions that are consistent with their needs and values. Here we will review the history of the role of the genetic counselor in the prenatal setting and will provide strategies and tools for supporting informed patient decision-making in the face of an increasingly complex reproductive genetic testing landscape.

Genetic counselors have played an important role in supporting patient decision-making in prenatal genetics since the inception of the profession. Recognizing that preferences related to prenatal genetic screening and diagnosis vary from person to person and are of great personal significance, genetic counseling practice has held the core value of supporting patient autonomy in the context of communicating balanced, current, and evidence-based information. Even with these priorities and goals in mind, decision-making in prenatal testing is complicated by numerous challenges such as the variety and complexity of available tests, reimbursement issues, and the possibility of uncertain findings (Fig. 1). Genetic test results as well as ultrasound findings that are difficult to interpret or for which there is insufficient information to predict clinical outcomes have long been a part of prenatal genetics. But as the number and scope of available genetic tests grow, so does the likelihood that an expectant parent will be faced

---

Figure 1. How to decide about prenatal genetic testing? The illustration is taken from the patient educational video, "How to Decide about Prenatal Genetic Testing?" This video is used to illustrate an example of a patient decision aid that was developed by genetic counselors. This and a series of patient educational videos on prenatal genetic testing were produced by the Washington State Department of Health and Genetic Support Foundation and are freely available online at www.geneticsupport.org/videos. (Figure reprinted courtesy of the Patient Library Genetic Support Foundation in conjunction with the Washington State Department of Health.)

with difficult decisions that must be made in the context of uncertain or incomplete information. Also complicating prenatal decision-making is the influence of agendas other than of the pregnant woman, including that of her family members and friends (Lewis et al. 2016), her reproductive partner, health-care providers (Muller and Cameron 2016), and commercial testing laboratories (Farrell et al. 2015). Deciphering the impact and importance of these external factors can be crucial in helping patients find their way to the choices that are best for them (Cernat et al. 2019).

Historically, genetic counselors have aspired to provide counseling that is "nondirective" (Kessler 1997). There is no single universally agreed upon definition of this concept, and approaches to operationalizing it vary (Bartels et al. 1997), but in the context of prenatal counseling the most fundamental issue is that genetic counselors seek to provide patients with information and support regarding their options without undue influence regarding their decision on what, if any, testing to choose or on how to act on the results (Benkendorf et al. 2001). The history behind the concept of non-directiveness in genetic counseling is complex (Stern 2012), but, ultimately, the goal is to support and promote patient autonomy in decision-making.

However, when a patient is presented with more than one potential pathway or treatment option, simply providing information about all options may not adequately enable patients to arrive at a decision that is right for them. Ideally, the genetic counselor is able to provide each patient with up-to-date, evidence-based information related to the risks, benefits, and limitations of each option and to elicit information about the patient's values, preferences, and personal situation. The goal of counseling should not be to impart all of the information the genetic counselor knows to the patient; rather, the goal should be for the patient and genetic counselor to engage in a meaningful dialogue and a two-way exchange of information that will best help the patient arrive at a decision that is right for her (Kessler 1997). Through this dialogue, the patient's questions and preferences may become clear and the genetic counselor has the opportunity to develop an understanding of factors at play in their decision-making process such that they can reflect this understanding back to the patient. Evidence suggests that genetic counselors successfully utilize these techniques in their sessions (Salema et al. 2019). Such a dialogue also provides the opportunity for the patient to reveal any possible factual misconceptions so that these may be corrected. Ultimately, patients value how genetic counselors provide affirmation for the weighty and uncomfortable decisions they are responsible for in the prenatal context, without making the decision on their behalf (Salema et al. 2019).

Herein, we will provide context and background on the importance of supporting informed decisions in prenatal care. Standardized tools that may be beneficial to the genetic counselor to facilitate more meaningful and active support of patient decision-making in a variety of clinical scenarios will be introduced, and we will consider the importance of involving genetic counselors in the development of decision aids for use outside of the genetic counseling office (see Box 1).

 Cite this article as *Cold Spring Harb Perspect Med* doi: 10.1101/cshperspect.a036509

BOX 1. STANDARDIZING INFORMATION FOR *ALL* WOMEN; SHARED DECISION-MAKING FOR *EVERY* WOMAN

by *Katie Stoll*

In prenatal care today, all women have the option to consider prenatal genetic testing including aneuploidy testing and carrier screening for recessive and X-linked conditions. Although I had been through conversations about prenatal testing decisions with patients thousands of times, I did not appreciate how much my approach to these conversations aligned with the shared decision-making (SDM) model. As with SDM, my process typically involved three steps when offering prenatal genetic testing. In each case, the patient or couple must understand that they have a choice, and that all possible choices that are available to them are reasonable to consider and pursue. They must then be provided with sufficient information to make an informed choice among the presented options, and then they must be supported to deliberate their options.

For every patient there may be multiple decision points in which these steps may be considered. For example, a patient first may be asked to decide if they want to undergo any testing for aneuploidy in the pregnancy. If they do, they may be asked to consider whether they would like to undergo diagnostic testing directly or consider screening tests with follow-up to diagnostic testing only if the screening tests indicate an increased probability of aneuploidy. If they indicate their preference is to undergo screening tests, they may need to decide between multiple available options. If they elect diagnostic testing, they may need to decide between karyotype analysis versus chromosomal microarray (CMA). With each of these questions, patients need sufficient information to understand the differences among the options, the potential benefits, drawbacks, and limitations of each test they are offered, and the possible outcomes that may come from accepting or declining any of these options.

As the number and complexity of the genetic tests offered to women in pregnancy continue to increase, and as the demand for genetic counselors in every area of medicine continues to grow, genetic counselors may be less available to utilize these SDM strategies with patients one-on-one. As we consider the optimal way to deliver information about genetic testing to pregnant women in a patient-centered way that promotes informed and value-consistent decisions, genetic counselors are uniquely suited to develop decision aids that will best support informed and autonomous patient decision-making. One existing decision tool that I had a role in developing along with other genetic counselors is a short informational video that was created for patients to help them consider their initial decisions on what if any prenatal genetic tests to consider (https://www.youtube.com/watch?v=-vIJGFWJquk) (see Fig. 1).

The first step in this video decision aid is to identify that the patient has a personal choice to make about prenatal testing. The narrator reads, "Which tests or whether to undergo any of these tests in pregnancy is your choice. One of the most important questions to consider when deciding about prenatal testing is what will this information mean for you?"

The second step is to provide the necessary information to support an informed decision. Through animation to illustrate key concepts, as well as written and spoken words, information is shared regarding the overall scope, as well as the potential benefits, risks, and limitations of the various screening and diagnostic testing options.

And the third step supports deliberation of their options by posing questions that women may wish to consider when weighing the options presented to them. The following is an excerpt from the video illustrates this:

Questions to consider regarding diagnostic testing include:

- If your baby had a genetic condition, would you want to know before birth?
  - Some women would want to know if their baby had a genetic condition or birth defect because they would want to be able to prepare before delivery.
  - Some women would consider making an adoption plan for their baby.
  - Some would consider ending the pregnancy if they knew their baby had a genetic condition.

- Do you feel like "yes or no" answers would help you feel less worried?
- Are you comfortable with the risk of miscarriage associated with these procedures?"

Genetic counselors can help support informed patient decisions regarding prenatal testing in a scalable way through the development of decision tools such as this video. Although such tools do not replace one-on-one consultation with a skilled genetic counselor, they can provide important information and empower patients to recognize they have an active decision to make about testing and provide some guided questions on what should be considered as one makes choices about these testing options.

HISTORY OF GENETIC COUNSELING AND SUPPORT OF INFORMED DECISIONS

The profession of genetic counseling emerged in tandem with the technological development and clinical availability of prenatal testing. The first prenatal diagnosis through fetal karyotype using amniocytes occurred in 1968 (Valenti et al. 1968), and the following year, the first genetic counseling training program at Sarah Lawrence College welcomed its initial cohort of genetic counseling students (Stern 2009). The 1970s brought many new developments in prenatal screening including the discovery of the association between elevated maternal serum alpha-fetoprotein and open neural tube defects (Wald and Cuckle 1977) and of reduced β-hexosaminidase A activity in Tay–Sachs disease carriers. Carrier screening for hemoglobinopathies such as sickle cell anemia and β-thalassemia also became possible in the 1970s (O'Brien et al. 1970; Markel 1997). As prenatal testing options grew, many genetic counselors began their careers in antenatal diagnostic centers throughout the country.

A number of social and political factors contributed to establishing the role of the genetic counselor in perinatal care in the final decades of the twentieth century. These included the women's reproductive and patients' rights movements and an increasing focus on bioethical principles in medicine. In addition to progressive political forces that shaped the profession's focus on reproductive autonomy, there was also a strong desire on the part of genetic professionals to distance themselves from the eugenics movement in the first half of the twentieth century. Genetic counselors approached reproduc-

tive counseling with the goal of providing information and support to help facilitate informed decision-making consistent with the patient's personal needs and values (Stern 2012).

CHALLENGES TO INFORMED DECISION-MAKING

For as long as prenatal genetic testing has been available, there have been concerns about whether or not all patients are given the opportunity to make informed decisions about their options (Seavilleklein 2009). The challenges faced in terms of patient education and counseling have only grown in number and complexity as prenatal genetic testing options have expanded.

Practice guidelines today dictate that all women should be offered a menu of genetic tests including carrier screening, screening for aneuploidy, and diagnostic testing (American College of Obstetricians and Gynecologists 2016, 2017). Typically, the discussion of these tests first occurs in the obstetric office, and the amount of time and attention dedicated to these discussions can vary tremendously, as do which tests are offered and under what circumstances. This variability may be attributed to differences in the preference of individual providers, a lack of awareness about current practice guidelines, site-specific screening protocols, and ever-changing insurance payer policies regarding prenatal genetic tests (Bernhardt et al. 1998; Farrell et al. 2011).

In some cases, patients may not be aware that they have the option of having prenatal genetic testing (Bryant et al. 2015). Or the genetic tests may be offered in such a way that they appear to be a routine part of prenatal care, and it is unclear to patients that they have an active choice to make with regards to which, if any testing, they would like to pursue; patients may not always be aware that they have the option to decline testing altogether (Seror and Ville 2009; Johnston et al. 2017).

There are many factors that may lead practitioners to favor testing over a nuanced discussion and informed consent process or documenting the patient's decision not to test. For one, the current health-care system requires prenatal care providers to cover a growing list of complex topics with their patients in an increasingly short clinical encounter. Additionally, many prenatal care providers are concerned that they may be vulnerable to a wrongful birth suit if a genetic condition is not diagnosed (Pergament and Ilijic 2014).

In October of 2011, prenatal cell-free DNA (cfDNA) screening became clinically available in parts of the world, including the United States, for women with an increased chance for aneuploidy in their pregnancy (Minear et al. 2015). The initial cfDNA screens evaluated for trisomy 21 exceeded the performance of previously available aneuploidy testing with both higher detection rates and lower false-positive rates. By early 2012, commercially available cfDNA screening was expanded to also include trisomy 13 and 18 (Palomaki et al. 2012), and later that year screening for sex chromosome aneuploidy was also available through many laboratories (Ramdaney et al. 2018). In screening the X and Y chromosomes, reporting of predicted fetal sex as early as 10 wk gestational age was also made possible through cfDNA screening. This feature of cfDNA screening and the intense marketing campaigns by the testing laboratories directed at both health-care providers and patients were factors that led to adoption of new technology into clinical practice at an unprecedented rate (Minear et al. 2015).

Currently, cfDNA screening is often offered to many women who are considered to have a low probability of a chromosome abnormality. Its utilization has been significantly expanded internationally as well. In addition to screening for aneuploidy, many laboratories now screen for common microdeletions, and there are cfDNA products that screen for copy number variants throughout the genome. Testing for single-gene disorders has also recently become available on a clinical basis (Zhang et al. 2019), and it seems possible and perhaps even likely that this technology will expand to genome sequencing in the clinical setting in the near future (Hayward and Chitty 2018).

With the increased sensitivity of screening tests for aneuploidy through prenatal cfDNA screening relative to previously available aneuploidy screens, fewer women are electing to have diagnostic testing (Huang et al. 2018). Given that consideration of diagnostic testing has historically been an entry point for the involvement of genetic counselors, fewer patients taking this path results in fewer patients who are given the opportunity to meet with a genetic counselor (Minkoff and Berkowitz 2014).

These factors are making the utilization of genetic testing in pregnancy more commonplace while at the same time making genetic counseling prior to testing rarer. Many women are only offered genetic counseling after testing has identified an increased probability of a genetic condition. This routinization of genetic testing in pregnancy represents a fundamental shift from the original goal of informed and autonomous decision-making on testing preferences.

Although many patients desire prenatal genetic testing, some individuals may not wish to face the decisions that may ensue, including whether or not to undergo diagnostic testing or how to respond to an abnormal prenatal diagnosis. If the blood test they have taken as a matter of course shows an increased probability for a genetic condition, they may feel pressured to undergo subsequent diagnostic testing, and depending on the results of the diagnostic testing, they may feel pressure to act on results in a specific way (Schoonen et al. 2012). This may occur when patients undergo genetic screening in pregnancy without fully anticipating the possible outcomes and possible next steps in advance of a testing decision.

INFORMED DECISIONS WITH UNCERTAIN INFORMATION

Initially prenatal genetic testing for Mendelian disease was focused on tests for a few specific conditions for which information about anticipated diagnoses could be relatively clear. For example, carrier screening was initially offered clinically for well-characterized recessive conditions in higher-risk populations based on ethnicity or family history. The variants selected for carrier screening were reviewed carefully to ensure that residual risk estimates and genotype–phenotype information were available to allow for genetic counseling with relatively unambiguous information about expected prognosis and relative risk.

Much more extensive genetic carrier screening is now available to some women preconception or during pregnancy, including screening for dozens to hundreds of variants associated with several genetic conditions. However, this opportunity for more information comes at the cost of a greater likelihood of uncertain or unanticipated findings. When screening for more genes, the probability of an abnormal result is higher, and many who undergo expanded carrier screening are carriers for a variant in at least one of the genes on the panel (Guo and Gregg 2019). And given that many laboratories now utilize next-generation sequencing of the entire gene rather than genotyping for known pathogenic variants, it is becoming increasingly common for laboratories to report out novel variants for which our ability to predict the clinical outcome is limited (Kraft et al. 2019).

With the expansion of cfDNA screening to rare microdeletions and even single-gene disorders, there is a higher possibility of a false-positive result. The predictive value of any screening test is dependent on the prevalence of the condition screened for, and thus, at a given level of accuracy, the rarer a condition is, the more likely the result is to be a false positive. Given that many women receive little pretest counseling regarding their prenatal cfDNA screening, people may fail to recognize the possibility that these abnormal results are incorrect as well as the limitations of the testing (Farrell et al. 2015).

Patients undergoing amniocentesis or chorionic villus sampling may now consider CMA in place of karyotype analysis for diagnostic testing. CMA identifies more clinically significant chromosomal abnormalities than traditional karyotype analysis, identifying a clinically significant chromosomal variant in 6% of pregnancies with a structural abnormality on ultrasound and 1.7% of pregnancies in cases of maternal age or abnormal aneuploidy screening. Additionally, 1%–2% of women who undergo testing with CMA in pregnancy will have a finding of uncertain significance (Wapner et al. 2012).

A study by Bernhardt et al. (2013) examining the effects of receiving abnormal or uncertain findings from CMA during pregnancy revealed that even with thorough pretest genetic counseling to help them understand and anticipate the possible results, participants still reported feeling blindsided by test results when they received an uncertain finding. They struggled with uncertain results and some described tests results as "toxic knowledge"—information that caused great anxiety that they regretted having received. Reflecting on their decision to undergo testing in the first place, many women expressed that they had felt it was an offer too good to pass up. Although many of these patients felt supported by their provider after receiving abnormal results, some reported feeling abandoned during the decision-making process (Bernhardt et al. 2013).

Dealing with uncertain information has always been a part of prenatal genetics. With genomic screening becoming both more accessible and more comprehensive, up to and including exome or genome sequencing, the potential for uncertain findings will only grow, creating new challenges in genetic counseling and informed consent. Approaches to counseling should recognize the unique needs and values of each couple and strive to help individuals understand the range of possible outcomes and the possibility of receiving results with unclear implications. Although many will find that the potential benefit of more information is worth the possibility of uncertainty, more information is not a plus in all circumstances.

Cite this article as *Cold Spring Harb Perspect Med* doi: 10.1101/cshperspect.a036509

SHARED DECISION-MAKING

Given the problems associated with the use of the descriptor "nondirectiveness" (Kessler 1992; Weil 2000), the more clearly defined concept of SDM has received increased recognition in recent years as a patient-centered approach to supporting informed choices in medical care (Beach and Sugarman 2019). SDM has been demonstrated to be an effective strategy for empowering patients to take an active role in making complex medical decisions when there is more than one reasonable treatment option available. The foundational principle for SDM is that patients are capable of making informed decisions that are consistent with their values through weighing new information and deliberating options with a supportive health-care provider (Elwyn et al. 2012). These principles of SDM are used in genetic counseling practice (Birch et al. 2018). SDM often lends itself very well to genetic counseling in that there is often more than one reasonable path or option available to patients, and the factors that shape patient decisions may rely heavily on individual values and the preferences of the patient.

A three-step process for SDM has been proposed (Elwyn et al. 2012). In the first step, the practitioner makes clear that there is a *choice* in how to proceed and that there are two or more viable treatment plans or options. The patient should understand that they are being asked to be actively engaged in a decision regarding their care with multiple reasonable possible options. This is an issue in the area of prenatal genetics, where patients may not always appreciate which tests and procedures are optional, and they may not always feel empowered to decide for themselves how they would like to proceed.

The second step in the SDM process is *providing information*. This process is intended to ensure that patients are provided with adequate information about their treatment options so that they are able to make an informed decision. The information provided to the patient should be unbiased, up-to-date, and evidence-based.

The third step is to *support deliberation*. During this step, patients are encouraged to explore what matters most to them as they consider their options and eventually arrive at an informed decision with the clinician to help guide them.

Decision Aids

In some clinical situations, SDM may be supported through the use of specific decision aids. In their simplest form, aids may communicate the probability of benefit and harm and describe the limitations of various testing options. More elaborate aids may incorporate value clarification exercises or review the experiences of patients who have faced similar choices. Some tools may be recommended for patients prior to meeting with the clinician to allow them to prepare for the visit. Other aids may be used during the session itself. Patient decision aids may be in the form of a booklet, a printed-out worksheet, a video, or an internet-based survey tool. Patient decision aids have become more available (O'Conner et al. 1999; Stacey et al. 2017), and genetic counselors have played a role in developing aids to support decisions in multiple clinical genetic areas (Box 2). Analysis of the effectiveness of patient decision aids indicates that the use of these tools can improve patient agency in decision-making and clarify the patient's values and needs. Additionally, these tools have been shown to reduce decisional conflict and regret, helping patients have a better understanding of outcome probabilities (Leinweber et al. 2019).

One SDM tool that may be used in a variety of genetic counseling sessions is the Ottawa Personal Decision Guide https://decisionaid.ohri.ca/decguide.html (Fig. 2). An example of its use in a clinical scenario involving decision-making in the face of uncertainty is provided (Box 2).

The case illustrated in Box 2 demonstrates the use of a shared decision aid by a genetic counselor working with a couple faced with a difficult decision in which the prognosis was uncertain. However, many people facing difficult decisions arrive at a choice without this formal process. The use of a formal decision aid such as the Ottawa Personal Decision Guide may be especially beneficial in cases in which a

Ottawa Personal Decision Guide
For People Making Health or Social Decisions

❶ Clarify your decision.

What decision do you face?

What are your reasons for making this decision?

When do you need to make a choice?

How far along are you with making a choice?
- ☐ Not thought about it
- ☐ Thinking about it
- ☐ Close to choosing
- ☐ Made a choice

❷ Explore your decision.

Knowledge
List the options and benefits and risks you know.

Values
Rate each benefit and risk using stars (★) to show how much each one matters to you.

Certainty
Choose the option with the benefits that matter most to you. Avoid the options with the risks that matter most to you.

	Reasons to Choose this Option Benefits / Advantages / Pros	How much it matters to you: 0★ not at all 5★ a great deal	Reasons to Avoid this Option Risks / Disadvantages / Cons	How much it matters to you: 0★ not at all 5★ a great deal
Option #1				
Option #2				
Option #3				

Which option do you prefer? ☐ Option #1 ☐ Option #2 ☐ Option #3 ☐ Unsure

Support

Who else is involved?			
Which option do they prefer?			
Is this person pressuring you?	☐ Yes ☐ No	☐ Yes ☐ No	☐ Yes ☐ No
How can they support you?			

What role do you prefer in making the choice?
- ☐ Share the decision with…
- ☐ Decide myself after hearing views of…
- ☐ Someone else decides…

Figure 2. Ottawa Personal Decision Guide. The Ottawa Personal Decision Guide (OPDG) and Ottawa Personal Decision Guide for Two (OPDGx2) are validated patient decision aids that are designed for any health-related or social decisions. These decision aids are freely available for use online (https://decisionaid.ohri.ca/decguide.html) and can be used by patients on their own or with the assistance of a health-care provider acting as a decision coach. (The pages from the guide are reprinted with permission from the authors, Stacey, O'Connnor, and Jacobsen at the Patient Decision Aids Research Group, Ottawa Hospital Research Institute and University of Ottawa, Canada, 2015.)

 Cite this article as *Cold Spring Harb Perspect Med* doi: 10.1101/cshperspect.a036509

❸ Identify your decision making needs.

	Knowledge	Do you know the benefits and risks of each option?	☐ Yes	☐ No
	Values	Are you clear about which benefits and risks matter most to you?	☐ Yes	☐ No
	Support	Do you have enough support and advice to make a choice?	☐ Yes	☐ No
	Certainty	Do you feel sure about the best choice for you?	☐ Yes	☐ No

If you answer 'no' to any question, you can work through steps two ❷ and four ❹, focusing on your needs.
People who answer "No" to one or more of these questions are more likely to delay their decision, change their mind, feel regret about their choice or blame others for bad outcomes.

❹ Plan the next steps based on your needs.

Decision making needs	✓ Things you could try
Knowledge If you feel you do NOT have enough facts	☐ Find out more about the options and the chances of the benefits and risks. ☐ List your questions. ☐ List where to find the answers (e.g. library, health professionals, counsellors):
Values If you are NOT sure which benefits and risks matter most to you	☐ Review the stars in step two ❷ to see what matters most to you. ☐ Find people who know what it is like to experience the benefits and risks. ☐ Talk to others who have made the decision. ☐ Read stories of what mattered most to others. ☐ Discuss with others what matters most to you.
Support If you feel you do NOT have enough support	☐ Discuss your options with a trusted person (e.g. health professional, counsellor, family, friends). ☐ Find help to support your choice (e.g. funds, transport, child care).
If you feel PRESSURE from others to make a specific choice	☐ Focus on the views of others who matter most. ☐ Share your guide with others. ☐ Ask others to fill in this guide. (See where you agree. If you disagree on facts, get more information. If you disagree on what matters most, consider the other person's views. Take turns to listen to what the other person says matters most to them.) ☐ Find a person to help you and others involved.
Certainty If you feel UNSURE about the best choice for you	☐ Work through steps two ❷ and four ❹, focusing on your needs.
Other factors making the decision DIFFICULT	List anything else you could try:

Fig. 2. *Continued.*

BOX 2. PATIENT COACHING WITH USE OF A DECISION AID

by *Judith Jackson*

After several months of trying to conceive, Sara and Brian were elated to learn they were expecting. Sara was 37 yr old, and this would be the first pregnancy and first baby for the couple. When Sara was 12 wk pregnant, she met with my colleague, another genetic counselor in our department. At this visit, the genetic counselor reviewed prenatal genetic screening and diagnostic testing options with Sara and Brian and they opted to proceed with combined screening. The first-trimester ultrasound including nuchal translucency measurement was normal, as were the analyte markers. The couple was reassured by the results of these tests which indicated that they had a very low chance for Down syndrome, trisomy 18, or a neural tube defect. At 18 wk 2 d gestation, Sara and Brian returned to the clinic for a detailed ultrasound examination. At this visit an abnormal curvature of the spine was noted in the fetus. The couple was referred to a local pediatric hospital for a fetal MRI, and a pediorthopedic consultation. The MRI detected significant kyphosis at L2-3 caused by block vertebrae or two levels of hemivertebrae. In addition, the spinal cord appeared tethered. It was thought that it was most likely an isolated abnormality and that is was surgically correctable, although there was a significant risk for paralysis after birth.

At 22 wk gestation, after multiple appointments with specialists, the couple contacted the MFM clinic and requested a same-day appointment with a genetic counselor. My colleague who had met with the couple previously was not in the office that day. I spoke with Brian and explained that they might prefer to come back the following day when they could meet with the genetic counselor they had previously seen. Brian expressed that they were feeling some urgency to speak with someone that day. He shared that they had consulted with many specialists regarding the prognosis and felt comfortable that they knew as much as there was to know. He said they needed help with the decision-making process, regarding whether to continue with the pregnancy or not. As I was new to meeting this family and they were so specific in their request for support in making this decision, which was clearly very difficult for them, I felt like a decision support tool could be useful in helping to guide our discussion.

I met with the couple later that afternoon, and in using the Ottawa Personal Decision Guide for Two (see Fig. 2), I helped guide Sara and Brian as they worked through their thoughts and feelings regarding this very difficult decision.

The first step is early defining the decision and the possible options. In this case, Sara and Brian were clear, "We are deciding whether to continue with the pregnancy or not." With this, we clarified the two options being considered with this decision: continuing the pregnancy or not continuing in the pregnancy.

Next, I asked the couple, in addition to the two of them, who was weighing in on this decision? In this case, this was a critical component of our discussion. Sara and Brian expressed that they had shared the information about the birth defect and prognosis only with their parents, and one set of parents had very strong opinions about what they believed was the best decision. From this, we explored Sara and Brian's preferences for what role they wanted in making the choice. The couple mutually agreed that they wanted to make the decision together and that, ultimately, it was only their opinion that mattered. They were asked to try to push aside the opinions of others regarding the impending decision to be made. I then asked Sara and Brian to consider the reasons to choose each option (benefits/advantages/pros) and also the reasons not to choose each option and (risks/disadvantages/cons). The couple then each answered yes or no to the following questions:

- Do you know the benefits and risks of each option?
- Are you clear about which benefits and risks matter most to you?
- Do you have enough support and advice to make a choice?
- Do you feel sure about the best choice for you?

For any "no" answers to the above, as the coach, I could help provide accurate information or direct them to resources that they further explore. The Ottawa Personal Decision Guide together with the context of what Brian and Sara had shared with me to that point prompted follow-up questions that I could ask and help answer to help them move closer to a decision.

Sara and Brian identified one piece of information that might make a more informed decision. They hoped to learn more about the experience of being a parent to a child who was paralyzed. I offered to connect them with a mother I knew who had a child with spina bifida who was in a wheelchair, although I also counseled the couple about the limitations of hearing just one story. The couple was interested in meeting or talking with this woman and eventually did. The guide helped to structure the discussion in such a way that ultimately the couple was able to reach a decision that was consistent with their own values and wishes.

Several years later I met with the couple again and they remained confident that they had made the decision that was right for them.

patient is struggling with arriving at a decision or at times when members of a couple are not initially in agreement.

In cases when a decision aid is not used, components of the guide and process can often be utilized informally in a genetic counseling session. A key step in this model is to ask the patient if there are any remaining questions that could be answered to help in their decision-making. Patients have expressed decisional regret when they recall unanswered questions prior to moving forward with a treatment plan (Bernhardt et al. 2013). Genetic counselors can ensure that misunderstandings are clarified whenever possible with the goal of providing accurate and up-to-date information. The patient's response may give an indication of the extent of their support system and facilitate the provision of resources when there are deficiencies in support. Genetic counselors can ask their patients what role they want to have in this decision. Additionally, they can ask who else is a part of this decision, and how much does their opinion matter.

CONCLUDING REMARKS

Genetic counselors have played a crucial role in supporting patients to make informed and value-consistent decisions in the prenatal genetic setting since prenatal screening and diagnosis were first possible. As complex tests such as cfDNA and expanded carrier testing become more routine in obstetrical care, the percentage of patients who work directly with a genetic counselor to help them navigate through the decision-making process will necessarily decrease. And with the growing use of these technologies taking place outside of the genetic counseling setting, new tools to support patient decision-making in prenatal genetics will be needed. Given the experience of genetic counselors in supporting patients through sometimes difficult decisions with regard to prenatal testing, genetic counselors are uniquely equipped to design programs and tools including decision aids that can support patient decisions even when they are not directly involved in their care.

Although genetic counselors have been practicing with elements of the SDM principles for decades, the application of the SDM model to the prenatal genetic counseling practice has not been extensively evaluated. Thus, additional research is needed on the effectiveness of using SDM principles in a genetic counseling setting. However, because in-person counseling is not a viable answer in light of the expansion of prenatal screening, genetic counselors should play an integral role in developing decision aids to be used by counselors and in OB offices. These decision aids allow genetic counselors to impact the quality of prenatal decisions even when they are not able to work directly with the patient.

REFERENCES

American College of Obstetricians and Gynecologists. 2016. Practice Bulletin No. 163 Summary: Screening for fetal aneuploidy. *Obstet Gynecol* **127:** 979–981. doi:10.1097/AOG.0000000000001439

American College of Obstetricians and Gynecologists. 2017. Committee Opinion No 691: Carrier screening for genetic conditions. *Obstet Gynecol* **129:** e41–e55. doi:10.1097/AOG.0000000000001952

Bartels DM, LeRoy BS, McCarthy P, Caplan AL. 1997. Nondirectiveness in genetic counseling: a survey of practitioners. *Am J Med Genet* **72:** 172–179. doi:10.1002/(SICI)1096-8628(19971017)72:2<172::AID-AJMG9>3.0.CO;2-X

Beach MC, Sugarman J. 2019. Realizing shared decision-making in practice. *J Am Med Assoc* doi:10.1001/jama.2019.9797

Benkendorf JL, Prince MB, Rose MA, De Fina A, Hamilton HE. 2001. Does indirect speech promote non-directive genetic counseling? Results of a sociolinguistic investigation. *Am J Med Genet* **106:** 199–207. doi:10.1002/ajmg.10012

Bernhardt BA, Geller G, Doksum T, Larson SM, Roter D, Holtzman NA. 1998. Prenatal genetic testing: content of discussions between obstetric providers and pregnant women. *Obstet Gynecol* **91:** 648–655.

Bernhardt BA, Soucier D, Hanson K, Savage MS, Jackson L, Wapner RJ. 2013. Women's experiences receiving abnormal prenatal chromosomal microarray testing results. *Genet Med* **15:** 139–145. doi:10.1038/gim.2012.113

Birch PH, Adam S, Coe RR, Port AV, Vortel M, Friedman JM, Légaré F. 2018. Assessing shared decision-making clinical behaviors among genetic counsellors. *J Genet Couns* **1:** 40–49.

Bryant AS, Norton ME, Nakagawa S, Bishop JT, Pena S, Gregorich SE, Kuppermann M. 2015. Variation in women's understanding of prenatal testing. *Obstet Gynecol* **125:** 1306-1312.

Cernat A, De Freitas C, Majid U, Trivedi F, Higgins C, Vanstone M. 2019. Facilitating informed choice about non-invasive prenatal testing (NIPT): a systematic review and qualitative meta-synthesis of women's experiences. *BMC Pregnancy Childbirth* **19:** 27. doi:10.1186/s12884-018-2168-4

Elwyn G, Frosch D, Thomson R, Joseph-Williams N, Lloyd A, Kinnersley P, Cording E, Tomson D, Dodd C, Rollnick S, et al. 2012. Shared decision making: a model for clinical practice. *J Gen Intern Med* **27:** 1361–1367. doi:10.1007/s11606-012-2077-6

Farrell RM, Dolgin N, Flocke SA, Winbush V, Mercer MB, Simon C. 2011. Risk and uncertainty: shifting decision making for aneuploidy screening to the first trimester of pregnancy. *Genet Med* **13:** 429–436. doi:10.1097/GIM.0b013e3182076633

Farrell RM, Agatisa P, Mercer M, Coleridge M. 2015. Online direct-to-consumer messages about non-invasive prenatal genetic testing. *Reprod Biomed Soc Online* **1:** 88–97. doi:10.1016/j.rbms.2016.02.002

Guo MH, Gregg AR. 2019. Estimating yields of prenatal carrier screening and implications for design of expanded carrier screening panels. *Genet Med* **21:** 1940–1947.

Hayward J, Chitty LS. 2018. Beyond screening for chromosomal abnormalities: advances in non-invasive diagnosis of single gene disorders and fetal exome sequencing. *Semin Fetal Neonatal Med* **23:** 94–101. doi:10.1016/j.siny.2017.12.002

Huang T, Dougan S, Walker M, Armour C, Okun N. 2018. Trends in the use of prenatal testing services for fetal aneuploidy in Ontario: a descriptive study. *CMAJ* **6:** E436–E444. doi:10.9778/cmajo.20180046

Johnston J, Farrell R, Parens E. 2017. Supporting women's autonomy in prenatal testing. *N Engl J Med* **377:** 505–507. doi:10.1056/NEJMp1703425

Kessler S. 1992. Psychological aspects of genetic counseling. VII. Thoughts on directiveness. *J Genet Couns* **1:** 9–17. doi:10.1007/BF00960080

Kessler S. 1997. Psychological aspects of genetic counseling. XI. Nondirectiveness revisited. *Am J Med Genet* **72:** 164–171. doi:10.1002/(SICI)1096-8628(19971017)72:2<164::AID-AJMG8>3.0.CO;2-V

Kraft SA, Duenas D, Wilfond BS, Goddard KAB. 2019. The evolving landscape of expanded carrier screening: challenges and opportunities. *Genet Med* **21:** 790–797. doi:10.1038/s41436-018-0273-4

Leinweber KA, Columbo JA, Kang R, Trooboff SW, Goodney PP. 2019. A review of decision aids for patients considering more than one type of invasive treatment. *J Surg Res* **235:** 350–366. doi:10.1016/j.jss.2018.09.017

Lewis C, Hill M, Chitty LS. 2016. A qualitative study looking at informed choice in the context of non-invasive prenatal testing for aneuploidy. *Prenat Diagn* **36:** 875–881. doi:10.1002/pd.4879

Markel H. 1997. Scientific advances and social risks: historical perspectives of genetic screening programs for sickle cell disease, Tay–Sachs disease, neural tube defects, and Down syndrome, 1970-1997. *NIH-DOE Working Group on Ethical, Legal and Social Implications of Human Genome Research.* NIH, Bethesda. https://www.ncbi.nlm.nih.gov/books/NBK231976

Minear MA, Lewis C, Pradhan S, Chandrasekharan S. 2015. Global perspectives on clinical adoption of NIPT. *Prenat Diagn* **35:** 959–967. doi:10.1002/pd.4637

Minkoff H, Berkowitz R. 2014. The case for universal prenatal genetic counseling. *Obstet Gynecol* **123:** 1335–1338. doi:10.1097/AOG.0000000000000267

Muller C, Cameron LD. 2016. *It's complicated*—factors predicting decisional conflict in prenatal diagnostic testing. *Health Expect* **19:** 388–402. doi:10.1111/hex.12363

O'Brien JS, Okada S, Chen A, Fillerup DL. 1970. Tay-Sachs disease. Detection of heterozygotes and homozygotes by serum hexosaminidase assay. *N Engl J Med* **283:** 15–20. doi:10.1056/NEJM197007022830104

O'Conner AM, Rostom A, Fiset V, Tetroe J, Entwistle V, Llewellyn-Thomas H, Holmes-Rovner M, Barry M, Jones J. 1999. Decision aids for patients facing health treatment or screening decisions: systematic review. *BMJ* **319:** 731–734. doi:10.1136/bmj.319.7212.731

Palomaki GE, Deciu C, Kloza EM, Lambert-Messerlian GM, Haddow JE, Neveux LM, Ehrich M, van den Boom D, Bombard AT, Grody WW, et al. 2012. DNA sequencing of maternal plasma reliably identifies trisomy 18 and trisomy 13 as well as Down syndrome: an international collaborative study. *Genet Med* **14:** 296–305. doi:10.1038/gim.2011.73

Pergament D, Ilijic K. 2014. The legal past, present and future of prenatal genetic testing: professional liability and other legal challenges affecting patient access to services. *J Clin Med* **3:** 1437–1465. doi:10.3390/jcm3041437

Ramdaney A, Hoskovec J, Harkenrider J, Soto E, Murphy L. 2018. Clinical experience with sex chromosome aneuploidies detected by noninvasive prenatal testing (NIPT): accuracy and patient decision-making. *Prenat Diagn* **38:** 841–848. doi:10.1002/pd.5339

Salema D, Townsend A, Austin J. 2019. Patient decision-making and the role of the prenatal genetic counselor: an exploratory study. *J Genet Couns* **28:** 155–163.

Schoonen M, van der Zee B, Wildschut H, de Beaufort I, de Wert G, de Koning H, Essink-Bot ML, Steegers E. 2012. Informing on prenatal screening for Down syndrome prior to conception. An empirical and ethical perspective. *Am J Med Genet A* **158A:** 485–497. doi:10.1002/ajmg.a .35213

Seavilleklein V. 2009. Challenging the rhetoric of choice in prenatal screening. *Bioethics* **23:** 68–77. doi:10.1111/j .1467-8519.2008.00674.x

Seror V, Ville Y. 2009. Prenatal screening for Down syndrome: women's involvement in decision-making and their attitudes to screening. *Prenat Diagn* **29:** 120–128. doi:10.1002/pd.2183

Stacey D, Légaré F, Lewis K, Barry MJ, Bennett CL, Eden KB, Holmes-Rovner M, Llewellyn-Thomas H, Lyddiatt A, Thomson R, et al. 2017. Decision aids for people facing health treatment or screening decisions. *Cochrane Database of Syst Rev* **1:** CD001431.

Stern AM. 2009. A quiet revolution: the birth of the genetic counselor at Sarah Lawrence College, 1969. *J Genet Couns* **18:** 1–11.

Stern AM. 2012. *Telling genes: the story of genetic counseling in America.* Johns Hopkins University Press, Baltimore.

Valenti C, Schutta EJ, Kehaty T. 1968. Prenatal diagnosis of Down's syndrome. *Lancet* **2:** 220. doi:10.1016/S0140-6736(68)92656-1

Wald NJ, Cuckle H. 1977. Maternal serum α-fetoprotein measurement in antenatal screening for anencephaly and spina bifida in early pregnancy. Report of the U.K. Collaborative Study on α-Fetoprotein in Relation to Neural-Tube Defects. *Lancet* 1323–1332.

Wapner RJ, Martin CL, Levy B, Ballif BC, Eng CM, Zachary JM, Savage M, Platt LD, Saltzman D, Grobman WA, et al. 2012. Chromosomal microarray versus karyotyping for prenatal diagnosis. *N Engl J Med* **367:** 2175–2184. doi:10.1056/NEJMoa1203382

Weil J. 2000. *Psychosocial genetic counseling.* Oxford, New York.

Zhang J, Li J, Saucier J, Feng Y, Jiang Y, Sinson J, McCombs A, Schmitt ES, Peacock S, Chen S, et al. 2019. Non-invasive prenatal sequencing for multiple Mendelian monogenic disorders using circulating cell-free fetal DNA. *Nat Med* **25:** 439–447. doi:10.1038/s41591-018-0334-x

Impact of Emerging Technologies in Prenatal Genetic Counseling

Blair Stevens

McGovern Medical School at UTHealth in Houston, Department of Obstetrics, Gynecology and Reproductive Sciences, Houston, Texas 77030, USA

Correspondence: blair.k.stevens@uth.tmc.edu

For decades, prenatal testing has been offered to evaluate pregnancies for genetic conditions. In recent years, the number of testing options and range of testing capabilities has dramatically increased. Because of the risks associated with invasive diagnostic testing, research has focused on the detection of genetic conditions through screening technologies such as cell-free DNA. Screening for aneuploidy, copy number variants, and monogenic disorders is clinically available using a sample of maternal blood, but limited data exist on the accuracy of some of these testing options. Additional research is needed to examine the accuracy and utility of screening for increasingly rare conditions. As the breadth of prenatal genetic testing options continues to expand, patients, clinical providers, laboratories, and researchers need to find collaborative means to validate and introduce new testing technologies responsibly. Adequate validation of prenatal tests and effective integration of emerging technologies into prenatal care will become even more important once prenatal treatments for genetic conditions become available.

In prenatal genetic counseling, counselors offer pregnant women and their partners information about genetic risks to their developing fetus and provide psychological counseling about test results and their implications. Prenatal genetic counseling involves understanding a patient's needs and values, assessing risk factors, discussing testing options, interpreting test results, and providing counseling to address the psychological responses of parents. Although genetic counseling in the prenatal setting may not result in testing, advances in prenatal testing technology have shaped the practice of prenatal counseling. As early as the 1950s, the ability to use amniocentesis with cytogenetic and biochemical analysis allowed couples to learn their fetus was affected by a genetic condition. By the 1970s, amniocentesis was routinely offered to women over 35 yr of age or with a family history of a genetic condition (Resta 2002). The availability of amniocentesis and its risks and limitations were the impetus for the development of the earliest iteration of prenatal genetic counseling, which is now evolving to face new challenges related to the development and expansion of cell-free DNA screening technologies.

Although technological advancements in genetics have revolutionized many facets of

Cite this article as *Cold Spring Harb Perspect Med* doi: 10.1101/cshperspect.a036517

health care, the impact in obstetrics has novel attributes that make genetic counseling critical to prenatal care. First, there are few treatments for genetic conditions prior to birth and a prenatal diagnosis may not impact the management of a pregnancy or delivery. Therefore, the decision to undergo prenatal genetic testing depends on a patient's values, needs, and desires. Further, the decision to undergo diagnostic genetic testing requires invasive testing, with a procedure-related risk of miscarriage. Finally, prenatal genetic testing for some families involves facing a decision about whether or not to terminate a wanted pregnancy if a genetic condition is identified. As a consequence of these issues, prenatal testing is notably different from other types of tests performed during pregnancy. Genetic counseling by an expert provider is warranted to properly address the potential benefits, risks, limitations, and consequences of testing.

This review aims to provide a snapshot of prenatal genetic counseling in 2019, an overview of the use of prenatal diagnostic and screening tests, the challenges of prenatal testing, and a glimpse into future directions of reproductive genetic testing and counseling.

GENETIC COUNSELING

Prenatal genetic counseling is relevant to all pregnancies as there is a 3%–5% baseline risk for a birth defect or genetic condition (Centers for Disease Control and Prevention 2008). Genetic counseling can be provided by a variety of health-care providers including genetic counselors, obstetricians, maternal–fetal medicine specialists, geneticists, and nurses. Historically, referrals for genetic counseling were limited to patients with advanced maternal age, abnormal screening results, a significant family history, or suggestive ultrasound findings. Although this referral practice continues to predominate, the new, more accurate, and lower-risk genetic tests have increased the number of pregnant women with fewer or none of these risk factors attending genetic counseling for risk assessment, discussion of genetic testing options, decision-making, and counseling centered on values and beliefs.

A typical prenatal genetic counseling session includes the following (the information is tailored to each patient based on the indication and delivered with sensitivity to her educational, emotional and cultural needs):

1. *Contracting*: Exploring the patient's needs and goals through a dialog called "contracting" is useful for understanding the purpose of the visit and the patient's values and needs. This understanding can be used by the counselor to tailor information and help facilitate patient decision-making. Contracting includes an assessment of the patient's concerns, psychological well-being, and understanding, all of which leads to shared goal setting.

2. *Risk Assessment*: To facilitate informed decision-making, a genetic counselor evaluates and explains risks to the fetus. Personal and family medical history is assessed by collecting health information for a three-generation pedigree. Medical records are also reviewed for maternal age, screening results, and a history of any teratogen exposures. Additionally, if an ultrasound has been done, certain findings may indicate a higher risk for the fetus to have a condition. Collectively, this information is used to determine a baseline fetal risk assessment.

3. *Testing Review*: General population testing is available for all pregnant patients. Additional genetic testing may be indicated based on the individualized risk assessment. There is often a range of testing options available and pretest counseling by a knowledgeable provider is important. Key components of pretest counseling include a discussion of the purpose of testing, how the test is performed, risks and benefits of testing, possible results, turnaround time, and communication of results (Janssens et al. 2017). Counseling and decision-making are discussed in Goldman (2019).

4. *Psychological Counseling*: In addition to roles as educator and testing facilitator, genetic counselors also explore the psychological needs of patients as they often experience

Cite this article as *Cold Spring Harb Perspect Med* doi: 10.1101/cshperspect.a036517

feelings of shock, anxiety, disbelief, guilt, grief, and mourning when a risk factor is identified or when results are abnormal and unexpected, particularly when significant uncertainty about the future health and well-being of the fetus exists (Raymer 2004; Bernhardt et al. 2013; Werner-Lin et al. 2016)

5. *Plan for Follow-Up*: Expectations for follow-up are communicated to the patient, if needed. This may include plans for results disclosure (in-person vs. over the phone), referrals to a specialist, and communication back to the referring provider. Written information and/or contact information for patient support groups and advocacy networks are also helpful resources, depending on the nature and outcome of the session.

OVERVIEW OF GENETIC TESTING OPTIONS

The discussion of prenatal genetic testing options with patients is essential as the number and types of testing options change often. A knowledgeable provider with expertise in the risks, benefits, and limitations of prenatal genetic testing should facilitate this discussion.

Prenatal Screening Tests

Screening tests evaluate the risk for a specific condition or group of conditions but cannot definitively diagnose or rule them out. In the prenatal context, screening tests provide information with minimal risk to the pregnancy as they are typically performed on maternal blood or via ultrasound. Prenatal screening tests are available for a variety of genetic conditions, but the most common tests screen for chromosome conditions (such as Down syndrome) or autosomal recessive conditions (such as sickle cell anemia and cystic fibrosis).

Screening for Fetal Aneuploidy

Chromosomal aneuploidy is extra or missing chromosomal material. Full aneuploidy refers to an extra or missing chromosome that results in a condition. Risk factors for full aneuploidy include increasing maternal age, previous child

with aneuploidy, and ultrasound abnormalities (Gardner et al. 2012). Because of their sporadic nature, screening for chromosome abnormalities should be available to all pregnant patients, and their decision to test or not to test should be supported. Various screening methods can be used depending on the patient preference, insurance coverage, and indication.

Multiple Marker Screening: Traditional screening tests utilize biochemical markers in maternal blood that are produced by the fetus and placenta. Patterns of abnormal analyte levels are associated with conditions such as Down syndrome, trisomy 18, and open neural tube defects. Additionally, less common conditions, such as Smith–Lemli–Opitz syndrome and triploidy, may cause abnormal maternal serum markers. Multiple marker screening is designed to provide a risk assessment with detection rates ranging from 69% to 96%, depending on the number of analytes assessed, the use of first- and/or second-trimester maternal blood samples, and the use of ultrasound markers evaluated in conjunction with maternal analytes. The false-positive rate for these various screening methods is typically ~5% (Practice Bulletin No. 163 2016).

Cell-Free DNA (cfDNA)/Noninvasive Prenatal Testing (NIPT)/Noninvasive Prenatal Screening (NIPS): First introduced in 2011, this novel screening test evaluates the likelihood of fetal chromosome abnormalities from cfDNA in maternal blood. After ~10 wk gestation, enough cfDNA from the trophoblast cells of the placenta is present in maternal blood and can be sequenced to determine the chromosome of origin (Committee Opinion No. 640 2015). Using a counting method, the number of fragments from the chromosome of interest can be evaluated and samples with a significantly greater or lesser number of these counts are reported out as increased risk. Alternatively, single-nucleotide polymorphisms (SNPs) can be analyzed on the chromosomes of interest to determine whether an aneuploidy is suspected. Both methods have shown very high detection rates and low false-positive rates (Table 1).

Initially, cfDNA screening was offered exclusively to women at high risk. However, it is in-

Table 1. Detection rate and false-positive rates of cell-free DNA screening

Condition	Detection rate (95% CI)	False-positive rate (95% CI)
Trisomy 21	>99% (99.1%–99.9%)	0.04% (0.02%–0.07%)
Trisomy 18	97.9% (94.9%–99.1%)	0.04% (0.03%–0.07%)
Trisomy 13	99% (65.8%–100%)	0.04% (0.02%–0.07%)
Monosomy X	95.8% (70.3%–99.5%)	0.14% (0.05%–0.38%)
Sex chromosomes abnormalities (other than monosomy X)	>99% (83.6%–100%)	0.004% (0%–0.08%)

Data in table adapted from Gil et al. 2017.

creasingly utilized in average-risk populations as it has been shown to have similar sensitivity and specificity in low-risk pregnancies (Pergament et al. 2014). Although the test performance is similar in low-risk populations, the meaning of a screen positive result can differ greatly because of differences in positive predictive value (PPV). Positive predictive value is the likelihood that any given positive screen is a true positive; PPV is impacted not only by sensitivity and specificity of the test but also by the prevalence of the condition at issue. For example, if NIPT has a 99.9% sensitivity and specificity, and a 45-yr-old woman gets a positive screen for Down syndrome, she has a 98% PPV. A 25-yr-old woman, on the other hand, has a 49% PPV even though the NIPT test she took has the same sensitivity and specificity. All other things being equal, women with a higher a priori risk for Down syndrome will have a higher PPV than women at low risk. To aid providers in calculating the positive and negative predictive values of screening tests, a calculator was created by the National Society of Genetic Counselors and the Perinatal Quality Foundation (PQF) (https://www.perinatalquality.org/Vendors/NSGC/NIPT/). Using the prevalence, sensitivity, and specificity of the condition, positive and negative predictive values can be generated to aid in posttest counseling.

ACOG and the Society of Maternal–Fetal Medicine (SMFM) recommend that laboratories report PPVs to assist with posttest counseling (Committee Opinion No. 640 2015). The PPV can also be calculated manually if the test sensitivity, specificity, and prevalence of the condition are known. Sensitivity and specificity data are readily available for conditions that have been well-studied, such as trisomy 21 and triso-

my 18, but limited data are available on the sensitivity and specificity for other conditions, such as sex chromosome abnormalities and microdeletion syndromes. Despite the existence of a PPV and NPV calculator, accurate postscreen risk assessment cannot be performed without reliable sensitivity and specificity data.

False-positive and false-negative cfDNA results do occur, albeit less commonly than with multiple marker screening (Gil et al. 2017). Several maternal and fetal factors can lead to false-positive or -negative screening results, including mosaicism (Grati et al. 2014; Hartwig et al. 2017). Mosaicism is defined as two or more cell lines with different genetic compositions (Gardner et al. 2012). NIPT results reflect the makeup of the placenta (Lo et al. 1997), and if the placenta contains abnormal cells, then it can lead to a positive NIPT result. If the mosaicism is only in the placenta, which is known as confined placental mosaicism (CPM), the NIPT results may not reflect the fetal karyotype (Van Opstal and Srebniak 2016). On the other hand, if the placenta contains cells with normal chromosomes, but the fetus is chromosomally abnormal or mosaic, NIPT results will yield a false-negative result.

Other causes of false-positive NIPT results include the demise of a twin with a chromosome abnormality, a maternal chromosome abnormality, maternal mosaicism for a chromosome abnormality, and a maternal condition such as cancer (Bianchi et al. 2015; Hartwig et al. 2017; Leonard et al. 2018; Kim et al. 2019; Yu et al. 2019).

A false-negative result, the failure to detect a fetal chromosome abnormality, is commonly due to low fetal fraction (FF). The FF is the

Cite this article as *Cold Spring Harb Perspect Med* doi: 10.1101/cshperspect.a036517

percentage of cfDNA in maternal blood derived from the placenta. Pregnant women have on average 10% FF, meaning that 90% of the cfDNA is maternal in origin (Ashoor et al. 2013). An insufficient level of cfDNA from the pregnancy make is less likely to detect a fetal abnormality. Low FF has been associated with many factors such as increased maternal weight and early gestational age, but also with maternal conditions such as lupus and maternal therapy with blood thinners (Ashoor et al. 2013; Burns et al. 2017; Dabi et al. 2018). Low FF has also been found to be associated with genetic conditions, such as trisomy 18, trisomy 13, and triploidy (McKanna et al. 2019). Therefore, when a low FF is found and results cannot be reported, diagnostic testing is recommended over repeat testing via NIPT, because a repeat test is not guaranteed to yield a result and may further delay obtaining diagnostic information (Committee Opinion No. 640 2015).

Screening for Microdeletions

NIPT has the capacity to screen for countless numbers of genetic conditions as the entire nuclear genome from trophoblastic cells exists in maternal blood. As previously described, aneuploidy is evaluated by detecting an abundance or deficiency of cfDNA from a particular chromosome or by observing a trisomic or monosomic SNP pattern. Using this same available data, pregnancies can also be evaluated for microdeletions and microduplications, such as 22q11 deletion syndrome. Some argue that it is appropriate to screen the general population for microdeletion syndromes as they are more prevalent than Down syndrome for women in their early and mid-20s.

Most clinical laboratories offer the option of microdeletion screening for a select number of microdeletion syndromes, such as 22q11, 1p36, 4p, 5p, 8q, 11q, and 15q deletions. Sensitivity and specificity data are limited because of the rarity of these conditions. For example, Helgeson et al. (2015) reported 100% PPV for detection of 4p and 11q deletions (95% CI 5.5%–100%), but only detected one case of each in their study population, whereas two of six cases

(95% CI 24.1%–94%) of 5p deletion cases were found to be false positives. Another study reports PPVs ranging from 0% to 66.7%, but 40% of cases had an unknown outcome (Martin et al. 2018). To address the limited evidence on detection rates for rare conditions, laboratories have created artificial plasma mixtures or simulated samples to evaluate screening accuracy (Wapner et al. 2015). It is unknown whether these spiked samples can accurately assess how the test will function with actual patients.

It is important to note that even if high sensitivity and specificity are confirmed, the PPV will be low in the absence of other risk factors (family history, ultrasound findings, etc.) because of the rarity of these conditions. For example, Wolf–Hirschhorn syndrome is a condition caused by a deletion of chromosome 4p. The prevalence of Wolf–Hirschhorn is ~1 in 20,000. Even with a 99.9% sensitivity and specificity, the PPV of a positive cfDNA screen would be only 5%.

Screening for Monogenic Disorders via Cell-Free DNA

Evidence regarding the accuracy of clinically available monogenic disorder screening is even harder to find (Zhang et al. 2019). Screening for autosomal dominant conditions via maternal blood is clinically available for select de novo or paternally inherited conditions, but not maternally inherited dominant conditions. Screening for de novo or paternally inherited conditions is hypothetically feasible because the testing evaluates the presence of pathogenic variants that are not present in the maternal genome. Thus, the presence of the variant in maternal blood indicates its presence in the fetus (Hudecova and Chiu 2017).

Data appear promising for the detection of de novo or paternal variants; however, outcome data were available in <50% of reported cases in the single peer-reviewed article available to date (Zhang et al. 2019). Furthermore, positive cfDNA results were detected only in pregnancies with abnormal ultrasound findings consistent with the screening result or those with a positive paternal family history. None of the pregnancies tested for other indications (abnormal screening results, advanced maternal age, advanced paternal age, other positive family history) was found to have abnormal results. In other words, all cases with positive results had a high a priori risk based on ultrasound findings or based on a 50% risk to inherit the paternal mutation. Proof of clinical utility in the general population requires further study and additional research is needed to confirm the accuracy of screening for monogenic conditions via cfDNA.

Screening for known autosomal recessive, autosomal dominant, and X-linked conditions via cfDNA is also clinically available, but there are no peer-reviewed publications that demonstrate the accuracy of this method for that purpose (https://www.progenity.com/tests/resura).

Reproductive Carrier Screening

Screening parents for heritable autosomal recessive and X-linked conditions they may carry is far simpler than screening the fetus for de novo conditions, as parental DNA can be sequenced or genotyped for inherited pathogenic variants. This type of reproductive carrier screening has been available for decades; however, the number of conditions for which carrier screening is available continues to expand. All individuals are thought to carry on average four to five autosomal recessive conditions (Morton et al. 1956), and a lack of family history is not an effective screening tool, as both parents must carry the same genetic condition to be at increased risk and therefore children with recessive conditions are typically born to families without a family history.

Traditionally, carrier screening has been offered according to ethnic group (Gregg and Edwards 2018). For example, African–American women were offered sickle cell anemia screening and Caucasian women were offered cystic fibrosis screening. With the rising affordability of genetic testing and the high rates of mixed or unknown ancestry, expanded carrier screening (ECS) is now offered on a routine basis. Expanded carrier screening entails screening for more conditions than are recommended by professional guidelines and enables the same screening test to be offered all patients, regardless of ethnicity or ancestry (Gregg and Edwards 2018).

If an individual is found to be a carrier of an autosomal recessive condition, protocol requires offering the partner carrier screening. The fetus is at risk of being affected only when both parents carry a variant in the same gene. If the partner also tests positive and both parents carry a disease-causing variant, there is a 25% risk to each fetus of being affected. With a negative test, the risk of the partner being a carrier is reduced but not eliminated because carrier screening does not detect 100% of carriers. The detection rate for carrier screening varies depending on the condition of the patient, ethnicity of the patient, and testing methodology. Carrier screening can utilize targeted genotyping panels or genome sequencing to identify pathogenic variants among many genes. The advantage of sequencing is a higher detection rate of pathogenic variants. Yet sequencing also identifies likely pathogenic variants and variants of uncertain significance, which are changes that are uninterpretable based on available evidence and thus have unclear or unknown pathogenicity. The reporting of likely pathogenic variants or variants of uncertain significance is inconsistent and at the discretion of the laboratory that conducts the screening.

Carrier screening panels have grown to include hundreds of autosomal recessive and X-linked recessive conditions. As of 2017, screening for cystic fibrosis and spinal muscular atrophy was recommended for all patients (Committee on Genetics 2017). Additional screening for hemoglobinopathies, thalassemia or conditions that occur more often among individuals of Ashkenazi Jewish background were recommended based on ethnicity. Fragile X syndrome screening was recommended for those with a positive family history (Committee on Genetics 2017). Expanded panels have increased the number of conditions available for screening but panels vary by laboratory. Multiple publications have put forth guidelines from which the selection of conditions on expanded panels should be based, but interpretation of these guidelines can be challenging and adher-

Cite this article as Cold Spring Harb Perspect Med doi: 10.1101/cshperspect.a036517

ence to these guidelines is not universal (Grody et al. 2013; Edwards et al. 2015; Stevens et al. 2017).

As panels expand, the chances of detecting carrier status in a patient increase. Although the ability to diagnose at-risk couples may be valued as a benefit, risks and limitations of expanded screening also exist. Possible drawbacks include the psychological impact of receiving positive carrier screen results, cost and limited availability of follow-up testing for the father, and the cost of time spent on follow-up. More evidence is needed to evaluate expanded panels for their added benefit to current standards for carrier screening.

Prenatal Diagnostic Tests

Diagnostic tests require an invasive procedure to obtain a direct fetal sample for testing. This direct testing allows for more definitive results and the ability to test for more conditions than most screening tests. The major disadvantage is that the procedures required to obtain a diagnostic sample come with risks to the pregnancy, including miscarriage.

There are two primary diagnostic procedure currently in use for diagnostic genetic testing: chorionic villus sampling (CVS) and amniocentesis (amnio). CVS entails a biopsy of the chorionic villi from the placenta in the first trimester of pregnancy (10–14 wk). The sample can be obtained by passing a catheter transcervically or inserting a needle transabdominally into the placenta. Amniocentesis is done by sampling amniotic fluid transabdominally and can be per-

formed later in pregnancy, from 16-wk gestation on. Both procedures are associated with high diagnostic accuracy and low procedure-related risks.

The types of tests that can be performed using CVS or amniocentesis samples are summarized in Table 2. In addition to cytogenetic and molecular genetic testing, amniotic fluid can also be evaluated for viral infections, elevated alpha-fetoprotein and acetylcholinesterase levels indicative of open neural tube defects and biochemical abnormalities.

Prenatal Diagnosis of Chromosome Aneuploidy

Karyotyping and fluorescence in situ hybridization (FISH) testing are routinely ordered to detect aneuploidy. Indications for this testing may include advanced maternal age, a previous child with aneuploidy, or a positive screening test. Regardless of the indication, testing for additional chromosome conditions such as microdeletion and microduplication syndromes may also be considered when invasive testing is pursued.

Prenatal Diagnosis of Microdeletions and Microduplications

Chromosome microarray is used to detect smaller chromosome deletions and duplications and is increasingly standard, particularly following the detection of fetal abnormalities on ultrasound (Durham et al. 2019). Studies have shown that women 35 yr of age and older and those

Table 2. Common prenatal diagnostic tests

Test	Evaluated conditions	Examples
Karyotype	Aneuploidy, large deletions and duplications (>5 Mb), chromosome rearrangements	Down syndrome, Wolff–Hirschhorn, translocations
Fluorescence in situ hybridization (FISH) for aneuploidy	Common aneuploidies (21, 18, 13, X, Y)	Down syndrome, Turner syndrome
Chromosome microarray (CMA)	Microdeletion and microduplication syndromes (<5 Mb)	22q11 deletion syndrome, Williams syndrome
Next-generation sequencing (single-gene testing, NGS panels, exome sequencing)	Monogenic disorders	Noonan syndrome, cystic fibrosis

(Mb) Megabase.

who have a positive screening test result, have a 1%–2% chance that they will learn their fetus is affected by a condition that would not be detected by routine karyotyping. If ultrasound abnormalities are detected, the chance of a significant finding in the fetus increases to 6%. It is important to note that a chromosome microarray also has a 1%–2% chance to detect a change that cannot be characterized as benign or pathogenic, also known as a variant of uncertain significance (Wapner et al. 2012). Patients who receive uncertain results report intense negative reactions, demonstrating the need for adequate pretest counseling regarding the possibility of abnormal or uncertain results (Bernhardt et al. 2013).

Prenatal Diagnosis of Monogenic Disorders

Although karyotyping and chromosome microarray results can effectively identify chromosome abnormalities, sequencing is needed to identify variants within genes. The majority of sequencing tests use next-generation sequencing (NGS), a quicker, cost-effective sequencing approach. A variety of sequencing tests are available prenatally depending on the indication. For example, a couple known to carry variants in the CFTR gene can undergo testing for the known variants. On the other hand, testing a fetus found to have congenital anomalies may require a multigene panel depending on the differential diagnosis. These panels could include a number of genes associated with a single condition, such as Noonan syndrome, or a number of genes related to a phenotype, such as a skeletal dysplasia panel.

When sequencing panels have not identified a cause or there is not an appropriate panel for the indication, exome sequencing may be considered. Exons are the translated parts of the genetic code and they make up the exome, which comprises ~1%–2% of the genome. The majority of disease-causing variants are found in the exome, so targeting the exome for sequencing is a more feasible method to detect sequence variants (Ng et al. 2009; Stenson et al. 2009; Bainbridge et al. 2010).

Contemporary Use of Prenatal Screening and Diagnostic Testing

Practice guidelines recommend that all pregnant women be offered prenatal screening and diagnostic testing, regardless of maternal age or risk factors. Previously, guidelines recommended invasive diagnostic testing only for women at increased risk. The change in recommendations is largely based on a belief that patients and their partners should weigh the risks and benefits of testing in light of their preferences and values to determine the use of prenatal genetic testing.

1. *Screening*: ACOG recommends that women be offered some form of screening for aneuploidy. There is no one superior prenatal screen for all circumstances, so the provider and patient together must make an individualized determination of the best screening option, if any (Cai et al. 2016). ACOG specifically states in a 2015 Committee Opinion (Committee Opinion No. 640 2015) and a 2019 Practice Advisory (https://www.acog.org/Clinical-Guidance-and-Publications/Practice-Advisories/Cell-free-DNA-to-Screen-for-Single-Gene-Disorders) that screening for microdeletion syndromes and monogenic disorders is not recommended because of lack of clinical validation data.

   Carrier screening for recessive conditions should also be offered all women who present for care, ideally prior to conceiving. There is no one carrier screening panel (ethnicity based vs. expanded) that is ideal for all patient populations, so the risks, benefits and limitations of each panel option should be reviewed with patients to facilitate informed decision-making.

2. *Diagnostic Testing*: Uptake depends on many factors, but the rate of diagnostic testing has dropped significantly since the introduction of cfDNA screening despite its diagnostic advantages over screening tests (Warsof et al. 2015; Stevens et al. 2019). When CVS or amniocentesis is desired, patients should routinely be offered both karyotyping and chromosome microarray along with a discussion of the risks, benefits, and limitations of each (Committee on Genetics and the Society for Maternal-Fetal Medicine 2016). Addi-

tional testing options may be offered depending on the testing indication, such as NGS panels for monogenic disorders, biochemical testing for open neural tube defects or biochemical testing for metabolic conditions.

Exome sequencing (ES) is not recommended by ACMG for routine testing of the fetus. However, when other recommended prenatal tests have not arrived at a diagnosis, ES may be considered. Limitations of ES should be emphasized, including longer turnaround times, false positives, false negatives, and variants of unknown significance (ACMG Board of Directors 2012).

Similarly, the International Society of Prenatal Diagnosis (ISPD), PQF, and SMFM do not recommend prenatal ES be used routinely, and they note that when it is performed, it should be done under a research protocol. However, use of prenatal ES can be considered when a genetic condition is suspected and the case is managed with expert guidance of a genetic professional (International Society for Prenatal Diagnosis 2018). Prenatal ES is used almost exclusively during pregnancy when ultrasound abnormalities are detected that increase suspicion for a genetic condition. Although this will likely change in the future, laboratories are not offering prenatal ES for apparently healthy pregnancies. The diagnostic yield of prenatal ES varies depending on the indication for testing, but ranges from 10% to 30%.

3. *Introduction of New or Expanded Clinical Testing*: It is not uncommon for national guidelines on genetic testing to lag behind clinical care. For example, cfDNA screening was introduced in clinical practice in 2011 after publication of a multicenter research study performed on more than 4000 pregnancies by 27 centers, including more than 200 pregnancies diagnosed with Down syndrome (Palomaki et al. 2011). ACOG did not endorse it as a screening option for high-risk pregnancies until December of 2012 (American College of Obstetricians and Gynecologists Committee on Genetics 2012). It is still not specifically endorsed for low-risk populations (Committee Opinion No. 640 2015).

Subsequent expansions of cfDNA, including screening for microdeletions and monogenic disorders, have been introduced clinically without sufficient validity testing. For example, screening for microdeletions, including 22q11, 5p, 4p, and 1p36 deletions, became clinically available in late 2013. Peer-reviewed validation data were published two years later in 2015 (Helgeson et al. 2015; Wapner et al. 2015; Zhao et al. 2015), and these studies used a combination of simulated or artificial samples as well as maternal plasma DNA with microdeletions. Similarly, genome-wide NIPT was introduced in 2015 by one company after showcasing data in a poster presentation at the ISPD conference. Clinical validation was subsequently reported in a peer-reviewed journal in August of 2016 (Lefkowitz et al. 2016), followed by data on the first 10,000 cases published December of 2017 (Ehrich et al. 2017). Screening for monogenic disorders became clinically available in 2017 and only a white paper published by the performing laboratory was made available (https://bcm.box .com/shared/static/eehibpd8d7bpobn7wypiwoe f3iv9z8hn.pdf) until 2019 when the first peer-reviewed research was published on the performance of cfDNA screening for these monogenic conditions (Zhang et al. 2019).

Up until the introduction of cfDNA for common aneuploidies, research was conducted and validation data was published prior to the clinical introduction of a screening test. The expansion of cfDNA screening to include microdeletions, genome-wide screening, and monogenic disorders have upended this standard of care. Clinical samples and patients undergoing clinical testing are now used to gather validation data. This industry-driven approach to prenatal screening presents major challenges for patients and providers, leaving the latter to offer screening tests whose risks and detection rates are poorly understood. If a provider offers nonvalidated tests, patients do not have sufficient information to make informed decisions and may not be aware that their results will be the source of data to validate screening.

Major obstacles to conducting prospective validation studies for very rare conditions are costs and time. A validation study for detection of trisomy 21, the most common chromosome abnormality, took 27 centers enrolling more than 4000 pregnant patients (Palomaki et al. 2011). Significantly more participants are required for a condition that occurs in 1 in 5000 pregnancies. One may argue that unlike Down syndrome and trisomy 18, there are no alternative screening methods for conditions like 22q11 deletion syndrome, and a somewhat accurate test may be better than no screening test. Therefore, as long as adequate pretest counseling is performed to ensure the limitations of expanded testing are understood, arguably patients seeking information should be able to obtain as much as possible while avoiding procedure-related risks. However, there are disadvantages and risks associated with offering inadequately validated screening tests to patients that informed consent may not be adequate to address.

As prenatal screening expands to rarer conditions using more advanced technology, patients, clinicians, laboratories, researchers, and policy makers must work together to integrate testing responsibly into clinical care. Missing outcome data and unknown sensitivity and specificity for clinically available tests is not appropriate and a solution is imperative as the pace of new technology and patient demand continue to accelerate. Infrastructure to facilitate data collection and pregnancy outcomes is needed as are patient-reported preferences for delivery of information. Ideally, nonvalidated tests should only be offered to patients under a research protocol in a prospective manner. If outcome data can be collected on participating patients, then sensitivity and specificity data can be determined. If the option of expanded testing is limited to research protocols through testing laboratories and testing is provided without additional cost to the patient, interested patients and professionals may be likely to participate.

Past validation studies of prenatal chromosome microarray demonstrate the success of such efforts (Wapner et al. 2012). Professional organizations will be unable to recommend expanded screening tests without evidence and insurance providers are unlikely to cover services that are not recommended by professional organizations. In addition to test accuracy,

data on patient experiences with newly introduced tests that assess understanding and psychological well-being can be used to guide the offer and disclosure of prenatal testing results in a sensitive and responsible way. Public policy on use of prenatal testing may consider allocation of funds for data collection to offset costs fronted by testing laboratories.

FUTURE DIRECTIONS IN PRENATAL GENETIC TESTING

Despite technological advances, it is not possible to diagnose all suspected genetic conditions using current genetic testing techniques. As tests improve, the ability to interpret genetic and genomic data must keep pace, because once these tests are established in the pediatric setting they will be rapidly introduced in the prenatal setting. In the prenatal setting, uncertainty generated by testing is further compounded by limited designation of a fetal phenotype, which is typically restricted to ultrasound imaging. Thus, identifying exome or genome sequencing data relevant to the fetal phenotype can be challenging. Advancements in fetal imaging are needed along with advancements in genetic testing and interpretation of results.

Studies have shown that women are interested in learning about the health of their fetus, but do not want to put their pregnancy at risk from the procedure (Kalynchuk et al. 2015; Sullivan et al. 2019). For this reason, much research has focused on the ability to test for genetic conditions using maternal blood. Some of the emerging areas of research are outlined below.

1. *Improved Cell-Free DNA Screening for Monogenic Diseases*: Screening tests for select de novo dominant conditions and known inherited variants are currently available. Although screening for the presence of a de novo or paternally inherited variant is relatively straightforward, cfDNA screening for the maternal variant requires more sophisticated technology as the fetal variant must be detected in the presence of maternal cfDNA containing the same variant. Screening for the paternal variant is more challenging if

the parental variants are identical, as may be the case in consanguineous couples or among those carrying founder mutations.

Two testing approaches, relative mutation dosage by droplet digital PCR and relative haplotype dosage by massively parallel sequencing (MPS), are techniques used to screen for monogenic disorders using cfDNA. Relative mutation dosage analyzes the mutant allele proportion to the wild-type allele proportion in maternal serum, expecting greater mutant allele proportion in an affected pregnancy, greater wild-type proportion for an unaffected pregnancy and a balanced ratio when the fetus is a carrier. Relative haplotype dosage involves determining maternal and paternal haplotypes from genomic regions linked to the mutation. Then SNPs heterozygous in the mother and homozygous in the father are selected to determine whether the mutation or wild-type allele was passed on by the mother. Both methods are highly dependent on a sufficient FF (Hudecova and Chiu 2017).

Several cases of prenatal detection of monogenic disorders using these two methods have been reported, including identification of β-thalassemia, sickle cell anemia, methylmalonic acidemia, hemophilia, cystic fibrosis, hearing loss, congenital adrenal hyperplasia, Duchenne muscular dystrophy, Gaucher, and α-thalassemia (Chiu et al. 2002; Ding et al. 2004; Li et al. 2005; Tung-wiwat et al. 2006, 2007; Papasavva et al. 2008, 2013; Ho et al. 2010; Galbiati et al. 2011, 2016; Tsui et al. 2011; Yan et al. 2011; Barrett et al. 2012; Phylipsen et al. 2012; Sirichotiya-kul et al. 2012; Ge et al. 2013; Gu et al. 2014; Ma et al. 2014; New et al. 2014; Xiong et al. 2015, 2018; Xu et al. 2015; Yoo et al. 2015; Zeevi et al. 2015; Parks et al. 2016; Vermeulen et al. 2017; Chang et al. 2018; Guissart et al. 2018).

A 2019 study by Luo et al. evaluated the feasibility of performing cfDNA screening for aneuploidy, copy number variants, and monogenic disorders to demonstrate that it is possible to have a 3 in 1 cfDNA screen that is cheaper and faster than doing separate analyses. One of the challenges of combining chromosome screening with screening for monogenic disorders is the greater depth of sequencing necessary for identification of sequencing variants. Luo and colleagues assessed chromosome aneuploidy, copy number variants and mutation hotspots in six genes of the three of the most common single-gene disorders in China: β-thalassemia, hearing-impairment, and phenylketonuria. Every aneuploidy (7/7) and deletion/duplication >20 Mb (3/3) was detected and 86.4% of the monogenic disorders were accurately detected (19/22) (Luo et al. 2019). Malcher et al. (2018) also developed a comprehensive cfDNA product that demonstrated 100% sensitivity and 98.53% specificity for trisomy 21, demonstrated 100% accuracy for fetal sex, and detected five of seven skeletal dysplasia cases.

2. *Exome/Genome Sequencing by Cell-Free DNA Screening*: Testing for targeted variants associated with a known family history is achievable by the methods previously described, but testing for multiple targets is less feasible using these methods. Genome sequencing of a fetus using maternal blood has been accomplished but with several limitations, primarily involving the amount of fetal DNA available for analysis. For this reason, some studies have been performed using genome sequencing in the second or third trimester when FF levels are higher (Chan et al. 2016).

Several approaches have also been used to enrich the fetal DNA analyzed for genome sequencing (Rabinowitz et al. 2019). Fetal fragments of cfDNA are typically shorter than maternal fragments because of differences in nucleosome position (Chan et al. 2004; Fan et al. 2010, Lo et al. 2010; Yu et al. 2014). This approach has been used to screen for chromosome abnormalities (Cirigliano et al. 2017; Sun et al. 2017) and genome sequencing (Rabinowitz et al. 2019). However, enrichment based on DNA strand size can lead to PCR amplification concerns that may skew allele ratios. A new method, circulating single-molecule amplification and resequencing technology (cSMART), has been introduced to remove amplification bias. The cSMART method was implemented in a 2015 study that correctly classified four pregnancies at risk for Wilson disease (Lv et al. 2015).

Other applications of exome or genome sequencing may be the identification of carriers of genetic conditions, particularly for couples at higher risk such as consanguineous couples or those who have had multiple children/pregnancies affected with a suspected genetic condition. Additionally, couples undergoing in vitro fertilization and testing of embryos via preimplantation genetic testing (PGT) may also consider GS/ES.

3. *Whole-Cell Identification*: Because of many situational challenges presented by cfDNA screening, such as assessment of twin pregnancies, inability to detect Robertsonian translocation versus trisomy, abnormal cfDNA from maternal malignancy, placental mosaicism, and low FF, efforts to identify intact fetal cells in maternal blood for analysis have continued. The biggest challenge to this approach is the low quantity of fetal cells in maternal circulation, estimated at 1–45 cells per 30 mL maternal blood (Beaudet 2016). Fetal cells are identified and isolated using unique fetal receptors and antibodies. Although studies have shown successful isolation and diagnosis of chromosome abnormalities and microdeletions, significant challenges remain (Feng et al. 2018).

Patient Perspectives

Multiple studies have demonstrated patient interest in learning more about the genetic health of their unborn baby. A population of more than 200 pregnant patients presenting for genetic testing were studied in 2018 and the majority of respondents (83%) felt ES should be offered prenatally and about half of participants were interested in pursuing it. However, only 17% reported they would undergo amniocentesis to obtain the information. This study demon-strates positive patient response to noninvasive screening for a wide variety of genetic conditions (Kalynchuk et al. 2015).

Another study of more than 500 pregnant women demonstrated the desire for information about their pregnancy primarily "to prepare financially, medically, or psychologically for a child with special needs." The majority of patients wanted information regarding serious treatable childhood-onset conditions (89.7%) and were least likely to want information about nonmedical traits (40%) (Sullivan et al. 2019).

It is important to note that these surveys were hypothetical offers for genetic testing and did not entail detailed counseling about the potential for variants of uncertain significance. Therefore, it is unclear if patients will demonstrate as much interest in genetic testing as these studies indicate.

If cfDNA and fetal cell isolation technologies continue to improve, the risk for miscarriage associated with invasive testing will no longer be a decision-making factor. However, patients must still contemplate the risks, benefits and limitations of information obtained from genetic testing. The risk to obtain variants of uncertain significance or discovery of incidental findings will continue to be decision-making factors. Additional barriers may also include cost and insurance coverage. In addition to research on the accuracy of noninvasive testing, research must focus on the psychological impact of the decision-making process, the experience of receiving positive results and the impact of genetic testing on outcomes.

Changing Paradigm: Emergence of Prenatal Treatments

Prenatal genetic testing is optional and requires pretest counseling because the risks and benefits of the information obtained from genetic testing are weighed differently in light of patients' values, needs, and desires. The goal of testing for most patients is to get reassurance from normal results, but learning of abnormal results can also be beneficial for some. Prenatal identification of a genetic condition can help decrease uncertainty during the pregnancy, and can allow for termination of pregnancy or mental preparation for a genetic condition, adjustments to the birth plan (i.e., monitoring during labor or plans for vaginal delivery vs. Caesarian section), changes in location and timing of delivery, and immediate neonatal management. For example, a 2017 study of ES performed on neonates <100-d-old revealed that medical management was affected in >50% of cases, which demonstrates the value of early diagnosis and highlights the benefit of prenatal diagnosis (Meng et al. 2017). Despite these potential benefits, the outcome for many individuals with genetic conditions will not change because of a prenatal diagnosis, and testing for the sake of knowing ahead of time may not be perceived as a benefit for all families. Some expectant parents find information obtained through prenatal testing more harmful than beneficial. The decision-making process with regard to prenatal testing is complex and is affected by many personal factors including a patient's "need to know," feelings about pregnancy termination, and availability of interventions.

As new treatments and preventive measures become available for genetic conditions, prenatal testing will become increasingly relevant during pregnancy as a means to improve health outcomes. Presumably, use of prenatal testing will increase, given that pregnant patients have been shown to be more motivated to pursue testing if a condition is treatable prenatally or in the immediate postnatal period (Kalynchuk et al. 2015).

Conditions that are ideal candidates for prenatal intervention include conditions that result in irreparable damage prior to delivery, that do not have effective postnatal treatment, when small fetal size is an advantage for treatment efficacy (i.e., higher treatment to target cell ratio), or when early delivery of a vector or transgene could produce a decreased immune response and/or a future immune tolerance which could improve outcomes (David and Waddington 2012; AJMG 2018).

Prenatal gene therapy in mouse models have shown promise in several conditions: hemophilia A and B, congenital blindness, Crigler–Najjar type 1 syndrome, and Pompe disease (glycogen storage disease type II). Other conditions that

are potential gene therapy candidates include cystic fibrosis, Duchenne muscular dystrophy, lysosomal storage disorders, spinal muscular atrophy, urea cycle defects, α-thalassemia, severe combined immunodeficiency, epidermolysis bullosa, and even intrauterine growth restriction and congenital diaphragmatic hernia (David and Waddington 2012).

Gene editing is an emerging gene therapy that can be utilized in the fetal period; however, much research is needed to determine the safety of such technology. Research in animal models is already underway and the β-thalassemia phenotype has apparently been cured in a mouse model (McClain and Flake 2016; AJMG 2018).

In the case of conditions with an effective postnatal treatment, prenatal diagnosis can allow for early diagnosis so that postnatal treatment can be initiated without the need to wait for testing results following delivery, such as either of the two FDA approved treatments for spinal muscular atrophy.

When prenatal genetic therapies are ready for clinical use, the impact on prenatal genetic testing will be profound. Multifaceted research is imperative, not only on efficacy and safety of treatments, but also on long term outcomes, provider education, determining effective patient counseling models, service delivery models for increased genetic testing needs and patient experiences and attitudes toward testing in light of treatment options.

CONCLUSION

We have witnessed prenatal testing evolve drastically as we have progressed from the discovery of trisomy 21 to ES on prenatal samples. Screening has also advanced significantly from the observation of abnormal maternal analyte levels in pregnancies with Down syndrome. Today, we have the ability to screen for a multitude of chromosome and monogenic disorders with cfDNA. Despite these advances, many genetic conditions are not yet diagnosed in the prenatal setting and many challenges exist when expanding screening to more rare conditions. Furthermore, for those that are diagnosed, a prenatal diagnosis does not often lead to major improvements in postnatal outcomes. Benefits of prenatal diagnosis are often limited, with only some perceiving as beneficial advance knowledge of the diagnosis or the option to terminate the pregnancy. As therapies and treatments for genetic testing emerge, the purpose and goals of prenatal testing will shift, and the need for the accurate prenatal diagnosis will be essential.

As genetic testing continues to expand and treatments are developed, risks and benefits to prenatal testing will still coexist. Ensuring patient understanding of these benefits, risks, and also limitations will be of utmost importance. The roles for genetic counselors are numerous as these advancements are made, including non-clinical and clinical research involvement, the introduction of new technologies into clinical care, provider education, patient advocacy, obtaining informed consent, facilitating of testing and treatment decisions, delivery of results, and psychological support of patients and their patients throughout the process.

REFERENCES

*Reference is also in this subject collection.

ACMG Board of Directors. 2012. Points to consider in the clinical application of genomic sequencing. *Genet Med* **14:** 759–761. doi:10.1038/gim.2012.74

AJMG Sequence: decoding news and trends for the medical genetics community by Roxanne Nelson. 2018. Novel Gene-editing technique cures β-thalassemia in utero: a novel peptide nucleic acid–based gene-editing technique using a nanoparticle delivery system seemingly cured β thalassemia in fetal mice. *Am J Med Genet A* **176:** 2052–2053.

American College of Obstetricians and Gynecologists Committee on Genetics. 2012. Committee Opinion No. 545: noninvasive prenatal testing for fetal aneuploidy. *Obstet Gynecol* **120:** 1532–1534. doi:10.1097/01.AOG.00004 23819.85283.f4

American College of Obstetricians and Gynecologists' Committee on Practice Bulletins—Obstetrics, Committee on Genetics, and the Society for Maternal Fetal Medicine. 2016. Practice Bulletin No. 163: Screening for Fetal Aneuploidy. *Obstet Gynecol* **127:** 979–981. doi:10.1097/AOG .0000000000001439

Ashoor G, Syngelaki A, Poon LC, Rezende JC, Nicolaides KH. 2013. Fetal fraction in maternal plasma cell-free DNA at 11–13 weeks' gestation: relation to maternal and fetal characteristics. *Ultrasoun Obstet Gyncol* **41:** 26–32. doi:10.1002/uog.12331

Bainbridge MN, Wang M, Burgess DL, Kovar C, Rodesch MJ, D'Ascenzo M, Kitzman J, Wu YQ, Newsham I, Rich-

mond TA, et al. 2010. Whole exome capture in solution with 3 Gbp of data. *Genome Biol* **11:** R62. doi:10.1186/gb-2010-11-6-r62

Barrett AN, McDonnell TC, Chan KC, Chitty LS. 2012. Digital PCR analysis of maternal plasma for noninvasive detection of sickle cell anemia. *Clin Chem* **58:** 1026–1032. doi:10.1373/clinchem.2011.178939

Beaudet AL. 2016. Using fetal cells for prenatal diagnosis: history and recent progress. *Am J Med Genet C Semin Med Genet* **172:** 123–127. doi:10.1002/ajmg.c.31487

Bernhardt BA, Soucier D, Hanson K, Savage MS, Jackson L, Wapner RJ. 2013. Women's experiences receiving abnormal prenatal chromosomal microarray testing results. *Genet Med* **15:** 139–145. doi:10.1038/gim.2012.113

Bianchi DW, Chudova D, Sehnert AJ, Bhatt S, Murray K, Prosen TL, Garber JE, Wilkins-Haug L, Vora NL, Warsof S, et al. 2015. Noninvasive prenatal testing and incidental detection of occult maternal malignancies. *J Am Med Assoc* **314:** 162–169. doi:10.1001/jama.2015.7120

Burns W, Koelper N, Barberio A, Deagostino-Kelly M, Mennuti M, Sammel MD, Dugoff L. 2017. The association between anticoagulation therapy, maternal characteristics, and a failed cfDNA test due to a low fetal fraction. *Prenat Diagn* **37:** 1125–1129. doi:10.1002/pd.5152

Cai L, Wang J, Du S, Zhu S, Wang T, Lu D, Chen H. 2016. Comparison of hybrid fixation to dual plating for both-bone forearm fractures in older children. *Am J Ther* **23:** e1391–e1396. doi:10.1097/MJT.0000000000000227

Centers for Disease Control and Prevention. 2008. Update on the overall prevalence of major birth defects—Atlanta, Georgia, 1978–2005. *MMWR Morb Mortal Wkly Rep* **57:** 1–5.

Chan KC, Zhang J, Hui AB, Wong N, Lau TK, Leung TN, Lo KW, Huang DW, Lo YM. 2004. Size distributions of maternal and fetal DNA in maternal plasma. *Clin Chem* **50:** 88-92. doi:10.1373/clinchem.2003.024893

Chan KC, Jiang P, Sun K, Cheng YK, Tong TY, Cheng SH, Wong AI, Hudecova I, Leung TY, Chiu RW, et al. 2016. Second generation noninvasive fetal genome analysis reveals de novo mutations, single-base parental inheritance, and preferred DNA ends. *Proc Natl Acad Sci* **113:** E8159–E8168. doi:10.1073/pnas.1615800113

Chang MY, Ahn S, Kim MY, Han JH, Park HR, Seo HK, Yoon J, Lee S, Oh DY, Kang C, et al. 2018. One-step noninvasive prenatal testing (NIPT) for autosomal recessive homozygous point mutations using digital PCR. *Sci Rep* **8:** 2877. doi:10.1038/s41598-018-21236-w

Chiu RW, Lau TK, Leung TN, Chow KC, Chui DH, Lo YM. 2002. Prenatal exclusion of β thalassaemia major by examination of maternal plasma. *Lancet* **360:** 998–1000. doi:10.1016/S0140-6736(02)11086-5

Cirigliano V, Ordoñez E, Rueda L, Syngelaki A, Nicolaides KH. 2017. Performance of the neoBona test: a new paired-end massively parallel shotgun sequencing approach for cell-free DNA-based aneuploidy screening. *Ultrasound Obstet Gynecol* **49:** 460-464. doi:10.1002/uog.17386

Committee on Genetics. 2017. Committee opinion No. 691: carrier screening for genetic conditions. *Obstet Gynecol* **129:** e41–e55. doi:10.1097/AOG.0000000000001952

Committee on Genetics and the Society for Maternal-Fetal Medicine. 2016. Committee Opinion No.682: Microarrays and next-generation sequencing technology: the use of advanced genetic diagnostic tools in obstetrics and gynecology. *Obstet Gynecol* **128**: e262–e268. doi:10.1097/AOG.0000000000001817

Committee Opinion No. 640. 2015. Cell-free DNA screening for fetal aneuploidy. *Obstet Gynecol* **126**: e31–37. doi:10.1097/AOG.0000000000001051

Dabi Y, Guterman S, Jani JC, Letourneau A, Demain A, Kleinfinger P, Lohmann L, Costa JM, Benachi A. 2018. Autoimmune disorders but not heparin are associated with cell-free fetal DNA test failure. *J Transl Med* **16**: 335. doi:10.1186/s12967-018-1705-2

David AL, Waddington SN. 2012. Candidate diseases for prenatal gene therapy. *Methods Mol Biol* **891**: 9–39.

Ding C, Chiu RW, Lau TK, Leung TN, Chan LC, Chan AY, Charoenkwan P, Ng IS, Law HY, Ma ES, et al. 2004. MS analysis of single-nucleotide differences in circulating nucleic acids: application to noninvasive prenatal diagnosis. *Proc Natl Acad Sci* **101**: 10762–10767. doi:10.1073/pnas.0403962101

Durham L, Papanna R, Stevens B, Noblin S, Rodriguez-Buritica D, Hashmi SS, Krstic N. 2019. The utilization of prenatal microarray: a survey of current genetic counseling practices and barriers. *Prenat Diag* **39**: 351–360. doi:10.1002/pd.5435

Edwards JG, Feldman G, Goldberg J, Gregg AR, Norton ME, Rose NC, Schneider A, Stoll K, Wapner R, Watson MS. 2015. Expanded carrier screening in reproductive medicine-points to consider: a joint statement of the American College of Medical Genetics and Genomics, American College of Obstetricians and Gynecologists, National Society of Genetic Counselors, Perinatal Quality Foundation, and Society for Maternal-Fetal Medicine. *Obstet Gynecol* **125**: 653–662. doi:10.1097/AOG.0000000000000666

Ehrich M, Tynan J, Mazloom A, Almasri E, McCullough R, Boomer T, Grosu D, Chibuk J. 2017. Genome-wide cfDNA screening: clinical laboratory experience with the first 10,000 cases. *Genet Med* **19**: 1332–1337. doi:10.1038/gim.2017.56

Fan HC, Blumenfeld YJ, Chitkara U, Hudgins L, Quake SR. 2010. Analysis of the size distributions of fetal and maternal cell-free DNA by paired-end sequencing. *Clin Chem* **56**: 1279-1286. doi:10.1373/clinchem.2010.144188

Feng C, He Z, Cai B, Peng J, Song J, Yu X, Sun Y, Yuan J, Zhao X, Zhang Y. 2018. Non-invasive prenatal diagnosis of chromosomal aneuploidies and microdeletion syndrome using fetal nucleated red blood cells isolated by nanostructure microchips. *Theranostics* **8**: 1301–1311. doi:10.7150/thno.21979

Galbiati S, Brisci A, Lalatta F, Seia M, Makrigiorgos GM, Ferrari M, Cremonesi L. 2011. Full COLD-PCR protocol for noninvasive prenatal diagnosis of genetic diseases. *Clin Chem* **57**: 136–138. doi:10.1373/clinchem.2010.155671

Galbiati S, Monguzzi A, Damin F, Soriani N, Passiu M, Castellani C, Natacci F, Curcio C, Seia M, Lalatta F, et al. 2016. COLD-PCR and microarray: two independent highly sensitive approaches allowing the identification of fetal paternally inherited mutations in maternal plasma. *J Med Genet* **53**: 481–487. doi:10.1136/jmedgenet-2015-103229

Gardner RJM, Sutherland GR, Shaffer LG. 2012. *Chromosome abnormalities and genetic counseling*, pp. 283, 403, 417. Oxford University Press, Oxford.

Ge H, Huang X, Li X, Chen S, Zheng J, Jiang H, Zhang C, Pan X, Guo J, Chen F, et al. 2013. Noninvasive prenatal detection for pathogenic CNVs: the application in α-thalassemia. *PLoS One* **8**: e67464. doi:10.1371/journal.pone.0067464

Gil MM, Accurti V, Santacruz B, Plana MM, Nicolaides KH. 2017. Analysis of cell-free DNA in maternal blood in screening for aneuploidies: updated meta-analysis. *Ultrasound Obstet Gynecol* **50**: 302–314. doi:10.1002/uog.17484

* Goldman JS. 2019. Predictive genetic counseling for neurodegenerative diseases: past, present, and future. *Cold Spring Harb Perspect Med* doi: 10.1101/cshperspect.a036525

Grati FR, Malvestiti F, Ferreira JC, Bajaj K, Gaetani E, Agrati C, Grimi B, Dulcetti F, Ruggeri AM, De Toffol S, et al. 2014. Fetoplacental mosaicism: potential implications for false-positive and false-negative noninvasive prenatal screening results. *Genet Med* **16**: 620–624. doi:10.1038/gim.2014.3

Gregg AR, Edwards JG. 2018. Prenatal genetic carrier screening in the genomic age. *Semin Perinatol* **42**: 303–306. doi:10.1053/j.semperi.2018.07.019

Grody WW, Thompson BH, Gregg AR, Bean LH, Monaghan KG, Schneider A, Lebo RV. 2013. ACMG position statement on prenatal/preconception expanded carrier screening. *Genet Med* **15**: 482–483. doi:10.1038/gim.2013.47

Gu W, Koh W, Blumenfeld YJ, El-Sayed YY, Hudgins L, Hintz SR, Quake SR. 2014. Noninvasive prenatal diagnosis in a fetus at risk for methylmalonic acidemia. *Genet Med* **16**: 564–567. doi:10.1038/gim.2013.194

Guissart C, Tran Mau Them F, Debant V, Viart V, Dubucs C, Pritchard V, Rouzier C, Boureau-Wirth A, Haquet E, Puechberty J, et al. 2018. A broad test based on fluorescent-multiplex PCR for noninvasive prenatal diagnosis of cystic fibrosis. *Fetal Diagn Ther* **17**: 1–10.

Hartwig TS, Ambye L, Sørensen S, Jørgensen FS. 2017. Discordant non-invasive prenatal testing (NIPT)—a systematic review. *Prenat Diagn* **37**: 527–539. doi:10.1002/pd.5049

Helgeson J, Wardrop J, Boomer T, Almasri E, Paxton WB, Saldivar JS, Dharajiya N, Monroe TJ, Farkas DH, Grosu DS, et al. 2015. Clinical outcome of subchromosomal events detected by whole-genome noninvasive prenatal testing. *Prenat Diagn* **35**: 999–1004. doi:10.1002/pd.4640

Ho SS, Chong SS, Koay ES, Ponnusamy S, Chiu L, Chan YH, Rauff M, Baig S, Chan J, Su LL, et al. 2010. Noninvasive prenatal exclusion of haemoglobin Bart's using foetal DNA from maternal plasma. *Prenat Diagn* **30**: 65–73.

Hudecova I, Chiu RW. 2017. Non-invasive prenatal diagnosis of thalassemias using maternal plasma cell free DNA. *Best Pract Res Clin Obstet Gynaecol* **39**: 63–73. doi:10.1016/j.bpobgyn.2016.10.016

International Society for Prenatal Diagnosis; Society for Maternal and Fetal Medicine; Perinatal Quality Foundation. 2018. Joint position statement from the international society for prenatal diagnosis (ISPD), the Society for Maternal Fetal Medicine (SMFM), and the Perinatal Quality

Foundation (PQF) on the use of genome-wide sequencing for fetal diagnosis. *Prenat Diag* **38:** 6–9. doi:10.1002/pd.5195

Janssens S, Chokoshvili D, Vears DF, De Paepe A, Borry P. 2017. Pre- and post-testing counseling considerations for the provision of expanded carrier screening: exploration of European geneticists' views. *BMC Med Ethics* **18:** 46. doi:10.1186/s12910-017-0206-9

Kalynchuk EJ, Althouse A, Parker LS, Saller DN Jr Rajkovic A. 2015. Prenatal whole-exome sequencing: parental attitudes. *Prenat Diagn* **35:** 1030–1036. doi:10.1002/pd.4635

Kim SC, Cha DH, Jeong HR, Lee J, Jang JH, Cho EH. 2019. Clinically significant maternal X chromosomal copy number variation detected by noninvasive prenatal test. *J Obstet Gynaecol Res* doi: 10.1111/jog.14033

Lefkowitz RB, Tynan JA, Liu T, Wu Y, Mazloom AR, Almasri E, Hogg G, Angkachatchai V, Zhao C, Grosu DS, et al. 2016. Clinical validation of a noninvasive prenatal test for genomewide detection of fetal copy number variants. *Am J Obstet Gynecol* **215:** 227.e1–227.e16. doi:10.1016/j.ajog.2016.02.030

Leonard F, Gueben R, Gueben R, Grisart B, Van Linthout C. 2018. Incidental findings of maternal genetic abnormalities during non-invasive prenatal screening. *Rev Med Liege* **73:** 125–128.

Li Y, Di Naro E, Vitucci A, Zimmermann B, Holzgreve W, Hahn S. 2005. Detection of paternally inherited fetal point mutations for β-thalassemia using size-fractioned cell-free DNA in maternal plasma. *J Am Med Assoc* **293:** 843–849. doi:10.1001/jama.293.7.843

Lo YMD, Corbetta N, Chamberlain PF, Rai V, Sargent IL, Redman CW, Wainscoat JS. 1997. Presence of fetal DNA in maternal plasma and serum. *Lancet* **350:** 485–487. doi:10.1016/S0140-6736(97)02174-0

Lo YM, Chan KC, Sun H, Chen EZ, Jiang P, Lun FM, Zheng YW, Leung TY, Lau TK, Cantor CR, et al. 2010. Maternal plasma DNA sequencing reveals the genome-wide genetic and mutational profile of the fetus. *Sci Transl Med* **2:** 61ra91. doi:10.1126/scitranslmed.3001720

Luo Y, Jia B, Yan K, Liu S, Song X, Chen M, Jin F, Du Y, Wang J, Hong Y, et al. 2019. Pilot study of a novel multi-functional noninvasive prenatal test on fetus aneuploidy, copy number variation, and single-gene disorder screening. *Mol Genet Genomic Med* **7:** e00597. doi: 10.1002/mgg3.597.

Lv W, Wei X, Guo R, Liu Q, Zheng Y, Chang J, Bai T, Li H, Zhang J, Song Z, et al. 2015. Noninvasive prenatal testing for Wilson disease by use of circulating single-molecule amplification and resequencing technology (cSMART). *Clin Chem* **61:** 172–181. doi:10.1373/clinchem.2014.229328

Ma D, Ge H, Li X, Jiang T, Chen F, Zhang Y, Hu P, Chen S, Zhang J, Ji X, et al. 2014. Haplotype-based approach for noninvasive prenatal diagnosis of congenital adrenal hyperplasia by maternal plasma DNA sequencing. *Gene* **544:** 252–258. doi:10.1016/j.gene.2014.04.055

Malcher C, Yamamoto GL, Burnham P, Ezquina SAM, Lourenço NCV, Balkassmi S, Antonio DSM, Hsia GSP, Gollop T, Pavanello RC, et al. 2018. Development of a comprehensive noninvasive prenatal test. *Genet*

Mol Biol **41:** 545–554. doi:10.1590/1678-4685-gmb-2017-0177

Martin K, Iyengar S, Kalyan A, Lan C, Simon AL, Stosic M, Kobara K, Ravi H, Truong T, Ryan A, et al. 2018. Clinical experience with a single-nucleotide polymorphism-based non-invasive prenatal test for five clinically significant microdeletions. *Clin Genet* **93:** 293–300. doi:10.1111/cge.13098

McClain LE, Flake AW. 2016. In utero stem cell transplantation and gene therapy: recent progress and the potential for clinical application. *Best Pract Res Clin Obstet Gynaecol* **31:** 88–98. doi:10.1016/j.bpobgyn.2015.08.006

McKanna T, Ryan A, Krinshpun S, Kareht S, Marchand K, Grabarits C, Ali M, McElheny A, Gardiner K, LeChien K, et al. 2019. Fetal fraction-based risk algorithm for non-invasive prenatal testing: screening for trisomies 13 and 18 and triploidy in women with low cell-free fetal DNA. *Ultrasound Obstet Gynecol* **53:** 73–79. doi:10.1002/uog.19176

Meng L, Pammi M, Saronwala A, Magoulas P, Ghazi AR, Vetrini F, Zhang J, He W, Dharmadhikari AV, Qu C, et al. 2017. Use of exome sequencing for infants in intensive care units: ascertainment of Sever single-gene disorders and effect on medical management. *JAMA Pediatr* **171:** e173438. doi:10.1001/jamapediatrics.2017.3438

Morton NE, Crow JF, Muller HJ. 1956. An estimate of the mutational damage in a man from data on consanguineous marriages. *Proc Natl Acad Sci* **42:** 855–863. doi:10.1073/pnas.42.11.855

New MI, Tong YK, Yuen T, Jiang P, Pina C, Chan KC, Khattab A, Liao GJ, Yau M, Kim SM, et al. 2014. Noninvasive prenatal diagnosis of congenital adrenal hyperplasia using cell-free fetal DNA in maternal plasma. *J Clin Endocrinol Metab* **99:** E1022–E1030. doi:10.1210/jc.2014-1118

Ng SB, Turner EH, Robertson PD, Flygare SD, Bigham AW, Lee C, Shaffer T, Wong M, Bhattacharjee A, Eichler EE, et al. 2009. Targeted capture and massively parallel sequencing of 12 human exomes. *Nature* **461:** 272–276. doi: 10.1038/nature08250.

Palomaki GE, Kloza EM, Lambert-Messerlian GM, Haddow JE, Neveux LM, Ehrich M, van den Boom D, Bombard AT, Deciu C, Grody WW, et al. 2011. DNA sequencing of maternal plasma to detect Down syndrome: an international clinical validation study. *Genet Med* **13:** 913–920. doi:10.1097/GIM.0b013e3182368a0e

Papasavva T, Kalikas I, Kyrri A, Kleanthous M. 2008. Arrayed primer extension for the noninvasive prenatal diagnosis of β-thalassemia based on detection of single nucleotide polymorphisms. *Ann N Y Acad Sci* **1137:** 302–308. doi:10.1196/annals.1448.029

Papasavva TE, Lederer CW, Traeger-Synodinos J, Mavrou A, Kanavakis E, Ioannou C, Makariou C, Kleanthous M. 2013. A minimal set of SNPs for the noninvasive prenatal diagnosis of β-thalassaemia. *Ann Hum Genet* **77:** 115–124. doi:10.1111/ahg.12004

Parks M, Court S, Cleary S, Clokie S, Hewitt J, Williams D, Cole T, MacDonald F, Griffiths M, Allen S. 2016. Non-invasive prenatal diagnosis of Duchenne and Becker muscular dystrophies by relative haplotype dosage. *Prenat Diag* **36:** 312–320. doi:10.1002/pd.4781

Pergament E, Cuckle H, Zimmermann B, Banjevic M, Sigurjonsson S, Ryan A, Hall MP, Dodd M, Lacroute P, Stosic M, et al. 2014. Single-nucleotide polymorphism-based noninvasive prenatal screening in a high-risk and low-risk cohort. *Obstet Gynecol* **124:** 210–218. doi:10.1097/AOG.0000000000000363

Phylipsen M, Yamsri S, Treffers EE, Jansen DT, Kanhai WA, Boon EM, Giordano PC, Fucharoen S, Bakker E, Harteveld CL. 2012. Non-invasive prenatal diagnosis of β-thalassemia and sickle-cell disease using pyrophosphorolysis-activated polymerization and melting curve analysis. *Prenat Diagn* **32:** 578–587. doi:10.1002/pd.3864

Rabinowitz T, Polsky A, Golan D, Danilevsky A, Shapira G, Raff C, Basel-Salmon L, Matar RT, Shomron N. 2019. Bayesian-based noninvasive prenatal diagnosis of single-gene disorders. *Genome Res* **29:** 428–438. doi:10.1101/gr.235796.118

Raymer K. 2004. Prenatal screening and the assessment of risk: the view from the other side. *J Obstet Gynaecol Can* **26:** 329–332. doi:10.1016/S1701-2163(16)30360-7

Resta R. 2002. Historical aspects of genetic counseling: why was maternal age 35 chosen as the cut-off for offering amniocentesis? *Med Secoli* **14:** 793–811.

Sirichotiyakul S, Charoenkwan P, Sanguansermsri T. 2012. Prenatal diagnosis of homozygous α-thalassemia-1 by cell-free fetal DNA in maternal plasma. *Prenat Diagn* **32:** 45–49. doi:10.1002/pd.2892

Stenson PD, Ball EV, Howells K, Phillips AD, Mort M, Cooper DN. 2009. The Human Gene Mutation Database: providing a comprehensive central mutation database for molecular diagnostics and personalized genomics. *Hum Genomics* **4:** 69–72. doi:10.1186/1479-7364-4-2-69

Stevens B, Krstic N, Jones M, Murphy L, Hoskovec J. 2017. Finding middle ground in constructing a clinically useful expanded carrier screening panel. *Obstet Gynecol* **130:** 279–284. doi:10.1097/AOG.0000000000002139

Stevens BK, Noblin SJ, Chen HY, Czerwinski J, Friel LA, Wagner C. 2019. Introduction of cell-free DNA screening is associated with changes in prenatal genetic counseling indications. *J Genet Couns* **28:** 692–699. doi:10.1002/jgc4.1095

Sullivan HK, Bayefsky M, Wakim PG, Huddleston K, Biesecker BB, Hull SC, Berkman BE. 2019. Noninvasive prenatal whole genome sequencing: pregnant women's views and preferences. *Obstet Gynecol* **133:** 525–532. doi:10.1097/AOG.0000000000003121

Sun K, Chan KC, Hudecova I, Chiu RW, Lo YM, Jiang P. 2017. COFFEE: control-free noninvasive fetal chromosomal examination using maternal plasma DNA. *Prenat Diagn* **37:** 336-340. doi:10.1002/pd.5016

Tsui NB, Kadir RA, Chan KC, Chi C, Mellars G, Tuddenham EG, Leung TY, Lau TK, Chiu RW, Lo YM. 2011. Noninvasive prenatal diagnosis of hemophilia by microfluidics digital PCR analysis of maternal plasma DNA. *Blood* **117:** 3684–3691. doi:10.1182/blood-2010-10-310789

Tungwiwat W, Fucharoen S, Fucharoen G, Ratanasiri T, Sanchaisuriya K. 2006. Development and application of a real-time quantitative PCR for prenatal detection of fetal α⁰-thalassemia from maternal plasma. *Ann NY Acad Sci* **1075:** 103–107. doi:10.1196/annals.1368.013

Tungwiwat W, Fucharoen G, Fucharoen S, Ratanasiri T, Sanchaisuriya K, Sae-Ung N. 2007. Application of mater-

nal plasma DNA analysis for noninvasive prenatal diagnosis of Hb E-β-thalassemia. *Transl Res* **150:** 319–325. doi:10.1016/j.trsl.2007.06.006

Van Opstal D, Srebniak MI. 2016. Cytogenetic confirmation of a positive NIPT result: evidence-based choice between chorionic villus sampling and amniocentesis depending on chromosome aberration. *Exp Rev Mol Diagn* **16:** 513–520. doi:10.1586/14737159.2016.1152890

Vermeulen C, Geeven G, de Wit E, Verstegen MJAM, Jansen RPM, van Kranenburg M, de Bruijn E, Pulit SL, Kruisselbrink E, Shahsavari Z, et al. 2017. Sensitive monogenic noninvasive prenatal diagnosis by targeted haplotyping. *Am J Hum Genet* **101:** 326–339. doi:10.1016/j.ajhg.2017.07.012

Wapner RJ, Martin CL, Levy B, Ballif BC, Eng CM, Zachary JM, Savage M, Platt LD, Saltzman D, Grobman WA, et al. 2012. Chromosomal microarray versus karyotyping for prenatal diagnosis. *N Engl J Med* **367:** 2175–2184. doi:10.1056/NEJMoa1203382

Wapner RJ, Babiarz JE, Levy B, Stosic M, Zimmermann B, Sigurjonsson S, Wayham N, Ryan A, Banjevic M, Lacroute P, et al. 2015. Expanding the scope of noninvasive prenatal testing: detection of fetal microdeletion syndromes. *Am J Obstet Gynecol* **212:** 332.e1–332.e9. doi:10.1016/j.ajog.2014.11.041

Warsof SL, Larion S, Abuhamad AZ. 2015. Overview of the impact of noninvasive prenatal testing on diagnostic procedures. *Prenat Diagn* **35:** 972–979. doi:10.1002/pd.4601

Werner-Lin A, Barg FK, Kellom KS, Stumm KJ, Pilchman L, Tomlinson AN, Bernhardt BA. 2016. Couple's narratives of communion and isolation following abnormal prenatal microarray testing results. *Qual Health Res* **26:** 1975–1987. doi:10.1177/1049732315603367

Xiong L, Barrett AN, Hua R, Tan TZ, Ho SS, Chan JK, Zhong M, Choolani M. 2015. Non-invasive prenatal diagnostic testing for β-thalassaemia using cell-free fetal DNA and next generation sequencing. *Prenat Diagn* **35:** 258–265. doi:10.1002/pd.4536

Xiong L, Barrett AN, Hua R, Ho S, Jun L, Chan K, Mei Z, Choolani M. 2018. Non-invasive prenatal testing for fetal inheritance of maternal β-thalassaemia mutations using targeted sequencing and relative mutation dosage: a feasibility study. *BJOG* **125:** 461–468. doi:10.1111/1471-0528.15045

Xu Y, Li X, Ge HJ, Xiao B, Zhang YY, Ying XM, Pan XY, Wang L, Xie WW, Ni L, et al. 2015. Haplotype-based approach for noninvasive prenatal tests of Duchenne muscular dystrophy using cell-free fetal DNA in maternal plasma. *Genet Med* **17:** 889–896. doi:10.1038/gim.2014.207

Yan TZ, Mo QH, Cai R, Chen X, Zhang C-M, Liu Y-H, Chen Y-J, Zhou W-J, Xiong F, Xu X-M 2011. Reliable detection of paternal SNPs within deletion breakpoints for non-invasive prenatal exclusion of homozygous α⁰-thalassemia in maternal plasma. *PLoS One* **6:** e24779. doi:10.1371/journal.pone.0024779

Yoo Sk, Lim BC, Byeun J, Hwang H, Kim KJ, Hwang YS, Lee J, Park JS, Lee YS, Namkung J, et al. 2015. Noninvasive prenatal diagnosis of duchenne muscular dystrophy: comprehensive genetic diagnosis in carrier, proband,

and fetus. *Clin Chem* **61:** 829–837. doi:10.1373/clinchem
.2014.236380

Yu SC, Chan KC, Zheng YW, Jiang P, Liao GJ, Sun H, Ako-
lekar R, Leung TY, Go AT, Vugt Jv, et al. 2014. Size-based
molecular diagnostics using plasma DNA for noninvasive
prenatal testing. *Proc Natl Acad Sci* **111:** 8583-8588.
doi:10.1073/pnas.1406103111

Yu T, Li S, Zhao W, Yu D. 2019. False positive non-invasive
prenatal testing results due to vanishing twins. *Zhonghua
Yi Xue Yi Chuan Xue Za Zhi* **36:** 327–330.

Zeevi DA, Altarescu G, Weinberg-Shukron A, Zahdeh F,
Dinur T, Chicco G, Herskovitz Y, Renbaum P, Elstein
D, Levy-Lahad E, et al. 2015. Proof-of-principle rapid
noninvasive prenatal diagnosis of autosomal recessive
founder mutations. *J Clin Invest* **125:** 3757–3765. doi:10
.1172/JCI79322

Zhang J, Li J, Saucier JB, Feng Y, Jiang Y, Sinson J, McCombs
AK, Schmitt ES, Peacock S, Chen S, et al. 2019. Noninvasive prenatal sequencing for multiple Mendelian
monogenic disorders using circulating cell-free fetal
DNA. *Nat Med* **25:** 439–447. doi:10.1038/s41591-018-
0334-x

Zhao C, Tynan J, Ehrich M, Hannum G, McCullough R,
Saldivar JS, Oeth P, van den Boom D, Deciu C. 2015.
Detection of fetal subchromosomal abnormalities by sequencing circulating cell-free DNA from maternal plasma. *Clin Chem* **61:** 608–616. doi:10.1373/clinchem.2014
.233312

Cite this article as *Cold Spring Harb Perspect Med* doi: 10.1101/cshperspect.a036517

Predictive Genetic Counseling for Neurodegenerative Diseases: Past, Present, and Future

Jill S. Goldman

Taub Institute for Research on Alzheimer's Disease and the Aging Brain, Department of Neurology, Columbia University Vagelos College of Medicine, New York, New York 10032, USA

Correspondence: jg2673@cumc.columbia.edu

Predictive genetic counseling for neurodegenerative diseases commenced with Huntington's disease (HD). Because the psychological issues and outcomes have been best studied in HD, the HD genetic counseling and testing protocol is still accepted as the gold standard for genetic counseling for these diseases. Yet, advances in genomic technology have produced an abundance of new information about the genetics of diseases such as Alzheimer's disease, frontotemporal dementia, amyotrophic lateral sclerosis, and Parkinson's disease. The resulting expansion of genetic tests together with the availability of direct-to-consumer testing and clinical trials for treatment of these diseases present new ethical and practical issues requiring modifications to the protocol for HD counseling and new demands on both physicians and genetic counselors. This work reviews the history of genetic counseling for neurodegenerative diseases, its current practice, and the future direction of genetic counseling for these conditions.

Predictive testing for neurodegenerative diseases began in the early 1990s and has been accompanied by much discussion and research from genetics, ethics, and neurology communities. This review will focus on how this history has resulted in the current practice of predictive genetic counseling and testing for neurodegenerative diseases and then examine how genetic counseling for predictive testing may change in the near future with the emergence of clinical treatment trials for these conditions.

HUNTINGTON'S DISEASE

HD Gene Discovery

Before the identification of the pathogenic variant that causes Huntington's disease (HD),

ethicists, geneticists, and other clinicians familiar with HD studied their patient populations and speculated about the impact of a future gene discovery. Policies were debated and protocols developed by the World Federation of Neurology Research Group on Huntington's Disease in 1990 (World Federation of Neurology Research Group on Huntington's Disease 1990) and the International Huntington Association and the World Federation of Neurology Research Group on Huntington's Chorea in 1994 (International Huntington Association and the World Federation of Neurology Research Group on Huntington's Chorea 1994). Although predictive testing using linkage markers was available for many years, the discovery of *HTT* in 1993 (Huntington's Disease Collaborative

Research Group 1993) introduced the possibility of providing individuals who were symptomatic or who were at risk for HD with a definitive diagnosis or risk status without the involvement of relatives. Many scholarly articles and qualitative research studies were published about the impact of predictive testing on the individual, the family, and society (Almqvist et al. 1999, 2003; Decruyenaere et al. 2004; Richards and Williams 2004; Larsson et al. 2006; Tibben 2007; Gargiulo et al. 2009). Thus, HD became the gold standard for predictive testing of neurodegenerative diseases.

The HD Predictive Counseling and Testing Protocol

The HD predictive testing protocol was developed in 1994 to facilitate the opportunity for at-risk individuals to make informed decisions about predictive testing and to provide genetic counseling and testing to those who choose it in the safest way possible (International Hunting-ton Association and the World Federation of Neurology Research Group on Huntington's Chorea 1994; Huntington's Disease Society of America 1994). Among the key points of the guidelines were not testing minors, using caution when testing informs about another individual who does not wish to know their status, delaying testing for those individuals with active major psychiatric conditions until their psychiatric symptoms could be stabilized, and having participants bring a support person with them to counseling and result sessions. The guidelines were very prescriptive about the pre- and post-testing appointments (Table 1): the timing of appointments in the testing protocol, the information that should be included in genetic counseling, and confidentiality. These guidelines were updated in 2013 by the European Huntington Disease Network to include the availability of genetic counseling to minors requesting testing, as well as additional recommendations for prenatal and preimplantation genetic diagnosis and research involvement. Remarkably, the

Table 1. HDSA 1994/2003 protocol versus 2016 revised protocol

1994/2003 protocol[a]	2016 revised protocol
Telephone screen	Telephone screen
Visit 1: In-person genetic counseling (with support person, strongly advised)	Visit 1: In-person (with support person, strongly advised)
	Genetic counseling
	Sign informed consent document
	Mental health assessment
	Neurological examination offered
	Draw blood
Visit 2: Psychiatric assessment	Visit 2:
	Disclosure of results in person
	Arrange postresult follow-up
Visit 3:	N/A
Neurological evaluation offered	
Review of potential impact of the test	
Informed consent	
Blood draw	
Visit 4: Disclosure of results in person (with support person strongly advised)	N/A
Telephone follow-up:	Telephone follow-up:
Additional visits for supportive counseling as needed	Additional visits for supportive counseling as needed
Neurological evaluation for those testing positive	Neurological evaluation for those testing positive

Data from Huntington's Disease Society of America 1994, 2016.
[a]The order of the genetic counseling, neurological evaluation, and psychological assessment may vary.

Cite this article as *Cold Spring Harb Perspect Med* doi: 10.1101/cshperspect.a036525

basic tenets of the original guidelines have persisted internationally (MacLeod et al. 2013).

In the United States, the Huntington's Disease Society of America (HDSA) has published its own guidelines, which were revised several times. Although original guidelines were similar to those of the World Federation of Neurology and International Huntington Association (Huntington's Disease Society of America 1994), the last revision had some significant differences. This last revision (Huntington's Disease Society of America 2016) was undertaken to address concerns about the paternalism in the approach to previous guidelines and improve access by reducing the time required for and expense associated with testing. The protocol allows for more flexibility, such as allowing the participant to test without a support person or a neurological evaluation if they so choose. However, the biggest change is the reduction in the number of required in-person appointments from four to two (see Table 1): (1) genetic counseling, mental health assessment, neurological exam, informed consent, and blood draw; and (2) return of results. This new protocol was received with both praise (Huntington's Disease Society of America 2016) for improving access to testing and criticism (Groves 2017) for lessening some safety features, including reducing the time for the participant to reflect on the genetic counseling discussion and no longer requiring the neurological examination to assess early symptoms. Presently, HD centers are assessing their own patient populations and resources and revising protocols accordingly. Perhaps the largest contribution of the latest HDSA protocol is to prompt centers to consider each request on a case-by-case basis.

Outcomes of Predictive Testing for HD

Well before the discovery of *HTT* and even before linkage testing, the potential for predictive testing for HD created concerns about discrimination, stigma, and psychological well-being (Rosenfeld 1984; Bird 1985; Craufurd and Harris 1986; Farrer et al. 1986; Kolata 1986; Bloch et al. 1987; Kessler 1987; Lamport et al. 1987; Quaid et al. 1987; Huggins et al. 1990; Tibben

et al. 1990). Whereas original estimates of the rate of uptake of testing were 55%–80% of those people at risk (Barette and Marsden 1979; Tyler and Harper 1983; Kessler et al. 1987), the actual percentage of at-risk individuals choosing *HTT* testing when it became available was only 4%–24% (Tibben 2007; Morrison et al. 2011; Baig et al. 2016). Thus, those people coming for testing are a self-selected group who may feel that the benefits of learning their results outweigh the fears of the devastating results.

Many studies looking at the outcomes of predictive testing have revealed that most of those within this self-selected group ultimately cope effectively with the information regardless of their result, yet some do experience significant psychological distress (Almqvist et al. 2003; Crozier et al. 2015). Some of the studies that reveal negative outcomes include a 1999 study, which found that ~1% of those who underwent testing experienced a catastrophic event (CE) such as suicide, suicide attempt, or psychiatric hospitalization (Almqvist et al. 1999). Approximately half of these participants had begun to manifest symptoms of HD. Those at risk for CEs were more likely to have had preexisting psychiatric problems, been unemployed, and recently diagnosed with HD.

A later study of people who had gone through HD testing found that 54% of those who tested positive had some suicidal ideation (Larsson et al. 2006). Yet, overall, the distress levels of both those testing positive and those testing negative returned to baseline levels after one year. Interestingly, most people testing positive experienced distress soon after the delivery of test results, whereas the distress of people testing negative peaked six months after the return of results (Lawson et al. 1996; Tibben 2007). Psychological distress was more common in those with positive results who felt guilty about risks to their children and those experiencing early signs of HD. The distress of the people testing negative was more common when they strongly identified with HD. Whether because of test results, prior marital strife, or early disease symptoms, many partners of those being tested also experienced distress and some relationships suffered (Decruyenaere et al. 2004; Richards and Williams 2004; Tibben

2007). Unsurprisingly, a prior history of depression was more often associated with posttest depression (Gargiulo et al. 2009). Of the 10% of participants in the Prospective Huntington At-Risk Observation Study (PHAROS) who chose testing, more than half of those testing negative experienced a reduction in depressive symptoms, whereas >60% of the positive participants had the same or higher levels of depressive symptoms than before testing (Quaid et al. 2017).

Overall, it seems that the predictive testing process generally attracts a self-selected sample of at-risk people who can cope better with results: Many people drop out of the process after counseling but before they receive results reporting reasons that included feeling unprepared to live with positive results (Ibisler et al. 2017; Mandich et al. 2017). Further, Tibben suggested that many people who have chosen to test are lost to follow-up and that those people may be experiencing psychological distress that is not captured by studies. Thus, research into the long-term impact of predictive testing may not fully reflect psychological well-being of those at risk for HD who chose to undergo testing (Tibben 2007). Another caveat is that the impact of testing has been studied only at centers that follow the international guidelines. Collectively, the evidence suggests that although many people proceed with predictive testing without long-term adverse outcomes, some experience negative outcomes, and attention to the HD guidelines may reduce the potential for negative repercussions.

The Predictive Testing Genetic Counseling Session

The international guidelines express the importance of genetic counseling and provide specific instructions about the content of a pretest predictive counseling session. The session is divided between giving and gathering information and anticipatory guidance about the posttest period. As an HD counselor, I have found that although a concrete set of information should be imparted, the patient will dictate when in the session they are prepared to hear specific information (Table 2). Beginning genetic counseling by inquiring about the patient's experience with the disease allows the counselor to assess the patient's understanding of the disease as well as any misperceptions and to learn their responses to experiencing HD in their family member(s) and their fears about the future. The genetic counselor can then use this information to explore feelings about being part of an HD family and living at risk for the disease and how the patient has managed living with the threat of the condition. The counselor may emphasize that there is significant variation in how each person with HD experiences the disease course. The counselor can facilitate patient understanding that in the event that she or he tests positive for HD, her or his personal experience may differ from that of their family member because of variability of clinical symptoms, personality, support, environment, or new treatments. Genetic counseling integrates this information into discourse about the patient's experiences. Often, misperceptions are expressed. By describing the symptoms of HD (psychiatric, motor, and cognitive) and reinforcing the disease variability, the counselor can address specific misperceptions. The counselor can refer to the patient's pedigree to discuss probable age of onset of symptoms and the concept of anticipation in which age of onset gets earlier with each generation because of an

Table 2. General HD information to include in genetic counseling

Explanation of autosomal-dominant inheritance and risk to the patient, relatives, and children
Description of DNA, the CAG expansion in *HTT*, and how the expansion produces a toxic protein that causes HD
Explanation of the four repeat ranges that can be seen in a test result: ≤ 26 = normal, 27–35 = intermediate–normal phenotype but expandable, 36–39 = reduced penetrance and expandable, ≥ 40 fully penetrant
Reverse correlation between age of onset and number of repeats
Anticipation (usually through the paternal line) and the possibility of juvenile HD

Cite this article as *Cold Spring Harb Perspect Med* doi: 10.1101/cshperspect.a036525

expansion of the pathogenic CAG repeat in *HTT*.

Before helping the patient to anticipate possible ramifications of testing, learning about the patient's motivation for pursuing predictive testing at this point in time may illuminate the context into which the information will be received and who is available to help support the patient. Common reasons for predictive testing include making reproductive decisions, life planning, and managing uncertainty. Communicating relevant information about HD facilitates patient engagement to think through and weigh the various consequences of following through with testing or not, to make an informed decision. The counselor may ask the patient to consider their responses to the following questions:

1. What do you anticipate your immediate reaction to a positive and a negative test result may be?

2. Over the long term, talk me through how you imagine you will manage living with knowing your risk status and coping with the stress.

3. What, if anything, about your future life plans may change?

4. With whom will you share your results? Have you discussed your testing decision with these people?

5. How are these results likely to affect these people?

6. What are your resources and who makes up your primary support system?

7. If your siblings are testing, how do you anticipate feeling if their results differ from yours?

During the session, the genetic counselor also questions the patient's support person about how she or he thinks the patient will cope with results and what they anticipate their own response will be. The counselor should explore the support person's feelings, especially when they differ from those of the patient. The genetic counselor may make the recommendation to delay testing until unresolved issues between a couple are addressed and/or may refer them for longer-term counseling.

In ending the session, the counselor communicates the next steps in the pretest protocol. If not already discussed, information can be provided about reproductive options, insurance, and resources such as the HDSA. Whether the patient decides to test or not, the opportunity to participate in observational studies is offered, which provides a way for at-risk individuals to stay connected to the medical community and, through helping further scientific understanding of HD, find some purpose for knowing they have a pathogenic variant for the disease.

How Genetic Counseling for HD May Change

The future of genetic counseling for HD will be impacted by the availability of drug trials that may provide additional hope to those found to be at risk, particularly if they lead to effective treatment and prevention. Even now, access to HD counseling in the United States is limited by the number and location of counselors sufficiently familiar with the novel characteristics of HD counseling. In many areas, access is further limited by the cost of completing the full protocol. As prevention trials become available, the demand for testing will likely overwhelm the existing resources, and new protocols will have to be established.

For example, in 2015, Ionis Pharmaceutical began a drug trial in Europe and Canada for Ionis-HTTRx, an antisense oligonucleotide (ASO) designed to reduce the abnormal Huntingtin protein through blocking translation of *HTT* mRNA. The drug showed significant success in mouse models and promising results in humans. The trial, which enrolls patients with early symptoms, has begun in the United States as a Phase 1b/2a trial. This trial—and others like it in the future—will study drug safety, tolerability, and biomarkers (Wild and Tabrizi 2017).

The availability of clinical trials of therapeutics specifically creates ethical issues for genetic counselors and physicians. Each trial will have strict inclusion/exclusion criteria—for example, the Ionis–Roche trial is only open to patients who are confirmed to have a pathogenic variant in *HTT* and are displaying early symptoms. Therefore, patients must understand that to qualify, they will need to know their genetic status and have confirmed symptoms. Given that some patients learn their genetic status without learning their neurological status, the lure of trial participation may prompt patients to seek out more information than they would have done otherwise. It is one thing to know that you are gene variant–positive and quite another to know that you have early HD, and some patients choose not to know. A possible scenario may be that patients elect to have genetic testing to participate, learn their status, and then are not given one of the limited participant slots. If the only motivation for genetic testing is study participation, there is the potential for seeing an increase in depressive symptoms and other adverse outcomes. Additionally, one family member may be accepted into the trial while another is not. It will be up to the genetic counselor to anticipate and explore such scenarios with the patients. Genetic counseling for people wanting testing for trial candidacy could potentially impact counseling for other people who are differently motivated. The genetic counselor will have to work with her professional HD group colleagues to determine equitable access. Regardless of motivation, genetic counseling will have to include an explanation of the trial and clarification about inclusion criteria. Patients will need to understand that if a trial is a double-blind randomized study, participation will not guarantee receiving the drug, and even if they do, there is no evidence to predict benefits or harms.

Complicated as they may be, clinical trials offer hope to people when there was none before. The genetic counselor can facilitate informing people about both observational research and clinical trials. Research can offer not only hope, but a way to find meaning in their genetic status and a way to stay connected to the medical community.

ALZHEIMER'S DISEASE

Predictive Counseling and Testing for Dominantly Inherited Alzheimer's Disease

As the genes for Alzheimer's disease (AD) were being mapped, concerns were raised about the legal, ethical, and social issues of predictive genetic testing. A workshop attended by scientists, geneticists, clinicians, and ethicists formulated guidelines similar to those for HD, but also pointed to some differences between AD and HD (Lennox et al. 1994). The first report of predictive testing for AD was in 1995, soon after the discovery of the amyloid precursor protein gene (*APP*) (Lannfelt et al. 1995). A protocol that included pre/posttest genetic counseling similar to that for predictive testing of HD was accepted for AD predictive testing. However, autosomal-dominant AD can be due to mutations in three different highly penetrant genes—*APP*, presenilin 1 (*PSEN1*), and presenilin 2 (*PSEN2*)—making testing more complicated than that for HD. Additionally, much of AD does not follow an autosomal-dominant inheritance pattern. Ruling out dominantly inherited AD (DIAD) does not reduce the risk for late-onset sporadic AD. Furthermore, although only ~1% of AD is due to autosomal-dominant genes, a much greater proportion is multifactorial and familial (Mayeux and Schupf 1995).

The reasons for interest in predictive testing for AD and other neurodegenerative diseases are similar to those for HD: assessing reproductive options, life planning, and managing uncertainty (Tibben et al. 1997), as well as clinical trial candidacy. As with HD, predictive testing for DIAD was found to be without significant adverse psychological consequences within the parameters of a predictive testing protocol (Steinbart et al. 2001; Molinuevo et al. 2005; Cassidy et al. 2008).

AD Susceptibility Testing and Genetic Counseling

Most of AD is not inherited as an autosomal-dominant trait, but instead is multifactorial. Through genetic counseling, individuals can better understand their empiric risk based on their family history. Still many people with a family member with AD want to undergo genetic testing and, therefore, seek out genetic counseling to learn about variants in genes that increase risk. Of the multiple genes that contribute to risk for AD discovered through genome-wide association studies (GWASs) and exome/genome sequencing (ES/GS), the apolipoprotein E gene (*APOE*) is recognized to have the greatest association with AD risk. *APOE* has three different allelic forms (ε2, ε3, and ε4) and, therefore, six different possible genotypes. Individuals heterozygous for ε4 are at two to three times higher risk for AD than those without an ε4, and homozygotes have a significantly higher lifetime risk. Estimates of the actual *APOE* ε4/ε4-associated lifetime risk vary considerably. New studies estimate risk of AD or mild cognitive impairment (MCI) as 30%–55%, which is significantly less than early studies (Qian et al. 2017). Despite this association, APOE ε4 is neither necessary nor sufficient for the development of AD. Numerous research papers have been published cautioning use of *APOE* for either diagnostic or predictive reasons. Statements against *APOE* testing cite possible misinterpretation of risk (too much risk reduction or too high a disease risk perception) and possible insurance discrimination (CMG/ASHG 1995; Mayeux and Schupf 1995; Relkin et al. 1996; Post et al. 1997; Goldman et al. 2011). In addition to its association with late onset AD, *APOE* has been reported to lower the age of onset of early-onset AD (Corder et al. 1993; Saunders et al. 1993). The *APOE*-associated risk for AD also varies with ethnicity and sex (Farrer et al. 1997). Another distinguishing factor is that it is also a risk factor for cardiovascular disease.

Despite the guidelines against *APOE* testing, substantial interest in susceptibility testing led to a series of National Institutes of Health (NIH)-funded randomized clinical studies called the Risk Evaluation and Education for Alzheimer's Disease (REVEAL) (Roberts et al. 2003, 2004; Marteau et al. 2005; Zick et al. 2005). In these studies, individuals at risk for AD (as a result of having a parent with the disease) were randomized to receive or not receive *APOE* results. Information concerning risk was given to both groups by genetic counselors, but only one group received their *APOE* results. The major limitation to these studies was the ascertainment bias of a largely Caucasian, female, highly educated sample. Motivation to test included anxiety about risk, desire to know as much as they

could about their health, and a need to take control (Gooding et al. 2006). In general, the outcome assessments found that participants experienced little psychological distress (Cassidy et al. 2008; Green et al. 2009). REVEAL found that perception of lifetime risk was reduced among individuals with ε4 negative but not positive results, and that perceived risk may not be accurate, even after counseling (Marteau et al. 2005; Linnenbringer et al. 2010). Another interesting finding from these studies was that *APOE* ε4 positive participants were significantly more likely to buy long-term care insurance. The REVEAL investigators suggested that their participants were protected by research confidentiality, whereas if the general public chose to be tested, their results would be accessible. Awareness of this testing behavior by the insurance industry could ultimately lead to higher rates for people with family histories of AD (Zick et al. 2005). One REVEAL study showed that *APOE* positive individuals were more likely to change health behaviors, including use of unproven supplements (Chao et al. 2008; Vernarelli et al. 2010). Learning that *APOE* ε4 is also a risk factor for coronary heart disease was tolerated well, possibly because participants perceived that they have more control over cardiac risk than AD risk (Christensen et al. 2016).

The main goals of genetic counseling for multifactorial AD should be to clarify individual risk by examining the empiric risk generated from family history and to explain the lack of clinical utility of *APOE* testing. Counseling should explore motivation for wanting testing and question how the patient might achieve their goal without testing. Genetic counseling can also include a discussion of some of the REVEAL study findings on health behaviors (e.g., choosing not to use supplements and to live a heart healthy life).

The New World of Alzheimer's Genetics: DTC, Clinic Trials, and Polygenic Risk Scores

Until recently, *APOE* testing for Alzheimer risk has been limited. Yet, now *APOE* testing is being offered without genetic counseling by direct-to-consumer (DTC) companies such as 23andMe and Helix. The majority of people choosing to test seem to accept their results without much distress, and even find positive ε4 results helpful for future planning and motivation for lifestyle change. Yet, in other studies, as many as 24% of individuals experienced psychological distress on finding that they are ε4 carriers (Zallen 2018; Marshe 2019). Without pretest counseling, these individuals had not understood the meaning of the test nor considered its consequences. Additionally, testing for *APOE* on 23andMe requires users to understand that *APOE*-associated risk is dependent on ethnicity and that their overall risk is both polygenic and multifactorial. Knowledge of having an *APOE ε4* allele has been shown to negatively influence subjective and objective performance on memory tests (Lineweaver et al. 2014). Those testing positive often seek advice and guidance from their doctors who may or may not be able to help (see alzforum.org/news/community-news/genetic-wild-west-23andme-raw-data-contains-75-alzheimers-mutations). With luck, some find their way to genetic counselors. Others turn to social media support groups (such as apoe4.info) for help.

The considerable fear of AD has fostered both semilegitimate and dubious prevention claims. Alzheimer's disease prevention centers have emerged, including at academic institutions. Although such centers can help people to change lifestyles that will reduce the risk of cardiovascular disease, whether this will translate to reduced AD risk is uncertain. Moreover, these centers target a well-educated clientele that can afford extensive elective appointments and testing. Such centers are also ordering *APOE* tests as well as biomarker tests such as amyloid positron emission tomography (PET) scans, which measure the amount of amyloid on the brain. Presently, no medication can reduce this amyloid and most asymptomatic people with amyloid will not develop AD (Hellmuth et al. 2019).

To date, drug trials for AD have failed because of safety concerns or lack of efficacy. These trials were designed for people with moderate to severe disease. Since long-term observational studies show that AD biomarkers start changing 15 or so years before diagnosis, treating people in mid to late disease stages may be too late (Bateman et al. 2012). Thus, trials have now started for asymptomatic people at high risk for developing AD (Bateman et al. 2017; Lopez et al. 2019). Genetic counseling is an important component of these studies to insure fully informed consent and to reduce the likelihood of coercion, such that someone would elect to learn their genetic status only because they want to be in the trial (Grill et al. 2015). Once again, being part of research instills hope, but participants should consider how knowing their genetic status will impact their lives whether or not the experimental drug is effective. Genetic counseling can help in this process.

Polygenic risk scores are currently used in research to assess overall risk of AD, risk of conversion from MCI to AD, and age of onset of AD (Escott-Price et al. 2015; Tan et al. 2017; Tosto et al. 2017; Xiao et al. 2017; Cruchaga et al. 2018). Polygenic risk scores provide a value based on the weighted risks of multiple variants at the same time. The utility of such scores is yet to be determined, but scientists are speculating that they can even be used to determine cognition abilities in early life (Axelrud et al. 2018). These polygenic risk scores need to be validated before they can be used clinically for risk prediction. Despite this, third-party websites already exist that will generate polygenic risk scores for conditions including AD using raw data from companies like 23andMe and ancestry.com (e.g., impute.me). If the use of such scores moves into the clinic, genetic counselors will become responsible for explaining the benefits and limitations of such tests and will have to interpret results. Already physicians are ordering predictive dementia panel tests on their patients without confirming the presence of a pathogenic variant in affected family members. Often, this is performed without genetic counseling, which can lead to ill-considered actions by the ordering physician and the patient. Without an established family pathogenic variant, the presence of one or more variant of unknown significance (VUS) may result in significant anxiety for the patient.

OTHER COMPLEX NEURODEGENERATIVE DISEASES

With the advent of next-generation sequencing, our knowledge of genetic components of disease has exploded and with it, the furthering of our understanding of disease mechanisms. In the world of neurodegenerative conditions, this proliferation of information is particularly apparent in the fields of frontotemporal dementia (FTD) and amyotrophic lateral sclerosis (ALS). These two seemingly different diseases are now perceived as being a disease spectrum. Excluding very few unique genes (particularly, *SOD1* for ALS and *MAPT*, *PGRN*, and *CHMP2B* for FTD), the two diseases overlap in causal genes. Additionally, 98% of ALS and 50% of FTD share the common pathological finding of TDP-43 neuronal and glial cytoplasmic inclusions (Turner et al. 2017). Given that up to 13.5% of ALS and 43% of FTD is familial (Crook et al. 2017), providing the most current and accurate information for families is crucial.

Unlike HD, which is monogenetic, or AD, which is either autosomal-dominant or inheritance as a single pathology, FTD/ALS can be caused by numerous genes resulting in heterogenous clinical presentations and pathologies. Of particular importance is a pathogenic hexanucleotide expansion in the *C9orf72* gene, which causes as much as 34% of familial ALS and 26% of familial FTD (van Blitterswijk et al. 2012). Clinical presentations include pure ALS, pure FTD, FTD/ALS, AD-like disease, HD-like disease, cerebellar ataxia, psychosis, and more. Age of onset and disease duration are also highly variable for this gene, as well as many of the other FTD/ALS genes. To add to the complexity, FTD/ALS genes display different disease mechanisms (e.g., dominant-negative: *MAPT*, *SOD1* vs. haploinsufficiency: *PGRN*; recessive inheritance [*SOD1* D90A variant]; and different types of variations: single-nucleotide changes, indels, repeat expansions [*C9orf72*]). Much about the penetrance of these genes remains elusive, but some appear to be highly penetrant (*MAPT*, *CHMP2B*), whereas others have variable penetrance depending on the variation (*SOD1*). Moreover, several case reports and research findings raise the possibility of oligogenic inheritance and discordant family results due to sporadic ALS phenocopy (Mandich et al. 2015; Giannoccaro et al. 2017). Last, *TMEM106B* and possible other risk genes have been unveiled that influence the risk of sporadic disease as well as the presentation of familial disease when in combination with various pathogenic variants (e.g., *C9orf72*, *PGRN*) (Nicholson and Rademakers 2016).

Previous studies have shown that, like HD and AD, predictive testing for FTD and ALS, when conducted under an HD-like protocol, results in relatively few adverse psychological or social outcomes (Fanos et al. 2011; Wagner et al. 2018). Genetic counseling includes information on multiple genes that may be causative, uncertainty about clinical presentation, penetrance, and age of onset. Presently, many ALS centers are performing *C9orf72* genetic testing on all consenting symptomatic patients, whether they have a family history or not. The rationale behind this testing is that *C9orf72* expansions have been found in a significant number of people with apparently sporadic disease (Majounie et al. 2012). Finding a genetic cause for disease can lead to better understanding of the etiology, but can also cause substantial anxiety among relatives of those with a pathogenic variant. Additionally, because many of the genes are pleiotropic, any at-risk individual should be helped to understand that if tested, they may learn whether they carry the family variant, but not how or when that gene will be expressed. Predictive testing without a known family variant involves use of a large panel of dementia and ALS genes. As a result, the possibility of multiple VUS results is significant. Moreover, 30%–40% of people with familial ALS do not have a pathogenic variant in any known ALS gene; thus, negative predictive testing will not rule out the possibility of carrying a yet-to-be identified gene (Benatar et al. 2016).

Like HD, new clinical drug studies for ALS and some other neurodegenerative diseases have commenced. Qualifying for such trials is complicated by the multifactorial etiology of these diseases. Some trials are specific to a particular genetic etiology. For example, in some situations, people will not qualify for a trial if they have sporadic disease or a genetic form of the disease other than that targeted by the trial (e.g., FTD caused by an *MAPT* variant vs. FTD-*C9orf72*). Thus, candidates undergo genetic screening but may not qualify for participation in a trial because they carry a variant in the "wrong" gene. Patients will, therefore, learn their genetic status but have no clinical recourse. Substantial pre- and posttest genetic counseling will be needed to prepare patients for these potential outcomes. Patients will need to consider how knowledge of their status will affect themselves and their relatives.

As neurologists increase the number of tests they order, clinically or within studies, the importance of genetic counseling expands. Although genetic research leads to increased understanding of disease mechanisms, it also leads to recognition of the greater complexity of the genetics of neurogenetic diseases. Genetic counseling can help patients to have realistic expectations about possible results and better understand the concepts of variable expression, pleiotropy, penetrance, genetic heterogeneity, and VUSs. Genetic counselors will continue to play a key role in informed consent and exploration of tolerance for the uncertainty that can be generated by test results. Genetic counselors will also be instrumental in ordering appropriate genomic tests.

MOVEMENT DISORDERS

Parkinson's disease (PD) is the second most common neurodegenerative disease after AD. Although 5%–10% of PD is monogenic, the majority of these cases are early-onset, recessive forms (Lill 2016). Thus, only 2%–3% of idiopathic PD can be attributed to monogenic etiology (Domingo and Klein 2018). The α-synuclein gene (*SNCA*) causes a rare, severe type of autosomal-dominant PD. Only four point-pathogenic variants are known, and most pathogenic variants associated with *SNCA* are duplications and triplications. The duplications have low penetrance and result in a generally milder disease. Duplications have also been found in patients with sporadic disease (Ahn et al. 2008).

The most common cause of autosomal-dominant PD is a pathogenic variant in the *LRRK2* gene. Again, *LRRK2* has been identified in patients with apparently sporadic disease. However, some *LRRK2* pathogenic variants have low penetrance so these cases may not in reality be sporadic. The G2019S variant with a penetrance of <30% is especially common in North Africans (with a frequency of 39% of sporadic and 36% of familial PD) and Ashkenazi Jews (with a frequency of 28% in familial and 10% in sporadic PD) (Schneider and Alcalay 2017). Although the phenotype overlaps with idiopathic PD, LRRK2-PD tends to be tremor-predominant (as opposed to having more posture instability) and have a lower rate of associated dementia.

Genetic counseling for PD is complicated not only by issues of penetrance, but also by the presence of variants that increase disease susceptibility. The gene most important in predicting risk for PD is *GBA*. Whereas homozygous *GBA* causes Gaucher disease, heterozygous carriers are at risk for PD. This risk is both age- and pathogenic variant–dependent. *GBA* variants are particularly common among the Ashkenazi Jewish population, in which ~15% of people with PD have a *GBA* variant, whereas only ~3% of non-Ashkenazi Jews with PD carry one. The most common pathogenic variant for Ashkenazi Jews is N370S, which has a lifetime penetrance of <10% (Alcalay et al. 2014). Other variants may have a higher penetrance. *GBA*-associated PD resembles idiopathic PD, but may have earlier onset and faster progression and has a higher risk of psychiatric symptoms and dementia. In fact, *GBA* variants are associated with risk of Lewy body dementia.

Until recently, because of the lack of clinical use of knowing genetic information on idiopathic PD, genetic testing was rarely performed. However, with the increase in prognostic information and both observational studies on and clinical trials for genetic forms of the disease, many physicians are encouraging testing and patients are asking for it (Brüggemann and Klein 2019). Additionally, DTC testing is available. As with AD, 23andMe is offering risk assessment for PD, and as with *APOE*, testing for

Cite this article as *Cold Spring Harb Perspect Med* doi: 10.1101/cshperspect.a036525

PD through 23andMe requires reading the small print. 23andMe tests only for the common Ashkenazi Jewish variants in *LRRK2* and *GBA*. If the user is not Ashkenazi Jewish, the results do little in predicting their risk for developing PD. Genetic counseling can help explain results.

Several observational studies and treatment trials for PD require knowing gene status. Most require genetic counseling for informed consent. Patients often are motivated to help PD research. However, some asymptomatic individuals participate primarily to receive their test results. Interestingly, because both *LRRK2* and *GBA* gene variants are relatively common in this population, testing an unaffected person occasionally revealed a variant in the unaffected parent. Additionally, people interested in learning their genetic status should consider their privacy and possible discrimination (Suter 2019). These issues again point to the importance of pretest genetic counseling.

SUMMARY AND THE FUTURE OF GENETIC COUNSELING FOR NEURODEGENERATIVE DISEASE

Genetic counseling for neurodegenerative diseases is evolving. Counseling for the small number of people at risk for autosomal-dominant diseases will likely remain similar to current practices. Additionally, genetic counseling will play an essential role in future clinical trials for this population. Through genetic counseling, potential participants can consider and weigh issues such as whether or not they want to learn their genetic status, whether family dynamics are causing any coercion with regards to trial participation, whether trial participation will increase or decrease stress, and whether to delay having children to participate in a trial. As clinical trials increase, new methods such as tele-genetic counseling may shorten or replace in-person counseling sessions.

As soon as any treatment for a neurogenetic disease is proven effective, however, genetic counseling will change. Pretest counseling for autosomal-dominant diseases will need to refocus on the possibility that genetic testing may still leave uncertainty because of VUSs, pheno-typic variability, and genetic heterogeneity. Individuals coming for testing may learn that they are positive for a gene for which the new therapeutic is not effective. And regardless of what genomic variant they carry, they will still need to consider reproductive options, impact on family dynamics, and communication of results to other family members. Genetic counseling will still be a key to informed decision-making. With any new treatment, the number of people wishing to be genetically tested will increase dramatically and outrun the number of neurogenetic counselors. At that point, testing for these autosomal-dominant diseases is likely to be performed routinely through physicians. Yet, genetic counseling will still be important for the reasons mentioned above. Thus, another important role for neurogenetic counseling will include novel delivery modes such as tele-genetic counseling, advising clinicians who are ordering tests, and advising on ethical issues generated by the new interventions, including efficacy.

Most neurodegenerative disease is multifactorial, not autosomal-dominant. For these cases, risk profiles can be generated. Although changes in lifestyle may reduce risk somewhat, the underlying genetic propensity may be too great to prevent disease. Trials are studying how the combination of genetic profiling and biomarker studies can determine who and when to treat. Genetic counseling will continue to play an active role in preparing participants for such studies and helping them to interpret results. Because of the great volume of people at risk for diseases such as late-onset AD or PD, in-person counseling with current practice models will be largely impractical. New strategies, whereby delivery of information is provided online, or in other mass dissemination ways, with genetic counseling focused on helping people adapt, need to be identified. Genetic counselors can play an essential part in designing and overseeing these tools. In summary, genetic counseling for neurodegenerative diseases will continue to be essential for people at risk to understand their risk, think about the impact of knowing their risk, and determining how to adapt to that risk.

REFERENCES

*Reference is also in this subject collection.

Ahn TB, Kim SY, Kim JY, Park SS, Lee DS, Min HJ, Kim YK, Kim SE, Kim JM, Kim HJ, et al. 2008. α-Synuclein gene duplication is present in sporadic Parkinson disease. *Neurology* **70:** 43–49. doi:10.1212/01.wnl.0000271080.53272 .c7

Alcalay RN, Dinur T, Quinn T, Sakanaka K, Levy O, Waters C, Fahn S, Dorovski T, Chung WK, Pauciulo M, et al. 2014. Comparison of Parkinson risk in Ashkenazi Jewish patients with Gaucher disease and *GBA* heterozygotes. *JAMA Neurol* **71:** 752–757. doi:10.1001/jamaneurol .2014.313

Almqvist EW, Bloch M, Brinkman R, Craufurd D, Hayden MR. 1999. A worldwide assessment of the frequency of suicide, suicide attempts, or psychiatric hospitalization after predictive testing for Huntington disease. *Am J Hum Genet* **64:** 1293–1304. doi:10.1086/302374

Almqvist EW, Brinkman RR, Wiggins S, Hayden MR; Canadian Collaborative Study of Predictive Testing. 2003. Psychological consequences and predictors of adverse events in the first 5 years after predictive testing for Huntington's disease. *Clin Genet* **64:** 300–309. doi:10.1034/j .1399-0004.2003.00157.x

American College of Medical Genetics/American Society of Human Genetics Working Group on ApoE and Alzheimer disease. 1995. Statement on use of apolipoprotein E testing for Alzheimer disease. *J Am Med Assoc* **274:** 1627–1629. doi:10.1001/jama.1995.03530200063039

Axelrud LK, Santoro ML, Pine DS, Talarico F, Gadelha A, Manfro GG, Pan PM, Jackowski A, Picon F, Brietzke E, et al. 2018. Polygenic risk score for Alzheimer's disease: implications for memory performance and hippocampal volumes in early life. *Am J Psychiatry* **175:** 555–563. doi:10.1176/appi.ajp.2017.17050529

Baig SS, Strong M, Rosser E, Taverner NV, Glew R, Miedzybrodzka Z, Clarke A, Craufurd D; UK Huntington's Disease Prediction Consortium, Quarrell OW. 2016. 22 Years of predictive testing for Huntington's disease: the experience of the UK Huntington's Prediction Consortium. *Eur J Hum Genet* **24:** 1396–1402. doi:10.1038/ejhg.2016.36

Barette J, Marsden CD. 1979. Attitudes of families to some aspects of Huntington's chorea. *Psychol Med* **9:** 327–336. doi:10.1017/S0033291700030841

Bateman RJ, Xiong C, Benzinger TL, Fagan AM, Goate A, Fox NC, Marcus DS, Cairns NJ, Xie X, Blazey TM, et al. 2012. Clinical and biomarker changes in dominantly inherited Alzheimer's disease. *N Engl J Med* **367:** 795–804. doi:10.1056/NEJMoa1202753

Bateman RJ, Benzinger TL, Berry S, Clifford DB, Duggan C, Fagan AM, Fanning K, Farlow MR, Hassenstab J, McDade EM, et al. 2017. The DIAN-TU Next Generation Alzheimer's Prevention trial: adaptive design and disease progression model. *Alzheimers Dement* **13:** 8–19. doi:10 .1016/j.jalz.2016.07.005

Benatar M, Stanislaw C, Reyes E, Hussain S, Cooley A, Fernandez MC, Dauphin DD, Michon SC, Andersen PM, Wuu J. 2016. Presymptomatic ALS genetic counseling and testing: experience and recommendations. *Neurology* **86:** 2295–2302. doi:10.1212/WNL.0000000000002773

Bird SJ. 1985. Presymptomatic testing for Huntington's disease. *J Am Med Assoc* **253:** 3286–3291. doi:10.1001/ jama.1985.03350460086027

Bloch M, Hayden MR, Opitz JM. 1987. Preclinical testing in Huntington disease. *Am J Med Genet* **27:** 733–734. doi:10 .1002/ajmg.1320270333

Brüggemann N, Klein C. 2019. Will genotype drive treatment options? *Mov Disord* doi: 10.1002/mds.27699.

Cassidy MR, Roberts JS, Bird TD, Steinbart EJ, Cupples LA, Chen CA, Linnenbringer E, Green RC. 2008. Comparing test-specific distress of susceptibility versus deterministic genetic testing for Alzheimer's disease. *Alzheimers Dement* **4:** 406–413. doi:10.1016/j.jalz.2008.04.007

Chao S, Roberts JS, Marteau TM, Silliman R, Cupples LA, Green RC. 2008. Health behavior changes after genetic risk assessment for Alzheimer disease: the REVEAL Study. *Alzheimer Dis Assoc Disord* **22:** 94–97. doi:10 .1097/WAD.0b013e31815a9dcc

Christensen KD, Roberts JS, Whitehouse PJ, Royal CD, Obisesan TO, Cupples LA, Vernarelli JA, Bhatt DL, Linnenbringer E, Butson MB, et al. 2016. Disclosing pleiotropic effects during genetic risk assessment for Alzheimer disease: a randomized trial. *Ann Intern Med* **164:** 155–163. doi:10.7326/M15-0187

Corder EH, Saunders AM, Strittmatter WJ, Schmechel DE, Gaskell PC, Small GW, Roses AD, Haines JL, Pericak-Vance MA. 1993. Gene dose of apolipoprotein E type 4 allele and the risk of Alzheimer's disease in late onset families. *Science* **261:** 921–923. doi:10.1126/science .8346443

Craufurd DI, Harris R. 1986. Ethics of predictive testing for Huntington's chorea: the need for more information. *Br Med J (Clin Res Ed)* **293:** 249–251. doi:10.1136/bmj.293 .6541.249

Crook A, Williams K, Adams L, Blair I, Rowe DB. 2017. Predictive genetic testing for amyotrophic lateral sclerosis and frontotemporal dementia: genetic counselling considerations. *Amyotroph Lateral Scler Frontotemporal Degener* **18:** 475–485. doi:10.1080/21678421.2017.1332079

Crozier S, Robertson N, Dale M. 2015. The psychological impact of predictive genetic testing for Huntington's disease: a systematic review of the literature. *J Genet Couns* **24:** 29–39. doi:10.1007/s10897-014-9755-y

Cruchaga C, Del-Aguila JL, Saef B, Black K, Fernandez MV, Budde J, Ibanez L, Deming Y, Kapoor M, Tosto G, et al. 2018. Polygenic risk score of sporadic late-onset Alzheimer's disease reveals a shared architecture with the familial and early-onset forms. *Alzheimers Dement* **14:** 205–214. doi:10.1016/j.jalz.2017.08.013

Decruyenaere M, Evers-Kiebooms G, Cloostermans T, Boogaerts A, Demyttenaere K, Dom R, Fryns JP. 2004. Predictive testing for Huntington's disease: relationship with partners after testing. *Clin Genet* **65:** 24–31. doi:10.1111/j. .2004.00168.x

Domingo A, Klein C. 2018. Genetics of Parkinson disease. *Handb Clin Neurol* **147:** 211–227. doi:10.1016/B978-0-444-63233-3.00014-2

Escott-Price V, Sims R, Bannister C, Harold D, Vronskaya M, Majounie E, Badarinarayan N; GERAD/PERADES; IGAP consortia, Morgan K, Passmore P, et al. 2015. Common polygenic variation enhances risk prediction for Alzheimer's disease. *Brain* **138:** 3673–3684. doi:10.1093/brain/awv268

Fanos JH, Gronka S, Wuu J, Stanislaw C, Andersen PM, Benatar M. 2011. Impact of presymptomatic genetic testing for familial amyotrophic lateral sclerosis. *Genet Med* **13:** 342–348. doi:10.1097/GIM.0b013e318204d004

Farrer LA, Opitz JM, Reynolds JF. 1986. Suicide and attempted suicide in Huntington disease: implications for preclinical testing of persons at risk. *Am J Med Genet* **24:** 305–311. doi:10.1002/ajmg.1320240211

Farrer LA, Cupples A, Haines JL, Hyman B, Kukull WA, Mayeux R, Myers RH, Pericak-Vance MA, Risch N, van Duijn CM. 1997. Effects of age, sex, and ethnicity on the association between apolipoprotein E genotype and Alzheimer disease. *J Am Med Assoc* **278:** 1349–1356. doi:10.1001/jama.1997.03550160069041

Gargiulo M, Lejeune S, Tanguy ML, Lahlou-Laforêt K, Faudet A, Cohen D, Feingold J, Durr A. 2009. Long-term outcome of presymptomatic testing in Huntington disease. *Eur J Hum Genet* **17:** 165–171. doi:10.1038/ejhg.2008.146

Giannoccaro MP, Bartoletti-Stella A, Piras S, Pession A, De Massis P, Oppi F, Stanzani-Maserati M, Pasini E, Baiardi S, Avoni P, et al. 2017. Multiple variants in families with amyotrophic lateral sclerosis and frontotemporal dementia related to *C9orf72* repeat expansion: further observations on their oligogenic nature. *J Neurol* **264:** 1426–1433. doi:10.1007/s00415-017-8540-x

Goldman JS, Hahn SE, Catania JW, LaRusse-Eckert S, Butson MB, Rumbaugh M, Strecker MN, Roberts JS, Burke W, Mayeux R, et al. 2011. Genetic counseling and testing for Alzheimer disease: joint practice guidelines of the American College of Medical Genetics and the National Society of Genetic Counselors. *Genet Med* **13:** 597–605. doi:10.1097/GIM.0b013e31821d69b8

Gooding HC, Linnenbringer EL, Burack J, Roberts JS, Green RC, Biesecker BB. 2006. Genetic susceptibility testing for Alzheimer disease: motivation to obtain information and control as precursors to coping with increased risk. *Patient Educ Couns* **64:** 259–267. doi:10.1016/j.pec.2006.03.002

Green RC, Roberts JS, Cupples LA, Relkin NR, Whitehouse PJ, Brown T, Eckert SL, Butson M, Sadovnick AD, Quaid KA, et al. 2009. Disclosure of *APOE* genotype for risk of Alzheimer's disease. *N Engl J Med* **361:** 245–254. doi:10.1056/NEJMoa0809578

Grill JD, Bateman RJ, Buckles V, Oliver A, Morris JC, Masters CL, Klunk WE, Ringman JM; Dominantly Inherited Alzheimer's Network. 2015. A survey of attitudes toward clinical trials and genetic disclosure in autosomal dominant Alzheimer's disease. *Alzheimers Res Ther* **7:** 50. doi:10.1186/s13195-015-0135-0

Groves M. 2017. The highly anxious individual presenting for Huntington disease-predictive genetic testing: the psychiatrist's role in assessment and counseling. *Handb Clin Neurol* **144:** 99–105. doi:10.1016/B978-0-12-801893-4.00008-0

Hellmuth J, Rabinovici GD, Miller BL. 2019. The rise of pseudomedicine for dementia and brain health. *J Am Med Assoc* **321:** 543–544. doi:10.1001/jama.2018.21560

Huggins M, Bloch M, Kanani S, Quarrell OW, Theilman J, Hedrick A, Dickens B, Lynch A, Hayden M. 1990. Ethical

and legal dilemmas arising during predictive testing for adult-onset disease: the experience of Huntington disease. *Am J Hum Genet* **47:** 4–12.

Huntington's Disease Collaborative Research Group. 1993. A novel gene containing a trinucleotide repeat that is expanded and unstable on Huntington's disease chromosomes. *Cell* **72:** 971–983. doi:10.1016/0092-8674(93)90585-E

Huntington's Disease Society of America. 1994. *Guidelines for predictive testing for Huntington's disease.* Huntington's Disease Society of America, New York.

Huntington's Disease Society of America. 2016. *Genetic testing protocol for Huntington's disease.* Huntington's Disease Society of America, New York.

Ibisler A, Ocklenburg S, Stemmler S, Arning L, Epplen JT, Saft C, Hoffjan S. 2017. Prospective evaluation of predictive DNA testing for Huntington's disease in a large German center. *J Genet Couns* **26:** 1029–1040. doi:10.1007/s10897-017-0085-8

International Huntington Association and the World Federation of Neurology Research Group on Huntington's Chorea. 1994. Guidelines for the molecular genetics predictive test in Huntington's disease. *J Med Genet* **31:** 555–559. doi:10.1136/jmg.31.7.555

Kessler S. 1987. Psychiatric implications of presymptomatic testing for Huntington's disease. *Am J Orthopsychiatry* **57:** 212–219. doi:10.1111/j.1939-0025.1987.tb03531.x

Kessler S, Field T, Worth L, Mosbarger H, Opitz JM, Reynolds JF. 1987. Attitudes of persons at risk for Huntington disease toward predictive testing. *Am J Med Genet* **26:** 259–270. doi:10.1002/ajmg.1320260204

Kolata G. 1986. Genetic screening raises questions for employers and insurers. *Science* **232:** 317–319. doi:10.1126/science.2938255

Lamport AT, Opitz JM, Shaw MW. 1987. Presymptomatic testing for Huntington chorea: ethical and legal issues. *Am J Med Genet* **26:** 307–314. doi:10.1002/ajmg.1320260208

Lannfelt L, Axelman K, Lilius L, Basun H. 1995. Genetic counseling in a Swedish Alzheimer family with amyloid precursor protein mutation. *Am J Hum Genet* **56:** 332–335.

Larsson MU, Luszcz MA, Bui TH, Wahlin TB. 2006. Depression and suicidal ideation after predictive testing for Huntington's disease: a two-year follow-up study. *J Genet Couns* **15:** 361–374. doi:10.1007/s10897-006-9027-6

Lawson K, Wiggins S, Green T, Adam S, Bloch M, Hayden MR. 1996. Adverse psychological events occurring in the first year after predictive testing for Huntington's disease. The Canadian Collaborative Study Predictive Testing. *J Med Genet* **33:** 856–862. doi:10.1136/jmg.33.10.856

Lennox A, Karlinsky H, Meschino W, Buchanan JA, Percy ME, Berg JM. 1994. Molecular genetic predictive testing for Alzheimer's disease: deliberations and preliminary recommendations. *Alzheimer Dis Assoc Disord* **8:** 126–147. doi:10.1097/00002093-199408020-00009

Lill CM. 2016. Genetics of Parkinson's disease. *Mol Cell Probes* **30:** 386–396. doi:10.1016/j.mcp.2016.11.001

Lineweaver TT, Bondi MW, Galasko D, Salmon DP. 2014. Effect of knowledge of APOE genotype on subjective and objective memory performance in healthy older adults.

Am J Psychiatry **171:** 201–208. doi:10.1176/appi.ajp.2013 .12121590

Linnenbringer E, Roberts JS, Hiraki S, Cupples LA, Green RC. 2010. "I know what you told me, but this is what I think:" perceived risk of Alzheimer disease among individuals who accurately recall their genetics-based risk estimate. *Genet Med* **12:** 219–227. doi:10.1097/GIM .0b013e3181cef9e1

Lopez Lopez C, Tariot PN, Caputo A, Langbaum JB, Liu F, Riviere ME, Langlois C, Rouzade-Dominguez ML, Zalesak M, Hendrix S, et al. 2019. The Alzheimer's Prevention Initiative Generation Program: study design of two randomized controlled trials for individuals at risk for clinical onset of Alzheimer's disease. *Alzheimers Dement (N Y)* **5:** 216–227.

MacLeod R, Tibben A, Frontali M, Evers-Kiebooms G, Jones A, Martinez-Descales A, Roos RA; Editorial Committee and Working Group 'Genetic Testing Counselling' of the European Huntington Disease Network. 2013. Recommendations for the predictive genetic test in Huntington's disease. *Clin Genet* **83:** 221–231. doi:10.1111/j.1399-0004 .2012.01900.x

Majounie E, Renton AE, Mok K, Dopper EG, Waite A, Rollinson S, Chiò A, Restagno G, Nicolaou N, Simon-Sanchez J, et al. 2012. Frequency of the *C9orf72* hexanucleotide repeat expansion in patients with amyotrophic lateral sclerosis and frontotemporal dementia: a cross-sectional study. *Lancet Neurol* **11:** 323–330. doi:10.1016/S1474-4422(12)70043-1

Mandich P, Mantero V, Verdiani S, Gotta F, Caponnetto C, Bellone E, Ferrandes G, Origone P. 2015. Complexities of genetic counseling for ALS: a case of two siblings with discordant genetic test results. *J Genet Couns* **24:** 553–557. doi:10.1007/s10897-015-9831-y

Mandich P, Lamp M, Gotta F, Gulli R, Iacometti A, Marchese R, Bellone E, Abbruzzese G, Ferrandes G. 2017. 1993–2014: two decades of predictive testing for Huntington's disease at the Medical Genetics Unit of the University of Genoa. *Mol Genet Genomic Med* **5:** 473–480. doi:10.1002/mgg3.238

Marshe VS, Gorbovskaya I, Kanji S, Kish M, Müller DJ. 2019. Clinical implications of APOE genotyping for late-onset Alzheimer's disease (LOAD) risk estimation: a review of the literature. *J Neural Transm* **126:** 65–85. doi:10.1007/s00702-018-1934-9

Marteau TM, Roberts S, LaRusse S, Green RC. 2005. Predictive genetic testing for Alzheimer's disease: impact upon risk perception. *Risk Anal* **25:** 397–404. doi:10.1111/j .1539-6924.2005.00598.x

Mayeux R, Schupf N. 1995. Apolipoprotein E and Alzheimer's disease: the implications of progress in molecular medicine. *Am J Public Health* **85:** 1280–1284. doi:10 .2105/AJPH.85.9.1280

Molinuevo JL, Pintor L, Peri JM, Lleó A, Oliva R, Marcos T, Blesa R. 2005. Emotional reactions to predictive testing in Alzheimer's disease and other inherited dementias. *Am J Alzheimers Dis Other Demen* **20:** 233–238. doi:10.1177/153331750502000408

Morrison PJ, Harding-Lester S, Bradley A. 2011. Uptake of Huntington disease predictive testing in a complete population. *Clin Genet* **80:** 281–286. doi:10.1111/j.1399-0004 .2010.01538.x

Nicholson AM, Rademakers R. 2016. What we know about *TMEM106B* in neurodegeneration. *Acta Neuropathol* **132:** 639–651. doi:10.1007/s00401-016-1610-9

Post SG, Whitehouse PJ, Binstock RH, Bird TD, Eckert SK, Farrer LA, Fleck LM, Gaines AD, Juengst ET, Karlinsky H, et al. 1997. The clinical introduction of genetic testing for Alzheimer disease. An ethical perspective. *J Am Med Assoc* **277:** 832–836. doi:10.1001/jama.1997.03540340066035

Qian J, Wolters FJ, Beiser A, Haan M, Ikram MA, Karlawish J, Langbaum JB, Neuhaus JM, Reiman EM, Roberts JS, et al. 2017. APOE-related risk of mild cognitive impairment and dementia for prevention trials: an analysis of four cohorts. *PLoS Med* **14:** e1002254. doi:10.1371/journal .pmed.1002254

Quaid KA, Brandt J, Folstein SE. 1987. The decision to be tested for Huntington's disease. *J Am Med Assoc* **257:** 3362. doi:10.1001/jama.1987.03390240068015

Quaid KA, Eberly SW, Kayson-Rubin E, Oakes D, Shoulson I; Huntington Study Group PHAROS Investigators and Coordinators. 2017. Factors related to genetic testing in adults at risk for Huntington disease: the prospective Huntington at-risk observational study (PHAROS). *Clin Genet* **91:** 824-831. doi:10.1111/cge.12893

Relkin NR, Kwon YJ, Tsai J, Gandy S. 1996. The National Institute on Aging/Alzheimer's Association recommendations on the application of apolipoprotein E genotyping to Alzheimer's disease. *Ann NY Acad Sci* **802:** 149–176. doi:10.1111/j.1749-6632.1996.tb32608.x

Richards F, Williams K. 2004. Impact on couple relationships of predictive testing for Huntington disease: a longitudinal study. *Am J Med Genet A* **126A:** 161–169. doi:10 .1002/ajmg.a.20582

Roberts JS, LaRusse SA, Katzen H, Whitehouse PJ, Barber M, Post SG, Relkin N, Quaid K, Pietrzak RH, Cupples LA, et al. 2003. Reasons for seeking genetic susceptibility testing among first-degree relatives of people with Alzheimer disease. *Alzheimer Dis Assoc Disord* **17:** 86–93. doi:10 .1097/00002093-200304000-00006

Roberts JS, Barber M, Brown TM, Cupples LA, Farrer LA, LaRusse SA, Post SG, Quaid KA, Ravdin LD, Relkin NR, et al. 2004. Who seeks genetic susceptibility testing for Alzheimer's disease? Findings from a multisite, randomized clinical trial. *Genet Med* **6:** 197–203. doi:10.1097/01 .GIM.0000132688.55591.77

Rosenfeld A. 1984. At risk for Huntington's disease: who should know what and when? *Hastings Cent Rep* **14:** 5–8. doi:10.2307/3561179

Saunders AM, Strittmatter WJ, Schmechel D, George-Hyslop PH, Pericak-Vance MA, Joo SH, Rosi BL, Gusella JF, Crapper-MacLachlan DR, Alberts MJ, et al. 1993. Association of apolipoprotein E allele ε4 with late-onset familial and sporadic Alzheimer's disease. *Neurology* **43:** 1467–1472. doi:10.1212/WNL.43.8.1467

Schneider SA, Alcalay RN. 2017. Neuropathology of genetic synucleinopathies with parkinsonism: review of the literature. *Mov Disord* **32:** 1504–1523. doi:10.1002/mds .27193

Steinbart EJ, Smith CO, Poorkaj P, Bird TD. 2001. Impact of DNA testing for early-onset familial Alzheimer disease and frontotemporal dementia. *Arch Neurol* **58:** 1828–1831. doi:10.1001/archneur.58.11.1828

Cite this article as *Cold Spring Harb Perspect Med* doi: 10.1101/cshperspect.a036525

* Suter S. 2019. Legal challenges in genetics, including duty to warn and genetic discriminations. *Cold Spring Harb Perspect Med* doi:10.1101/cshperspect.a036665

Tan CH, Hyman BT, Tan JJX, Hess CP, Dillon WP, Schellenberg GD, Besser LM, Kukull WA, Kauppi K, McEvoy LK, et al. 2017. Polygenic hazard scores in preclinical Alzheimer disease. *Ann Neurol* 82: 484–488. doi:10.1002/ana.25029

Tibben A. 2007. Predictive testing for Huntington's disease. *Brain Res Bull* 72: 165–171. doi:10.1016/j.brainresbull.2006.10.023

Tibben A, Vegter-vd Vlis M, vd Niermeijer MF, Kamp JJ, Roos RA, Rooijmans HG, Frets PG, Verhage F. 1990. Testing for Huntington's disease with support for all parties. *Lancet* 335: 553. doi:10.1016/0140-6736(90)90796-8

Tibben A, Stevens M, de Wert GM, Niermeijer MF, van Duijn CM, van Swieten JC. 1997. Preparing for presymptomatic DNA testing for early onset Alzheimer's disease/cerebral haemorrhage and hereditary Pick disease. *J Med Genet* 34: 63–72. doi:10.1136/jmg.34.1.63

Tosto G, Bird TD, Tsuang D, Bennett DA, Boeve BF, Cruchaga C, Faber K, Foroud TM, Farlow M, Goate AM, et al. 2017. Polygenic risk scores in familial Alzheimer disease. *Neurology* 88: 1180–1186. doi:10.1212/WNL.0000000000003734

Turner MR, Al-Chalabi A, Chio A, Hardiman O, Kiernan MC, Rohrer JD, Rowe J, Seeley W, Talbot K. 2017. Genetic screening in sporadic ALS and FTD. *J Neurol Neurosurg Psychiatry* 88: 1042–1044. doi:10.1136/jnnp-2017-315995

Tyler A, Harper PS. 1983. Attitudes of subjects at risk and their relatives towards genetic counselling in Huntington's chorea. *J Med Genet* 20: 179–188. doi:10.1136/jmg.20.3.179

van Blitterswijk M, DeJesus-Hernandez M, Rademakers R. 2012. How do *C9ORF72* repeat expansions cause amyotrophic lateral sclerosis and frontotemporal dementia: can we learn from other noncoding repeat expansion disorders? *Curr Opin Neurol* 25: 689–700. doi:10.1097/WCO.0b013e32835a3efb

Vernarelli JA, Roberts JS, Hiraki S, Chen CA, Cupples LA, Green RC. 2010. Effect of Alzheimer disease genetic risk disclosure on dietary supplement use. *Am J Clin Nutr* 91: 1402–1407. doi:10.3945/ajcn.2009.28981

Wagner KN, Nagaraja HN, Allain DC, Quick A, Kolb SJ, Roggenbuck J. 2018. Patients with sporadic and familial amyotrophic lateral sclerosis found value in genetic testing. *Mol Genet Genomic Med* 6: 224–229. doi:10.1002/mgg3.360

Wild EJ, Tabrizi SJ. 2017. Therapies targeting DNA and RNA in Huntington's disease. *Lancet Neurol* 16: 837–847. doi:10.1016/S1474-4422(17)30280-6

World Federation of Neurology Research Group on Huntington's Disease. 1990. Ethical issues policy statement on Huntington's disease molecular genetics predictive test. *J Med Genet* 27: 34–38. doi:10.1136/jmg.27.1.34

Xiao E, Chen Q, Goldman AL, Tan HY, Healy K, Zoltick B, Das S, Kolachana B, Callicott JH, Dickinson D, et al. 2017. Late-onset Alzheimer's disease polygenic risk profile score predicts hippocampal function. *Biol Psychiatry Cogn Neurosci Neuroimaging* 2: 673–679. doi:10.1016/j.bpsc.2017.08.004

Zallen DT. 2018. "Well, good luck with that": reactions to learning of increased genetic risk for Alzheimer disease. *Genet Med* 20: 1462–1467. doi:10.1038/gim.2018.13

Zick CD, Mathews CJ, Roberts JS, Cook-Deegan R, Pokorski RJ, Green RC. 2005. Genetic testing for Alzheimer's disease and its impact on insurance purchasing behavior. *Health Aff (Millwood)* 24: 483–490. doi:10.1377/hlthaff.24.2.483

Genetic Counseling in Neurodevelopmental Disorders

Alyssa Blesson[1] and Julie S. Cohen[2,3]

[1]Center for Autism and Related Disorders, Kennedy Krieger Institute, Baltimore, Maryland 21211, USA

[2]Department of Neurology and Developmental Medicine, Kennedy Krieger Institute, Baltimore, Maryland 21205, USA

[3]Department of Neurology, Johns Hopkins University School of Medicine, Baltimore, Maryland 21205, USA

Correspondence: cohenju@kennedykrieger.org

Neurodevelopmental disorders (NDDs), including global developmental delay (GDD), intellectual disability (ID), and autism spectrum disorder (ASD), represent a continuum of developmental brain dysfunction. Although the etiology of NDD is heterogeneous, genetic variation represents the largest contribution, strongly supporting the recommendation for genetic evaluation in individuals with GDD/ID and ASD. Technological advances now allow for a specific genetic diagnosis to be identified in a substantial portion of affected individuals. This information has important ramifications for treatment, prognosis, and recurrence risk, as well as psychological and social benefits for the family. Genetic counseling is a vital service to enable patients and their families to understand and adapt to the genetic contribution to NDDs. As the demand for genetic evaluation for NDDs increases, genetic counselors will have a predominant role in the ongoing evaluation of NDDs, especially as identification of genetic etiologies has the potential to lead to targeted treatments for NDDs in the future.

Genetic counselors are increasingly becoming integral members of the team working with families facing a diagnosis of a neurodevelopmental disorder (NDD). Genetic counselors have many roles in this setting—helping families understand the NDD diagnosis and its implications for their child and family, providing education about potential etiologies and the role of genetics, and participating in the diagnostic evaluation. In addition, genetic counselors help families with making decisions related to genetics and adapting to the disability or genetic condition.

Herein, we will begin with a review of the definitions and etiologies of NDDs. We will provide an overview of diagnostic genetic testing for NDDs and then delve deeper into the numerous counseling issues that arise. Overall, this work will focus on genetic evaluation and counseling for NDDs that have the greatest likelihood to yield an identifiable genetic cause—namely, global developmental delay (GDD), intellectual disability (ID), and autism spectrum disorder (ASD)—highlighting the importance of genetic counseling and ongoing follow-up.

DEFINITIONS

NDDs are a common group of conditions that manifest as impairments in development and function that begin in childhood. As defined in the *Diagnostic and Statistical Manual of Mental Disorders*, 5th edition (DSM-5), NDDs include ID, ASD, attention deficit hyperactivity disorder, communication disorders, motor disorders such as cerebral palsy, and specific learning disorders (American Psychiatric Association 2013). These conditions are typically diagnosed after comprehensive assessments with a developmental pediatrician, pediatric neurologist, neuropsychologist, psychiatrist, or other specialist.

As defined by the American Association on Intellectual and Developmental Disability, ID (formerly called mental retardation) is "characterized by significant limitations both in intellectual functioning and in adaptive behavior as expressed in conceptual, social, and practical adaptive skills. The disability originates before age 18 years" (Schalock et al. 2007). Standardized measures of adaptive functioning and intelligence are typically reliable in children over age 5 years. Before that time, a diagnosis of GDD is often given when a child presents with significant delays in two or more developmental domains, including gross or fine motor, speech/language, cognitive, social/personal, and activities of daily living (Moeschler et al. 2014). Not all children diagnosed with GDD will meet diagnostic criteria for ID, particularly those who present with mild delays and who are expected to meet age-appropriate developmental milestones. The prevalence of GDD/ID in the United States is estimated to be 1%–3% (Moeschler et al. 2014).

ASD is characterized by deficits in social communication and interactions, along with restrictive, repetitive patterns of behaviors, interests, and activities (American Psychiatric Association 2013). This definition encompasses previously used terminology such as pervasive

developmental disorder and its subgroupings (autistic disorder, Asperger's disorder, childhood disintegrative disorder, and pervasive developmental disorder not otherwise specified). ASD is four times more common in males than females (Baio et al. 2018). The severity of the ASD diagnosis is based on level of support: Level 1 indicates "requiring support," Level 2 indicates "requiring substantial support," or Level 3 "requiring very substantial support" (American Psychiatric Association 2013). According to the Autism and Developmental Disabilities Monitoring (ADDM) Network study from surveillance year 2014, one in 59 children (1.7%) aged 8 years has been identified with a diagnosis of ASD (Baio et al. 2018).

NDDs often present with complex patterns of impairment across motor, cognitive, and neurobehavioral domains. For example, individuals with GDD/ID may show autistic features, which may or may not meet criteria for an ASD diagnosis. Likewise, approximately one-third of individuals with ASD have ID (Baio et al. 2018). Although cerebral palsy (CP) is often secondary to prenatal and/or perinatal brain injury, there are many genetic disorders that masquerade as CP and may also have ID and/or ASD as a feature (Lee et al. 2014b). Other medical comorbidities seen in NDDs include epilepsy, congenital anomalies, and psychiatric conditions (Schalock et al. 2007).

Ultimately, these various NDDs can be represented on a larger continuum of an underlying developmental brain dysfunction (DBD) (Moreno-De-Luca et al. 2013; Finucane and Myers 2016). Research has shown a shared genetic basis for ID and related disorders such as epilepsy, ASD, schizophrenia, and others that can be considered part of the spectrum of DBD (Vissers et al. 2016). There can be significant diagnostic overlap and co-occurrence of these brain-based diagnoses within an affected individual and/or among family members, which further points toward an underlying genetic contribution.

ETIOLOGY OF NDDs

The etiology of NDD is heterogeneous, including genetic as well as nongenetic factors. The genetic contribution to NDDs is variable, but reports of increased rates of concordance for NDDs among twins and family members support a high degree of heritability (Chaste and Leboyer 2012). There is substantial genetic heterogeneity, as there are an enormous number of individually rare genetic conditions that present with an NDD as a prominent feature. These include chromosome copy number variants (CNVs) and single-gene disorders. Nongenetic factors may modify the manifestation of features, thus resulting in a multifactorial condition. Variable expressivity and reduced penetrance also impact the diagnosis and presence of medical comorbidities with NDD, especially within the same family. For example, the pathogenic 16p11.2 microdeletion has been reported in individuals with ASD, GDD/ID, and/or neuropsychiatric conditions, as well as apparently unaffected individuals (Ho et al. 2016). Other mechanisms include epigenetic alternations or a polygenic contribution with multiple common variants or polymorphisms collectively resulting in an NDD (Siu and Weksberg 2017).

Epidemiologic studies have attempted to establish the association of various environmental factors with NDDs, such as maternal health conditions, perinatal infection, malnutrition, trauma, and many others (Gentner and Leppert 2019). For instance, prenatal exposure to alcohol and environmental exposure to lead are confirmed to be associated with an increased risk for GDD/ID. On the other hand, any association between vaccines and ASD has been thoroughly debunked by many studies, including those conducted by the Centers for Disease Control (DeStefano et al. 2013).

The high diagnostic yield from genetic testing indicates that genetic variation has a larger causal contribution to the development of NDDs when compared with environmental risk factors. Indeed, modern techniques now result in the identification of genetic etiologies in more than half of individuals with severe GDD/ID (Gilissen et al. 2014), with that number expected to increase with improvements in technology and disease-gene discovery. Therefore, in the absence of a clear and substantial environ-

mental factor that explains the clinical presentation, a genetic workup is appropriate to offer to individuals with GDD/ID and ASD.

GENETIC TESTING IN NDDs

Professional societies have published guidelines on genetic evaluation of GDD/ID and ASD. The most recent published guidelines were authored by the American College of Medical Genetics and Genomics (ACMG) for ASD (Schaefer et al. 2013) and by the American Academy of Pediatrics (AAP) for GDD and ID (Moeschler et al. 2014), and their recommendations are largely consistent.

As always, the evaluation should begin with review of the medical, developmental, and antenatal history, physical and neurologic exam, and review of three-generation family history. Keeping in mind the concepts of DBD as well as variable expressivity, family history queries should include conditions such as epilepsy, psychiatric disorders, and learning disabilities in addition to ASD and GDD/ID, especially with regard to parental abilities (Finucane and Myers 2016). If a specific diagnosis is suspected based on the history and exam, targeted testing could be performed for that disorder. If no specific diagnosis is suspected, both AAP and ACMG recommend a tiered approach to genetic testing.

There is widespread consensus that chromosomal microarray (CMA) testing should be the first-tier genetic test for any patient with unexplained GDD, ID, and/or ASD. The expected yield is ~10%, or possibly up to 15%–20% in patients with multiple congenital anomalies (Miller et al. 2010). G-banded karyotype is no longer recommended as a first-tier test unless there is suspicion of aneuploidy (e.g., Down syndrome) or a history of recurrent miscarriages (Miller et al. 2010). *FMR1* trinucleotide repeat analysis for fragile X syndrome is also recommended as a first-tier test for patients with nonspecific GDD, ID, and/or ASD. The expected yield is 1%–5% in males and lower in females.

The AAP and ACMG guidelines highlighted several other tests that could be considered as second-tier in patients with negative CMA and fragile X testing. These included analysis of the *MECP2* gene in females (expected yield 1%–4%), analysis of the *PTEN* gene in patients with macrocephaly (expected yield 5%), and X-linked ID gene panels in males with positive family history. The guidelines also addressed screening for inborn errors of metabolism (IEMs). There remains a lack of consensus about the utility of routine IEM screening in patients with nonspecific NDD. The overall expected yield is relatively low, <5%, but IEMs are considered high-impact as some have targeted treatment (van Karnebeek et al. 2014). Indications that should prompt consideration of IEM include developmental regression (beyond the typical loss of speech that may occur at 18–24 mo in ASD), multisystem involvement, failure to thrive, specific dietary triggers, and seizures, among others. Neuroimaging via brain magnetic resonance imaging (MRI) is an important part of the workup in patients with NDDs who have an abnormal neurologic exam, micro- or macrocephaly, regression, and/or seizures. Otherwise, brain MRI is not performed routinely in individuals with NDDs because of the low yield in absence of specific indicators, as well as the need for sedation.

At the time these guidelines were published, next-generation sequencing (NGS) techniques were still relatively new and, accordingly, the AAP and ACMG did not recommend exome sequencing (ES) in the diagnostic evaluation of patients with NDDs. ES became clinically available in late 2011, and, initially, it was typically ordered for patients as a last resort after extensive other investigations failed to uncover the diagnosis. However, in recent years, substantial literature has been published regarding the diagnostic yield, clinical utility, and cost effectiveness of ES in individuals with NDDs. Numerous studies have found a high diagnostic yield of ES—potentially >40% in patients with ID and other NDDs, especially when performed as a trio including both biological parents—with frequent changes to medical management (Lee et al. 2014a; Srivastava et al. 2014, 2019; Kuperberg et al. 2016; Nolan and Carlson 2016; Clark et al. 2018; Wright et al. 2018). Several cost-effectiveness studies have found that using ES

early in the diagnostic pathway markedly increases the diagnostic rate while reducing the cost per diagnosis, resulting in an overall cost savings compared with traditional diagnostic trajectory of sequential testing with or without ES as last resort (Monroe et al. 2016; Stark et al. 2017). Given the rapid pace of disease gene discovery, it is recommended to periodically reanalyze the exome data for individuals without a definitive diagnosis, because evidence suggests that doing so may increase the diagnostic yield by 10% or more (Wenger et al. 2017; Al-Nabhani et al. 2018).

These data strongly support the use of ES as a second-tier test in patients with NDD (see Box 1). Indeed, this has become standard practice in many centers across the United States, and it is anticipated that the next updates to the guidelines will reflect this. One caveat is that the expected diagnostic yield of ES in patients with isolated ASD (i.e., without cognitive impairment or morphologic abnormalities) is much lower (3% in one study) than in patients with ID or more complex presentations (Tammimies et al. 2015); therefore, at this time it is reasonable to offer CMA and fragile X, but not ES, to patients with isolated ASD.

Although genome sequencing (GS) is primarily used in research settings at this time, it is expected that this too will become a routine clinical test. GS has advantages over both CMA and ES, including more even coverage across the genome, analysis of noncoding regions, and ability to detect CNVs and structural rearrangements with specific breakpoints (Hehir-Kwa et al. 2015). GS may also be able to detect pathogenic short tandem repeat expansions, which underlie a number of Mendelian diseases including some NDDs such as fragile X syndrome (Dashnow et al. 2018).

Notably, there is emerging evidence based on two recent meta-analyses that ES and possibly GS should be considered as a first-tier test for the diagnostic evaluation of individuals with NDDs. Clark et al. (2018) conducted a meta-analysis of 37 studies encompassing 20,068 children with a range of indications including NDDs and found that the diagnostic yields of GS (41%) and ES (36%) were substantially

---

BOX 1. GENETIC TESTING IN NDDs

The following tests are frequently offered to individuals with NDDs in whom a specific diagnosis is not suspected after history and exam:

- Chromosomal microarray (CMA)
- *FMR1* trinucleotide repeat analysis for fragile X syndrome
- Exome sequencing (ES)

The current practice in most centers is to begin with CMA and *FMR1* testing, followed by ES as second tier; however emerging evidence supports the use of ES as a first-tier test.
Other tests to consider as indicated based on clinical presentation:

- Screening for inborn errors of metabolism
- Brain MRI for patients with abnormal neurologic exam
- Testing for specific Mendelian disorders

---

greater than CMA (10%) in these children with suspected genetic diseases. Furthermore, in this meta-analysis, the clinical utility, defined as change in clinical management, of GS (27%) and ES (17%) were higher than of CMA (6%). Srivastava et al. (2019) conducted a meta-analysis of ES in patients with isolated NDDs and NDDs plus associated conditions (e.g., syndromic features, systemic findings, neurologic features). Based on 30 studies comprising 3350 patients, the diagnostic yield of ES (31% in isolated NDD, 53% in NDD plus associated conditions, and 36% overall) was substantially higher than the generally accepted diagnostic yield of CMA (10%–20%). Srivastava et al. (2019) proposed a tiered workup beginning with ES and followed by CMA as second tier if ES analysis did not include CNV detection (Pfundt et al. 2017); if still no diagnosis, the investigators suggested additional tests to consider as indicated based on clinical presentation.

Some important genetic disorders are still not readily detected by NGS techniques, such as trinucleotide repeat expansions in fragile X syndrome and methylation defects in a subset of cases of Angelman syndrome (Biesecker and Green 2014); therefore, careful phenotyping and thoughtful consideration of a differential diagnosis are still crucial. Nonetheless, with decreasing cost and the superior diagnostic yield, in the future ES/GS will likely become the first-tier test

for patients with suspected monogenic disorders including NDDs.

GENETIC COUNSELING FOR NDDs TO PROMOTE DECISION-MAKING AND ADAPTATION

Every family who has a child diagnosed with an NDD has unique expectations and interest in pursuing a genetic evaluation. Thus, genetic counseling entails understanding what these expectations and interests are and how they originated. Addressing them directly can lead to providing an invaluable service to enable patients and their families to understand the contribution of genetics to an individual's diagnosis of NDD and to make informed decisions about testing and management. In addition, even for families who elect not to pursue testing, genetic counseling can still be beneficial to help them understand the role of genetics in NDDs, cope with the challenges surrounding NDDs, and adapt to their child's disability.

Parental Attitudes toward Diagnosis

Parents have many reactions after their child has been diagnosed with an NDD. They often experience acute grief and sorrow over the loss of the hoped-for child, as well as the loss of their expected future (Bruce et al. 1996; Nelson Goff et al. 2013). Parents are faced with the prospect of caring for a child through adulthood who may never be able to live independently. Parents worry about the impact on their day-to-day life and finances, as well as impact on their other typically developing children (Schuntermann 2007). Receiving an NDD diagnosis brings with it anxiety and uncertainty about the future (Makela et al. 2009). Parents frequently experience social isolation and stigmatization (Ali et al. 2012). They grapple with the question of why this has happened to them and their child, and this often triggers feeling of guilt and blame (Ali et al. 2012). Parents may also worry about the possibility of subsequent children having an NDD and/or if they did something to have caused the diagnosis to occur in their child. All of this naturally leads to a search for an explanation.

In a qualitative study of parents of children with unidentified multiple congenital anomaly syndromes, Rosenthal et al. (2001) defined dimensions to parents' views on the importance of diagnostic information: labeling (e.g., to help other people understand their child's condition), causation, prognosis, treatment, acceptance, and social support. Prognosis included wanting to know future health risks, mortality, how much progress their child is expected to make, and boundaries of their child's capabilities. Similarly, in a qualitative study of parents of children with and without a diagnosis for their child's ID, Makela et al. (2009) found that parents' interest in a diagnosis stemmed from seeking validation, information to guide expectations and treatment, assistance with accessing services, social support, and satisfying their curiosity.

The drive to find a diagnosis and etiology may be intense in close proximity to the time when an NDD diagnosis is made, especially for parents of younger children (Rosenthal et al. 2001; Makela et al. 2009). This drive may lessen somewhat as the child ages and parents have adapted to their child's condition over time. However, when given the opportunity, many parents of older children remain interested in finding the cause of their child's NDD. Renewed interest in a diagnosis and etiology may be triggered by learning of advances in genetic testing or prompting by typically developing siblings who are approaching adulthood and facing reproductive decisions, among other reasons.

Exploring the Relationship between Genetics and NDDs

After a child has been diagnosed with an NDD, it may not be intuitive to parents when their child's health-care provider raises the issue of genetics. For example, parents may ask questions like "My child has autism, what do you mean he/she could have a genetic problem?" and "So, if you diagnose a genetic disorder, does that mean my child doesn't have autism?" They may not realize that genetic variation has been implicated in NDDs and, instead, have other suspicions as to its etiology. Often parents attribute their child's NDD to a concrete event or tangible factor, such as nuchal cord at birth or maternal stress during pregnancy. The genetic counselor should respectfully explore these suspicions to understand and appreciate what use they serve, and only when needed, correct misconceptions with information supported by the medical literature. Research has shown that parents accept multiple attributions for their child's condition (Dong 2005; Vetsch et al. 2019). As such, they can accept a spiritual cause alongside a biological one.

When broaching the topic of genetics, it can be helpful to distinguish the NDD diagnosis from the potential etiologic diagnosis. For example, the genetic counselor can explain that NDD diagnoses such as ID and ASD describe the developmental concern, which stems from differences in brain functioning; genes influence how a person's brain develops and functions, so changes in genes (a genetic disorder) may result in differences in brain functioning, which leads to the symptoms that are labeled with an NDD diagnosis. By introducing the concept of multifactorial causes, this acknowledges that other factors including the environment may contribute to the complexities seen in NDDs. In our experience, separating these out—cause of brain difference (etiology) leading to NDD diagnosis—may help parents understand the role of genetics and why genetic evaluation and testing are being offered. Making this distinction may also help parents understand why establishing a specific genetic diagnosis usually does not change their child's NDD diagnosis.

Facilitating Decision-Making about Genetic Testing

Much of the treatment for NDDs, such as therapies and educational supports, is aimed at maximizing developmental potential and managing symptoms irrespective of etiology. Yet, it can also be important for the individual's care and to the family to determine the underlying cause. Establishing a specific genetic diagnosis may impact an individual's medical management and treatment, allow provision of prognostic information, enable family counseling with a specific recurrence risk, and reproductive options, as well result in psychological and social benefits to the individual and the family. On the other hand, there are important limitations of genetic testing and potential risks. Although genetic testing is a standard recommendation in the evaluation of patients with an NDD, it is ultimately a choice made by the family and, therefore, the genetic counselor must present a balanced discussion of potential benefits, limitations, and possible downsides of genetic testing (Cohen et al. 2013). Key benefits, limitations, and risks of genetic testing in NDD are summarized in Box 2 and described below.

Regarding the potential benefits of genetic testing in NDD, identifying a genetic etiology for an NDD through genetic testing may offer an end to a diagnostic odyssey and provide closure by ruling out other causes, as well as provide a "name" that unifies the individual's symptoms and challenges (Rosenthal et al. 2001). A genetic diagnosis may also direct medical management by providing information about other associated medical problems or possible future complications that need to be evaluated, supporting

---

BOX 2. KEY BENEFITS, RISKS, AND LIMITATIONS OF GENETIC TESTING IN NDDs

Potential benefits

- End diagnostic odyssey
- Provide a name that unifies child's symptoms
- Enable provision of prognostic information
- Enable tailoring of medical management
- Result in targeted treatment
- Alleviation of negative emotions such as guilt or blame
- Increase access to services and condition-specific support groups
- Enable counseling with specific recurrence risk and reproductive options

Potential risks and limitations

- Failure to identify definitive etiologic diagnosis
- Genetic diagnosis may not alter medical management or treatment
- Genetic diagnosis associated with limited or no prognostic information
- Possibility of variants of unknown significance, incidental/secondary findings, or unexpected information about familial relationships
- Negative emotional responses to results
- Unexpected diagnosis of parent or relative based on inherited variant
- Concerns about genetic discrimination and privacy of data

---

evidence for early intervention and educational supports, and using prognostic information in order to better prepare a family for their child's future. When discussing the benefit of learning prognostic information, it is important to advise families that the genetic result may not be fully predictive of future (dis)ability because there is often variability in outcomes. Furthermore, even for well-described conditions, there can be limited information available on outcomes in adulthood. Rarely, a targeted treatment is available, although this may change over time with advances in research. Knowledge of inheritance may alleviate feelings of guilt or blame for parents if the causative variant occurred de novo and enable specific recurrence risk counseling and family planning options as discussed below.

On the other hand, there are several reasons a family may decline genetic testing. First, there is no guarantee that genetic testing will identify the etiologic diagnosis. For GDD/ID, currently genetic testing identifies the cause about half of the time, taking into account the ~10% yield of CMA and ~40% yield of ES, and the yield is substantially lower for isolated ASD without comorbidities (Lee et al. 2014a; Srivastava et al. 2014, 2019; Kuperberg et al. 2016; Nolan and Carlson 2016; Clark et al. 2018; Wright et al. 2018). The possibility of uncovering variants of unknown significance (VUSs) may be a deterrent to some families, as a VUS could trigger worry and/or frustration due to lack of definitive information to interpret the pathogenicity of the variant. Similarly, identification of a genetic etiology may not result in immediate changes to medical management or prognosis. As a result, some families may defer testing in the hope that technology and knowledge will improve over time, reduce the likelihood of learning uncertain results, and increase the probability of learning medically useful information. Parents also consider the potential emotional impact of results, such as fear that they will learn that one or both parents "caused" their child's condition in the case of an inherited disorder. Additionally, because NDD represents a spectrum of DBD, identification of an inherited genetic variant in a family may suggest the presence of related diagnoses in family members. Genetic testing may

also reveal unexpected familial relationships such as consanguinity or misattributed parentage. Finally, some families are concerned about genetic discrimination and privacy of genetic data because of the limitations of the Genetic Information Nondiscrimination Act and other legislation, especially as they make plans for long-term care for their child.

Counseling about Recurrence Risk for NDDs

One of the major motivators for families to pursue genetic testing is knowledge of a specific recurrence risk (Finucane and Myers 2016). Many parents express great fear about the possibility of the condition happening again in their future children and concern about the potential risks to other family members, thus creating the desire for accurate recurrence risk counseling. Frequently, in cases in which the family history is negative, genetic disorders that present with NDD are due to de novo variants (Gilissen et al. 2014), which are associated with low recurrence risk for future siblings. On the other hand, autosomal dominant, autosomal recessive, and X-linked disorders are associated with substantially higher recurrence risks. Identifying a specific genetic diagnosis and determining the respective inheritance pattern allows for counseling with a specific risk figure and enables the use of reproductive testing options. This outcome may be realized by family members, as well as by individuals with milder NDDs who may have children of their own in the future. For siblings and other at-risk relatives, targeted testing for a causal variant may allow for diagnosis and early intervention. It is important to emphasize in the case of an inherited disorder, there is nothing the parent(s) did that caused the condition and to help them appreciate that they had no control over its occurrence.

In the absence of an identifiable genetic etiology, empiric recurrence risk estimates are used. There are several limitations in using empiric recurrence risks for NDDs. Over time, diagnostic criteria for certain NDDs change and study populations may be limited in size, compromising the accuracy of the risk estimates. In certain clinical scenarios, these estimates may not effectively account for the possibility of an affected parent, as variable expressivity and co-morbidities can be present in NDDs, which suggests a higher recurrence risk. The 2013 ACMG guidelines (Schaefer et al. 2013) stated that the accepted recurrence risk for full siblings of a child with ASD is in the range of 3% to 10%. More recently, larger studies have been published to update and refine empiric risks that take into account new diagnostic criteria and advances in genetic testing (Ozonoff et al. 2011). As reviewed by Schaefer (2016), it is now recommended to provide a 10%–20% recurrence risk for full siblings of a child with ASD and a 30%–50% risk if there is more than one affected sibling. Empiric recurrence risk for a full sibling of a child with ID is ~3% and increases to 10% if a parent is also diagnosed with ID (Harper 2010). Genetic counselors have an especially important role in helping parents make informed family planning decisions and cope with the fear of recurrence when there is not a single definitive probability of recurrence.

Delivering the Genetic Diagnosis

Genetic counselors play a key role in delivery of a genetic diagnosis. Although the experience of receiving a genetic diagnosis is usually difficult for parents, research has shown that how the diagnosis is delivered impacts whether the experience is perceived by parents as positive or negative (Nelson Goff et al. 2013; Waxler et al. 2013; Ashtiani et al. 2014).

Genetic counselors should be cognizant of how recently the parents received their child's NDD diagnosis. Parents of children who were recently diagnosed with an NDD may still be coming to terms with the practical and emotional ramifications of their child's developmental disability and experiencing acute grief, so receiving a genetic diagnosis could be further overwhelming or possibly devastating. In contrast, parents of children who are older or have had a long-standing NDD diagnosis may be better adapted to their child's disability and capabilities, so they may have a more positive experience

of receiving the diagnosis and may experience a sense of relief (Waxler et al. 2013).

Two studies specifically examined parental experiences of receiving a genetic diagnosis for their child's NDD, and the findings were remarkably consistent (Waxler et al. 2013; Ashtiani et al. 2014). Themes that emerged from these studies included the importance of communicating hope, emotional support, up-to-date information (verbal and written), resources, and follow-up plans. In addition, both studies underscored the importance of engaging parents in a discussion during which their emotions, questions, and concerns are elicited and addressed. Box 3 summarizes suggestions for delivery of a genetic diagnosis in NDD based on these studies.

Managing Uncertainty and Increasing Parental Sense of Control

Adaptation can be defined as the dynamic and multidimensional process of coming to terms with the implications of a health threat and the observable outcomes of that process (Biesecker and Erby 2008). As reviewed by these investigators, the appraisals individuals make in response to a health threat (in this context, parenting a child with an NDD), such as perceived uncertainty and personal control, are important predictors of adaptation to living with a genetic condition or risk.

It can be disappointing and frustrating to families when all available genetic testing fails to identify an etiologic diagnosis. The uncertainty generated by the absence of a diagnosis may lead to lower perceived personal control and less optimism (Madeo et al. 2012). Even when a genetic etiology is identified, there remain levels of uncertainty for medical management and prognosis, particularly for rare conditions and those with substantial phenotypic variability (Han et al. 2017), and parents may perceive a low sense of control (Lipinski et al. 2006). Uncertainty may be appraised by parents as negative/bad or positive/good (Whitmarsh et al. 2007). As an example of the latter, some parents may feel that "an uncertain outcome leaves open the possibilities for

---

BOX 3. SUGGESTIONS FOR DELIVERY OF A GENETIC DIAGNOSIS IN NDDs

---

- Attend to parents' emotions and provide emotional support
- Offer messages of hope and perspective.
- Engage the parents in a dialogue and encourage parents to talk (avoid verbal dominance).
- Check in with parents throughout the discussion and reengage as necessary.
- Limit the use of difficult medical terminology.
- Elicit parental preferences (e.g., asking whether they would like to see a picture of other individuals with the same condition).
- Provide the most up-to-date information possible.
- Provide balanced information (e.g., in addition to describing the features of the condition, point out aspects of the child's health that are not expected to be affected, if appropriate).
- Give written information about the diagnosis and an outline of follow-up plans.
- Give resources such as condition-specific support groups, when available.

---

a positive outcome in their child" (Madeo et al. 2012). Depending on parental appraisal, perceived uncertainty could aid or impede parental adaptation.

For parents struggling with uncertainty, genetic counselors can help by providing pieces of relatively certain information. For example, when appropriate, the parents can be told that the child's condition is not expected to be degenerative or life-limiting, other medical complications are not anticipated, and the child will still benefit from therapies. It may also be helpful to reassure parents that all available testing has been pursued and an exhaustive search for information has been performed, and then the genetic counselor can assist them with coping with the residual uncertainty, stress, and related challenges of caring for their child with NDD. Genetic counselors can use interventions that aid parents in using more effective coping strategies (e.g., Haakonsen Smith et al. 2018). To enhance parents' sense of personal control, genetic counselors can help parents recognize aspects of their situation over which they do have control (e.g., making decisions about therapies and other treatments) and identify tangible ways they may be able to gain some control (e.g., joining a support group, advocating for research into the condition).

Fostering Hope and Meaning-Making

As discussed above, messages of hope are important to share with parents during the process of delivering a genetic diagnosis. In addition, research has shown a relationship between hope, uncertainty, and psychological adaptation. Hope can be conceptualized as a cognitive process that is comprised of a sense of agency (goal-directed determination) and pathways (planning of ways to meet goals) (Snyder et al. 1991). In a study of caregivers of children with Down syndrome, hope and perceived uncertainty were significantly associated with caregiver adaptation (Truitt et al. 2012). Based on their findings, the investigators proposed that genetic counselors can explore what parents are hoping for and evaluate the degree to which that is adaptive. In addition, genetic counselors can help parents identify new opportunities and ways to work toward those goals as well as reinforce the steps parents are already taking.

Meaning-making is a key component of the adaptation process. In a study of family caregivers of individuals with developmental disabilities, Werner and Shulman (2013) found that identifying positive meaning in caregiving was a strong predictor of caregivers' subjective well-being. Genetic counselors can facilitate a

Cite this article as *Cold Spring Harb Perspect Med* doi: 10.1101/cshperspect.a036533

discussion with parents about their experience of being a caregiver and encourage them to tell their story, verbally or through a writing exercise. This allows parents an opportunity for "cognitive restructuring" that may help develop a sense of meaning and also may help them integrate the genetic/biological explanation for their child's NDD with other causal beliefs (Biesecker and Erby 2008).

Addressing Stigma, Shame, and Guilt

Parents of children with NDDs frequently experience affiliate stigma, defined as the internalized stigma experienced by an unaffected individual, because of his or her association with a person who bears a stigma (an NDD in this case), and affiliate stigma may have a negative impact on their well-being (Ali et al. 2012). Feelings of shame and guilt are also common and may be associated with poor psychological outcomes for parents such as anxiety and depression (Gallagher et al. 2008; Ali et al. 2012). Parental self-esteem and social support are further predictors of well-being and may protect against the negative effects of affiliate stigma (Werner and Shulman 2013). Genetic counselors can explore with parents their experience with stigma, the ways in which they currently cope, and the extent to which those strategies are effective. Genetic counselors can also assess the parent's social support network. All parents should be referred to relevant support groups, although some parents may need additional professional support or long-term counseling.

Finding Opportunities for Social Support

Genetic counselors can help families with NDDs by connecting them with resources and support groups. The ability to connect with support groups and research organizations is helpful as there are other individuals who understand their unique situation. There are multiple organizations dedicated to genetic disorders with NDD such as Unique (rarechromo .org), MyGene2 (mygene2.org), Genome Connect (genomeconnect.org), and Simons Searchlight (simonsearchlight.org). Even for rare conditions for which there may not be an established foundation, the approachability and ease of social media allows for small groups to form that can grow in size, as more individuals pursue genetic testing and are identified with the same diagnosis (Rocha et al. 2018). The growth of these groups can provide research opportunities into specific treatments and natural history studies as a potential study population is easily accessible. For individuals who do not have a known genetic diagnosis, there are a number of general support organizations centered on the NDD diagnosis, such as The Arc (thearc .org) and Autism Speaks (autismspeaks.org).

CONCLUDING REMARKS

We are on the cusp of precision medicine for individuals with NDDs. ID and related disorders have always been considered incurable, with treatment instead focused on symptom management and therapies to support development. Genomic advances have led to an increased understanding of the pathophysiology and neurobiology of NDDs, and it is expected that this will eventually lead to targeted treatments based on underlying genetic etiology. There are more than 80 NDD-associated IEMs that have established treatments (Sayson et al. 2015). Moreover, efforts are underway to develop novel therapeutics for genetic conditions with NDDs. For example, everolimus, a mammalian target of rapamycin (mTOR) inhibitor, is used for treatment of tumor manifestations in patients with tuberous sclerosis complex. Impressively, emerging evidence suggests that this drug is also beneficial for controlling seizures and improving neuropsychiatric symptoms (Kilincaslan et al. 2017). As another example, in a mouse model of Kabuki syndrome, postnatal treatment with a histone deacetylase inhibitor improved neurogenesis and ameliorated functional deficits (Bjornsson et al. 2014). This shows great promise for treatment of ID in Mendelian disorders of the epigenetic machinery and potentially other related conditions in the future. The vast potential for targeted treatment is perhaps the strongest impetus for genetic evaluation and testing in individuals with NDDs.

Genetic counselors are now poised to take a prominent role in the field of neurodevelopmental medicine. Similar to specialties such as oncology and cardiology in which genetic counselors provide services outside of the traditional medical genetics clinic, genetic counselors can create and develop their own positions in pediatric neurology and developmental pediatrics clinics. By partnering our genetics knowledge and counseling skills with the clinical expertise of these specialists, we can greatly increase access to genetic evaluation and testing for individuals with NDD. Finally, although further evidence is needed from studies of interventions in this population, research thus far has identified potential targets for genetic counseling to facilitate adaptation among parents of individuals with NDDs.

ACKNOWLEDGMENTS

We thank the patients and families of Kennedy Krieger Institute for inspiring us with their resilience, as well as Gillian Hooker, Alec Hoon, and Mahim Jain for their thoughtful review of the manuscript.

REFERENCES

Ali A, Hassiotis A, Strydom A, King M. 2012. Self stigma in people with intellectual disabilities and courtesy stigma in family carers: a systematic review. *Res Dev Disabil* 33: 2122–2140. doi:10.1016/j.ridd.2012.06.013

Al-Nabhani M, Al-Rashdi S, Al-Murshedi F, Al-Kindi A, Al-Thihli K, Al-Saegh A, Al-Futaisi A, Al-Mamari W, Zadjali F, Al-Maawali A. 2018. Reanalysis of exome sequencing data of intellectual disability samples: yields and benefits. *Clin Genet* 94: 495–501. doi:10.1111/cge.13438

American Psychiatric Association. 2013. *Diagnostic and statistical manual of mental disorders*, 5th ed. American Psychiatric Association, Arlington, VA.

Ashtiani S, Makela N, Carrion P, Austin J. 2014. Parents' experiences of receiving their child's genetic diagnosis: a qualitative study to inform clinical genetics practice. *Am J Med Genet A* 164: 1496–1502. doi:10.1002/ajmg.a.36525

Baio J, Wiggins L, Christensen DL, Maenner MJ, Daniels J, Warren Z, Kurzius-Spencer M, Zahorodny W, Robinson Rosenberg C, White T, et al. 2018. Prevalence of autism spectrum disorder among children aged 8 years—Autism and Developmental Disabilities Monitoring Network, 11 Sites, United States, 2014. *MMWR Surveill Summ* 67: 1–23. doi:10.15585/mmwr.ss6706a1

Biesecker BB, Erby L. 2008. Adaptation to living with a genetic condition or risk: a mini-review. *Clin Genet* 74: 401–407. doi:10.1111/j.1399-0004.2008.01088.x

Biesecker LG, Green RC. 2014. Diagnostic clinical genome and exome sequencing. *N Engl J Med* 370: 2418–2425. doi:10.1056/NEJMra1312543

Bjornsson HT, Benjamin JS, Zhang L, Weissman J, Gerber EE, Chen Y-C, Vaurio RG, Potter MC, Hansen KD, Dietz HC. 2014. Histone deacetylase inhibition rescues structural and functional brain deficits in a mouse model of Kabuki syndrome. *Sci Transl Med* 6: 256ra135. doi:10.1126/scitranslmed.3009278

Bruce EJ, Schultz CL, Smyrnios KX. 1996. A longitudinal study of the grief of mothers and fathers of children with intellectual disability. *Br J Med Psychol* 69: 33–45. doi:10.1111/j.2044-8341.1996.tb01848.x

Chaste P, Leboyer M. 2012. Autism risk factors: genes, environment, and gene–environment interactions. *Dialogues Clin Neurosci* 14: 281–292.

Clark MM, Stark Z, Farnaes L, Tan TY, White SM, Dimmock D, Kingsmore SF. 2018. Meta-analysis of the diagnostic and clinical utility of genome and exome sequencing and chromosomal microarray in children with suspected genetic diseases. *NPJ Genom Med* 3: 16. doi:10.1038/s41525-018-0053-8

Cohen J, Hoon A, Wilms Floet AM. 2013. Providing family guidance in rapidly shifting sand: informed consent for genetic testing. *Dev Med Child Neurol* 55: 766–768.

Dashnow H, Lek M, Phipson B, Halman A, Sadedin S, Lonsdale A, Davis M, Lamont P, Clayton JS, Laing NG, et al. 2018. STRetch: detecting and discovering pathogenic short tandem repeat expansions. *Genome Biol* 19: 121. doi:10.1186/s13059-018-1505-2

DeStefano F, Price CS, Weintraub ES. 2013. Increasing exposure to antibody-stimulating proteins and polysaccharides in vaccines is not associated with risk of autism. *J Pediatr* 163: 561–567. doi:10.1016/j.jpeds.2013.02.001

Dong DA. 2005. "Causal attributions for autistic spectrum disorders: influences on perceived personal control." MS Thesis, Bloomberg School of Public Health of Johns Hopkins University, Baltimore.

Finucane B, Myers SM. 2016. Genetic counseling for autism spectrum disorder in an evolving theoretical landscape. *Curr Genet Med Rep* 4: 147–153. doi:10.1007/s40142-016-0099-9

Gallagher S, Phillips AC, Oliver C, Carroll D. 2008. Predictors of psychological morbidity in parents of children with intellectual disabilities. *J Pediatr Psychol* 33: 1129–1136. doi:10.1093/jpepsy/jsn040

Gentner MB, Leppert MLO. 2019. Environmental influences on health and development: nutrition, substance exposure, and adverse childhood experiences. *Dev Med Child Neurol* 61: 1008–1014. doi:10.1111/dmcn.14149

Gilissen C, Hehir-Kwa JY, Thung DT, van de Vorst M, van Bon BWM, Willemsen MH, Kwint M, Janssen IM, Hoischen A, Schenck A, et al. 2014. Genome sequencing identifies major causes of severe intellectual disability. *Nature* 511: 344–347. doi:10.1038/nature13394

Haakonsen Smith C, Turbitt E, Muschelli J, Leonard L, Lewis KL, Freedman B, Muratori M, Biesecker BB. 2018. Feasibility of coping effectiveness training for caregivers of children with autism spectrum disorder: a genetic counseling intervention. *J Genet Couns* **27**: 252–262. doi:10.1007/s10897-017-0144-1

Han PKJ, Umstead KL, Bernhardt BA, Green RC, Joffe S, Koenig B, Krantz I, Waterston LB, Biesecker LG, Biesecker BB. 2017. A taxonomy of medical uncertainties in clinical genome sequencing. *Genet Med* **19**: 918–925. doi:10.1038/gim.2016.212

Harper PS. 2010. *Practical genetic counselling*, 7th ed. CRC Press, Boca Raton, FL.

Hehir-Kwa JY, Pfundt R, Veltman JA. 2015. Exome sequencing and whole genome sequencing for the detection of copy number variation. *Expert Rev Mol Diagn* **15**: 1023–1032. doi:10.1586/14737159.2015.1053467

Ho K, Wassman E, Baxter A, Hensel C, Martin M, Prasad A, Twede H, Vanzo R, Butler M. 2016. Chromosomal microarray analysis of consecutive individuals with autism spectrum disorders using an ultra-high resolution chromosomal microarray optimized for neurodevelopmental disorders. *Int J Mol Sci* **17**: 2070. doi:10.3390/ijms.17122070

Kilincaslan A, Kok BE, Tekturk P, Yalcinkaya C, Ozkara C, Yapici Z. 2017. Beneficial effects of everolimus on autism and attention-deficit/hyperactivity disorder symptoms in a group of patients with tuberous sclerosis complex. *J Child Adolesc Psychopharmacol* **27**: 383–388. doi:10.1089/cap.2016.0100

Kuperberg M, Lev D, Blumkin L, Zerem A, Ginsberg M, Linder I, Carmi N, Kivity S, Lerman-Sagie T, Leshinsky-Silver E. 2016. Utility of whole exome sequencing for genetic diagnosis of previously undiagnosed pediatric neurology patients. *J Child Neurol* **31**: 1534–1539. doi:10.1177/0883073816664836

Lee H, Deignan JL, Dorrani N, Strom SP, Kantarci S, Quintero-Rivera F, Das K, Toy T, Harry B, Yourshaw M, et al. 2014a. Clinical exome sequencing for genetic identification of rare Mendelian disorders. *JAMA* **312**: 1880–1887. doi:10.1001/jama.2014.14604

Lee RW, Poretti A, Cohen JS, Levey E, Gwynn H, Johnston MV, Hoon AH, Fatemi A. 2014b. A diagnostic approach for cerebral palsy in the genomic era. *Neuromolecular Med* **16**: 821–844. doi:10.1007/s12017-014-8331-9

Lipinski SE, Lipinski MJ, Biesecker LG, Biesecker BB. 2006. Uncertainty and perceived personal control among parents of children with rare chromosome conditions: the role of genetic counseling. *Am J Med Genet C Semin Med Genet* **142C**: 232–240. doi:10.1002/ajmg.c.30107

Madeo AC, O'Brien KE, Bernhardt BA, Biesecker BB. 2012. Factors associated with perceived uncertainty among parents of children with undiagnosed medical conditions. *Am J Med Genet A* **158A**: 1877–1884. doi:10.1002/ajmg.a.35425

Makela NL, Birch PH, Friedman JM, Marra CA. 2009. Parental perceived value of a diagnosis for intellectual disability (ID): a qualitative comparison of families with and without a diagnosis for their child's ID. *Am J Med Genet A* **149A**: 2393–2402. doi:10.1002/ajmg.a.33050

Miller DT, Adam MP, Aradhya S, Biesecker LG, Brothman AR, Carter NP, Church DM, Crolla JA, Eichler EE, Epstein CJ, et al. 2010. Consensus statement: chromosomal microarray is a first-tier clinical diagnostic test for individuals with developmental disabilities or congenital anomalies. *Am J Hum Genet* **86**: 749–764. doi:10.1016/j.ajhg.2010.04.006

Moeschler JB, Shevell M; Committee on Genetics. 2014. Comprehensive evaluation of the child with intellectual disability or global developmental delays. *Pediatrics* **134**: e903–e918. doi:10.1542/peds.2014-1839

Monroe GR, Frederix GW, Savelberg SMC, de Vries TI, Duran KJ, van der Smagt JJ, Terhal PA, van Hasselt PM, Kroes HY, Verhoeven-Duif NM, et al. 2016. Effectiveness of whole-exome sequencing and costs of the traditional diagnostic trajectory in children with intellectual disability. *Genet Med* **18**: 949–956. doi:10.1038/gim.2015.200

Moreno-De-Luca A, Myers SM, Challman TD, Moreno-De-Luca D, Evans DW, Ledbetter DH. 2013. Developmental brain dysfunction: revival and expansion of old concepts based on new genetic evidence. *Lancet Neurol* **12**: 406–414. doi:10.1016/S1474-4422(13)70011-5

Nelson Goff BS, Springer N, Foote LC, Frantz C, Peak M, Tracy C, Veh T, Bentley GE, Cross KA. 2013. Receiving the initial Down syndrome diagnosis: a comparison of prenatal and postnatal parent group experiences. *Intellect Dev Disabil* **51**: 446–457. doi:10.1352/1934-9556-51.6.446

Nolan D, Carlson M. 2016. Whole exome sequencing in pediatric neurology patients: clinical implications and estimated cost analysis. *J Child Neurol* **31**: 887–894. doi:10.1177/0883073815627880

Ozonoff S, Young GS, Carter A, Messinger D, Yirmiya N, Zwaigenbaum L, Bryson S, Carver LJ, Constantino JN, Dobkins K, et al. 2011. Recurrence risk for autism spectrum disorders: a Baby Siblings Research Consortium study. *Pediatrics* **128**: e488–e495.

Pfundt R, Del Rosario M, Vissers LELM, Kwint MP, Janssen IM, de Leeuw N, Yntema HG, Nelen MR, Lugtenberg D, Kamsteeg EJ, et al. 2017. Detection of clinically relevant copy-number variants by exome sequencing in a large cohort of genetic disorders. *Genet Med* **19**: 667–675. doi:10.1038/gim.2016.163

Rocha HM, Savatt JM, Riggs ER, Wagner JK, Faucett WA, Martin CL. 2018. Incorporating social media into your support tool box: points to consider from genetics-based communities. *J Genet Couns* **27**: 470–480. doi:10.1007/s10897-017-0170-z

Rosenthal ET, Biesecker LG, Biesecker BB. 2001. Parental attitudes toward a diagnosis in children with unidentified multiple congenital anomaly syndromes. *Am J Med Genet* **103**: 106–114. doi:10.1002/ajmg.1527

Sayson B, Popurs MAM, Lafek M, Berkow R, Stockler-Ipsiroglu S, van Karnebeek CDM. 2015. Retrospective analysis supports algorithm as efficient diagnostic approach to treatable intellectual developmental disabilities. *Mol Genet Metab* **115**: 1–9. doi:10.1016/j.ymgme.2015.03.001

Schaefer GB. 2016. Clinical genetic aspects of autism spectrum disorders. *Int J Mol Sci* **17**: 180. doi:10.3390/ijms17020180

Schaefer GB, Mendelsohn NJ; Professional Practice and Guidelines Committee. 2013. Clinical genetics evaluation in identifying the etiology of autism spectrum disorders: 2013 guideline revisions. *Genet Med* **15**: 399–407. doi:10.1038/gim.2013.32

Schalock RL, Luckasson RA, Shogren KA, Borthwick-Duffy S, Bradley V, Buntinx WH, Coulter DL, Craig EM, Gomez SC, Lachapelle Y, et al. 2007. The renaming of "mental retardation": understanding the change to the term "intellectual disability". *Intellect Dev Disabil* **45:** 116–124. doi:10.1352/1934-9556(2007)45[116:TROMRU]2.0.CO;2

Schuntermann P. 2007. The sibling experience: growing up with a child who has pervasive developmental disorder or mental retardation. *Harv Rev Psychiatry* **15:** 93–108. doi:10.1080/10673220701432188

Siu MT, Weksberg R. 2017. Epigenetics of autism spectrum disorder. *Adv Exp Med Biol* **978:** 63–90. doi:10.1007/978-3-319-53889-1_4

Snyder CR, Harris C, Anderson JR, Holleran SA, Irving LM, Sigmon ST, Yoshinobu L, Gibb J, Langelle C, Harney P. 1991. The will and the ways: development and validation of an individual-differences measure of hope. *J Pers Soc Psychol* **60:** 570–585. doi:10.1037/0022-3514.60.4.570

Srivastava S, Cohen JS, Vernon H, Barañano K, McClellan R, Jamal L, Naidu S, Fatemi A. 2014. Clinical whole exome sequencing in child neurology practice. *Ann Neurol* **76:** 473–483. doi:10.1002/ana.24251

Srivastava S, Love-Nichols J, Dies K, Ledbetter D, Martin C, Chung W, Firth H, Frazier T, Hansen R, Prock L, et al. 2019. Meta-analysis and multidisciplinary consensus statement: exome sequencing is a first-tier clinical diagnostic test for individuals with neurodevelopmental disorders. *Genet Med* doi:10.1038/s41436-019-0554-6.

Stark Z, Schofield D, Alam K, Wilson W, Mupfeki N, Macciocca I, Shrestha R, White SM, Gaff C. 2017. Prospective comparison of the cost-effectiveness of clinical whole-exome sequencing with that of usual care overwhelmingly supports early use and reimbursement. *Genet Med* **19:** 867–874. doi:10.1038/gim.2016.221

Tammimies K, Marshall CR, Walker S, Kaur G, Thiruvahindrapuram B, Lionel AC, Yuen RKC, Uddin M, Roberts W, Weksberg R, et al. 2015. Molecular diagnostic yield of chromosomal microarray analysis and whole-exome sequencing in children with autism spectrum disorder. *JAMA* **314:** 895–903. doi:10.1001/jama.2015.10078

Truitt M, Biesecker B, Capone G, Bailey T, Erby L. 2012. The role of hope in adaptation to uncertainty: the experience of caregivers of children with Down syndrome. *Patient Educ Couns* **87:** 233–238. doi:10.1016/j.pec.2011.08.015

Van Karnebeek CDM, Shevell M, Zschocke J, Moeschler JB, Stockler S. 2014. The metabolic evaluation of the child with an intellectual developmental disorder: diagnostic algorithm for identification of treatable causes and new digital resource. *Mol Genet Metab* **111:** 428–438. doi:10.1016/j.ymgme.2014.01.011

Vetsch J, Wakefield CE, Doolan EL, Signorelli C, McGill BM, Moore L, Techakesari P, Pieters R, Patenaude AF, McCarthy M, et al. 2019. 'Why us?' Causal attributions of childhood cancer survivors, survivors' parents and community comparisons—a mixed methods analysis. *Acta Oncol* **58:** 209–217. doi:10.1080/0284186X.2018.1532600

Vissers LE, Gilissen C, Veltman JA, 2016. Genetic studies in intellectual disability and related disorders. *Nat Rev Genet* **17:** 9–18. doi:10.1038/nrg3999

Waxler JL, Cherniske EM, Dieter K, Herd P, Pober BR. 2013. Hearing from parents: the impact of receiving the diagnosis of Williams syndrome in their child. *Am J Med Genet A* **161:** 534–541. doi:10.1002/ajmg.a.35789

Wenger AM, Guturu H, Bernstein JA, Bejerano G. 2017. Systematic reanalysis of clinical exome data yields additional diagnoses: implications for providers. *Genet Med* **19:** 209–214. doi:10.1038/gim.2016.88

Werner S, Shulman C. 2013. Subjective well-being among family caregivers of individuals with developmental disabilities: the role of affiliate stigma and psychosocial moderating variables. *Res Dev Disabil* **34:** 4103–4114. doi:10.1016/j.ridd.2013.08.029

Whitmarsh I, Davis AM, Skinner D, Bailey DB Jr. 2007. A place for genetic uncertainty: parents valuing an unknown in the meaning of disease. *Soc Sci Med* **65:** 1082–1093. doi:10.1016/j.socscimed.2007.04.034

Wright CF, McRae JF, Clayton S, Gallone G, Aitken S, FitzGerald TW, Jones P, Prigmore E, Rajan D, Lord J, et al. 2018. Making new genetic diagnoses with old data: iterative reanalysis and reporting from genome-wide data in 1,133 families with developmental disorders. *Genet Med* **20:** 1216–1223. doi:10.1038/gim.2017.246

Cancer Genetic Counseling—Current Practice and Future Challenges

Jaclyn Schienda and Jill Stopfer

Division of Cancer Genetics and Prevention, Dana Farber Cancer Institute, Boston, Massachusetts 02215, USA

Correspondence: JillE_Stopfer@dfci.harvard.edu

Cancer genetic counseling practice is rapidly evolving, with services being provided in increasingly novel ways. Pretest counseling for cancer patients may be abbreviated from traditional models to cover the elements of informed consent in the broadest of strokes. Genetic testing may be ordered by a cancer genetics professional, oncology provider, or primary care provider. Increasingly, direct-to-consumer testing options are available and utilized by consumers anxious to take control of their genetic health. Finally, genetic information is being used to inform oncology care, from surgical decision-making to selection of chemotherapeutic agent. This review provides an overview of the current and evolving practice of cancer genetic counseling as well as opportunities and challenges for a wide variety of indications in both the adult and pediatric setting.

CANCER GENETIC COUNSELING—WHERE WE HAVE BEEN AND WHERE WE ARE GOING

Historically, individuals with known or suspected risk for an inherited mutation in a cancer predisposition gene were likely to be referred for genetic counseling if their care was received at an academic medical center or Comprehensive Cancer Center. As part of a genetic counseling session lasting ~1 h, a three-generation pedigree would be constructed and the prior probability that the patient had inherited a germline mutation calculated. Genetic testing was offered for those having an ~5%–10% chance of having a mutation detected and ordered from a handful of specialty commercial or academic labs. Prior to the 2013 Supreme Court ruling overturning gene patenting, many genetic tests cost $3000 or more, and only the rare patient could afford the cost without meeting a strict set of insurance-based criteria. Genetic counseling and testing, including disclosure of results, was typically provided in person as part of a multivisit, multidisciplinary encounter (Hoskins et al. 1995; Resta et al. 2006).

The practice of genetic counseling has been undergoing rapid change in recent years, as increasing numbers of patients require access to genetic testing for treatment decisions, and broader categories of individuals are now considered appropriate candidates for genetic services. This work reviews traditional models for the practice of cancer genetic counseling, and preliminary evidence supporting the inclusion of novel interventions.

WHO IS A CANDIDATE FOR CANCER GENETIC COUNSELING AND TESTING?

According to SEER estimates from 2014 to 2016, ~39.3% of men and women will develop cancer in their lifetime, and most will be over the age of 55 yr (https://SEER.cancer.gov/statfacts/html/all.html). Germline pathogenic variants (PVs) in highly penetrant genes play a major role in at least 5%–10% of all cancers and in more than 50 hereditary cancer syndromes. Genetic testing for these syndromes provides cancer risk information that can be used to personalize medical management options to help mitigate the risks. Specific features of a person's medical or family history, or "red flags," that increase the likelihood of having a hereditary cancer predisposition are summarized in Table 1. Having cancer at an earlier than typical age for the tumor type is a possible indicator of inherited risk and is particularly true for a child diagnosed with an adult-onset tumor, such as colon cancer (Lynch et al. 1979; Cannon-Albright et al. 1989). Cer-

tain rare tumors also have a strong association with cancer predisposition. For example, ~40% of individuals diagnosed with a paraganglioma will have a hereditary PV in one of at least six different genes (Fishbein et al. 2013). An overview of selected pediatric and adult tumors that warrant referral for genetics evaluation regardless of family history are highlighted in Table 2.

Any person who develops multifocal, bilateral, or multiple primary tumors has a higher likelihood of having a cancer predisposition (Offit and Brown 1994). However, the likelihood is reduced if the person has had a known environmental exposure such as radiation or certain chemotherapeutic agents to treat a prior cancer. Well-established examples of associations between cancers and environmental exposures include breast cancer after mantle radiation to the chest wall to treat Hodgkin's lymphoma as a teenager (Conway et al. 2017). Certain cancer predisposition syndromes are also associated with significant medical issues, developmental delays, or physical differences, so assessment

Table 1. Personal and family history "red flags" suggestive of an underlying hereditary cancer predisposition

History	Examples
Personal	
Early age at diagnosis	Colon cancer at 30
Rare tumor	Male breast cancer, rhabdoid tumor
Tumor associated with known syndrome	Pheochromocytoma, retinoblastoma
Multifocal or bilateral tumors	Bilateral Wilms tumor
Multiple primary tumors	Breast cancer and ovarian cancer
Lack of known environmental factor	Lung cancer in nonsmoker
Excessive toxicity to treatment	Severe toxicity to chemotherapy in Fanconi anemia
Other developmental and physical differences	Large head size, asymmetry, congenital heart defects, etc. (for comprehensive examples, see Kesserwan et al. 2016; Coury et al. 2018)
Pathogenic variant detected in tumor/ abnormal tumor testing	*ALK* mutation in neuroblastoma, MSH6 absent (IHC) colon cancer
Family	
Multiple generations affected with tumors or cancers (on same side of family)	
Multiple first-degree relatives affected (parent, child, sibling)	
Pattern of cancers suggestive of known syndrome	Osteosarcoma, brain tumor, adrenocortical carcinoma
Known mutation in family member	*SDHB* mutation in father
Consanguinity	Parents of affected child are first cousins
Ethnicity	Ashkenazi Jewish, Icelandic founder mutations in *BRCA1/2*

Cite this article as *Cold Spring Harb Perspect Med* doi: 10.1101/cshperspect.a036541

Table 2. Select tumors that warrant genetics evaluation

Tumor	Highly associated syndromes	Gene	Commonly associated tumor/cancer risks
Adrenocortical carcinoma	Li–Fraumeni syndrome	TP53	Breast, sarcoma, brain tumor, adrenocortical carcinoma, many cancer types
Anaplastic rhabdomyosarcoma	Li–Fraumeni syndrome	TP53	See above
Cerebellar hemangioblastoma	von Hippel–Lindau syndrome	VHL	Endolymphatic sac tumor, pancreatic islet cell carcinoma, hemangioblastoma of CNS or retina, renal cell carcinoma, cysts in the liver, kidney, pancreas, or spleen
Choroid plexus carcinoma	Li–Fraumeni syndrome	TP53	See above
Diffuse gastric cancer	Hereditary diffuse gastric cancer	CDH1	Lobular breast cancer, diffuse gastric cancer
Hypodiploid ALL	Li–Fraumeni syndrome	TP53	See above
Medullary thyroid cancer	MEN2	RET	Medullary thyroid cancer, pheochromocytoma, hyperparathyroidism
Medulloblastoma	Familial adenomatous polyposis	APC	Colon, colon polyps, ampullary, small bowel, small bowel polyps, thyroid, desmoid
	Gorlin syndrome	SUFU, PTCH1	Basal cell carcinoma, ovarian fibroma, jaw keratocyst
	Hereditary breast/ovarian cancer	BRCA2, PALB2	Breast, ovarian, pancreatic, prostate, melanoma
	Li–Fraumeni syndrome	TP53	See above
Ovarian cancer	Hereditary breast/ovarian cancer	BRCA1, BRCA2	Breast, ovarian, pancreatic, prostate
	Lynch syndrome	MLH1, MSH2, MSH6, PMS2, EPCAM	Colon, endometrial, colon polyps, small bowel, urinary tract, ovarian
	Epithelial ovarian cancer	BRIP1, RAD51C, RAD51D	Possibly breast
Ovarian Sertoli–Leydig cell tumor	DICER1 syndrome	DICER1	Pleuropulmonary blastoma, Sertoli–Leydig cell tumor of the ovary, cystic nephroma, thyroid nodules/cancer
Pancreatic cancer	Familial atypical multiple mole melanoma syndrome	CDKN2A	Melanoma, pancreatic, dysplastic nevi
	Hereditary breast/ovarian cancer	BRCA1, BRCA2, PALB2	Breast, ovarian, pancreatic, prostate, melanoma
	Lynch syndrome	MLH1, MSH2, MSH6, PMS2, EPCAM	Colon, endometrial, colon polyps, small bowel, urinary tract, ovarian

Continued

Table 2. *Continued*

Tumor	Highly associated syndromes	Gene	Commonly associated tumor/cancer risks
	Peutz–Jeghers syndrome	*STK11*	Breast, gastrointestinal hamartomatous polyps, mucocutaneous pigmentation, colon, lung, small bowel, stomach, cervix, ovarian, testicular
Paraganglioma/ pheochromocytoma	Hereditary paraganglioma/ pheochromocytoma syndrome	*SDHA, SDHB, SDHC, SDHD, SDHAF2, MAX, TMEM127*	Paraganglioma, pheochromocytoma, GIST, renal cell cancer, papillary thyroid cancer
	MEN2	*RET*	
	von Hippel–Lindau syndrome	*VHL*	Endolymphatic sac tumor, pancreatic islet cell carcinoma, hemangioblastoma of CNS or retina, renal cell carcinoma, cysts in the liver, kidney, pancreas, or spleen
Pleuropulmonary blastoma	*DICER1* syndrome	*DICER1*	See above
Retinoblastoma	Hereditary retinoblastoma	*RB1*	Melanoma, sarcoma, pineoblastoma
Rhabdoid tumor, atypical teratoid/ rhabdoid tumor	Rhabdoid tumor predisposition syndrome	*SMARCB1, SMARCA4*	Schwannoma, meningioma

(CNS) Central nervous system, (GIST) gastrointestinal stromal tumor.

of additional medical history via targeted questioning or review of medical records may be necessary. For example, individuals with DNA damage repair syndromes such as Fanconi anemia may experience significantly increased toxicity from chemotherapy and display physical manifestations such as multiple congenital anomalies and café au lait spots (de Latour and Soulier 2016). There are many good review articles and guidelines that provide further detail about tumors and nononcologic features of cancer predisposition syndromes and make suggestions for referral and evaluation (Hampel et al. 2015; Kesserwan et al. 2016; Desai et al. 2017; Scollon et al. 2017; Coury et al. 2018; Kennedy and Shimamura 2019).

Determination of who will benefit from genetic testing, tailored screening, and risk-reducing interventions historically began with collection of family history information, and this is still integral to genetics practice today (Pyeritz 2012). Family history and risk factor questionnaires were provided in advance of the consultation to allow patients time to speak with their relatives to optimize the accuracy of the family history. Various software packages are now available to streamline this process. Examples include CRA Health (www.crahealth.com), CancerIQ (www.canceriq.com), Cancer Gene Connect (www.invitae.com/en/cancergeneconnect), and Progeny (Ozanne et al. 2009) (www.progenygenetics.com). Some of these software packages were designed to screen general populations for cancer genetics referral and may also provide quantitative risk assessments. These risk assessment tools are summarized in Table 3.

A number of professional societies, including the National Comprehensive Cancer Network (NCCN), have published guidelines to help clinicians determine who should be offered genetic risk assessment and testing. These guidelines, along with additional resources for

Table 3. Risk assessment tools

Tool	Cancer risks assessed	Mutation probabilities	Notable features	Website
Breast Cancer Risk Assessment Tool, aka Gail Model	Breast cancer 5 yr and lifetime risks	None	Used to determine eligibility for chemoprevention Family history and other risk factors	bcrisktool.cancer.gov
Claus Tables	Breast cancer 10 yr and lifetime risks	None	Risk based on family history, including ages at diagnosis	Breast Cancer Risk Assessment Application (BRisk APP)
Tyrer–Cuzick	Breast cancer 10 yr and lifetime risks	BRCA1, BRCA2	Family history and other risk factors including prior negative genetic testing	www.ems-trials.org/ riskevaluator
Penn II	None	BRCA1, BRCA2	Based on summary of family history features	pennmodel2.pmacs. upenn.edu/penn2
BOADICEA	Breast and ovarian cancer risks	BRCA1, BRCA2, PALB2, CHEK2, ATM	Family history–based Can calculate cancer risks for those with negative genetic testing results	ccge.medschl.cam.ac. uk/boadicea/ boadicea-model
BRCAPRO	Breast and ovarian cancer risks	BRCA1, BRCA2	Family history and other risk factors, includes contralateral breast cancer risk	projects.iq.harvard. edu/bayesmendel/ brcapro
MMRpro	Colon and endometrial cancer risks	MLH1, MSH2, MSH6	Risk for Lynch syndrome	projects.iq.harvard. edu/bayesmendel/ brcapro
MelaPRO	Melanoma	CDKN2A	Risk for CDKN2A-associated melanoma	projects.iq.harvard. edu/bayesmendel/ brcapro
PancPRO	Pancreatic cancer	Dominant pancreatic cancer risk gene	Risk for a putative AD gene for pancreatic cancer risk	projects.iq.harvard. edu/bayesmendel/ brcapro
PREMM5	None	MLH1, MSH2, MSH6, PMS2, EPCAM	Based on personal and family history of Lynch-associated cancers	premm.dfci.harvard. edu
ASK2ME	Cancer risks provided in age intervals for mutation carriers	32 high- and moderate-penetrance genes	Risks and summaries of existing management guidelines	ask2me.org/

(AD) Autosomal dominant.

cancer genetics providers, are summarized in Table 4. As with the risk assessment models, there is variability between professional guidelines. For example, the American Society of Breast Surgeons guidelines recommend genetic testing be offered to all individuals with breast cancer (Plichta et al. 2019), whereas the guideline published by NCCN clearly delineates personal and family history criteria for testing. Insurance companies typically stipulate criteria for coverage of genetic testing, which may differ from professional guidelines. This has led to considerable variation in provider practice, with insurers at times the ultimate arbiters of access based on individual policy coverage decisions.

Further complicating testing and referral decisions are data indicating that personal and family history often fall short in accurately predicting who will test positive for a PV in a clinically actionable gene. Some argue that guidelines, although useful, will miss too many individuals who would test positive (Yurgelun et al. 2015a; Rosenthal et al. 2017; Beitsch et al. 2019). Thus, in many practices the clinical threshold for testing has become less stringent. In fact, some are promoting the merits of population-based screening for cancer predisposition genes (King et al. 2014). Mary-Claire King, PhD, who was the first to localize the *BRCA1* gene, has advocated for population-wide screening for *BRCA1* and *BRCA2*, stating that "many women with mutations in these genes are identified as carriers only after their first cancer diagnosis because their family history of cancer was not sufficient to suggest genetic testing. To identify a woman as a carrier only *after* she develops cancer is a failure of cancer prevention" (King et al. 2014). Others are more measured in their views, but cite population screening as a laudable long-term goal, once the public health impact is better determined (Yurgelun et al. 2015b).

As direct-to-consumer (DTC) genetic testing is becoming more widely available, cancer genetic testing has expanded via this resource as well. Commercial cancer genetic testing companies are offering a patient-driven model of testing, in which a person requests testing on-

line, is paired with a physician or genetic counselor who will order the test, and is offered genetic counseling (McGowan et al. 2014; Covolo et al. 2015). These additional opportunities for consumer-initiated genetic testing will improve access but could also lead to lower testing quality or inappropriate ordering (Phillips et al. 2019). Increasingly, commercial labs are providing bundled services, including ancestry and health information, which may include assessment of cancer risk. One such company, 23andMe, currently offers testing for the three common Ashkenazi Jewish *BRCA1* and *BRCA2* PVs, as well as analysis of raw data at third-party companies. These approaches have stirred controversy because of concerns that users may not realize this limited genetic testing has little bearing on their chance of having a PV in *BRCA1* or *BRCA2*, unless they are of Ashkenazi Jewish ancestry (Gill et al. 2018). Further concern has been raised about the accuracy of certain DTC tests and third-party interpretations, with one study showing a false-positive result in 40% of samples subsequently sent for clinical confirmation at a CLIA-certified testing lab (Tandy-Connor et al. 2018, 2019).

Cohort studies such as the *All of Us* Research Program (allofus.nih.gov), The Partners Biobank Registry (biobank.partners.org), and the CSER (Clinical Sequencing Exploratory Research; cser-consortium.org) consortium, are offering genomic sequencing to individuals, unselected for any specific diagnosis, in a variety of clinical settings, and providing results for clinically actionable variants. Initial concerns about these large studies focused on the potential for participants' adverse psychological reactions and decisional regret after receiving unexpected results. However, so far, results from large-cohort studies have been reassuring, with no reports of clinically significant psychological harm or use of inappropriate medical services from return of results (Hart et al. 2019; Robinson et al. 2019). As access to genetic testing expands, and reassuring data accumulate about psychological outcomes, the number of suitable candidates for cancer genetic services is expected to continue to broaden.

Cite this article as *Cold Spring Harb Perspect Med* doi: 10.1101/cshperspect.a036541

Table 4. Cancer and genetics resources and societal guidelines for hereditary cancer predisposition evaluation

Category	Resources	Website	Societal guidelines/ policies	Reference	Content summary
Genetics	American College of Medical Genetics and Genomics (ACMG)	www.acmg.net	A practice guideline from the ACMG and the NSGC: referral indications for cancer predisposition assessment	10.1038/gim.2014.147 (Hampel et al. 2015)	Recommendations for referral for cancer genetics evaluation—not guidelines for testing; includes tables of cancer types with personal and family history; features suggestive of specific syndromes; provides description of syndromes and rationale for referral
	National Society of Genetic Counselors (NSGC)	www.nsgc.org			
	GeneReviews	www.ncbi.nlm.nih .gov/books/ NBK1116/			Comprehensive reviews of genetic conditions
	Genetics Home Reference	ghr.nlm.nih.gov			Information about genes and genetic conditions
General Cancer	American Cancer Society (ACS)	www.cancer.org			Comprehensive information about cancer and support for families
	American Society of Clinical Oncology (ASCO)	www.asco.org	ASCO Policy Statement Update: Genetic and Genomic Testing for Cancer Susceptibility	10.1200/JCO.2015.63.0996 (Robson et al. 2015)	Recommendations for use of multigene panels, somatic testing, and direct-to-consumer testing in clinical care; includes detailed table of informed consent/pretest education
			ASCO policy statement update: genetic and genomic testing for cancer susceptibility	10.1200/JCO.2009.27.0660 (Robson et al. 2010)	Recommendations address inclusion of moderate- and low-penetrance genes and direct-to-consumer testing; not in updated version
	National Comprehensive Cancer Network (NCCN)	www.nccn.org	NCCN Guidelines: Breast/ Ovarian, Colon, Gastric, Neuroendocrine and Adrenal Tumors, Myelodysplastic Syndromes	www.nccn.org/professionals/ physician_gls/default.aspx	Guidelines for referral for cancer genetics evaluation; includes criteria for genetic testing, criteria for clinical diagnosis, and management guidelines

Continued

Table 4. *Continued*

Category	Resources	Website	Societal guidelines/policies	Reference	Content summary
	National Cancer Institute (NCI)	www.cancer.gov			Comprehensive information about cancer for patients and providers; directory for cancer genetics services
			NCI Cancer Genetics Risk Assessment and Counseling (PDQ)–Health Professional Version. www.cancer.gov/about-cancer/causes-prevention/genetics/risk-assessment-pdq	www.cancer.gov/about-cancer/causes-prevention/genetics/risk-assessment-pdq	Very detailed review of process and content for cancer genetic risk assessment and counseling
	Familial Cancer Database Online (FaCD)	www.facd.info			Database that associates cancer types and syndromes; useful for creating differential diagnosis
Breast/ovarian	The American College of Obstetricians and Gynecologists (ACOG)	www.acog.org	Committee Opinion: Hereditary Cancer Syndromes and Risk Assessment	10.1097/01. AOG.0000466373.71146.51 (2015)	Recommendations for obtaining a family history, medical history, and referral for a cancer genetics evaluation; review of common hereditary syndromes with gynecologic cancers
			Practice Bulletin No 182: Hereditary Breast and Ovarian Cancer Syndrome	10.1097/AOG.0000000000002296 (Committee on Practice Bulletins–Gynecology 2017)	Guidelines for genetic counseling, testing, and management for hereditary breast and ovarian cancer syndrome
	The American Society of Breast Surgeons (ASBrS)	www .breastsurgeons .org	ASBrS: Consensus Guideline on Genetic Testing for Hereditary Breast Cancer	www.breastsurgeons.org/docs/statements/Consensus-Guideline-on-Genetic-Testing-for-Hereditary-Breast-Cancer.pdf	Recommendations for genetic testing for hereditary breast cancer

Cite this article as *Cold Spring Harb Perspect Med* doi: 10.1101/cshperspect.a036541

Category	Organization	Website	Guideline	Reference	Description
	Society of Gynecologic Oncology (SGO)	www.sgo.org	SGO statement on risk assessment for inherited gynecologic cancer predispositions	10.1016/j.ygyno.2014.09.009 (Lancaster et al. 2015)	Recommendations for referral criteria for genetic counseling and offering testing for hereditary breast and ovarian cancer and Lynch syndrome
	U.S. Preventive Services Task Force (USPSTF)	www.uspreventiveservicestaskforce.org	Draft Recommendation Statement BRCA-Related Cancer: Risk Assessment, Genetic Counseling, and Genetic Testing	www.uspreventiveservicestaskforce.org/Page/Document/draft-recommendation-statement/brca-related-cancer-risk-assessment-genetic-counseling-and-genetic-testing1	Recommendations for risk assessment, genetic counseling, and genetic testing for *BRCA1* and *BRCA2*
Endocrine	American Thyroid Association (ATA)	www.thyroid.org	Revised ATA for the Management of Medullary Thyroid Carcinoma Guidelines	10.1089/thy.2014.0335 (Wells et al. 2015)	Recommendations for *RET* genetic testing and management of hereditary MTC and MEN2 based on genotype
	Endocrine Society (ENDO)	www.endocrine.org	Pheochromocytoma and Paraganglioma: An Endocrine Society Clinical Practice Guideline	10.1210/jc.2014-1498 (Lenders et al. 2014)	Guidelines for genetic testing and management for hereditary paraganglioma and pheochromocytoma
Gastro-intestinal	American College of Gastroenterology (ACG)	www.gi.org	ACG Clinical Guideline: Genetic Testing and Management of Hereditary Gastrointestinal Cancer Syndromes	10.1038/ajg.2014.435 (Syngal et al. 2015)	Guidelines for genetic testing and management for hereditary gastrointestinal cancer syndromes
	American Gastroenterological Association (AGA)	www.gastro.org	AGA Institute Guideline on the Diagnosis and Management of Lynch Syndrome	10.1053/j.gastro.2015.07.036 (Rubenstein et al. 2015)	Guidelines for diagnosis and management of Lynch syndrome

Continued

Table 4. *Continued*

Category	Resources	Website	Societal guidelines/policies	Reference	Content summary
	American Society of Clinical Oncology (ASCO)	www.asco.org	Hereditary Colorectal Cancer Syndromes: ASCO Clinical Practice Guideline Endorsement of the Familial Risk–Colorectal Cancer: European Society for Medical Oncology Clinical Practice Guidelines	10.1200/JCO.2014.58.1322 (Stoffel et al. 2015)	Guidelines for genetic testing and management for hereditary colon cancer syndromes
			Evaluating Susceptibility to Pancreatic Cancer: ASCO Provisional Clinical Opinion	10.1200/JCO.18.01489 (Stoffel et al. 2019a)	Recommendations for genetic testing and management for hereditary pancreatic cancer syndromes
Hematologic	American Society of Hematology (ASH)	www.hematology.org			
Pediatric	American Association for Cancer Research (AACR)	www.aacr.org	Clinical Cancer Research: Pediatric Oncology Series by AACR	clincancerres.aacrjournals.org/pediatricseries	Series of articles summarizing cancer predisposition syndromes in childhood and providing expert consensus guidelines for management
	American Academy of Pediatrics (AAP)	www.aap.org	POLICY STATEMENT: Ethical and Policy Issues in Genetic Testing and Screening of Children	10.1542/peds.2012-3680 (BIOETHICS et al. 2013)	Recommendations about when genetic testing should be offered in different clinical scenarios
	The American Society of Pediatric Hematology/Oncology (ASPHO)	www.aspho.org			
	Children's Oncology Group (COG)	www.childrensoncologygroup.org			Clinical trials group for childhood cancer; provides support for families and connects researchers and clinicians

Cite this article as *Cold Spring Harb Perspect Med* doi: 10.1101/cshperspect.a036541

THE PROCESS OF CANCER GENETIC COUNSELING

Historical Context

Genetic counseling is "the process of helping people understand and adapt to the medical, psychological, and familial implications of genetic contributions to disease" (Resta et al. 2006). Genetic counseling was proposed as a key component of the cancer risk assessment process in the 1990s (Peters and Stopfer 1996; Stopfer 2000). Among the first families to receive genetic counseling for cancer risk were those participating in linkage studies aimed at localizing the *BRCA1* gene (Biesecker et al. 1993). Empowering patients to make informed decisions regarding screening, prevention, and genetic testing through provision of pertinent genetic and medical information and tailored psychological counseling remain core goals. The Cancer Genetics Studies Consortium (CSGC) Task Force was among the first groups in the oncology setting to develop consensus guidelines for the process and content of informed consent (Geller et al. 1997). This multidisciplinary group first considered why informed consent for genetic testing requires special consideration and acknowledged the following issues: (1) Genetic information affects an entire family, (2) genetic information can present unique challenges for medical professionals as it is probabilistic in nature, and (3) genetic information can lead to the reclassification of patients from healthy to high risk. At that time, the authors described the primary risks and benefits as psychological and social rather than physical or medical because the efficacy of preventive and therapeutic strategies in PV carriers had not yet been substantiated.

In 2004, the National Society of Genetic Counselors published their first set of recommendations for cancer genetic risk assessment and counseling (Trepanier et al. 2004). These guidelines were based on a literature review and the professional expertise of cancer genetic counselors. The Huntington's disease model of pre- and posttest counseling served as an initial template, paying significant attention to the possibility of psychological harm. This model advocated for three separate visits that could each last

an hour or more and would sometimes incorporate assessments from mental health professionals before disclosure of results (Biesecker and Garber 1995; Almqvist et al. 2003). Core components of traditional cancer genetic counseling sessions have included (Resta 2006; Madlensky et al. 2017):

- Collection and interpretation of family and medical histories.

- Risk assessment for a cancer diagnosis or for an inherited PV.

- Education about inheritance, testing options, management strategies, resources, prevention, and research opportunities.

- Counseling to facilitate informed decision-making, identification of psychological needs, and provision of appropriate support.

A number of practice guidelines for the provision of cancer genetic counseling and indications for genetic testing in oncology have been published and are summarized in Table 4.

Recent Developments

The evolution of cancer genetic counseling practice has been driven in part by advances in evidence-based medical interventions, social and behavioral research, and expansion of testing options (Athens et al. 2017). The field has moved from a presumption of medical benefit for carriers of cancer susceptibility variants to evidence-based approaches for risk reduction and specialized screening that lead to reduced morbidity and mortality. For example, bilateral salpingo-oophorectomy for women with inherited PVs in *BRCA1* or *BRCA2* reduces overall mortality, including for those who have already had a breast cancer diagnosis (Domchek et al. 2010; Metcalfe et al. 2015). Patients with Lynch syndrome adhering to screening endoscopy and hysterectomy recommendations demonstrated similar mortality rates to their relatives with negative genetic testing (Järvinen et al. 2009). Testing for *RET* mutations in families with multiple endocrine neoplasia and age-appropriate prophylactic thyroidectomy based on mutational status improves disease-free survival (Shepet et al. 2013).

Somatic and germline characterization of malignancy is becoming a standard part of the clinical evaluation for an increasing number of oncology patients. Recent studies suggest that nearly 10% of patients with advanced cancer may have actionable PVs that would not have been identified under existing guidelines for clinical testing (Mandelker et al. 2017). Knowledge about the presence of a hereditary PV can influence an oncologist's decision about drug of choice (Goyal et al. 2016; Cohen et al. 2017; Robson et al. 2017; Zhang et al. 2018; Abida et al. 2019; Bast et al. 2019). Therefore, broad categories of cancer patients are now candidates for genetic testing at the time of diagnosis. Because of the relatively high proportion and clinical actionability of PVs found in unselected patients with pancreatic and ovarian cancer, all patients with these diagnoses should be offered germline genetic testing (Stoffel et al. 2019b; Telli et al. 2019).

We now know that information obtained from genetic testing can be lifesaving, and for this reason, genetic testing may appropriately be recommended, rather than simply offered, for certain patients. As expected, when offered a test with implications for cancer treatment, most patients elect to undergo testing (Nilsson et al. 2019). Furthermore, because clinical decision-making will hinge on the outcome of genetic testing, providers and patients are anxious to get the genetic testing process underway and have results available as soon as possible. Therefore, the hesitancy to offer genetic testing in the absence of intensive discussion or a face-to-face encounter has also shifted, as there is a moral imperative to provide equitable access to testing in which a positive result has actionable consequences.

Psychological well-being among those undergoing cancer gene testing has been studied, and the available data do not support initial concerns. Numerous rigorous studies over the years have shown that genetic testing information does not increase the risk for true psychopathology such as suicidality, major depression, or anxiety in someone who does not otherwise

carry such a diagnosis (Lerman et al. 1998; Lammens et al. 2010b; Halbert et al. 2011; Kattentidt-Mouravieva et al. 2014). Patients with a prior psychiatric diagnosis, such as a history of depression or who require the use of psychotropic medicines, have demonstrated enhanced psychological vulnerability to genetic testing outcomes in several studies (Murakami et al. 2004; van Oostrom et al. 2007). This highlights the value of assessing prior history of psychiatric diagnoses, and individualizing care when possible, to address psychological support needs when they arise.

Finally, testing options have expanded significantly over the last 10 years, from single-site and single-gene testing to multigene panel tests (Kurian et al. 2018; Phillips et al. 2018). The demand for cancer genetic services has increased because of promising medical management options, lack of serious psychological burden for most individuals, and expansion of laboratory testing options. The need to provide these services to ever-increasing numbers of patients has fostered demand for novel and more efficient models for offering genetic counseling and testing (Hoskovec et al. 2018; McCuaig et al. 2018).

Pretest Counseling Models

The emergence of new models for the provision of pretest counseling challenges providers to ensure the core goals and values of genetic counseling are maintained. A major shift in clinical practice affecting the informed consent discussion has been the transition to multigene panel testing. The "tiered-binned model" of genetic counseling and informed consent was developed to address the content of a genetic counseling sessions for this purpose (Bradbury et al. 2015, 2016). Traditional models often included a detailed discussion about the clinical manifestations and management options of each gene tested. The tiered-binned model provides a layered approach, with broad concepts and common denominator elements required for all patients considered tier 1 "indispensable" information. More specific tier 2 information is provided as needed, to support the informational needs and specific preferences among diverse patient populations. "Binning" refers to the approach of organizing discussion about genes into categories, defined by variables such as high or moderate risk genes, or based on clinical utility (Bunnik et al. 2013).

Alternative pretest service delivery models can increase access to genetic testing by minimizing the number of required in-person appointments and reducing waiting times for results while maintaining patient satisfaction (McCuaig et al. 2018). Alternative models may be led by a genetic counselor, physician, or other provider (Schwartz et al. 2014; Senter et al. 2017; Colombo et al. 2018). Some patients are being offered direct access to genetic testing with minimal to no pretest discussion, with pretest education offered via a video, Web-based education, written information, a chatbot, or interactive relational agent (Sie et al. 2014; McCuaig et al. 2019). These models are summarized in Table 5. Typically, these alternative service delivery models rely on the availability of genetic counseling and further consultation for those found to have a PV. A critical part of posttest counseling will continue to be facilitation of cascade

testing for individuals with hereditary risk, as identifying family members with PVs can lead to targeted cancer prevention for these at-risk relatives (Caswell-Jin et al. 2019).

Although these models have been found noninferior to in-person genetic counseling in areas such as satisfaction, distress, and knowledge about genetic testing, many patients have preferred in-person genetic counseling (Schwartz et al. 2014; Sie et al. 2014). In addition, it is important to note that the physician-led approaches may not be generalizable to all settings, as many physicians are not comfortable ordering and interpreting genetic test results (Eccles et al. 2015; Kurian et al. 2018). Most studies published to date were conducted as research initiatives within academic medical centers and have focused heavily on more common indications such as hereditary breast and ovarian cancer and Lynch syndrome, potentially limiting the generalizability of conclusions to community-based settings (Manchanda et al. 2016; Bednar et al. 2017; Tutty et al. 2019).

As evidence-based novel models of pretest counseling and education are implemented, care should be taken to identify patients who are un-

Table 5. Alternative service delivery models for genetic counseling and testing

	Method	Potential benefits	Potential drawbacks
Genetic counselor–driven	Telephone	No travel needed, widely available interface	Limited reimbursement, no visual cues in the counseling session
	Videoconference	No travel needed, visual cues, may feel more personalized	Technical difficulties more common and may not be billable
	Group counseling	Group setting allows learning from other's questions, may feel supported	Lack of privacy, scheduling opportunities may be limited
	Embedded genetic counseling	Access to in-person counseling same day receiving treatment, increased likelihood of referral	Counseling may not be private if done in infusion area
Nongenetic counselor–driven	Tumor testing flags possible germline	Automated, enriches population of patients likely to have germline mutations	Tumor-only testing will miss detectable germline mutations
	Direct genetic testing	Rapid access	Unexpected results and lack of understanding
	Direct to consumer	Increases patient autonomy and privacy	Unexpected results and no relationship with local provider

decided about genetic testing after a brief intervention or who prefer a more detailed or in-person session because of psychological or informational concerns. If a genetic counselor is not available locally, remote access via telegenetic services is an option. Like so much of the practice of genetic counseling, using a flexible approach tailored to the patient facilitates optimal care, balancing efficiency with attention to specific psychological needs and preferences.

LABORATORY, TEST SELECTION, AND INTERPRETATION ISSUES

Germline genetic testing has been a valuable tool to elucidate inherited susceptibility to cancer for several decades (Ponder 1994). Genetic counselors are often at the forefront of test selection and are tasked with ensuring individuals make an informed choice about testing. The expansion of testing options has led to increased debates within the field and variability in practice.

Biosample Selection

It is important to understand the characteristics of the biological sample used for testing and its limitations. Blood or saliva is commonly used for germline genetic testing. However, if the patient has a hematologic malignancy, the test results may reflect somatic PV in the cancer. A somatic-only PV could be misinterpreted as germline, or a germline mutation could be missed. Similarly, a PV may be detected in *TP53* or another gene at a <0.50 variant allelic fraction in a person without a known hematologic malignancy and may represent a somatic change in the blood with unknown significance, referred to as clonal hematopoiesis of indeterminant potential or aberrant clonal expansions (Steensma 2018; Weitzel et al. 2018). A result like this may be misinterpreted as constitutional somatic mosaicism. In these cases, DNA testing from cultured skin fibroblasts is recommended, and eyebrow plucks are another potential source of DNA. A cheek swab or saliva sample may be contaminated with white blood cells and should be avoided if possible (Weitzel et al. 2018).

Laboratory Selection

An equally important aspect of the hereditary cancer evaluation is the selection of a laboratory. Whether single-gene analysis or a large multigene panel is indicated, it is critical to consider the capabilities and expertise of the lab performing the test. Even if the same basic technology is used, the amount of coverage, such as inclusion of deep intronic PVs, rearrangements, or pseudogene regions, may vary and significantly impact variant detection rates. Ensuring the gene or condition highest on the patient differential is thoroughly assessed provides the best sensitivity and detection rate (Richards et al. 2015).

Another key consideration is the laboratory's experience with interpreting germline variants within cancer genes. Variant interpretation incorporates subjective assessment of available data and may vary depending on a lab's experience (Amendola et al. 2016; Kim et al. 2019). The lack of diversity of those tested in large laboratory cohorts and within publicly accessible databases can lead to errors in classification and interpretation of findings (Manrai et al. 2016). Increasing representation of minority communities in testing cohorts will eventually lead to more accurate test interpretation and fewer variants of uncertain significance (VUSs); however, this will take time.

Test Selection

As with any new development, there is excitement, fear, reservation, and controversy surrounding multigene panel testing. Health-care providers typically seek interventions that maximize benefits while minimizing harm to the patient. Test selection and panel size are vigorously debated in the cancer genetics professional community (Lynce and Isaacs 2016; Lee et al. 2019b). On one extreme, there are those who believe a minimal amount of testing is most appropriate, such as single-site testing for a person with a known family history of a PV. On the other hand, some support the use of broad multigene panel testing that includes high and moderate penetrance genes for which the patient may or may not be at risk based on personal and family history (Douglas et al. 1999; Beadles et al. 2014; Desmond et al. 2015; Tung et al. 2015; Zhang et al. 2015; Parsons et al. 2016).

Even those who use multigene panel testing struggle with determining the optimal panel size and the degree to which a provider should steer patient choice for test selection (Domchek et al. 2013). It may not be appropriate to task patients with this burden when they are in crisis or under significant stress because of a new cancer diagnosis or family history concern. A potential middle ground may be to generally offer a broad multigene panel and help the patient decide if a smaller test is better for them based on their values, beliefs, and goals.

Germline Genetic Testing beyond Panels

Additional germline testing options are on the horizon. Exome and genome sequencing are becoming more widely available and may be helpful if the cancer patient also has congenital anomalies or developmental delay (Diets et al. 2018) or when multigene cancer panels are negative (Powis et al. 2018). Other technologies available include RNA sequencing paired with DNA analysis that can identify the presence and quantity of RNA transcripts, which may help in classification of VUS results and identify deep intronic PVs (Farber-Katz et al. 2018). We can expect that these and other enhancements to laboratory testing techniques will continue to increase the sensitivity of variant detection and clarity of interpretation over time.

Polygenic Risk Scores

Polygenic risk scores (PRSs) incorporate the combined effects of single-nucleotide polymorphisms (SNPs) identified from large-scale genome-wide association studies (GWASs) to estimate a person's risk to develop cancer. Generally, the magnitude of association of each individual SNP with disease risk is small (e.g., conferring a relative risk of <2). There is hope that the cumulative effect of 100 or more informative SNPs on their own, or in combination with a risk model such as Tyrer-Cuzick, can be used to stratify individuals into higher and lower risk categories that may lead to differential

screening recommendations. In addition, models such as BOADICEA are working to incorporate SNP data to further refine lifetime risk estimates for those who carry PVs in known cancer susceptibility genes (Lee et al. 2019a). However, many PRSs to date contribute relatively small improvements in risk prediction after other known risk factors are considered. In addition, commercial laboratories are offering SNP-based PRS genetic testing before clinical utility has been proven. Finally, there is differential access to this risk information, because GWAS data are less available for populations other than those of European ancestry, potentially creating or exacerbating disparities in health care. Efforts to study more diverse populations are needed to determine clinical utility and achieve equity in this area of personalized medicine (Martin et al. 2019).

Integration and Interpretation of Genetic Test Results

Cancer risk management recommendations should be based on the integration of the genetic test results with the patient's personal and family history (Robson et al. 2015). For example, a 40-yr-old woman who had breast cancer and has a family history meeting clinical criteria for Li–Fraumeni syndrome (LFS) would receive LFS guidelines even if her testing was negative or identified a VUS in *TP53*, whereas a similar woman without supporting family history would not receive such guidelines. Posttest risk integration is essential for any streamlined counseling and testing model to ensure that high-risk patients, like the former, are not not missed.

A continuing challenge for cancer genetics providers is managing VUS results. Approximately 90% of VUSs are later reclassified as benign, whereas a much smaller proportion are upgraded to pathogenic (Slavin et al. 2019; So et al. 2019; Tsai et al. 2019). Unfortunately, there is evidence (Kurian et al. 2018) that some providers are making inappropriate surgical or surveillance recommendations based on a lack of understanding of a VUS result, which are common in multigene panels (LaDuca et al. 2014; Kapoor et al. 2015).

It can also be challenging to interpret the significance of a PV in a less well-characterized gene, because cancer risks may not be clearly delineated and consensus management guidelines for follow-up may be lacking. In addition, when multigene panel testing is used, a challenge arises when PVs are found in genes considered highly penetrant, such as *TP53* or *CDH1*, for patients without suggestive personal or family history (Lynce and Isaacs 2016). These findings challenge penetrance estimates and decisions regarding appropriate medical management (Rana et al. 2019; Xicola et al. 2019).

SPECIAL ISSUES IN PEDIATRIC ONCOLOGY

Pediatric cancer genetics is an emerging subspecialty within cancer genetics that combines traditional cancer genetic counseling and pediatric genetics. There are unique aspects to pediatric oncology counseling that result in further complexity in the sessions.

When gathering a medical history for a child who has had cancer, it is important to complete a birth history, full medical review of systems, and a dysmorphology exam if possible, because increased cancer risk could be part of a broader Mendelian syndrome (Kesserwan et al. 2016). As is seen in the adult cancer clinic, there is a trend toward increased use of multigene panel testing, with debate over panel size. However, genetic testing in children raises additional ethical considerations. Genetic testing is typically offered when there is potential medical benefit to the child, such as a clear increased risk for pediatric cancer for which screening is available, along with the expectation of minimal harm (Committee on Bioethics et al. 2013).

Use of an expanded cancer panel in the pediatric population that includes adult-onset cancer genes is controversial. The identification of a PV in a gene known only to be associated with adult-onset tumors may or may not be the cause of a pediatric malignancy (Sylvester et al. 2018). As there are limited medical benefits to learning a child has a PV for an adult-onset cancer and potential harm (lack of future autonomy and potential psychological harm), children are not typically offered testing for adult-onset cancer predisposition genes, such as *BRCA1*, *BRCA2*, or Lynch syndrome (Kesserwan et al. 2016). However, this perspective is being challenged, and the testing approach is evolving (Druker et al. 2017; Sylvester et al. 2018; Kuhlen et al. 2019). Although not currently the standard practice in pediatrics, it is possible that knowing germline status of such genes may impact treatment opportunities, especially in well-advanced cancers. Tumor testing is increasingly being performed that may provide insight into the germline, and the information could benefit other family members. In addition, exome analysis is routinely performed in clinical genetics programs, with genetic variants reported as secondary findings according to the recommended ACMG list of actionable genes (Kalia et al. 2017). There are multiple large-panel tests and exome sequencing research studies ongoing that will assess the clinical and psychological impact of this testing in pediatric oncology.

It can be very challenging to involve a child in the pre- and posttest discussions, because parents can have strong opinions about what information should be shared and when. The benefit of open communication with children has been demonstrated in multiple studies, which show enhanced coping and adaptation, decreased anxiety, and increased family cohesion (Rowland and Metcalfe 2013; Valdez et al. 2018). Providers can offer to meet with parents before bringing the child into the visit to allow a more open discussion about the risk assessment and testing options, obtain full consent, and make a plan for how the child will be involved. Assent for genetic testing is recommended when possible (Committee on Bioethics et al. 2013), and some institutions have policies regarding the age of assent, which is commonly around age 11 yr. The amount of information and level of detail provided in these discussions will vary based on the developmental age of the child (Werner-Lin et al. 2018).

Management of pediatric cancer risk can be challenging because of the paucity of empiric data on surveillance or prevention in childhood. Existing guidelines are mostly based on expert consensus opinion (Brodeur et al. 2017). Studies have shown that parents often have an interest in pursuing genetic testing that could lead to enhanced medical management for their child or family (Lammens et al. 2010a; Rasmussen et al. 2010; Alderfer et al. 2015; McCullough et al. 2016; Brozou et al. 2018; Desrosiers et al. 2019). However, there are limited data on the emotional and financial burden that families face with ongoing surveillance and future studies are needed (Weber et al. 2019).

For the parents of a child with cancer, there is a high risk for psychological distress due to the cancer diagnosis (Pai et al. 2007), and the possibility of a hereditary component can be an additionally overwhelming (Lammens et al. 2010a). Parents fear for their child's future, worry about other children or parents, and may feel guilty or responsible for their child's diagnosis, but the benefits of testing often outweigh the potential downsides for parents who choose testing (McCullough et al. 2016; McGill et al. 2019; Scollon et al. 2019).

CONCLUSION

Cancer genetic services are evolving as they become more fully integrated into oncology care. In the past, these services were primarily provided at academic medical centers for those meeting specific personal and family history criteria. Testing to assess cancer risk is offered more broadly now through incorporation of novel service delivery models to reach increasingly diverse adult and pediatric patient populations. Opportunities to reduce morbidity and mortality for those with inherited cancer risk, as well as to stratify targeted therapy for cancer patients, requires equitable and appropriate access. Whereas numerous robust studies have provided promising results about the psychological well-being of those who have had genetic testing for inherited cancer risk, further studies will be critical as this testing extends to more diverse patient populations tested with genomic sequencing, who may have received only a brief educational intervention prior to testing. Genetic counselors will continue to facilitate interpretation of complex test results for both patients and providers and will remain critical to the delivery of high-quality, compassionate genetic services. Those with identifiable gene variants or significant family histories will continue to benefit from genetic counseling practices that help people understand and adapt to the clinical, psychological, and familial implications of genetic contributions to their disease.

REFERENCES

2015. Committee opinion no. 634: hereditary cancer syndromes and risk assessment. *Obstet Gynecol* **125:** 1538–1543. doi:10.1097/01.AOG.0000466373.71146.51

Abida W, Cheng ML, Armenia J, Middha S, Autio KA, Vargas HA, Rathkopf D, Morris MJ, Danila DC, Slovin SF, et al. 2019. Analysis of the prevalence of microsatellite instability in prostate cancer and response to immune checkpoint blockade. *JAMA Oncol* **5:** 471–478. doi:10.1001/jamaoncol.2018.5801

Alderfer MA, Zelley K, Lindell RB, Novokmet A, Mai PL, Garber JE, Nathan D, Scollon S, Chun NM, Patenaude AF, et al. 2015. Parent decision-making around the genetic testing of children for germline *TP53* mutations. *Cancer* **121:** 286–293. doi:10.1002/cncr.29027

Almqvist EW, Brinkman RR, Wiggins S, Hayden MR, Canadian Collaborative Study of Predictive Testing. 2003. Psychological consequences and predictors of adverse events in the first 5 years after predictive testing for Huntington's disease. *Clin Genet* **64:** 300–309. doi:10.1034/j.1399-0004.2003.00157.x

Amendola LM, Jarvik GP, Leo MC, McLaughlin HM, Akkari Y, Amaral MD, Berg JS, Biswas S, Bowling KM, Conlin LK, et al. 2016. Performance of ACMG-AMP variant-interpretation guidelines among nine laboratories in the Clinical Sequencing Exploratory Research Consortium. *Am J Hum Genet* **99:** 247. doi:10.1016/j.ajhg.2016.06.001

Athens BA, Caldwell SL, Umstead KL, Connors PD, Brenna E, Biesecker BB. 2017. A systematic review of randomized controlled trials to assess outcomes of genetic counseling. *J Genet Couns* **26:** 902–933. doi:10.1007/s10897-017-0082-y

Bast RC, Matulonis UA, Sood AK, Ahmed AA, Amobi AE, Balkwill FR, Wielgos-Bonvallet M, Bowtell DDL, Brenton JD, Brugge JS, et al. 2019. Critical questions in ovarian cancer research and treatment: report of an American Association for Cancer Research Special Conference. *Cancer* **125:** 1963–1972. doi:10.1002/cncr.32004

Beadles CA, Ryanne Wu R, Himmel T, Buchanan AH, Powell KP, Hauser E, Henrich VC, Ginsburg GS, Orlando LA. 2014. Providing patient education: impact on quantity and quality of family health history collection. *Fam Cancer* **13:** 325–332. doi:10.1007/s10689-014-9701-z

Bednar EM, Oakley HD, Sun CC, Burke CC, Munsell MF, Westin SN, Lu KH. 2017. A universal genetic testing initiative for patients with high-grade, non-mucinous epithelial ovarian cancer and the implications for cancer treatment. *Gynecol Oncol* **146:** 399–404. doi:10.1016/j.ygyno.2017.05.037

Beitsch PD, Whitworth PW, Hughes K, Patel R, Rosen B, Compagnoni G, Baron P, Simmons R, Smith LA, Grady I, et al. 2019. Underdiagnosis of hereditary breast cancer: are genetic testing guidelines a tool or an obstacle? *J Clin Oncol* **37:** 453–460. doi:10.1200/JCO.18.01631

Biesecker BB, Garber JE. 1995. Testing and counseling adults for heritable cancer risk. *J Natl Cancer Inst Monogr* **17:** 115–118.

Biesecker BB, Boehnke M, Calzone K, Markel DS, Garber JE, Collins FS, Weber BL. 1993. Genetic counseling for families with inherited susceptibility to breast and ovarian cancer. *J Am Med Assoc* **269:** 1970–1974. doi:10.1001/jama.1993.03500150082032

Bradbury AR, Patrick-Miller L, Long J, Powers J, Stopfer J, Forman A, Rybak C, Mattie K, Brandt A, Chambers R, et al. 2015. Development of a tiered and binned genetic counseling model for informed consent in the era of multiplex testing for cancer susceptibility. *Genet Med* **17:** 485–492. doi:10.1038/gim.2014.134

Bradbury AR, Patrick-Miller LJ, Egleston BL, DiGiovanni L, Brower J, Harris D, Stevens EM, Maxwell KN, Kulkarni A, Chavez T, et al. 2016. Patient feedback and early outcome data with a novel tiered-binned model for multiplex breast cancer susceptibility testing. *Genet Med* **18:** 25–33. doi:10.1038/gim.2015.19

Brodeur GM, Nichols KE, Plon SE, Schiffman JD, Malkin D. 2017. Pediatric cancer predisposition and surveillance: an overview, and a tribute to Alfred G. Knudson Jr. *Clin Cancer Res* **23:** e1–e5. doi:10.1158/1078-0432.CCR-17-0702

Brozou T, Taeubner J, Velleuer E, Dugas M, Wieczorek D, Borkhardt A, Kuhlen M. 2018. Genetic predisposition in children with cancer-affected families' acceptance of Trio-WES. *Eur J Pediatr* **177:** 53–60. doi:10.1007/s00431-017-2997-6

Bunnik EM, Janssens AC, Schermer MH. 2013. A tiered-layered-staged model for informed consent in personal genome testing. *Eur J Hum Genet* **21:** 596–601. doi:10.1038/ejhg.2012.237

Cannon-Albright LA, Thomas TC, Bishop DT, Skolnick MH, Burt RW. 1989. Characteristics of familial colon cancer in a large population data base. *Cancer* **64:** 1971–1975. doi:10.1002/1097-0142(19891101)64:9<1971::AID-CNCR2820640935>3.0.CO;2-L

Caswell-Jin JL, Zimmer AD, Stedden W, Kingham KE, Zhou AY, Kurian AW. 2019. Cascade genetic testing of relatives for hereditary cancer risk: results of an online initiative. *J Natl Cancer Inst* **111:** 95–98. doi:10.1093/jnci/djy147

Cohen R, Buhard O, Cervera P, Hain E, Dumont S, Bardier A, Bachet JB, Gornet JM, Lopez-Trabada D, Kaci R, et al. 2017. Clinical and molecular characterisation of hereditary and sporadic metastatic colorectal cancers harbouring microsatellite instability/DNA mismatch repair deficiency. *Eur J Cancer* **86:** 266–274. doi:10.1016/j.ejca.2017.09.022

Colombo N, Huang G, Scambia G, Chalas E, Pignata S, Fiorica J, Van Le L, Ghamande S, González-Santiago S, Bover I, et al. 2018. Evaluation of a streamlined oncologist-led *BRCA* mutation testing and counseling model for patients with ovarian cancer. *J Clin Oncol* **36:** 1300–1307. doi:10.1200/JCO.2017.76.2781

Committee on Bioethics, Committee on Genetics, and, the American College of Medical Genetics and, Genomics Social, Ethical, and Legal Issues Committee. 2013. Ethical and policy issues in genetic testing and screening of children. *Pediatrics* **131:** 620–622. doi:10.1542/peds.2012-3680

Committee on Practice Bulletins-Gynecology, Committee on Genetics, Society of Gynecologic Oncology. 2017. Practice bulletin No 182: hereditary breast and ovarian cancer syndrome. *Obstet Gynecol* **130:** e110–e126. doi:10.1097/AOG.0000000000002296

Conway JL, Connors JM, Tyldesley S, Savage KJ, Campbell BA, Zheng YY, Hamm J, Pickles T. 2017. Secondary breast cancer risk by radiation volume in women with Hodgkin lymphoma. *Int J Radiat Oncol Biol Phys* **97:** 35–41. doi:10.1016/j.ijrobp.2016.10.004

Coury SA, Schneider KA, Schienda J, Tan WH. 2018. Recognizing and managing children with a pediatric cancer predisposition syndrome: a guide for the pediatrician. *Pediatr Ann* **47:** e204–e216. doi:10.3928/19382359-20180424-02

Covolo L, Rubinelli S, Ceretti E, Gelatti U. 2015. Internet-based direct-to-consumer genetic testing: a systematic review. *J Med Internet Res* **17:** e279. doi:10.2196/jmir.4378

de Latour RP, Soulier J. 2016. How I treat MDS and AML in Fanconi anemia. *Blood* **127:** 2971–2979. doi:10.1182/blood-2016-01-583625

Desai AV, Perpich M, Godley LA. 2017. Clinical assessment and diagnosis of germline predisposition to hematopoietic malignancies: the University of Chicago experience. *Front Pediatr* **5:** 252. doi:10.3389/fped.2017.00252

Desmond A, Kurian AW, Gabree M, Mills MA, Anderson MJ, Kobayashi Y, Horick N, Yang S, Shannon KM, Tung N, et al. 2015. Clinical actionability of multigene panel testing for hereditary breast and ovarian cancer risk assessment. *JAMA Oncol* **1:** 943–951. doi:10.1001/jamaoncol.2015.2690

Desrosiers LR, Quinn E, Cramer S, Dobek W. 2019. Integrating genetic counseling and testing in the pediatric oncology setting: parental attitudes and influencing factors. *Pediatr Blood Cancer* **66:** e27907.

Diets IJ, Waanders E, Ligtenberg MJ, van Bladel DAG, Kamping EJ, Hoogerbrugge PM, Hopman S, Olderode-Berends MJ, Gerkes EH, Koolen DA, et al. 2018. High yield of pathogenic germline mutations causative or likely causative of the cancer phenotype in selected children with cancer. *Clin Cancer Res* **24:** 1594–1603. doi:10.1158/1078-0432.CCR-17-1725

Domchek SM, Friebel TM, Singer CF, Evans DG, Lynch HT, Isaacs C, Garber JE, Neuhausen SL, Matloff E, Eeles R, et al. 2010. Association of risk-reducing surgery in *BRCA1* or *BRCA2* mutation carriers with cancer risk and mortality. *J Am Med Assoc* **304:** 967–975. doi:10.1001/jama.2010.1237

Domchek SM, Bradbury A, Garber JE, Offit K, Robson ME. 2013. Multiplex genetic testing for cancer susceptibility: out on the high wire without a net? *J Clin Oncol* **31:** 1267–1270. doi:10.1200/JCO.2012.46.9403

Douglas FS, O'Dair LC, Robinson M, Evans DG, Lynch SA. 1999. The accuracy of diagnoses as reported in families with cancer: a retrospective study. *J Med Genet* **36:** 309–312.

Druker H, Zelley K, McGee RB, Scollon SR, Kohlmann WK, Schneider KA, Wolfe Schneider K. 2017. Genetic counselor recommendations for cancer predisposition evaluation and surveillance in the pediatric oncology patient. *Clin Cancer Res* **23:** e91–e97. doi:10.1158/1078-0432.CCR-17-0834

Eccles BK, Copson E, Maishman T, Abraham JE, Eccles DM. 2015. Understanding of BRCA VUS genetic results by breast cancer specialists. *BMC Cancer* **15:** 936. doi:10.1186/s12885-015-1934-1

Farber-Katz S, Hsuan V, Wu S, Landrith T, Vuong H, Xu D, Li B, Hoo J, Lam S, Nashed S, et al. 2018. Quantitative analysis of *BRCA1* and *BRCA2* germline splicing variants using a novel RNA-massively parallel sequencing assay. *Front Oncol* **8:** 286. doi:10.3389/fonc.2018.00286

Fishbein L, Merrill S, Fraker DL, Cohen DL, Nathanson KL. 2013. Inherited mutations in pheochromocytoma and paraganglioma: why all patients should be offered genetic testing. *Ann Surg Oncol* **20:** 1444–1450. doi:10.1245/s10434-013-2942-5

Geller G, Botkin JR, Green MJ, Press N, Biesecker BB, Wilfond B, Grana G, Daly MB, Schneider K, Kahn MJ. 1997. Genetic testing for susceptibility to adult-onset cancer. The process and content of informed consent. *J Am Med Assoc* **277:** 1467–1474. doi:10.1001/jama.1997.0354042006 3031

Gill J, Obley AJ, Prasad V. 2018. Direct-to-consumer genetic testing: the implications of the US FDA's first marketing authorization for BRCA mutation testing. *J Am Med Assoc* **319:** 2377–2378. doi:10.1001/jama.2018.5330

Goyal G, Fan T, Silberstein PT. 2016. Hereditary cancer syndromes: utilizing DNA repair deficiency as therapeutic target. *Fam Cancer* **15:** 359–366. doi:10.1007/s10689-016-9883-7

Halbert CH, Stopfer JE, McDonald J, Weathers B, Collier A, Troxel AB, Domchek S. 2011. Long-term reactions to genetic testing for *BRCA1* and *BRCA2* mutations: does time heal women's concerns? *J Clin Oncol* **29:** 4302–4306. doi:10.1200/JCO.2010.33.1561

Hampel H, Bennett RL, Buchanan A, Pearlman R, Wiesner GL; Guideline Development Group, American College of Medical Genetics and Genomics Professional Practice and Guidelines Committee and National Society of Genetic Counselors Practice Guidelines Committee. 2015. A practice guideline from the American College of Medical Genetics and Genomics and the National Society of Genetic Counselors: referral indications for cancer predisposition assessment. *Genet Med* **17:** 70–87. doi:10.1038/gim.2014.147

Hart MR, Biesecker BB, Blout CL, Christensen KD, Amendola LM, Bergstrom KL, Biswas S, Bowling KM, Brothers KB, Conlin LK, et al. 2019. Secondary findings from clinical genomic sequencing: prevalence, patient perspectives, family history assessment, and health-care costs from a multisite study. *Genet Med* **21:** 1100–1110. doi:10.1038/s41436-018-0308-x

Hoskins KF, Stopfer JE, Calzone KA, Merajver SD, Rebbeck TR, Garber JE, Weber BL. 1995. Assessment and counseling for women with a family history of breast cancer. A guide for clinicians. *J Am Med Assoc* **273:** 577–585. doi:10.1001/jama.1995.03520310075033

Hoskovec JM, Bennett RL, Carey ME, DaVanzo JE, Dougherty M, Hahn SE, LeRoy BS, O'Neal S, Richardson JG, Wicklund CA. 2018. Projecting the supply and demand for certified genetic counselors: a workforce study. *J Genet Couns* **27:** 16–20. doi:10.1007/s10897-017-0158-8

Järvinen HJ, Renkonen-Sinisalo L, Aktán-Collán K, Peltomäki P, Aaltonen LA, Mecklin JP. 2009. Ten years after mutation testing for Lynch syndrome: cancer incidence and outcome in mutation-positive and mutation-negative family members. *J Clin Oncol* **27:** 4793–4797. doi:10.1200/JCO.2009.23.7784

Kalia SS, Adelman K, Bale SJ, Chung WK, Eng C, Evans JP, Herman GE, Hufnagel SB, Klein TE, Korf BR, et al. 2017. Recommendations for reporting of secondary findings in clinical exome and genome sequencing, 2016 update (ACMG SF v2.0): a policy statement of the American College of Medical Genetics and Genomics. *Genet Med* **19:** 249–255. doi:10.1038/gim.2016.190

Kapoor NS, Curcio LD, Blakemore CA, Bremner AK, McFarland RE, West JG, Banks KC. 2015. Multigene panel testing detects equal rates of pathogenic *BRCA1/2* mutations and has a higher diagnostic yield compared to limited *BRCA1/2* analysis alone in patients at risk for hereditary breast cancer. *Ann Surg Oncol* **22:** 3282–3288. doi:10.1245/s10434-015-4754-2

Kattentidt-Mouravieva AA, den Heijer M, van Kessel I, Wagner A. 2014. How harmful is genetic testing for familial adenomatous polyposis (FAP) in young children; the parents' experience. *Fam Cancer* **13:** 391–399. doi:10.1007/s10689-014-9724-5

Kennedy AL, Shimamura A. 2019. Genetic predisposition to MDS: clinical features and clonal evolution. *Blood* **133:** 1071–1085. doi:10.1182/blood-2018-10-844662

Kesserwan C, Friedman Ross L, Bradbury AR, Nichols KE. 2016. The advantages and challenges of testing children for heritable predisposition to cancer. *Am Soc Clin Oncol Educ Book* **35:** 251–269. doi:10.1200/EDBK_160621

Kim YE, Ki CS, Jang MA. 2019. Challenges and considerations in sequence variant interpretation for Mendelian disorders. *Ann Lab Med* **39:** 421–429. doi:10.3343/alm.2019.39.5.421

King MC, Levy-Lahad E, Lahad A. 2014. Population-based screening for *BRCA1* and *BRCA2*: 2014 Lasker Award. *J Am Med Assoc* **312:** 1091–1092. doi:10.1001/jama.2014.12483

Kuhlen M, Taeubner J, Brozou T, Wieczorek D, Siebert R, Borkhardt A. 2019. Family-based germline sequencing in children with cancer. *Oncogene* **38:** 1367–1380. doi:10.1038/s41388-018-0520-9

Kurian AW, Ward KC, Hamilton AS, Deapen DM, Abrahamse P, Bondarenko I, Li Y, Hawley ST, Morrow M, Jagsi R, et al. 2018. Uptake, results, and outcomes of germline multiple-gene sequencing after diagnosis of breast cancer. *JAMA Oncol* **4:** 1066–1072. doi:10.1001/jamaoncol.2018.0644

LaDuca H, Stuenkel AJ, Dolinsky JS, Keiles S, Tandy S, Pesaran T, Chen E, Gau CL, Palmaer E, Shoaepour K, et al. 2014. Utilization of multigene panels in hereditary cancer predisposition testing: analysis of more than 2,000 patients. *Genet Med* **16:** 830–837. doi:10.1038/gim.2014.40

Lammens CR, Aaronson NK, Wagner A, Sijmons RH, Ausems MG, Vriends AH, Ruijs MW, van Os TA, Spruijt L, Gómez García EB, et al. 2010a. Genetic testing in Li-Fraumeni syndrome: uptake and psychosocial conse-

quences. *J Clin Oncol* **28**: 3008–3014. doi:10.1200/JCO
.2009.27.2112

Lammens CR, Bleiker EM, Aaronson NK, Wagner A, Sijmons RH, Ausems MG, Vriends AH, Ruijs MW, van Os TA, Spruijt L, et al. 2010b. Regular surveillance for Li-Fraumeni syndrome: advice, adherence and perceived benefits. *Fam Cancer* **9**: 647–654. doi:10.1007/s10689-010-9368-z

Lancaster JM, Powell CB, Chen LM, Richardson DL, SGO Clinical Practice Committee. 2015. Society of Gynecologic Oncology statement on risk assessment for inherited gynecologic cancer predispositions. *Gynecol Oncol* **136**: 3–7. doi:10.1016/j.ygyno.2014.09.009

Lee A, Mavaddat N, Wilcox AN, Cunningham AP, Carver T, Hartley S, Babb de Villiers C, Izquierdo A, Simard J, Schmidt MK, et al. 2019a. BOADICEA: a comprehensive breast cancer risk prediction model incorporating genetic and nongenetic risk factors. *Genet Med* **21**: 1708–1718. doi:10.1038/s41436-018-0406-9

Lee K, Seifert BA, Shimelis H, Ghosh R, Crowley SB, Carter NJ, Doonanco K, Foreman AK, Ritter DI, Jimenez S, et al. 2019b. Clinical validity assessment of genes frequently tested on hereditary breast and ovarian cancer susceptibility sequencing panels. *Genet Med* **21**: 1497–1506. doi:10.1038/s41436-018-0361-5

Lenders JW, Duh QY, Eisenhofer G, Gimenez-Roqueplo AP, Grebe SK, Murad MH, Naruse M, Pacak K, Young WF Jr; Endocrine Society. 2014. Pheochromocytoma and paraganglioma: an endocrine society clinical practice guideline. *J Clin Endocrinol Metab* **99**: 1915–1942. doi:10.1210/jc.2014-1498

Lerman C, Hughes C, Lemon SJ, Main D, Snyder C, Durham C, Narod S, Lynch HT. 1998. What you don't know can hurt you: adverse psychologic effects in members of *BRCA1*-linked and *BRCA2*-linked families who decline genetic testing. *J Clin Oncol* **16**: 1650–1654. doi:10.1200/JCO.1998.16.5.1650

Lynce F, Isaacs C. 2016. How far do we go with genetic evaluation? Gene, panel, and tumor testing. *Am Soc Clin Oncol Educ Book* **35**: e72–e78. doi:10.1200/EDBK_ 160391

Lynch HT, Follett KL, Lynch PM, Albano WA, Mailliard JL, Pierson RL. 1979. Family history in an oncology clinic. Implications for cancer genetics. *J Am Med Assoc* **242**: 1268–1272. doi:10.1001/jama.1979.03300120022017

Madlensky L, Trepanier AM, Cragun D, Lerner B, Shannon KM, Zierhut H. 2017. A rapid systematic review of outcomes studies in genetic counseling. *J Genet Couns* **26**: 361–378. doi:10.1007/s10897-017-0067-x

Manchanda R, Burnell M, Loggenberg K, Desai R, Wardle J, Sanderson SC, Gessler S, Side L, Balogun N, Kumar A, et al. 2016. Cluster-randomised non-inferiority trial comparing DVD-assisted and traditional genetic counselling in systematic population testing for *BRCA1/2* mutations. *J Med Genet* **53**: 472–480. doi:10.1136/jmedgenet-2015-103740

Mandelker D, Zhang L, Kemel Y, Stadler ZK, Joseph V, Zehir A, Pradhan N, Arnold A, Walsh MF, Li Y, et al. 2017. Mutation detection in patients with advanced cancer by universal sequencing of cancer-related genes in tumor and normal DNA vs guideline-based germline testing. *J Am Med Assoc* **318**: 825–835. doi:10.1001/jama.2017.11137

Manrai AK, Funke BH, Rehm HL, Olesen MS, Maron BA, Szolovits P, Margulies DM, Loscalzo J, Kohane IS. 2016. Genetic misdiagnoses and the potential for health disparities. *N Engl J Med* **375**: 655–665. doi:10.1056/NEJMsa1507092

Martin AR, Kanai M, Kamatani Y, Okada Y, Neale BM, Daly MJ. 2019. Clinical use of current polygenic risk scores may exacerbate health disparities. *Nat Genet* **51**: 584–591. doi:10.1038/s41588-019-0379-x

McCuaig JM, Armel SR, Care M, Volenik A, Kim RH, Metcalfe KA. 2018. Next-generation service delivery: a scoping review of patient outcomes associated with alternative models of genetic counseling and genetic testing for hereditary cancer. *Cancers (Basel)* **10**: E435. doi:10.3390/cancers10110435

McCuaig JM, Tone AA, Maganti M, Romagnuolo T, Ricker N, Shuldiner J, Rodin G, Stockley T, Kim RH, Bernardini MQ. 2019. Modified panel-based genetic counseling for ovarian cancer susceptibility: a randomized non-inferiority study. *Gynecol Oncol* **153**: 108–115. doi:10.1016/j.ygyno.2018.12.027

McCullough LB, Slashinski MJ, McGuire AL, Street RL Jr, Eng CM, Gibbs RA, Parsons DW, Plon SE. 2016. Is whole-exome sequencing an ethically disruptive technology? Perspectives of pediatric oncologists and parents of pediatric patients with solid tumors. *Pediatr Blood Cancer* **63**: 511–515. doi:10.1002/pbc.25815

McGill BC, Wakefield CE, Vetsch J, Lim Q, Warby M, Metcalfe A, Byrne JA, Cohn RJ, Tucker KM. 2019. "I remember how I felt, but I don't remember the gene": families' experiences of cancer-related genetic testing in childhood. *Pediatr Blood Cancer* **66**: e27762. doi:10.1002/pbc.27762

McGowan ML, Fishman JR, Settersten RA, Lambrix MA, Juengst ET. 2014. Gatekeepers or intermediaries? The role of clinicians in commercial genomic testing. *PLoS ONE* **9**: e108484. doi:10.1371/journal.pone.0108484

Metcalfe K, Lynch HT, Foulkes WD, Tung N, Kim-Sing C, Olopade OI, Eisen A, Rosen B, Snyder C, Gershman S, et al. 2015. Effect of oophorectomy on survival after breast cancer in *BRCA1* and *BRCA2* mutation carriers. *JAMA Oncol* **1**: 306–313. doi:10.1001/jamaoncol.2015.0658

Murakami Y, Okamura H, Sugano K, Yoshida T, Kazuma K, Akechi T, Uchitomi Y. 2004. Psychologic distress after disclosure of genetic test results regarding hereditary nonpolyposis colorectal carcinoma. *Cancer* **101**: 395–403. doi:10.1002/cncr.20363

Nilsson MP, Nilsson ED, Borg Å, Brandberg Y, Silfverberg B, Loman N. 2019. High patient satisfaction with a simplified *BRCA1/2* testing procedure: long-term results of a prospective study. *Breast Cancer Res Treat* **173**: 313–318. doi:10.1007/s10549-018-5000-y

Offit K, Brown K. 1994. Quantitating familial cancer risk: a resource for clinical oncologists. *J Clin Oncol* **12**: 1724–1736. doi:10.1200/JCO.1994.12.8.1724

Ozanne EM, Loberg A, Hughes S, Lawrence C, Drohan B, Semine A, Jellinek M, Cronin C, Milham F, Dowd D, et al. 2009. Identification and management of women at high risk for hereditary breast/ovarian cancer syndrome. *Breast J* **15**: 155–162. doi:10.1111/j.1524-4741.2009.00690.x

Pai AL, Greenley RN, Lewandowski A, Drotar D, Youngstrom E, Peterson CC. 2007. A meta-analytic review of

Cite this article as *Cold Spring Harb Perspect Med* doi: 10.1101/cshperspect.a036541

the influence of pediatric cancer on parent and family functioning. *J Fam Psychol* **21:** 407–415. doi:10.1037/0893-3200.21.3.407

Parsons DW, Roy A, Yang Y, Wang T, Scollon S, Bergstrom K, Kerstein RA, Gutierrez S, Petersen AK, Bavle A, et al. 2016. Diagnostic yield of clinical tumor and germline whole-exome sequencing for children with solid tumors. *JAMA Oncol* **2:** 616–624. doi:10.1001/jamaoncol.2015.5699

Peters JA, Stopfer JE. 1996. Role of the genetic counselor in familial cancer. *Oncology (Williston Park)* **10:** 159–166, 175; discussion 176-156, 178.

Phillips KA, Deverka PA, Hooker GW, Douglas MP. 2018. Genetic test availability and spending: where are we now? Where are we going? *Health Aff (Millwood)* **37:** 710–716. doi:10.1377/hlthaff.2017.1427

Phillips KA, Trosman JR, Douglas MP. 2019. Emergence of hybrid models of genetic testing beyond direct-to-consumer or traditional labs. *J Am Med Assoc* **321:** 2403–2404. doi:10.1001/jama.2019.5670

Plichta JK, Sebastian ML, Smith LA, Menendez CS, Johnson AT, Bays SM, Euhus DM, Clifford EJ, Jalali M, Kurtzman SH, et al. 2019. Germline genetic testing: what the breast surgeon needs to know. *Ann Surg Oncol* **26:** 2184–2190. doi:10.1245/s10434-019-07341-8

Ponder BA. 1994. Setting up and running a familial cancer clinic. *Br Med Bull* **50:** 732–745. doi:10.1093/oxfordjournals.bmb.a072921

Powis Z, Farwell Hagman KD, Speare V, Cain T, Blanco K, Mowlavi LS, Mayerhofer EM, Tilstra D, Vedder T, Hunter JM, et al. 2018. Exome sequencing in neonates: diagnostic rates, characteristics, and time to diagnosis. *Genet Med* **20:** 1468–1471. doi:10.1038/gim.2018.11

Pyeritz RE. 2012. The family history: the first genetic test, and still useful after all those years? *Genet Med* **14:** 3–9. doi:10.1038/gim.0b013e3182310bcf

Rana HQ, Clifford J, Hoang L, LaDuca H, Helen Black M, Li S, McGoldrick K, Speare V, Dolinsky JS, Gau CL, et al. 2019. Genotype–phenotype associations among panel-based TP53[+] subjects. *Genet Med.* doi:10.1038/s41436-019-0541-v

Rasmussen A, Alonso E, Ochoa A, De Biase I, Familiar I, Yescas P, Sosa AL, Rodríguez Y, Chávez M, López-López M, et al. 2010. Uptake of genetic testing and long-term tumor surveillance in von Hippel–Lindau disease. *BMC Med Genet* **11:** 4. doi:10.1186/1471-2350-11-4

Resta RG. 2006. Defining and redefining the scope and goals of genetic counseling. *Am J Med Genet C Semin Med Genet* **142C:** 269–275. doi:10.1002/ajmg.c.30093

Resta R, Biesecker BB, Bennett RL, Blum S, Hahn SE, Strecker MN, Williams JL, National Society of Genetic Counselors' Definition Task Force. 2006. A new definition of genetic counseling: National Society of Genetic Counselors' Task Force report. *J Genet Couns* **15:** 77–83. doi:10.1007/s10897-005-9014-3

Richards S, Aziz N, Bale S, Bick D, Das S, Gastier-Foster J, Grody WW, Hegde M, Lyon E, Spector E, et al. 2015. Standards and guidelines for the interpretation of sequence variants: a joint consensus recommendation of the American College of Medical Genetics and Genomics and the Association for Molecular Pathology. *Genet Med* **17:** 405–424. doi:10.1038/gim.2015.30

Robinson JO, Wynn J, Biesecker B, Biesecker LG, Bernhardt B, Brothers KB, Chung WK, Christensen KD, Green RC, McGuire AL, et al. 2019. Psychological outcomes related to exome and genome sequencing result disclosure: a meta-analysis of seven Clinical Sequencing Exploratory Research (CSER) Consortium studies. *Genet Med.* doi:10.1038/s41436-019-0565-3

Robson ME, Storm CD, Weitzel J, Wollins DS, Offit K; American Society of Clinical Oncology. 2010. American Society of Clinical Oncology policy statement update: genetic and genomic testing for cancer susceptibility. *J Clin Oncol* **28:** 893–901. doi:10.1200/JCO.2009.27.0660

Robson ME, Bradbury AR, Arun B, Domchek SM, Ford JM, Hampel HL, Lipkin SM, Syngal S, Wollins DS, Lindor NM. 2015. American Society of Clinical Oncology Policy Statement Update: genetic and genomic testing for cancer susceptibility. *J Clin Oncol* **33:** 3660–3667. doi:10.1200/JCO.2015.63.0996

Robson M, Im SA, Senkus E, Xu B, Domchek SM, Masuda N, Delaloge S, Li W, Tung N, Armstrong A, et al. 2017. Olaparib for metastatic breast cancer in patients with a germline *BRCA* mutation. *N Engl J Med* **377:** 523–533. doi:10.1056/NEJMoa1706450

Rosenthal ET, Bernhisel R, Brown K, Kidd J, Manley S. 2017. Clinical testing with a panel of 25 genes associated with increased cancer risk results in a significant increase in clinically significant findings across a broad range of cancer histories. *Cancer Genet* **218-219:** 58–68. doi:10.1016/j.cancergen.2017.09.003

Rowland E, Metcalfe A. 2013. Communicating inherited genetic risk between parent and child: a meta-thematic synthesis. *Int J Nurs Stud* **50:** 870–880. doi:10.1016/j.ijnurstu.2012.09.002

Rubenstein JH, Enns R, Heidelbaugh J, Barkun A; Clinical Guidelines Committee. 2015. American Gastroenterological Association Institute guideline on the diagnosis and management of Lynch syndrome. *Gastroenterology* **149:** 777–782; quiz e716-777. doi:10.1053/j.gastro.2015.07.036

Schwartz MD, Valdimarsdottir HB, Peshkin BN, Mandelblatt J, Nusbaum R, Huang AT, Chang Y, Graves K, Isaacs C, Wood M, et al. 2014. Randomized noninferiority trial of telephone versus in-person genetic counseling for hereditary breast and ovarian cancer. *J Clin Oncol* **32:** 618–626. doi:10.1200/JCO.2013.51.3226

Scollon S, Anglin AK, Thomas M, Turner JT, Wolfe Schneider K. 2017. A comprehensive review of pediatric tumors and associated cancer predisposition syndromes. *J Genet Couns* **26:** 387–434. doi:10.1007/s10897-017-0077-8

Scollon S, Majumder MA, Bergstrom K, Wang T, McGuire AL, Robinson JO, Gutierrez AM, Lee CH, Hilsenbeck SG, Plon SE, et al. 2019. Exome sequencing disclosures in pediatric cancer care: patterns of communication among oncologists, genetic counselors, and parents. *Patient Educ Couns* **102:** 680–686. doi:10.1016/j.pec.2018.11.007

Senter L, O'Malley DM, Backes FJ, Copeland LJ, Fowler JM, Salani R, Cohn DE. 2017. Genetic consultation embedded in a gynecologic oncology clinic improves compliance with guideline-based care. *Gynecol Oncol* **147:** 110–114. doi:10.1016/j.ygyno.2017.07.141

Shepet K, Alhefdhi A, Lai N, Mazeh H, Sippel R, Chen H. 2013. Hereditary medullary thyroid cancer: age-appropriate thyroidectomy improves disease-free survival. *Ann Surg Oncol* **20**: 1451–1455. doi:10.1245/s10434-012-2757-9

Sie AS, van Zelst-Stams WA, Spruijt L, Mensenkamp AR, Ligtenberg MJ, Brunner HG, Prins JB, Hoogerbrugge N. 2014. More breast cancer patients prefer BRCA-mutation testing without prior face-to-face genetic counseling. *Fam Cancer* **13**: 143–151.

Slavin TP, Manjarrez S, Pritchard CC, Gray S, Weitzel JN. 2019. The effects of genomic germline variant reclassification on clinical cancer care. *Oncotarget* **10**: 417–423. doi:10.18632/oncotarget.26501

So MK, Jeong TD, Lim W, Moon BI, Paik NS, Kim SC, Huh J. 2019. Reinterpretation of *BRCA1* and *BRCA2* variants of uncertain significance in patients with hereditary breast/ovarian cancer using the ACMG/AMP 2015 guidelines. *Breast Cancer (Auckl)* **26**: 510–519. doi:10.1007/s12282-019-00951-w

Steensma DP. 2018. Clinical consequences of clonal hematopoiesis of indeterminate potential. *Blood Adv* **2**: 3404–3410. doi:10.1182/bloodadvances.2018020222

Stoffel EM, Mangu PB, Gruber SB, Hamilton SR, Kalady MF, Lau MW, Lu KH, Roach N, Limburg PJ; American Society of Clinical Oncology, et al. 2015. Hereditary colorectal cancer syndromes: American Society of Clinical Oncology Clinical Practice Guideline endorsement of the familial risk-colorectal cancer: European Society for Medical Oncology Clinical Practice Guidelines. *J Clin Oncol* **33**: 209–217. doi:10.1200/JCO.2014.58.1322

Stoffel EM, McKernin SE, Brand R, Canto M, Goggins M, Moravek C, Nagarajan A, Petersen GM, Simeone DM, Yurgelun M, et al. 2019a. Evaluating susceptibility to pancreatic cancer: ASCO provisional clinical opinion. *J Clin Oncol* **37**: 153–164. doi:10.1200/JCO.18.01489

Stoffel EM, McKernin SE, Khorana AA. 2019b. Evaluating susceptibility to pancreatic cancer: ASCO Clinical Practice Provisional Clinical Opinion Summary. *J Oncol Pract* **15**: 108–111. doi:10.1200/JOP.18.00629

Stopfer JE. 2000. Genetic counseling and clinical cancer genetics services. *Semin Surg Oncol* **18**: 347–357. doi:10.1002/(SICI)1098-2388(200006)18:4<347::AID-SSU10>3.0.CO;2-D

Sylvester DE, Chen Y, Jamieson RV, Dalla-Pozza L, Byrne JA. 2018. Investigation of clinically relevant germline variants detected by next-generation sequencing in patients with childhood cancer: a review of the literature. *J Med Genet* **55**: 785–793. doi:10.1136/jmedgenet-2018-105488

Syngal S, Brand RE, Church JM, Giardiello FM, Hampel HL, Burt RW; American College of Gastroenterology. 2015. ACG clinical guideline: genetic testing and management of hereditary gastrointestinal cancer syndromes. *Am J Gastroenterol* **110**: 223–262; quiz 263. doi:10.1038/ajg.2014.435

Tandy-Connor S, Guiltinan J, Krempely K, LaDuca H, Reineke P, Gutierrez S, Gray P, Tippin Davis B. 2018. False-positive results released by direct-to-consumer genetic tests highlight the importance of clinical confirmation testing for appropriate patient care. *Genet Med* **20**: 1515–1521. doi:10.1038/gim.2018.38

Tandy-Connor S, Krempely K, Pesaran T, LaDuca H, Guiltinan J, Davis BT. 2019. Advocating for the consumer: clinical confirmation of all direct-to-consumer raw data

alterations remains critical. *Genet Med* **21**: 760–761. doi:10.1038/s41436-018-0095-4

Telli ML, Gradishar WJ, Ward JH. 2019. NCCN guidelines updates: breast cancer. *J Natl Compr Canc Netw* **17**: 552–555.

Trepanier A, Ahrens M, McKinnon W, Peters J, Stopfer J, Grumet SC, Manley S, Culver JO, Acton R, Larsen-Haidle J, et al. 2004. Genetic cancer risk assessment and counseling: recommendations of the National Society of Genetic Counselors. *J Genet Couns* **13**: 83–114. doi:10.1023/B:JOGC.0000018821.48330.77

Tsai GJ, Rañola JMO, Smith C, Garrett LT, Bergquist T, Casadei S, Bowen DJ, Shirts BH. 2019. Outcomes of 92 patient-driven family studies for reclassification of variants of uncertain significance. *Genet Med* **21**: 1435–1442. doi:10.1038/s41436-018-0335-7

Tung N, Battelli C, Allen B, Kaldate R, Bhatnagar S, Bowles K, Timms K, Garber JE, Herold C, Ellisen L, et al. 2015. Frequency of mutations in individuals with breast cancer referred for *BRCA1* and *BRCA2* testing using next-generation sequencing with a 25-gene panel. *Cancer* **121**: 25–33. doi:10.1002/cncr.29010

Tutty E, Petelin L, McKinley J, Young MA, Meiser B, Rasmussen VM, Forbes Shepherd R, James PA, Forrest LE. 2019. Evaluation of telephone genetic counselling to facilitate germline *BRCA1/2* testing in women with high-grade serous ovarian cancer. *Eur J Hum Genet* **27**: 1186–1196. doi:10.1038/s41431-019-0390-9

Valdez JM, Walker B, Ogg S, Gattuso J, Alderfer MA, Zelley K, Ford CA, Baker JN, Mandrell BN, Nichols KE. 2018. Parent-child communication surrounding genetic testing for Li–Fraumeni syndrome: living under the cloud of cancer. *Pediatr Blood Cancer* **65**: e27350. doi:10.1002/pbc.27350

van Oostrom I, Meijers-Heijboer H, Duivenvoorden HJ, Bröcker-Vriends AH, van Asperen CJ, Sijmons RH, Seynaeve C, Van Gool AR, Klijn JG, Tibben A. 2007. Prognostic factors for hereditary cancer distress six months after *BRCA1/2* or HNPCC genetic susceptibility testing. *Eur J Cancer* **43**: 71–77. doi:10.1016/j.ejca.2006.08.023

Weber E, Shuman C, Wasserman JD, Barrera M, Patenaude AF, Fung K, Chitayat D, Malkin D, Druker H. 2019. "A change in perspective": exploring the experiences of adolescents with hereditary tumor predisposition. *Pediatr Blood Cancer* **66**: e27445. doi:10.1002/pbc.27445

Weitzel JN, Chao EC, Nehoray B, Van Tongeren LR, LaDuca H, Blazer KR, Slavin T, Pesaran T, Rybak C, Solomon I, et al. 2018. Somatic *TP53* variants frequently confound germ-line testing results. *Genet Med* **20**: 809–816. doi:10.1038/gim.2017.196

Wells SA, Asa SL, Dralle H, Elisei R, Evans DB, Gagel RF, Lee N, Machens A, Moley JF, Pacini F, et al. 2015. Revised American Thyroid Association guidelines for the management of medullary thyroid carcinoma. *Thyroid* **25**: 567–610. doi:10.1089/thy.2014.0335

Werner-Lin A, Merrill SL, Brandt AC, Barnett RE, Matloff ET. 2018. Talking with children about adult-onset hereditary cancer risk: a developmental approach for parents. *J Genet Couns* **27**: 533–548. doi:10.1007/s10897-017-0191-7

Xicola RM, Li S, Rodriguez N, Reinecke P, Karam R, Speare V, Black MH, LaDuca H, Llor X. 2019. Clinical features and cancer risk in families with pathogenic *CDH1* vari-

Cite this article as *Cold Spring Harb Perspect Med* doi: 10.1101/cshperspect.a036541

ants irrespective of clinical criteria. *J Med Genet.* doi:10
.1136/jmedgenet-2019-105991

Yurgelun MB, Allen B, Kaldate RR, Bowles KR, Judkins T, Kaushik P, Roa BB, Wenstrup RJ, Hartman AR, Syngal S. 2015a. Identification of a variety of mutations in cancer predisposition genes in patients with suspected Lynch syndrome. *Gastroenterology* **149:** 604–613.e20. doi:10 .1053/j.gastro.2015.05.006

Yurgelun MB, Hiller E, Garber JE. 2015b. Population-wide screening for germline *BRCA1* and *BRCA2* mutations: too

much of a good thing? *J Clin Oncol* **33:** 3092–3095. doi:10 .1200/JCO.2015.60.8596

Zhang J, Walsh MF, Wu G, Edmonson MN, Gruber TA, Easton J, Hedges D, Ma X, Zhou X, Yergeau DA, et al. 2015. Germline mutations in predisposition genes in pediatric cancer. *N Engl J Med* **373:** 2336–2346. doi:10.1056/ NEJMoa1508054

Zhang L, Peng Y, Peng G. 2018. Mismatch repair-based stratification for immune checkpoint blockade therapy. *Am J Cancer Res* **8:** 1977–1988.

Tumor-Based Genetic Testing and Familial Cancer Risk

Andrea Forman[1] and Jilliane Sotelo[2]

[1]Department of Clinical Genetics, Risk Assessment Program, Fox Chase Cancer Center, Philadelphia, Pennsylvania 19111, USA

[2]Center for Cancer Genetics and Prevention, Dana Farber Cancer Institute, Boston, Massachusetts 02215, USA

Correspondence: AFormanCGC@gmail.com

As genetic testing on somatic tumor tissue becomes a more routine part of personalized cancer treatment, a growing opportunity arises to identify hereditary germline variants within those results. These germline results can affect future cancer screening for both patients and their family members. Finding this germline information can be complicated as a result of differences between somatic and germline testing processes, nomenclature, and outcome goals (e.g., treatment impact). The goal of this review is to highlight differences between somatic and germline testing and outline a potential guide to allow for appropriate clinical interpretation of somatic testing results in order to better facilitate genetic counseling referrals and confirmatory germline testing.

Tumor characteristics have long guided cancer treatment and follow-up care. As awareness of hereditary cancer risk has grown, correlations between histopathologic tumor features and prevalence of germline cancer risk variants have also been noted. Triple-negative breast cancers are more likely to carry *BRCA1* variants compared to estrogen-positive tumors (Peshkin et al. 2010; Krammer et al. 2017). Specific renal cancer histology may suggest particular hereditary renal cancer risk genes, such as oncocytic chromophobe tumors associated with variants in the *FLCN* gene (Peng and Chen 2018). Markers of mismatch repair deficiency were found to include microsatellite instability and loss of expression of certain proteins on immunohistochemistry (IHC), leading to recommendations for universal screening protocols for colon and endometrial tumors to screen for Lynch syndrome (de la Chapelle and Hampel 2010).

Recent technologies allow for genomic sequencing of tumor tissue with an increased understanding of the tumor genome and its impact on personalized cancer treatment. A growing number of therapies can target certain somatic variants, and tumor genetic testing is becoming a more integral part of cancer treatment. As this testing becomes more widely used, clinicians must be familiar with the potential for somatic tumor testing to identify hereditary cancer risks and recognize the need to facilitate confirmatory testing through genetic counseling. Knowledge of hereditary cancer risk can help patients determine future screening and provide family members with valuable information (Jain et al. 2016).

THE MISMATCH REPAIR PATHWAY AND LYNCH SYNDROME

Detection of hereditary cancer risk often begins with a family history–based assessment and various criteria assist in determining personal and family history most likely to yield informative genetic testing results. However, such criteria can miss at-risk families. For example, in Lynch syndrome, also called hereditary nonpolyposis colorectal cancer syndrome (HNPCC), family history–based guidelines may miss up to 39% of at-risk families (Syngal et al. 2000). Pathogenic variants in these genes (*MLH1*, *MSH2*, *MSH6*, *PMS2*, and *EPCAM*) cause mismatch repair deficiency (MMRD) and lead to distinctive tumor characteristics including microsatellite instability (MSI) and IHC-detected loss of protein expression from the relevant genes (de la Chapelle and Hampel 2010), which provide tumor-based avenues to identify at-risk patients, including those missed by family history.

Microsatellite Instability

Microsatellites are common regions of chromosomal DNA with repeating sequences of DNA nucleotides. These repeat sequences can expand in the setting of MMRD, such as those caused by Lynch syndrome, and present with high levels of MSI. Approximately 20% of uterine cancers will be MSI-high, with <10% of those associated with a diagnosis of Lynch syndrome (Hampel et al. 2006). Approximately 15% of colorectal tumors will show high MSI, with about one-third of those having Lynch syndrome (de la Chapelle and Hampel 2010). Not all MSI-high tumors will have Lynch syndrome as other events, particularly somatic hypermethylation of the *MLH1* promoter region and biallelic somatic inactivation of one of the genes associated with Lynch syndrome, can also lead to MSI-high findings (Salvador et al. 2019).

MSI testing has historically been a polymerase chain reaction (PCR)-based panel focusing on five loci identified by the National Cancer Institute (Boland et al. 1998). Variations have been implemented in many pathology departments as universal screening protocols have become more prevalent. As next-generation sequencing (NGS) technologies have expanded in both the tumor and germline genetic testing setting, the ability to detect microsatellite repeats using NGS methodology has been confirmed to have comparable sensitivity and specificity, allowing this information to be incorporated into some tumor sequencing test reports (Salipante et al. 2014).

Immunohistochemistry

Protein expression testing by IHC is also a highly sensitive screening tool for Lynch syndrome. Cancers that develop because of inactivation of the MMR genes typically show a complete lack of protein expression for those genes within the tumor (de la Chapelle and Hampel 2010). In Lynch syndrome, the patient inherits a single pathogenic variant on one of two alleles inherited from the parents. As part of the tumorigenesis process, the wild-type allele will characteristically become nonfunctional (through an acquired somatic variant or loss of heterozygosity) leading to a lack of protein expression in tumor tissue, whereas normal tissue will retain detectable protein levels because of one functional wild-type allele. Abnormal IHC results can also be acquired, similar to high microsatellite instability, through acquired biallelic somatic variants and hypermethylation of the *MLH1* promoter region (Salvador et al. 2019).

IHC is often the preferred method of routine Lynch screening as it is convenient, inexpensive, and makes use of supplies and expertise readily available in most pathology departments. Also, unlike MSI, IHC allows for more specificity in identifying the MMR gene likely affected. MMR proteins act as heterodimers, with MLH1 proteins typically pairing with PMS2 and MSH2 working with MSH6 (Boland et al. 2008). Nonfunctioning *MLH1* genes (whether through promoter hypermethylation, biallelic somatic variants, or germline *MLH1* variants) will typically lead to loss of both MLH1 and PMS2 protein expression within tumor cells. Inactivation

Cite this article as *Cold Spring Harb Perspect Med* doi: 10.1101/cshperspect.a036590

Table 1. Simplified summary of tumor microsatellite instability (MSI) and immunohistochemistry (IHC) assessment for Lynch Syndrome (LS)

IHC				MSI	Possible explanation
MLH1	MSH2	MSH6	PMS2		
+	+	+	+	MSS or low	Sporadic cancer
					Possibly non-Lynch syndrome cancer syndrome
N/a	N/a	N/a	N/a	High	Germline mutation in any LS gene
					Sporadic cancer (MPH, DSM, or other etiology)
−	+	+	−	N/a	Germline mutation—typically *MLH1*, rarely *PMS2*
					BRAF mutation (colon tumor only) or MPH
					DSM in *MLH1* and/or *PMS2*
+	−	−	+	N/a	Germline mutation—typically *MSH2/EPCAM*, rarely *MSH6*
					DSM in *MSH2* and/or *MSH6*
+	+	+	−	N/a	Germline mutation in *PMS2*, rarely in *MLH1*
					DSM in *PMS2* or (rarely) *MLH1*
+	+	−	+	N/a	Germline mutation in *MSH6*, rarely in *MSH2*
					DSM in *MSH6* or (rarely) *MSH2*

(MPH) MLH1 promoter hypermethylation, (DSMs) double somatic mutations.

of only the *PMS2* gene will often allow expression of the MLH1 protein and PMS2 protein expression alone absent on IHC. Similarly, *MSH2* inactivation will lead to loss of both MSH2 and MSH6 protein expression, whereas inactivation of *MSH6* will only affect MSH6 protein expression. It is important to note that MSH6 protein expression can be decreased compared to pretreatment levels in tissue that has undergone chemoradiation (Goldstein et al. 2017), leading to a false report of absence of MSH6 protein in a biopsy or resection of the treated tissue. The National Comprehensive Cancer Network has summarized common explanations for loss of IHC and follow-up recommendations based on IHC and MSI results, a modified form of which is available in Table 1 (NCCN 1.2018).

Universal MMR Screening

Routine screening of colorectal and endometrial tumors for microsatellite stability and IHC has been shown to increase detection of Lynch syndrome beyond family history–based guidelines and has been encouraged as part of routine care in the treatment of colorectal and uterine cancer (Hampel et al. 2008; EGAPP 2009). Universal screening, and subsequent identification

of Lynch syndrome, also allows for "cascade testing" for family members unaffected by cancer, increasing cost-effectiveness of such testing and providing health benefits to extended relatives (Hampel 2016). MMRD also has a growing impact on cancer treatment with the recent discovery of beneficial immunotherapies in MMRD tumors of various types (Le et al. 2015), supporting the need for even broader MMRD screening.

As both tumor gene sequencing and identification of MMRD have a growing impact on cancer treatment, it has been posited that up-front tumor sequencing may effectively replace screening programs focusing only on IHC and/or MSI testing. Identifying Lynch syndrome through IHC/MSI screening programs can require multiple steps over an extended period of time (i.e., tumor screening, *BRAF* testing, *MLH1* hypermethylation testing, and finally germline testing). A recent study found that test sensitivity improved with up-front tumor testing and allowed for identification of additional treatment-related information, such as *KRAS* status and tumor mutational burden, compiled within one test (Hampel et al. 2018). Feasibility and cost-effectiveness of this method compared to in-house IHC/MSI testing has yet to be determined.

TUMOR SEQUENCING ANALYSIS AND IMPLICATIONS FOR GERMLINE ALTERATION DETECTION

Somatic gene sequencing analyses are most frequently performed by NGS or massively parallel sequencing. This technology replaced single-target detection assays (e.g., Sanger sequencing) with simultaneous evaluation of many genes, utilizing millions of short nucleic acids sequencing in parallel (Bentley et al. 2008; Shendure and Ji 2008). However, the application of this technology has been inconsistent. In a 2015 survey by a clinical laboratory–focused working group from the Association for Molecular Pathology, responders reported using NGS testing in panels ranging from 1 to 10 genes to >100. A small number of laboratories reported performing exome (12%) or genome (5%) analysis. Although all participants reported small-nucleotide variants (SNVs), or the change of a single nucleotide, there was considerably more variability in reporting of copy number variants (CNVs). Only 35%–37% of participating laboratories analyzed for CNVs (Li et al. 2017).

METHODS OF ANALYSIS (TUMOR GENE SEQUENCING)

Analysis is typically carried out using amplicon-based methods, either through vendor-created tools with varying levels of customization or through targeted enrichment with hybridization capture (Cheng et al. 2015). Amplicon methods use PCR to select for desired sequences by using primers that target a particular gene or stretch of DNA. This process creates short segments of DNA for analysis, called amplicons. Amplicon-based methods without hybridization capture are typically more affordable and require less clinical laboratory investment. They are convenient, particularly for smaller gene panels, but susceptible to artifacts of random sequence mismatch and uneven sequence coverage across areas of interest (Nikiforova et al. 2013; Singh et al. 2013; Lin et al. 2014; Luthra et al. 2014; Tsongalis et al. 2014). Additional challenges arise as certain variants may not be detected due to poor coverage in an amplicon-based system or not reported because of quality control methods.

Some analyses use hybridization capture. For these assays, amplicon-based NGS is performed, and then biotinylated probes are used to specifically bind to areas of interest within the desired sequences. These areas may be known sequences with typically poor coverage, areas with pseudogene homology, or any other stretch of DNA for which additional amplicons for analysis may be desired. Once, the probes are bound and pulled down, one is left with a more specific library of sequences of interest. This methodology can create more even sequence coverage and reduce the amount of mismatch artifact. For these reasons, large panels have better performance when utilizing a hybridization capture tool with which more accurate estimates of CNVs can be made. However, this frequently involves significant infrastructure and bioinformatics investment (Cheng et al. 2015). The additive value of this investment may not be viewed as important in the treatment setting based on the goals of the institution and the analysis.

Deletion and duplication analysis using an NGS platform can be challenging. Probes must be selected appropriately to allow for even coverage and reduction of pseudogene homology, which is sometimes very challenging. The ability to detect a single exon deletion is quite good with higher levels of sequencing in the flanking intronic regions. However, most NGS assays used for somatic sequencing focus on exonic regions only. Depending on the bioinformatic tool used, deletions of up to 25 base pairs are easy to analyze provided the whole surrounding area has been sequenced. However, larger insertions and deletions can lead to errors in mapping because the aligner tool, the bioinformatics tool that matches the reported sequence to the expected sequence, must contend with large extra or missing portions of DNA when comparing the two sequences. This can lead to lack of alignment or misalignment of the sequence and therefore a possible missed variant call (Cheng et al. 2015, 2017). Understanding which methodology was used for testing is an important distinction when considering the

Cite this article as *Cold Spring Harb Perspect Med* doi: 10.1101/cshperspect.a036590

ability of a somatic analysis to identify a germline variant.

GERMLINE IMPLICATIONS FROM SOMATIC TESTING

For the clinical genetic counselor or other health-care provider, a primary concern when considering a somatic gene report is whether there are germline implications. The IMPACT program at Memorial Sloan Kettering, which offers paired somatic and germline analysis to participants, found ~15% of unselected individuals harbor a pathogenic germline variant. Discordance between tumor type and cancer susceptibility gene were identified in 60% of individuals, and about one-fourth had loss of the second allele (Schrader et al. 2016), suggesting that paired analysis may be a tool for opportunistic hereditary cancer screening in cancer patients who might not otherwise be identified through personal and family history assessment.

However, most cancer patients in the United States do not receive paired analysis when undergoing somatic testing. Using a somatic report as a screening tool for germline alterations can be complicated. Utilization is made even more overwhelming when considering that the mutational burden of these tumors can be high; more than 100 sequencing alterations and CNVs can be identified within a single sample. Studies have suggested that 20% of patients with ovarian cancer have a somatic gene variant associated with the Fanconi anemia/DNA repair pathway (Kanchi et al. 2014), and *TP53* variants were identified in >40% of all tumors analyzed in multiple organs (Zehir et al. 2017). Considering that the recent American Society of Clinical Oncology (ASCO) guidelines recommend genetic testing for all somatic *BRCA1* and *BRCA2* carriers, the ability to parse out germline alterations is both important and timely.

SOMATIC ANALYSIS COMPARED TO GERMLINE

When looking at a somatic report for germline implications, it is important to identify a few key items: intentional germline filtering, transcript utilization, tumor heterogeneity and purity, allelic fraction, possible founder variants, genes atypical to the tumorigenesis process in the organ, and tumor mutational burden. Differences in the variant classification process between germline and somatic variants should also be kept in mind.

Filtration

In some cases, tumor analysis intentionally filters out germline variants. Previously, there have been instances in which laboratories felt that patients were not adequately consented to receive incidental germline findings and therefore excluded these variants from the analysis (Mandelker and Zhang 2018). Other laboratories intentionally filter out suspected or known germline variants so that they can report on pure somatic alterations, thought to be most active in the tumor and therefore the best targets for therapy (Raymond et al. 2016; Mandelker and Zhang 2018). Therefore, it is possible that a known pathogenic germline variant may not appear on the somatic report if not thought to be clinically significant for the treatment of the cancer.

Transcript Use

Joint Commission Guidelines recommend that the canonical transcript (determined by consensus, longest coding sequence [CDS] or longest cDNA as listed in Ensembl [Zerbino et al. 2018] and University of California, Southern California (UCSC) Genome Browser [Kent et al. 2002]) be used for somatic as well as germline analysis (Li et al. 2017). However, because of the large number of aberrations in the somatic gene setting, there may be discrepancies between nucleotide and amino acid nomenclature when compared to HGVS nomenclature for that transcript at that location (Dalgleish et al. 2010). For example, a recent somatic report found a *MUTYH* pathogenic variant called c.1138delC (p.A382fs) at an allele frequency of 57%. The variant was confirmed to be present in the germline, but the nomenclature was different, c.1147del (p.Ala385Profs*23). Many publicly available databases, such as ClinVar (Landrum et al. 2018), will list aliases, and alternate locations may be included as a known alias. However, if discrepancy continues to exist, it may be beneficial to request genomic coordinates for comparison against existing databases.

In some research settings, an annotation, known as "Best Effect" annotation, may be given. This annotation considers all the transcriptional outcomes of all known transcripts at that location and reports the most deleterious (McLaren et al. 2016). As an example, if 15 known transcripts were identified for a single gene, and these were anticipated to result in 10 different transcriptional outcomes including intronic, missense, and splice site variants, the Best Effect annotation would report the splice site alteration because that would most affect translation. As the canonical transcript is sometimes chosen arbitrarily, and different transcripts of the same gene are transcribed in different organs, it is not currently clear which transcript is active at any given time in the tumor, so reporting the most severe effect may be desirable (Barrera et al. 2008; Singer et al. 2008; Wang et al. 2008; Gupta et al. 2010). The Best Effect annotation will frequently not match listed alterations in germline databases and genomic coordinates should be retrieved from the testing laboratory.

Tumor Heterogeneity

Human cancer cells in a given tumor are known to display differences in cellular morphology, metabolism, motility, proliferation, and metastatic potential (Fidler and Hart 1982; Heppner 1984; Nicolson 1984; Dick 2008). This extends to gene alteration and expression. Each tumor is a small universe controlled with Darwinian evolution of tumorigenesis, in which a high level of cell divisions creates a perfect environment for multiple mutants, potentially acting in response to a variety of intratumor and extraorganism pressures. As the analyzed tumor in most cases represents just a small locus of this universe, the analysis may not capture the germline alteration, either because of loss through CNVs or reversion, and therefore the germline alteration

does not appear on the report (Cheng et al. 2015; Cortes-Ciriano et al. 2017; Bailey et al. 2018; Knijnenburg et al. 2018).

Allelic Fraction

Allelic fraction, or the percent of the reads identifying the given alteration, has been posited as a method for determination of possible germline result. However, allelic fraction can be biased by multiple factors (Cheng et al. 2017; Knijnenburg et al. 2018). Indel alterations have an allelic fraction that frequently falls outside of anticipated 40%–60% ranges for germline variants (Cheng et al. 2017). If the given sample for analysis has a high level of tumor content, the allelic fraction may be biased by the heterogeneity of the sample itself. If the tumor purity has been compromised and the sample contains a high level of normal cells, the allelic fraction may also be impacted (Bailey et al. 2018). Homopolymer repeat regions may falsely elevate the amount of reads at any given location, therefore causing an imbalance of allelic fraction between the alteration and wild type (Cheng et al. 2017). Additionally, if a sample contains a high level of CNVs, allelic dropout in some cells of either the wild-type allele or altered allele could lead to a misleading allelic fraction that varies widely from the anticipated 50% (Knijnenburg et al. 2018; Sun et al. 2018). Last, reversion is a well-documented phenomenon, and loss of the mutant allele in samples from previously treated disease may mask an underlying germline variant (Domchek 2017; Mayor et al. 2017; Weigelt et al. 2017).

Presence of Founder Variants

The presence of a common founder variant is one of the best predictors that a somatic alteration is germline. Founder variants are well-typified in many genes including the *BRCA1* and *BRCA2* Ashkenazi Jewish founder variants (*BRCA1* c.68_69delAG and c.5266dupC, *BRCA2* c.5946delT) and Eastern European variants in *CHEK2* (c.1100delC). The presence of one of these founder variants in a tumor specimen supports the need for germline analysis (Knijnenburg et al. 2018). Conversely, a com-

Cite this article as *Cold Spring Harb Perspect Med* doi: 10.1101/cshperspect.a036590

mon founder variant may be filtered out of a somatic testing report if the filters on the somatic analysis are such that an alteration with population frequency >1% (typically derived from Gnomad [Karczewski et al. 2019] or ExAc [Lek et al. 2016]) are automatically filtered out (Cheng et al. 2015; Li et al. 2017).

Discordant Tumor Types

Although defining typical alterations within categories of tumors is still underway, many of the genes that increase empiric risk of cancer in certain organs are frequently involved in tumorigenesis in those same organs (Bailey et al. 2018). However, for most genes, their presence as prominent somatic alterations in discordant tumors should be a flag for possible germline involvement. For example, the presence of a *BRCA2* variant in papillary thyroid cancer should be considered for germline analysis. *TP53* and *PTEN* would be notable exceptions to this rule because of their constitutive nature and relatively common somatic association of multiple tumor types. (Bailey et al. 2018) MSK-IMPACT has found that among tumor paired normal samples with clinically actionable germline alterations, discordance between tumor type and cancer susceptibility gene was identified in 60% of individuals (Cheng et al. 2017). This suggests that an alteration identified in a gene not typical to the tumorigenesis process of that organ, particularly one that would be pathogenic in the germline setting, should be considered for germline genetic analysis.

Tumor Mutational Burden

High tumor mutational burden (TMB) is another scenario in which one should consider genetic testing. Although TMB and MSI have a high level of correlation (Huang et al. 2018), they do not assess for the same somatic phenomena. TMB is the number of relative variants within tumor cells versus the "typical," and MSI refers specifically to the number of repeat expansions of microsatellites of DNA. For tumors primarily driven by the mismatch repair (MMR) genes, high TMB will be detected (Alexandrov et al.

2013; Kim et al. 2013; Cortes-Ciriano et al. 2017; Bailey et al. 2018; Knijnenburg et al. 2018). In cancers known to be associated with the Lynch spectrum (e.g., colon, gastric, uterine, ovarian, urothelial, and kidney), referral to genetics should be considered if a high level of TMB exists (Cortes-Ciriano et al. 2017; Bailey et al. 2018). It is important to note that although the germline variant may be identified in the somatic analysis, allelic frequency may be an unreliable flag because of the hypermutated nature, and allelic loss of the germline alteration may have happened if large numbers of CNVs are present (Cortes-Ciriano et al. 2017; Bailey et al. 2018).

Variant Classification

Joint Consensus Recommendations for the Standards and Guidelines in the Interpretation and Reporting of Sequence Variants is made by the Association for Molecular Pathology (AMP), American Society of Clinical Oncology (ASCO), and the College of American Pathologists (CAP). Classification based on their recommendation is based on a four-tier system (Table 2): tier I, variants with strong clinical significance; tier II, variants with potential clinical significance; tier III, variants of unknown clinical significance; and tier IV, variants deemed benign or likely benign (Li et al. 2017). Unlike germline variants, in which measurable increased risk of cancer over a lifetime is an indicator of pathogenicity, somatic variants are interpreted based on their impact on clinical care. Therefore, those alterations that predict sensitivity, resistance, or toxicity to a specific therapy can be targeted as part of standard or investigational therapy (i.e., clinical trials). Those alterations that affect prognosis or diagnosis will be tiered in the highest tier. Alterations known to be pathogenic in the germline setting that do not fit the above criteria may therefore be placed in a lower tier. As an example, if a breast cancer patient were to carry a germline *EGFR* T790M variant, this may not be classified as a tier I alteration, as it has no bearing on the patient's breast cancer treatment. It is important to remember that the primary goal of germline

Table 2. Somatic variant classification system

Tier I	Variants of strong clinical significance	Level A evidence	FDA-approved therapy Included in professional guidelines
		Level B evidence	Well-powered studies with consensus from leaders in the field
Tier II	Variants of potential clinical significance	Level C evidence	FDA-approved therapies for different tumor types or investigational therapies
			Multiple small published studies with some consensus
		Level D evidence	Preclinical trials or a few case reports without consensus
Tier III	Variants of unknown clinical significance		Not observed at a significant allele frequency in the general or specific subpopulation databases, or pan-cancer or tumor-specific variant databases
			No convincing published evidence of cancer association
Tier IV	Benign or likely benign variants		Observed at significant allele frequency in the general or specific subpopulation databases
			No existing published evidence of cancer association

annotation is to identify alterations causing measurable changes in risk at the whole-body level. Annotation in the somatic setting, however, is used to identify those alterations that create a clinically actionable change in that type of tumor in that particular organ. This classification is by necessity not static and relies on the currently available technologies, therapeutic targets, and clinical trials. Although the joint consensus recommendations do indicate that those variants at high likelihood of being germline, either because of identification in a paired normal sample or because of other features (i.e., 50% allelic fraction in alterations identified as pathogenic in germline databases, known founder variants), should be classified as tier 1, germline alterations can be hard to detect or have possible ambiguous annotation and therefore may be relegated to a lower tier (Li et al. 2017).

In contrast, guidelines for germline analysis are put forth by the American College of Medical Genetics (ACMG). They include a five-basket system that classifies an alteration as Pathogenic (P), Likely Pathogenic (LP), Variant of Uncertain Significance (VUS), Likely Benign (LB), and Benign (B). Variants are scored with two scales; one "Evidence of Benign Impact" and the other "Evidence of Pathogenicity". The lines of evidence that support pathogenicity are classified as Very Strong, Strong, Moderate, Supporting. The lines of evidence that support benign impacts are classified as Stand Alone,

Strong, and Supporting. Set combinations of this classification correlate to the variant basket and lead to the final classification (Richards et al. 2015).

Case 1

A 37-year-old man presented to the emergency room with rectal bleeding. Colonoscopy identified a 3-cm mass in the cecum. Adenocarcinoma is confirmed by biopsy. He underwent a right hemicolectomy. Immunohistochemistry was performed with all four proteins (MLH1, MSH2, MSH6 and PMS2) present and MSI detected at 0:10 markers. On somatic analysis, the tumor showed a high level of TMB. The specimen was 60% tumor. Two *MSH6* tier III variants were detected: c.3938_3939insTCAAAAGG GACATAGAAAA (p.A1320Sfs*5) in 6% of reads and c.1309C>G (p.H437D) in 5% of reads. Germline large gene panel analysis was performed, and the patient was found to carry the *MSH6* p.A1320Sfs*5 pathogenic variant.

This case illustrates the sometimes-increased sensitivity of somatic analysis. The particular variant identified is a well-known, pathogenic, truncating variant that causes loss of 41 terminal amino acids, leading to a nonfunctioning protein that can escape nonsense mediated decay. Therefore, a false negative was identified on IHC. An important note is that the allelic fraction in this sample was low, likely

   Cite this article as *Cold Spring Harb Perspect Med* doi: 10.1101/cshperspect.a036590

driven down by the overall hypermutability of this tumor. These identified somatic alterations were classified as tier III variants based on the low allelic frequency and lack of actionability in regard to treatment (immunotherapy was not available at the time of this diagnosis).

Germline genetic testing would have been indicated for this patient based on age at diagnosis. Additionally, high TMB in absence of loss of IHC could indicate a polymerase gene variant (i.e., *POLE*, *POLD1*), and these genes were included when the patient's germline testing was ordered.

CLONAL HEMATOPOEISIS OF INDETERMINATE POTENTIAL (CHIP)

The advent of NGS technologies brought an increased sensitivity to detecting genetic variants in blood cells at low allele frequencies. Sanger sequencing can detect mosaic alleles at a 15%–20% allele frequency (Rohlin et al. 2009), whereas NGS can detect low-level mosaicism down to a 2% allele frequency or lower (Li and Stoneking 2012). Variants found at low allele frequencies can suggest the presence of mosaicism acquired during embryogenesis or clonal hematopoiesis (CH) (i.e., the expansions of cells with somatic variants) (Steensma et al. 2015). Clonal hematopoiesis of indeterminate potential (CHIP) generally refers to CH that involves driver mutations associated with myelodysplastic syndromes, such as *DNMT3A*, *ASXL1*, or *TET2* (Steensma 2018). Patients with CHIP have an increased risk of progressing to a heme malignancy, of ~0.5%–1% per year (Sperling et al. 2017), and CHIP has also been associated with acute cardiovascular events (Jaiswal et al. 2017). CHIP may also be a risk factor in the setting of bone marrow transplant as there have been cases of donor cell leukemia's arising from CHIP (Gondek et al. 2016), but other studies have suggested CHIP has no impact on overall survival of donor recipients (Frick et al. 2019). CH can be found in >10% of the population over age 65 (Jaiswal et al. 2014), representing significant implications for clinicians interpreting germline genetic testing results. How to follow patients with age-related CH or CHIP, given the potential impact on risks for cardiovascular events and heme malignancies, has not been determined.

CELL-FREE TUMOR DNA (ctDNA)

Cell-free DNA (cfDNA) is free-floating DNA in many different body fluids including plasma, saliva, lymph, breast milk, bile, urine, and spinal fluid (Thierry et al. 2016; Wan et al. 2017). Cell-free tumor DNA (ctDNA) refers specifically to short pieces of DNA derived from a tumor and appears in these fluids. ctDNA develops from either cellular breakdown, apoptosis and necrosis, or through active release mechanisms, encapsulated in vesicles, or associated with protein complexes (Jahr et al. 2001; Rykova et al. 2012). Although cfDNA is common even in healthy people, patients with cancer have higher blood levels of cfDNA (Leon et al. 1977; Stroun et al. 1987; Schwarzenbach et al. 2011). The average proportion of mutated DNA in plasma is very low, 0.4% in even advanced cancers (Barbany et al. 2019). Analysis of ctDNA is known as liquid biopsy. Liquid biopsy is performed using NGS to capture tumor-specific genetic variants. Data suggests that the length of ctDNA is shorter than naturally occurring cfDNA, so liquid biopsy utilizes size selection for shorter DNA fragment lengths to increase sensitivity of the analysis (Jiang et al. 2015; Underhill et al. 2016; Hellwig et al. 2018). Ideally, these represent a better summary of the heterogeneity of the genomic landscape of a patient's tumor than perhaps a localized biopsy would. This technology could also increase early stage detection and provide genetic information about a tumor in a hard to biopsy location (Hiemcke-Jiwa et al. 2018; Zill et al. 2018).

A large-scale study in 21,807 patients with advanced cancers in more than 50 cancer types utilized a panel of 70 defined cancer genes to determine the percentage of alterations in these patients. Somatic alterations were identified in 85% of the patients overall, with a range of 51% to 93% represented across various tumor types (Zill et al. 2018). The posited applications of ctDNA include identification of disease progression (i.e., ductal carcinoma in situ to invasive ductal breast cancer), cancer screening in high-risk populations (i.e., prostate cancer in men over 70), and distinguishing benign from malignant disease (Barbany et al. 2019). In addition to single-gene variants, it is possible that large CNVs or chromosomal disorders may be identified (Barbany et al. 2019). Additionally, there are a number of clinical trials using liquid biopsy in known variant carriers to detect early onset of tumors (Barbany et al. 2019).

IDENTIFYING GERMLINE VARIANTS THROUGH SOMATIC ANALYSIS

It is important to emphasize that there are no consensus recommendations for assessing which somatic variants should be confirmed with germline hereditary cancer testing other than recent National Comprehensive Cancer Network (NCCN) guidelines that recommend testing for all somatic pathogenic variants found in *BRCA1* or *BRCA2* (NCCN 3.2019). Insurance coverage for germline confirmatory testing varies. Recommendations for such testing are likely to evolve significantly as our understanding of this testing and technologies evolve. However, here we hope to begin the development of a framework for assessment based on the information previously reviewed (see Fig. 1).

Step 1. Assess all cases for clinical testing criteria for hereditary cancer risk based on personal and family history. Given the potential for germline variants to be lost in somatic tumor testing, even cases with no suggestive somatic variants should be offered germline testing based on appropriate clinical indications.

Step 2. Determine if tumor testing shows high TMB or MSI-high features (and *BRAF*-negative if colorectal tumor). These features are highly suggestive of variants in the genes associated with Lynch syndrome (*MLH1, MSH2, MSH6, PMS2,* and *EPCAM*) as well as *POLE* and *POLD1*, and subsequent germline testing should be done that includes these genes, at a minimum. Although these genes are typically included in somatic testing, the genes involve difficult to sequence regions, such as the pseudogene region of *PMS2*, which may be excluded from somatic analysis.

Step 3. Ensure that the somatic tumor result has not filtered out germline findings. If so, any decision to proceed with germline testing will be based solely on personal and family history guidelines or other germline-variant prediction models.

Step 4. Review somatic results for common founder variants. These are highly likely to be confirmed in the germline. Examples include the three common variants in the *BRCA1* and *BRCA2* genes associated with Jewish ancestry. It is less clear if testing is appropriate for moderate penetrance founder variants, although many of them impact cancer screening recommendations. These could include monoallelic variants in *MUTYH* (c.Y179C and c.G396D), *APC* c.I1307K, *NBN* c.657del5, and CHEK2 c.1100delc.

Step 5. At this point, the reviewer may be left with several variants in genes known to be associated with hereditary cancer risk and can begin the process of separating variants using the five-basket system: Benign (B), Likely Benign (LB), of Uncertain Significance (VUS), Pathogenic (P), and Likely Pathogenic (LP). Some variants will simply be noted as "detected" or similar, and further investigation may be necessary (Table 3). Variants confirmed to be B, LB, or VUS in the germline setting will generally not need germline testing unless other clinical indications are met.

Step 6. Some genes have a higher likelihood of being confirmed in the germline than others. NCCN now recommends confirmatory germline testing for all pathogenic and likely pathogenic variants in the *BRCA1* and *BRCA2* genes (NCCN 3.2019). Based on data from paired somatic/germline testing programs, other genes with a significant (>20%) chance of testing positive in the germline when seen in somatic tissue include the Lynch genes (*MLH1, MSH2, MSH6, PMS2,* and *EPCAM*) and *PALB2* (Meric-Bernstam et al. 2016). A reasonable argument could be made to proceed with germline testing that includes pathogenic or likely pathogenic somatic variants in any of these genes, but there is limited infor-

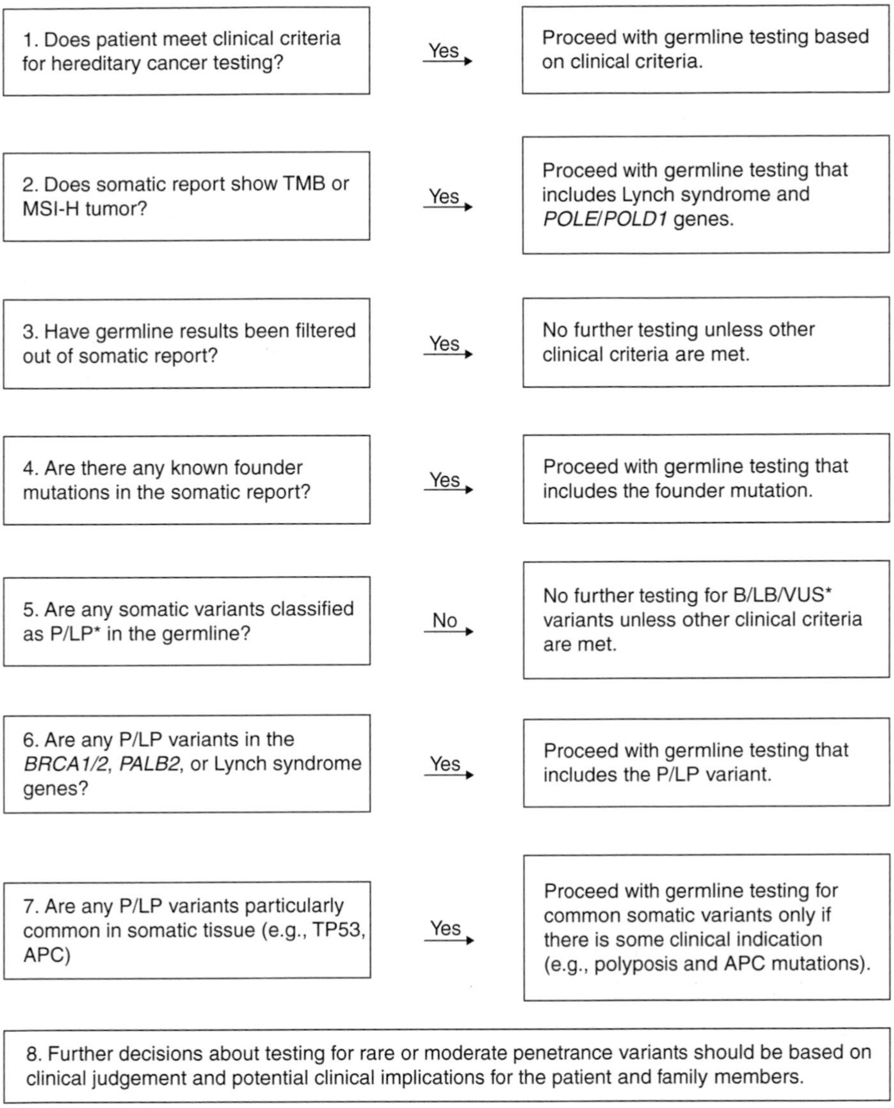

1. Does patient meet clinical criteria for hereditary cancer testing?	Yes →	Proceed with germline testing based on clinical criteria.
2. Does somatic report show TMB or MSI-H tumor?	Yes →	Proceed with germline testing that includes Lynch syndrome and *POLE/POLD1* genes.
3. Have germline results been filtered out of somatic report?	Yes →	No further testing unless other clinical criteria are met.
4. Are there any known founder mutations in the somatic report?	Yes →	Proceed with germline testing that includes the founder mutation.
5. Are any somatic variants classified as P/LP* in the germline?	No →	No further testing for B/LB/VUS* variants unless other clinical criteria are met.
6. Are any P/LP variants in the *BRCA1/2*, *PALB2*, or Lynch syndrome genes?	Yes →	Proceed with germline testing that includes the P/LP variant.
7. Are any P/LP variants particularly common in somatic tissue (e.g., TP53, APC)	Yes →	Proceed with germline testing for common somatic variants only if there is some clinical indication (e.g., polyposis and APC mutations).

8. Further decisions about testing for rare or moderate penetrance variants should be based on clinical judgement and potential clinical implications for the patient and family members.

*B = Benign, LB = Likely Benign, VUS = Variant of Uncertain Significance, LP = Likely Pathogenic, P = Pathogenic.

Figure 1. Flowchart to assess somatic results for potential germline testing.

mation regarding incidence of moderate-risk genes, such as *ATM*, *CHEK2*, *RAD51C*, and others. The provider may wish to consider the potential clinical impact of a pathogenic variant for various genes (Table 4).

Step 7. Conversely, many genes common in tumor cells are only rarely confirmed in the germline. These include, but are not limited to, *APC*, *TP53*, *MEN1*, *NF2*, *PTEN RET*, *STK11*, and *VHL* (Meric-Bernstam et al. 2016). The syndromes associated with these genes tend to have distinctive phenotypes. Although clinical assessment is not always a perfect predictor of germline variants, given the low overall proportion of germline variants in these genes, it may be appropriate to proceed with testing of P/LP variants only when there are additional clinical indications. Sometimes, the nonquantifiable "clinical judgment" may come into play.

Table 3. Resources to help assess somatic variants for germline testing

Transcript determination	Ensembl genome browser	useast.ensembl.org/index.html
	UCSC Genome Browser	www.genome.ucsc.edu
	NCBI Genome	www.ncbi.nlm.nih.gov/genome
	RefSeqGene	www.ncbi.nlm.nih.gov/refseq/rsg
Somatic variant classification	My Cancer Genome	www.mycancergenome.org
	cBioPortal, Memorial Sloan Kettering	www.cbioportal.org
	IARC TP53 mutation database	p53.iarc.fr
	International Cancer Genome Consortium	www.icgc.org
	Catalog of Somatic Mutations in Cancer (COSMIC)	cancer.sanger.ac.uk/cosmic
Germline variant classification	Clinvar	www.ncbi.nlm.nih.gov/clinvar
	Clinvitae	clinvitae.invitae.com
	International Society for Gastrointestinal Hereditary Tumors (InSiGHT)	www.insight-database.org
	Human Gene Mutation Database (HGMD)	www.hgmd.cf.ac.uk/ac/index.php
	Leiden Open Variant Database	www.lovd.nl
Gene tumor frequencies	cBio Portal	www.cbioportal.org
	COSMIC	cancer.sanger.ac.uk/cosmic
Germline population frequencies	ExAC	exac.broadinstitute.org/
	Genome Aggregation Database (gnomAD)	gnomad.broadinstitute.org/
	1000 Genomes	browser.1000genomes.org
	Exome variant server	evs.gs.washington.edu/EVS
	dbSNP	www.ncbi.nlm.nih.gov/snp
	dbVar	www.ncbi.nlm.nih.gov/dbvar
Gene association	GeneReviews	www.ncbi.nlm.nih.gov/books/ NBK1116
	GeneCards	www.genecards.org
	Genetics Home Reference	www.ghr.nlm.nih.gov
	Online Medelian Inheritance in Man (OMIM)	www.omim.org

Step 8. Debatably, other genes should be considered for confirmatory germline testing. The ACMG has developed a list of 25 cancer genes (among others) representing highly penetrant genetic disorders in an attempt to reduce the morbidity and mortality associated with these genes (Kalia et al. 2017) and this list may serve as a baseline for some clinicians. Many of the publications examining results from paired somatic and germline testing to date have not included moderate-penetrance genes such as *CHEK2* and *ATM*, yet pathogenic variants from these genes have the potential to affect cancer risk screening, given recommendations from NCCN and others (NCCN 1.2018; NCCN 3.2019). See Table 4.

As a final step, there will need to be a process developed to follow up with the ordering provider or the patient in order to ensure that co-ordination of germline testing is appropriately facilitated.

Case 2

A 54-year-old man was recently diagnosed with Stage IV gastric adenocarcinoma with signet ring cells, also known as diffuse gastric cancer (DGC). After progressing through several lines of therapy, he underwent testing with a commercial multiplatform, solid tumor biomarker analysis. The tumor was found to be Her2/Neu-negative by IHC and MSI-stable. Two somatic variants were noted, a pathogenic variant in *TP53* and a variant of uncertain significance in *CDH1*. The patient was referred to a genetic counselor for consideration of germline genetic testing.

The genetic counselor involved with this case noted that ~12% of gastric cancers will

Cite this article as *Cold Spring Harb Perspect Med* doi: 10.1101/cshperspect.a036590

Table 4. Hereditary cancer risk genes and screening implications

Hereditary cancer risk genes	ACMG[a]	Screening guidelines	Hereditary cancer risk genes	ACMG[a]	Screening guidelines
APC	✓	NCCN- GI	NF1		NCCN- HBOC
APC I1307K		NCCN- GI	NF2	✓	Evans et al. 2005
ATM		NCCN- HBOC	NTHL1 (biallelic only)		NCCN- GI
AXIN2		NCCN- GI	PALB2		NCCN- HBOC
BAP1		Rai et al. 2016	POLD1		NCCN- GI
BARD1		N/a	POLE		NCCN- GI
BMPR1A	✓	NCCN- GI	PTEN	✓	NCCN- HBOC
BRCA1	✓	NCCN- HBOC	RAD50		N/a
BRCA2	✓	NCCN- HBOC	RAD51C		NCCN- HBOC
BRIP1		NCCN- HBOC	RAD51D		NCCN- HBOC
CDH1		NCCN- HBOC	RB1	✓	Skalet et al. 2018
CDKN2A		Rossi et al. 2019	RET	✓	Kloos et al. 2009
CHEK2[b]		NCCN- HBOC	SDHA	✓	Lenders et al. 2014
CTNNA1		N/a	SDHB	✓	
DICER1		Schultz et al. 2018	SDHC	✓	
GREM1		NCCN- GI	SDHD	✓	
HOXB13		Giri et al. 2018	SMAD4	✓	NCCN- GI
MEN1	✓	Thakker et al. 2012	STK11	✓	NCCN- HBOC
Lynch syndrome/MMR genes (MLH1, MSH2, MSH6, PMS2, EPCAM)	✓	NCCN- GI	TP53	✓	NCCN- HBOC
MSH3 (biallelic only)		NCCN- GI	TSC1	✓	Krueger et al. 2013
MUTYH (monoallelic)		NCCN- GI	TSC2	✓	
MUTYH (biallelic)	✓	NCCN- GI	VHL	✓	Poulsen et al. 2010
NBN		NCCN- HBOC	WT1	✓	Lee et al. 2016

Genetic/Familial High-Risk Assessment-Gastrointestinal (NCCN-GI): Colorectal. Version 1.2018

Genetic/Familial High-Risk Assessment-Hereditary Breast and Ovarian Cancer (NCCN-HBOC): Breast and Ovarian. Version 3.2019

[a]ACMG Secondary Finding Genes (Kalia et al. 2017).

[b]Low-penetrance variants, such as S428F and I157T, have unclear clinical implications.

have somatic variants in *CDH1*, and there is limited data as to the proportion of somatic variants that are later confirmed in the germline. The reported somatic variant, c.D257Y, was classified as a variant of uncertain significance in the germline. In hereditary diffuse gastric cancer (HDGC) syndrome the average age of onset of DGC is 38 yr (range: 14–69 yr) and testing is recommended for anyone diagnosed at age 40 or younger. When diagnosed at older ages, additional family history of gastric cancer or lobular breast cancer is part of testing criteria (van der Post et al. 2015). Although the patient did not have known breast or gastric cancer in his family, there was some question about his father

having esophageal versus gastric cancer. Whereas the counselor and patient discussed the low likelihood of finding an actionable variant, the patient was interested in pursuing testing and willing to pay out-of-pocket costs, as he did not meet clinical criteria for testing. One potential benefit of testing would be to have the VUS status confirmed by a clinical laboratory, which would allow for future reclassification updates. Testing was pursued through a multigene cancer panel from a commercial laboratory.

Germline results showed a new pathogenic variant in *CDH1*, called deletion exons 14–16, which had not been previously identified in the somatic test report. This new variant was noted to involve a large deletion that may not have been detectible through the somatic testing laboratory processes. The discovery of this germline variant has allowed for cascade testing of several family members.

This case illustrates the limitations of somatic genetic testing in regards to certain types of variants and also highlights some of the remaining challenges when determining which somatic tumor findings justify follow-up germline testing.

DISCUSSION

Although this review speaks to the current state of practice, many new changes are on the horizon. It is likely that most somatic analysis will become paired with germline analysis, and the ambiguity surrounding whether a somatic gene change is also present in the germline will no longer exist. These changes will hopefully lead to higher rates of insurance coverage for comprehensive germline analysis in cancer patients, likely extending testing to cancer patients that do not fit current guidelines for analysis. Additionally, somatic analysis has already led to increased understanding of how genes typically classified as "cardiac" or "general genetics" may contribute to cancer risk (i.e., NOTCH1, Rasopathies, RECQL1). The continued identification of crossover risks may lead to greater fluidity between the traditionally siloed specialties within genetics.

Significant limitations exist at this time to this summary of the science. Although it is known that methylation, structural variants, and variation in the introns and promoters are likely involved in tumorigenesis, comprehensive analysis of these areas and types of genomic variation is uncommon, and we do not understand the implication of these alterations. Certainly, as our understanding of these variations deepens, the extent to which we acknowledge the interplay between somatic and germline variants will also broaden. The current data regarding somatic testing is limited by the reliance on sequencing and deletion and duplication of the DNA strand only.

As somatic technologies have primarily been performed in countries of European descent and among well-insured individuals in the United States, there is a significant lack of data about tumor profiles in individuals from non-White and Hispanic racial and ethnic groups. Until genomic research, both germline and somatic, is intentionally designed to capture genomic data from individuals that represent the global population, its applicability to understanding and treating cancer is hampered. There are likely many genetic nuances in each population that affect prognosis, response, baseline risk, and recurrence risk, and the current extrapolation of existing research to populations not included in the source data is limiting.

Further research will be needed to understand how other genomic changes in the germline (methylation, variable expression) effect tumorigenesis, how the tumorigenesis process may differ in racial and ethnic communities outside of European ancestry and how best, or if at all, to silo identified variants into areas of specialty.

REFERENCES

Alexandrov LB, Nik-Zainal S, Wedge DG, Aparicio SA, Behjati S, Biankin AV, Bignell AG, Bolli N, Borg A, Børresen-Dale AL, et al. 2013. Signatures of mutational processes in human cancer. Nature 500: 415–421. doi:10.1038/nature12477

Bailey MH, Tokheim C, Porta-Pardo E, Sengupta S, Betrand D, Weerasinghe A, Colaprico A, Wendi MC, Kim J, Reardon B, et al. 2018. Comprehensive characterization of cancer driver genes and mutations. Cell 173: 371–385. e18. doi:10.1016/j.cell.2018.02.060

Barbany G, Arthur C, Leiden A, Nordenskjold M, Rosenquist R, Tesi B, Wallander K, Tham E. 2019. Cell free tumour DNA testing for early detection of cancer—a potential future tool. J Int Med 286: 118–136. doi:10.1111/joim.12897

Barrera LO, Li Z, Smith AD, Arden KC, Cavenee WK, Zhang MQ, Green RD, Ren B. 2008. Genome-wide mapping and analysis of active promoters in mouse embryonic stem cells and adult organs. Genome Res 18: 46–59. doi:10.1101/gr.6654808

Bentley DR, Balasubramanian S, Swerdlow HP, Smith GP, Milton J, Brown CG, Hall KP, Evers DJ, Barnes CL, Bignell HR, et al. 2008. Accurate whole human genome sequencing using reversible terminator chemistry. Nature 456: 53–59. doi:10.1038/nature07517

Boland CR, Thibodeau SN, Hamilton SR, Sidransky D, Eshleman JR, Burt RW, Meltzer SJ, Rodriguez-Bigas MA, Fodde

Cite this article as Cold Spring Harb Perspect Med doi: 10.1101/cshperspect.a036590

R, Ranzani GN, et al. 1998. A National Cancer Institute Workshop on Microsatellite Instability for cancer detection and familial predisposition: development of international criteria for the determination of microsatellite instability in colorectal cancer. *Cancer Res* **58:** 5248–5257.

Boland CR, Koi M, Chang DK, Carethers JM. 2008. The biochemical basis of microsatellite instability and abnormal immunohistochemistry and clinical behavior in Lynch syndrome: from bench to bedside. *Fam Cancer* **7:** 41–52. doi:10.1007/s10689-007-9145-9

Cheng DT, Mitchell TN, Zehir A, Shah RH, Benayed R, Syed A, Chandramohan R, Liu ZY, Won HH, Scott SN, et al. 2015. Memorial Sloan Kettering-Integrated Mutation Profiling of Actionable Cancer Targets (MSK-IMPACT): a hybridization capture-based next-generation sequencing clinical assay for solid tumor molecular oncology. *J Mol Diagn* **17:** 251–264. doi:10.1016/j.jmoldx.2014.12 .006

Cheng DT, Prasad M, Chekaluk Y, Benayed R, Sadowska J, Zehir A, Syed A, Wang YE, Somar J, Li Y, et al. 2017. Comprehensive detection of germline variants by MSK-IMPACT, a clinical diagnostic platform for solid tumor molecular oncology and concurrent cancer predisposition testing. *BMC Med Genomics* **10:** 33. doi:10.1186/ s12920-017-0271-4

Cortes-Ciriano I, Lee S, Park WY, Kim TM, Park PJ. 2017. A molecular portrait of microsatellite instability across multiple cancers. *Nat Commun* **8:** 15180. doi:10.1038/ ncomms15180

Dalgleish R, Flicek P, Cunningham F, Astashyn A, Tully RE, Proctor G, Chen Y, McLaren WM, Larsson P, Vaughan BW, et al. 2010. Locus reference genomic sequences: an improved basis for describing human DNA variants. *Genome Med* **2:** 24. doi:10.1186/gm145

de la Chapelle A, Hampel H. 2010. Clinical relevance of microsatellite instability in colorectal cancer. *J Clin Oncol* **28:** 3380–3387. doi:10.1200/JCO.2009.27.0652

Dick JE. 2008. Stem cell concepts renew cancer research. *Blood* **112:** 4793–4807. doi:10.1182/blood-2008-08-077941

Domchek SM. 2017. Reversion mutations with clinical use of PARP inhibitors: many genes, many versions. *Cancer Discov* **7:** 937–939. doi:10.1158/2159-8290.CD-17-0734

Evaluation of Genomic Applications in Practice and Prevention (EGAPP) Working Group. 2009. Recommendations from the EGAPP Working Group: genetic testing strategies in newly diagnosed individuals with colorectal cancer aimed at reducing morbidity and mortality from Lynch syndrome in relatives. *Genet Med* **11:** 35–41. doi:10.1097/ GIM.0b013e31818fa2ff

Evans DG, Baser ME, O'Reilly B, Rowe J, Gleeson M, Saeed S, King A, Huson SM, Kerr R, Thomas N, et al. 2005. Management of the patient and family with neurofibromatosis 2: a consensus conference statement. *Br J Neurosurg* **19:** 5–12. doi:10.1080/02688690500081206

Fidler IJ, Hart IR. 1982. Biological diversity in metastatic neoplasms: origins and implications. *Science* **217:** 998–1003. doi:10.1126/science.7112116

Frick M, Chan W, Arends CM, Hablesreiter R, Halik A, Heuser M, Michonneau D, Blau O, Hoyer K, Christen F, et al. 2019. Role of donor clonal hematopoiesis in allo-geneic hematopoietic stem-cell transplantation. *J Clin Oncol* **37:** 375–385. doi:10.1200/JCO.2018.79.2184

Giri VN, Knudsen KE, Kelly WK, Abida W, Andriole GL, Bangma CH, Bekelman JE, Benson MC, Blanco A, Burnet A, et al. 2018. Role of genetic testing for inherited prostate cancer risk: Philadelphia Prostate Cancer Consensus Conference 2017. *J Clin Oncol* **36:** 414–424. doi:10 .1200/JCO.2017.74.1173

Goldstein JB, Wu W, Borras E, Masand G, Cuddy A, Mork ME, Bannon SA, Lynch PM, Rodriguez-Bigas M, Taggart MW, et al. 2017. Can microsatellite status of colorectal cancer be reliably assessed after neoadjuvant therapy? *Clin Cancer Res* **23:** 5246–5254. doi:10.1158/1078-0432 .CCR-16-2994

Gondek LP, Zheng G, Ghiaur G, DeZern AE, Matsui W, Yegnasubramanian S, Lin MT, Levis M, Eshleman JR, Varadhan R, et al. 2016. Donor cell leukemia arising from clonal hematopoiesis after bone marrow transplantation. *Leukemia* **30:** 1916–1920. doi:10.1038/leu.2016.63

Gupta R, Wikramasinghe P, Bhattacharyya A, Perez FA, Pal S, Davuluri RV. 2010. Annotation of gene promoters by integrative data-mining of ChIP-seq Pol-II enrichment data. *BMC Bioinformatics* **11:** S65. doi:10.1186/1471-2105-11-S1-S65

Hampel H. 2016. Genetic counseling and cascade genetic testing in Lynch syndrome. *Fam Cancer* **15:** 423–427. doi:10.1007/s10689-016-9893-5

Hampel H, Frankel W, Panescu J, Lockman J, Sotamaa K, Fix D, Comeras I, La Jeunesse J, Nakagawa H, Westman JA, et al. 2006. Screening for Lynch syndrome (hereditary nonpolyposis colorectal cancer) among endometrial cancer patients. *Cancer Res* **66:** 7810–7817. doi:10.1158/0008-5472.CAN-06-1114

Hampel H, Frankel WL, Martin E, Arnold M, Khanduja K, Kuebler P, Clendenning M, Sotamaa K, Prior T, Westman JA, et al. 2008. Feasibility of screening for Lynch syndrome among patients with colorectal cancer. *J Clin Oncol* **26:** 5783–5788. doi:10.1200/JCO.2008.17.5950

Hampel H, Pearlman R, Beightol M, Zhao W, Jones D, Frankel WL, Goodfellow PJ, Yilmaz A, Miller K, Bacher J, et al. 2018. Assessment of tumor sequencing as a replacement for Lynch syndrome screening and current molecular tests for patients with colorectal cancer. *JAMA Oncol* **4:** 806–813. doi:10.1001/jamaoncol.2018 .0104

Hellwig S, Nix DA, Gligorich KM, O'Shea JM, Thomas A, Fuertes CL, Bhetariya PJ, Marth GT, Bronner MP, Underhill HR. 2018. Automated size selection for short cell-free DNA fragments enriches for circulating tumor DNA and improves error correction during next generation sequencing. *PLoS ONE* **13:** e0197333. doi:10.1371/journal .pone.0197333

Heppner GH. 1984. Tumor heterogeneity. *Cancer Res* **44:** 2259–2265.

Hiemcke-Jiwa LS, ten Dam-van Loon NH, Leguit RJ, Nierkens S, Ossewaarde-van Norel J, de Boer JH, Roholl FF, de Weger RA, Huibers MMH, de Groot-Mijnes JDF, et al. 2018. Potential diagnosis of vitreoretinal lymphoma by detection of *MYD88* mutation in aqueous humor with ultrasensitive droplet digital polymerase chain reaction. *JAMA Ophthalmol* **136:** 1098–1104. doi:10.1001/ja maophthalmol.2018.2887

Huang X, Tse JY, Protopopov A, Russell M, Weeraratne D, Ring JE, Bjonnes A, Pei S, Sun R, Lvova M, et al. 2018. Characterization of tumor mutation burden (TMB) and microsatellite instability (MSI) interplay for gastroesophageal adenocarcinoma (GA) and colorectal carcinoma (CRC). *J Clin Oncol* **36:** 22. doi:10.1200/JCO.2018.36.5_suppl.22

Jahr S, Hentze H, Englisch S, Hardt D, Fackelmayer FO, Hesch RD, Knippers R. 2001. DNA fragments in the blood plasma of cancer patients: quantitations and evidence for their origin from apoptotic and necrotic cells. *Cancer Res* **61:** 1659–1665.

Jain R, Savage MJ, Forman AD, Mukherji R, Hall MJ. 2016. The relevance of hereditary cancer risks to precision oncology: what should providers consider when conducting tumor genomic profiling? *J Natl Compr Canc Netw* **14:** 795–806.

Jaiswal S, Fontanillas P, Flannick J, Manning A, Grauman PV, Mar BG, Lindsley RC, Mermel CH, Burtt N, Chavez A, et al. 2014. Age-related clonal hematopoiesis associated with adverse outcomes. *N Engl J Med* **371:** 2488–2498. doi:10.1056/NEJMoa1408617

Jaiswal S, Natarajan P, Silver AJ, Gibson CJ, Bick AG, Shvartz E, McConkey M, Gupta N, Gabriel S, Ardissino D, et al. 2017. Clonal hematopoiesis and risk of atherosclerotic cardiovascular disease. *N Engl J Med* **377:** 111–121. doi:10.1056/NEJMoa1701719

Jiang P, Chan CW, Chan KC, Cheng SH, Wong J, Wong VW, Wong GL, Chan SL, Mok TS, Chan HL, et al. 2015. Lengthening and shortening of plasma DNA in hepatocellular carcinoma patients. *Proc Natl Acad Sci* **112:** E1317–E1325. doi:10.1073/pnas.1500076112

Kalia SS, Adelman K, Bale SJ, Chung WK, Eng C, Evans JP, Herman GE, Hufnagel SB, Klein TE, Korf BR, et al. 2017. Recommendations for reporting of secondary findings in clinical exome and genome sequencing, 2016 update (ACMG SF v2.0): a policy statement of the American College of Medical Genetics and Genomics. *Genet Med* **19:** 249–255. doi:10.1038/gim.2016.190

Kanchi KL, Johnson KJ, Lu C, McLellan MD, Leiserson MD, Wendl MC, Zhang Q, Koboldt DC, Xie M, Kandoth C, et al. 2014. Integrated analysis of germline and somatic variants in ovarian cancer. *Nat Commun* **5:** 3156. doi:10.1038/ncomms4156

Karczewski KJ, Francioli LC, Tiao G, Collins RL, Cummings BB, Alföldi J, Wang Q, Laricchia KM, Kosmicki JA, Ganna A, et al. 2019. Variation across 141,456 human exomes and genomes reveals the spectrum of loss-of-function intolerance across human protein-coding genes. Cold Spring Harbor Laboratory: ASHG Platform Session. bioRxiv doi:10.1101/531210

Kent WJ, Sugnet CW, Furey TS, Roskin KM, Pringle TH, Zahler AM, Haussler D. 2002. The human genome browser at UCSC. *Genome Res* **12:** 996–1006. doi:10.1101/gr.229102

Kim TM, Laird PW, Park PJ. 2013. The landscape of microsatellite instability in colorectal and endometrial cancer genomes. *Cell* **155:** 858–868. doi:10.1016/j.cell.2013.10.015

Kloos RT, Eng C, Evans DB, Francis GL, Gagel RF, Gharib H, Moley JF, Pacini F, Ringel MD, Schlumberger M, et al. 2009. Medullary thyroid cancer: management guidelines of the American Thyroid Association. *Thyroid* **19:** 565–612. doi:10.1089/thy.2008.0403

Knijnenburg TA, Wang L, Zimmermann MT, Chambwe N, Gao GF, Cherniack AD, Fan H, Shen H, Way GP, Greene CS, et al. 2018. Genomic and molecular landscape of DNA damage repair deficiency across the Cancer Genome Atlas. *Cell Rep* **23:** 239–254.e6. doi:10.1016/j.celrep.2018.03.076

Krammer J, Pinker-Domenig K, Robson ME, Gönen M, Bernard-Davila B, Morris EA, Mangino DA, Jochelson MS. 2017. Breast cancer detection and tumor characteristics in *BRCA1* and *BRCA2* mutation carriers. *Breast Cancer Res Treat* **163:** 565–571. doi:10.1007/s10549-017-4198-4

Krueger DA, Northrup H; International Tuberous Sclerosis Complex Consensus Group. 2013. Tuberous sclerosis complex surveillance and management: recommendations of the 2012 International Tuberous Sclerosis Complex Consensus Conference. *Pediatr Neurol* **49:** 255–265. doi:10.1016/j.pediatrneurol.2013.08.002

Landrum MJ, Lee JM, Benson M, Brown GR, Chao C, Chitipiralla S, Gu B, Hart J, Hoffman D, Jang W, et al. 2018. ClinVar: improving access to variant interpretations and supporting evidence. *Nucleic Acids Res* **46:** D1062–D1067. doi:10.1093/nar/gkx1153

Le DT, Uram JN, Wang H, Bartlett BR, Kemberling H, Eyring AD, Skora AD, Luber BS, Azad NS, Laheru D, et al. 2015. PD-1 blockade in tumors with mismatch-repair deficiency. *N Engl J Med* **372:** 2509–2520. doi:10.1056/NEJMoa1500596

Lee PA, Nordenström A, Houk CP, Ahmed SF, Auchus R, Baratz A, Baratz Dalke K, Liao LM, Lin-Su K, Looijenga L III, et al. 2016. Global disorders of sex development update since 2006: perceptions, approach and care. *Horm Res Paediatr* **85:** 158–180. doi:10.1159/000442975

Lek M, Karczewski K, Minikel E, Samocha K, Banks E, Fennell T, O'Donnell-Luria AH, Ware J, Hill A, Cummings BB. 2016. Analysis of protein-coding genetic variation in 60,706 humans. *Nature* **536:** 285–291. doi:10.1038/nature19057

Lenders JW, Duh Q, Eisenhofer G, Gimenez-Roqueplo A, Grebe SK, Murad SM, Naruse M, Pacak K, Young WF. 2014. Pheochromocytoma and paraganglioma: an Endocrine Society Clinical Practice Guideline. *J Clin Endocrinol Metab* **99:** 1915–1942. doi:10.1210/jc.2014-1498

Leon SA, Shapiro B, Sklaroff DM, Yaros MJ. 1977. Free DNA in the serum of cancer patients and the effect of therapy. *Cancer Res* **37:** 646–650.

Li M, Stoneking M. 2012. A new approach for detecting low-level mutations in next-generation sequence data. *Genome Biol* **13:** R34. doi:10.1186/gb-2012-13-5-r34

Li M, Datto M, Duncavage EJ, Kulkarni S, Lindeman NI, Roy S, Tsimberidou AM, Vnencak-Jones CL, Wolff DJ, Younes A, et al. 2017. Standards and guidelines for the interpretation and reporting of sequence variants in cancer: A Joint Consensus Recommendation of the Association for Molecular Pathology, American Society of Clinical Oncology, and College of American Pathologists. *J Mol Diagn* **19:** 4–23. doi:10.1016/j.jmoldx.2016.10.002

Lin MT, Mosier SL, Thiess M, Beierl KF, Debeljak M, Tseng LH, Chen G, Yegnasubramanian S, Ho H, Cope L, et al. 2014. Clinical validation of *KRAS, BRAF*, and *EGFR* mutation detection using next-generation sequencing. *Am J Clin Pathol* **141**: 856–866. doi:10.1309/AJCPMWGWGO34EGOD

Luthra R, Patel KP, Reddy NG, Haghshenas V, Routbort MJ, Harmon MA, Barkoh BA, Kanagal-Shamanna R, Ravandi F, Cortes JE, et al. 2014. Next-generation sequencing-based multigene mutational screening for acute myeloid leukemia using MiSeq: applicability for diagnostics and disease monitoring. *Haematologica* **99**: 465–473. doi:10.3324/haematol.2013.093765

Mandelker D, Zhang L. 2018. The emerging significance of secondary germline testing in cancer genomics. *J Pathol* **244**: 610–615. doi:10.1002/path.5031

Mayor P, Gay LM, Lele S, Elvin JA. 2017. *BRCA1* reversion mutation acquired after treatment identified by liquid biopsy. *Gynecol Oncol Rep* **21**: 57–60. doi:10.1016/j.gore.2017.06.010

McLaren W, Gil L, Hunt SE, Riat HS, Ritchie GR, Thormann A, Flicek P, Cunningham F. 2016. The Ensembl variant effect predictor. *Genome Biol* **17**: 122. doi:10.1186/s13059-016-0974-4

Meric-Bernstam F, Brusco L, Daniels M, Wathoo C, Bailey AM, Strong L, Shaw K, Lu K, Qi Y, Zhao H, et al. 2016. Incidental germline variants in 1000 advanced cancers on a prospective somatic genomic profiling protocol. *Ann Oncol* **27**: 795–800. doi:10.1093/annonc/mdw018

National Comprehensive Cancer Network (NCCN). NCCN Clinical Practice Guidelines in Oncology. Genetic/Familial High-Risk Assessment: Colorectal Version 1.2018. July 12, 2018. Available at https://www.nccn.org/professionals/physician_gls/pdf/genetics_colon.pdf.

National Comprehensive Cancer Network (NCCN). NCCN Clinical Practice Guidelines in Oncology. Genetic/Familial High-Risk Assessment: Breast and Ovarian. Version 3.2019. January 18, 2019. Available at https://www.nccn.org/professionals/physician_gls/pdf/genetics_screening.pdf.

Nicolson GL. 1984. Generation of phenotypic diversity and progression in metastatic tumor cells. *Cancer Metastasis Rev* **3**: 25–42. doi:10.1007/BF00047691

Nikiforova MN, Wald AI, Roy S, Durso MB, Nikiforov YE. 2013. Targeted next-generation sequencing panel (ThyroSeq) for detection of mutations in thyroid cancer. *J Clin Endocrinol Metab* **98**: E1852–E1860. doi:10.1210/jc.2013-2292

Peng YC, Chen YB. 2018. Recognizing hereditary renal cancers through the microscope: a pathology update. *Surg Pathol Clin* **11**: 725–737. doi:10.1016/j.path.2018.07.010

Peshkin BN, Alabek ML, Isaacs C. 2010. BRCA1/2 mutations and triple negative breast cancers. *Breast Dis* **32**: 25–33. doi:10.3233/BD-2010-0306

Poulsen ML, Budtz-Jørgensen E, Bisgaard ML. 2010. Surveillance in von Hippel–Lindau disease (vHL). *Clin Genet* **77**: 49–59. doi:10.1111/j.1399-0004.2009.01281.x

Rai K, Pilarski R, Cebulla CM, Abdel-Rahman MH. 2016. Comprehensive review of *BAP1* tumor predisposition syndrome with report of two new cases. *Clin Genet* **89**: 285–294. doi:10.1111/cge.12630

Raymond VM, Gray SW, Roychowdhury S, Joffe S, Chinnaiyan AM, Parsons DW, Plon SE. 2016. Germline findings in tumor-only sequencing: points to consider for clinicians and laboratories. *J Natl Cancer Inst* **108**: djv351. doi:10.1093/jnci/djv351

Richards S, Aziz N, Bale S, Bick D, Das S, Gastier-Foster J, Grody WW, Hegde M, Lyon E, Spector E, et al. 2015. Standards and guidelines for the interpretation of sequence variants: a joint consensus recommendation of the American College of Medical Genetics and Genomics and the Association for Molecular Pathology. *Genet Med* **17**: 405–424. doi:10.1038/gim.2015.30

Rohlin A, Wernersson J, Engwall Y, Wiklund L, Björk J, Nordling M. 2009. Parallel sequencing used in detection of mosaic mutations: comparison with four diagnostic DNA screening techniques. *Hum Mutat* **30**: 1012–1020. doi:10.1002/humu.20980

Rossi M, Pellegrini C, Cardelli L, Ciciarelli V, Di Nardo L, Fargnoli MC. 2019. Familial melanoma: diagnostic and management implications. *Dermatol Pract Concept* **9**: 10–16. doi:10.5826/dpc.0901a03

Rykova EY, Morozkin ES, Ponomaryova AA, Loseva EM, Zaporozhchenko IA, Cherdyntseva NV, Vlassov VV, Laktionov PP. 2012. Cell-free and cell-bound circulating nucleic acid complexes: mechanisms of generation, concentration and content. *Expert Opin Biol Ther* **12**: S141–153. doi:10.1517/14712598.2012.673577

Salipante SJ, Scroggins SM, Hampel HL, Turner EH, Pritchard CC. 2014. Microsatellite instability detection by next generation sequencing. *Clin Chem* **60**: 1192–1199. doi:10.1373/clinchem.2014.223677

Salvador MU, Truelson MRF, Mason C, Souders B, LaDuca H, Dougall B, Black MH, Fulk K, Profato J, Gutierrez S, et al. 2019. Comprehensive paired tumor/germline testing for Lynch syndrome: bringing resolution to the diagnostic process. *J Clin Oncol* **37**: 647–657. doi:10.1200/JCO.18.00696

Schrader KA, Cheng DT, Joseph V, Prasad M, Walsh M, Zehir A, Ni A, Thomas T, Benayed R, Ashraf A, et al. 2016. Germline variants in targeted tumor sequencing using matched normal DNA. *JAMA Oncol* **2**: 104–111. doi:10.1001/jamaoncol.2015.5208

Schultz KP, Williams GM, Kamihara J, Stewart DR, Harris AK, Bauer AJ, Turner J, Shah R, Schneider K, Schneider KW, et al. 2018. *DICER1* and associated conditions: identification of at-risk individuals and recommended surveillance strategies. *Clin Cancer Res* **24**: 2251–2261. doi:10.1158/1078-0432.CCR-17-3089

Schwarzenbach H, Müller V, Milde-Langosch K, Steinbach B, Pantel K. 2011. Evaluation of cell-free tumour DNA and RNA in patients with breast cancer and benign breast disease. *Mol Biosyst* **7**: 2848–2854. doi:10.1039/c1mb05197k

Shendure J, Ji H. 2008. Next-generation DNA sequencing. *Nat Biotechnol* **26**: 1135–1145. doi:10.1038/nbt1486

Singer GA, Wu J, Yan P, Plass C, Huang TH, Davuluri RV. 2008. Genome-wide analysis of alternative promoters of human genes using a custom promoter tiling array. *BMC Genomics* **9**: 349. doi:10.1186/1471-2164-9-349

Singh RR, Patel KP, Routbort MJ, Reddy NG, Barkoh BA, Handal B, Kanagal-Shamanna R, Greaves WO, Medeiros LJ, Aldape KD, et al. 2013. Clinical validation of a next-generation sequencing screen for mutational hotspots in 46 cancer-related genes. *J Mol Diagn* **15**: 607–622. doi:10.1016/j.jmoldx.2013.05.003

Skalet AH, Gombos DSM, Gallie BL, Kim JW, Shields CL, Marr BP, Plon SE, Chévez-Barrios P. 2018. Screening children at risk for retinoblastoma: consensus report from the American Association of Ophthalmic Oncologists and Pathologists. *Ophthalmology* **125**: 453–458. doi:10.1016/j.ophtha.2017.09.001

Sperling AS, Gibson CJ, Ebert BL. 2017. The genetics of myelodysplastic syndrome: from clonal haematopoiesis to secondary leukaemia. *Nat Rev Cancer* **17**: 5–19. doi:10.1038/nrc.2016.112

Steensma DP. 2018. Clinical consequences of clonal hematopoiesis of indeterminate potential. *Blood Adv* **2**: 3404–3410. doi:10.1182/bloodadvances.2018020222

Steensma DP, Bejar R, Jaiswal S, Lindsley RC, Sekeres MA, Hasserjian RP, Ebert BL. 2015. Clonal hematopoiesis of indeterminate potential and its distinction from myelodysplastic syndromes. *Blood* **126**: 9–16. doi:10.1182/blood-2015-03-631747

Stroun M, Anker P, Lyautey J, Lederrey C, Maurice PA. 1987. Isolation and characterization of DNA from the plasma of cancer patients. *Eur J Cancer Clin Oncol* **23**: 707–712. doi:10.1016/0277-5379(87)90266-5

Sun JX, He Y, Sanford E, Montesion M, Frampton GM, Vignot S, Soria JC, Ross JS, Miller VA, Stephens PJ, et al. 2018. A computational approach to distinguish somatic vs. germline origin of genomic alterations from deep sequencing of cancer specimens without a matched normal. *PLoS Comput Biol* **14**: e1005965.

Syngal S, Fox EA, Eng C, Kolodner RD, Garber JE. 2000. Sensitivity and specificity of clinical criteria for hereditary non-polyposis colorectal cancer associated mutations in *MSH2* and *MLH1*. *J Med Genet* **37**: 641–645. doi:10.1136/jmg.37.9.641

Thakker RV, Newey PJ, Walls GV, Bilezikian J, Dralle H, Ebeling PR, Melmed S, Sakurai A, Tonelli F, Brandi ML. 2012. Clinical practice guidelines for multiple endocrine neoplasia type 1 (MEN1). *J Clin Endocrinol Metab* **97**: 2990–3011. doi:10.1210/jc.2012-1230

Thierry AR, El Messaoudi S, Gahan PB, Anker P, Stroun M. 2016. Origins, structures, and functions of circulating DNA in oncology. *Cancer Metastasis Rev* **35**: 347–376. doi:10.1007/s10555-016-9629-x

Tsongalis GJ, Peterson JD, de Abreu FB, Tunkey CD, Gallagher TL, Strausbaugh LD, Wells WA, Amos CI. 2014. Routine use of the Ion Torrent AmpliSeq™ Cancer Hotspot Panel for identification of clinically actionable somatic mutations. *Clin Chem Lab Med* **52**: 707–714. doi:10.1515/cclm-2013-0883

Underhill HR, Kitzman JO, Hellwig S, Welker NC, Daza R, Baker DN, Gligorich KM, Rostomily RC, Bronner MP, Shendure J. 2016. Fragment length of circulating tumor DNA. *PLoS Genet* **12**: e1006162. doi:10.1371/journal.pgen.1006162

van der Post RS, Vogelaar IP, Carneiro F, Guilford P, Huntsman D, Hoogerbrugge N, Caldas C, Schreiber KE, Hardwick RH, Ausems MG, et al. 2015. Hereditary diffuse gastric cancer: updated clinical guidelines with an emphasis on germline *CDH1* mutation carriers. *J Med Genet* **52**: 361–374. doi:10.1136/jmedgenet-2015-103094

Wan JCM, Massie C, Garcia-Corbacho J, Mouliere F, Brenton JD, Caldas C, Pacey S, Baird R, Rosenfeld N. 2017. Liquid biopsies come of age: towards implementation of circulating tumour DNA. *Nat Rev Cancer* **17**: 223–238. doi:10.1038/nrc.2017.7

Wang ET, Sandberg R, Luo S, Khrebtukova I, Zhang L, Mayr C, Kingsmore SF, Schroth GP, Burge CB. 2008. Alternative isoform regulation in human tissue transcriptomes. *Nature* **456**: 470–476. doi:10.1038/nature07509

Weigelt B, Comino-Méndez I, de Bruijn I, Tian L, Meisel JL, García-Murillas I, Fribbens C, Cutts R, Martelotto LG, Ng CKY, et al. 2017. Diverse *BRCA1* and *BRCA2* reversion mutations in circulating cell-free DNA of therapy-resistant breast or ovarian cancer. *Clin Cancer Res* **23**: 6708–6720. doi:10.1158/1078-0432.CCR-17-0544

Zehir A, Benayed R, Shah RH, Syed A, Middha S, Kim HR, Srinivasan P, Gao J, Chakravarty D, Devlin SM, et al. 2017. Mutational landscape of metastatic cancer revealed from prospective clinical sequencing of 10,000 patients. *Nat Med* **23**: 703–713. doi:10.1038/nm.4333

Zerbino DR, Achuthan P, Akanni W, Amode MR, Barrell D, Bhai J, Billis K, Cummins C, Gall A, Girón CG, et al. 2018. Ensembl 2018. *Nucleic Acids Res* **46**: D754–D761. doi:10.1093/nar/gkx1098

Zill OA, Banks KC, Fairclough SR, Mortimer SA, Vowles JV, Mokhtari R, Gandara DR, Mack PC, Odegaard JI, Nagy RJ, et al. 2018. The landscape of actionable genomic alterations in cell-free circulating tumor DNA from 21,807 advanced cancer patients. *Clin Cancer Res* **24**: 3528–3538. doi:10.1158/1078-0432.CCR-17-3837

Cite this article as *Cold Spring Harb Perspect Med* doi: 10.1101/cshperspect.a036590

Evidence-Based Genetic Counseling for Psychiatric Disorders: A Road Map

Jehannine C. Austin[1,2]

[1]Departments of Psychiatry and Medical Genetics, University of British Columbia, Vancouver, British Columbia V5Z 4H4, Canada

[2]BC Mental Health and Substance Use Services Research Institute, Vancouver, British Columbia V6Z 2A9, Canada

Correspondence: jehannine.austin@ubc.ca

Psychiatric disorders, such as schizophrenia, depression, anxiety, and bipolar disorder, are common conditions that arise as a result of complex and heterogeneous combinations of genetic and environmental factors. In contrast to childhood neurodevelopmental conditions such as autism and intellectual disability, there are no clinical practice guidelines for applying genetic testing in the context of these conditions. But genetic counseling and genetic testing are not synonymous, and people who live with psychiatric disorders and their family members are often interested in what psychiatric genetic counseling can offer. Further, research shows that it can improve outcomes like empowerment for this population. Despite this, psychiatric genetic counseling is not yet routinely or widely offered. This review describes the state of the evidence about the process and outcomes of psychiatric genetic counseling, focusing on its clinical implications and remaining research gaps.

Psychiatric disorders, such as schizophrenia, bipolar, depression, and anxiety, are common conditions that have a complex and heterogeneous etiology and arise in individuals as a result of varied contributions of genetic and environmental factors (Schmitt et al. 2014; Psychiatric Genetics Consortium 2018). Psychiatric disorders are not entirely genetically determined, and genetic testing cannot establish, confirm, or refine a psychiatric diagnosis. Whereas genetic testing is considered a first-tier approach for childhood neurodevelopmental disorders (e.g., autism, intellectual disability) (Miller et al. 2010) to diagnose underlying genetic syndromes whose manifestations include these phenotypes, genetic testing for psychiatric disorders is not standard clinical practice, nor is it suggested by any clinical practice guidelines in the context of psychiatric disorders like schizophrenia, bipolar, depression, anxiety, obsessive–compulsive disorder (OCD), eating disorders, etc. As an apparent consequence, genetic counseling is also not routinely provided for these conditions. However, benefits of genetic counseling do not depend on the use of genetic testing, and in most scenarios, neither necessitates the other.

Many clinical specialties within genetic counseling have emerged as a result of intuitive appreciation of patient needs in a particular area and/or the availability of genetic tests. In contrast, the emergence of genetic counseling for psychiatric disorders is driven by evidence that shows a clinical need (Quaid et al. 2001; Meiser et al. 2005, 2008; Austin et al. 2006; DeLisi and Bertisch 2006; Lyus 2007; Wilhelm et al. 2009; Erickson et al. 2014; Quinn et al. 2014). Further, studies have found that even in the absence of genetic testing, psychiatric genetic counseling has benefits for people with psychiatric disorders and their families (Austin and Honer 2008; Hippman et al. 2013, 2016; Inglis et al. 2015; Moldovan et al. 2017). Herein, I review this literature, and present the psychiatric genetic counseling practice model established at the world's first specialist psychiatric genetic counseling clinic, which has been shown to produce positive patient outcomes.

PSYCHIATRIC GENETIC COUNSELING: THE NEED

Psychiatric genetic counseling for conditions such as schizophrenia, bipolar, eating disorders, depression, anxiety—like genetic counseling for any other conditions—seeks to help people make personal meaning of the factors that contribute to the development of the condition that they have or that runs in their family (Resta et al. 2006). Even without knowing the full range of issues that psychiatric genetic counseling can address, people with psychiatric disorders and their families are interested in what the service can offer (Quaid et al. 2001; Lyus 2007; Kalb et al. 2017). Commonly, people with psychiatric diagnoses and their family members overestimate chances for others in their family to develop similar conditions (and this influences childbearing decisions) (Austin et al. 2006); they feel guilty about personal choices they believe may have caused their illness, or feel powerless to do anything about their illness; and parents feel guilty about their children's psychiatric illnesses (either for passing on "bad" genes or perceived responsibility for suboptimal parenting or having failed to prevent their child's condition). Despite these cognitive and affective burdens, people with psychiatric disorders and their families are not routinely referred for genetic counseling (Hunter et al. 2009; Leach et al. 2016).

THE HISTORY OF PSYCHIATRIC GENETIC COUNSELING RESEARCH

Speculation in the research literature about the potential application and outcomes of psychiatric genetic counseling span several decades (Kumar 1968; Stancer and Wagener 1984; Reveley 1985; Hodgkinson et al. 2001; Austin and Honer 2004; DeLisi and Bertisch 2006; Finn and Smoller 2006; Lyus 2007; Bennett et al. 2008; Gershon and Alliey-Rodriguez 2013). Some of these writings reflect common misconceptions about psychiatric genetic counseling—for example, that it is primarily concerned with helping people to understand risk for recurrence in the family (Gershon and Alliey-Rodriguez 2013), or it could involve presenting psychiatric disorders as "genetic conditions" (Bennett et al. 2008). Although the moniker "genetic counseling" implies an exclusive focus on genetics, in the context of complex disorders like psychiatric conditions, genetic counseling should always include a holistic interactive discussion of both genetic and environmental factors that contribute to the condition, and, a skilled counselor is alert and prepared not only to address emergent thoughts but also emotional consequences of the exchange of information. The studies actually exploring the provision of psychiatric genetic counseling that have emerged subsequent to early speculation about its potential outcomes in the research literature have all embraced this model for delivery of the intervention. Specifically, the psychiatric genetic counseling outcomes research literature is based on a model of genetic counseling that involves discussion of the genetic and environmental contributors in a holistic manner. Practitioners approach psychiatric genetic counseling from the perspective of valuing the interaction as more existential than risk-specific and focusing on emotional issues that can arise during the discussion.

Results from an initial pilot study (Austin and Honer 2008), observational studies (Costain et al. 2012a,b), qualitative work (Hippman et al. 2013), and a randomized controlled trial (Hippman et al. 2016) of psychiatric genetic counseling provide collective evidence of positive outcomes for patients with psychiatric disorders and their family members. These data catalyzed the founding of the first specialist psychiatric genetic counseling clinic—the Adapt Clinic in Vancouver. A study assessing the outcomes of patients of the Adapt Clinic showed significant increases in measures of empowerment and self-efficacy after psychiatric genetic counseling with large effect sizes for empowerment (Inglis et al. 2015). Empowerment was assessed using the genetic counseling outcomes scale (GCOS) (McAllister et al. 2011). A subsequent qualitative study showed that for people with psychiatric diagnoses, psychiatric genetic counseling can help patients accept their illness at a deeper level and integrate it more fully into their sense of self in a manner that they experience as empowering (Semaka and Austin 2019).

Psychiatric genetic counseling is an emerging specialty within clinical genetics (Moldovan et al. 2019). Given the evidence for positive patient outcomes associated with psychiatric genetic counseling as practiced in the studies that formed the foundation for the Adapt Clinic, the structure and process of the clinic is described below. The clinic continues to collect patient outcome data and is exploring how patient- and session-related variables (e.g., specific psychiatric diagnosis, whether or not risk information is discussed) impact patient outcomes. In the following sections, when data are available regarding how patient- and/or session-related variables impact patient outcomes, these are also discussed.

STRUCTURAL OVERVIEW OF THE ADAPT CLINIC

The Team, the Referral Base

The Adapt Clinic is a specialist psychiatric genetic counseling clinic. The name was developed in consultation with local psychiatrists who understood the value of the service and wished to, but were reluctant to, refer their patients for genetic counseling fearing their patients may interpret the referral as an indicator that their psychiatrist was implying that they should not have children.

The Adapt Clinic is housed within the existing general medical genetics clinic that services the province of British Columbia and uses two board-certified genetic counselors that are supported by a clinical secretary. A medical geneticist consults on cases in which a genetic syndrome is a possibility, based on medical/family history and signs off on clinical reports. A psychiatrist is available to consult on current mental health concerns (e.g., suicidality) that may be identified during a genetic counseling session. The clinic accepts referrals from both health-care providers (generally, psychiatrists who refer when patients ask specific questions about recurrence risks and/or etiology that they feel unequipped to address) (Leach et al. 2016) and self-referrals to ensure that barriers to access are as low as possible.

The clinic is available to men and women of any age who have a personal or family history of any psychiatric disorder. The service is covered by health insurance, and individuals are eligible for psychiatric genetic counseling if they reside within the geographical area served by the medical genetics center and have a personal and/or family history of a psychiatric disorder or genetic variation that confers increased risk for psychiatric illness.

Preappointment Procedures

On establishing initial contact, a phone call is booked to document family history in advance of the appointment. Initially, this protocol was established based on perceived advantages of this approach over that of documenting family history during the session. Subsequently, evidence has suggested advantages to this approach as greater increases in self-efficacy have been reported for patients whose family history was documented before their appointment than for those for whom it was documented in-session (Slomp et al. 2018). The phone call during which family history is documented takes, on average, about 45 minutes.

Confirmation of Psychiatric Diagnoses

Psychiatric diagnoses are established based on a clinical interview with a psychiatrist or other psychiatric health-care provider. There are no genetic tests with which to establish or confirm a psychiatric diagnosis. The Adapt Clinic counselors do not seek psychiatric records to confirm diagnoses for pragmatic reasons (e.g., challenges associated with obtaining consent to access records, psychiatric diagnoses can change over time). Rather, confidence in reported psychiatric diagnoses is established through the counselors' skills in eliciting information about psychiatric symptoms. Specifically, psychiatric interviewing skills are developed through a training that combines the use of the Structured Clinical Interview for Diagnostic Statistical Manual [DSM]) (First et al. 2015), the Positive and Negative Syndrome scale (Kay et al. 1987), and the Family Interview for Genetic studies (Maxwell 2018). The first two of these instruments focus on eliciting information about psychiatric symptomatology experienced by the interviewee, and the latter is an instrument used to assess psychiatric symptoms in a family member.

Appointment and Follow-Up

Psychiatric genetic counseling sessions (on average, 90 minutes in length) are provided in-person or by telephone or telehealth according to patient needs. Genetic testing is typically considered after consultation with the clinical geneticist when family/medial history is suggestive of a genetic syndrome that involves psychiatric manifestations as part of the phenotype (e.g., 22q11.2 deletion syndrome); outside of these relatively rare cases, no genetic testing is typically provided. After the session (described below), a consult report detailing the family history and issues discussed is sent to the referring health-care provider, a physician of the patient's choosing, and/or the patient. A routine follow-up phone call (on average, ~45 minutes) is conducted approximately one month after the appointment to see if the patient wishes to return to discuss any additional issues—to date, ~5% of patients have returned for another appointment.

Case Load, Supervision, Training

In total, each patient requires four hours of counselor time (preappointment procedures, appointment, follow-up, charting, and letter

writing). The clinic prioritizes ensuring that the counselor is able to provide quality care; therefore, participation in a one hour peer supervision session every two weeks is expected. To build capacity for the delivery of psychiatric genetic counseling, the clinic provides month-long training rotations for about five genetic counseling students and two or three observers (typically health-care professionals from other countries) per year. Finally, to build relationships with the community to ensure awareness of our service, in addition to all these duties, the counselors engage in outreach and teaching.

Evaluation of Patient Outcomes

Three scales are used as clinical assessment tools: the GCOS (McAllister et al. 2011), the self-stigma in relatives of people with mental illness (SSRMI) scale (Morris et al. 2018), and the illness management self-efficacy scale (IMSES) (Lorig et al. 1996). Specifically, all are used during contracting at the beginning of the genetic counseling appointment (see Initial Contracting, below). The same instruments are also administered to structure the check in on psychological and emotional issues during the follow-up telephone call that is standard for all patients one month after their appointment. A clinical database containing this information, as well as demographic and referral information, is maintained by trained volunteers, supervised by the counselors. As these scales are administered both before and after genetic counseling, they can be used to assess change in outcomes for quality assurance/quality improvement processes, and, with appropriate ethical approvals to conduct retrospective chart reviews, they can then be used in addressing research questions about the impact of psychiatric genetic counseling on patient outcomes (e.g., Inglis et al. 2015; Morris et al. 2018; Slomp et al. 2018; Borle et al. 2018).

PSYCHIATRIC GENETIC COUNSELING: CONTENT AND PROCESS

The process and content of the psychiatric genetic counseling delivered in the Adapt Clinic is guided by and based on the research that preceded the establishment of the clinic. The manual for psychiatric genetic counseling provided (see Table 1) summarizes the psychiatric genetic counseling as performed both in the context of the foundational research on which the Adapt Clinic is based and as provided in the Adapt Clinic. The manual is included as a quick refer-

ence guide for others who may want to ensure that the process and content of their psychiatric genetic counseling is evidence based. Specific suggestions for how to raise some of the most important issues and key messages to explicitly state for patients are listed elsewhere (see Inglis et al. 2017).

Table 1. Psychiatric genetic counseling manual

Item number	Session objective/content	Observed/completed (check)
1	Establish agenda with client—focus on what they hope to achieve from the session.	———
2	Elucidate (a) client's perspective on causes of mental illness (ask if they've thought about whether others in the family have similar conditions, and what was going on for them around the time they got sick), and (b) uncover associated emotions (e.g., guilt, shame, fear).	———
3	Review client's family history: Focus on psychiatric disorders.	———
4	Engage client in broad discussion of how mental illness develops (including roles of genes and environment) using jar model.	———
5	Personalize vulnerability factors (including genetic and environmental factors) based on family history and experiences that they shared during history taking—use family history as visual aid.	———
6	Use jar model to engage in a discussion of recovery and protective factors, helping client to articulate the self-management strategies (also known as protective factors) they use to reduce the impact of their mental illness (e.g., exercise), and patterns they have noticed that could help them to try new strategies (e.g., notice mood worsening after poor sleep).	———
7	Reinforce ideas raised in No. 6, and address any gaps—sleep, nutrition, exercise, and social support are good for everyone, but some things will be more personal/individual (e.g., meditation, attending church). Focus on and address emotions attached to these issues (e.g., guilt at not eating perfectly all the time).	———
8	Engage in explicit discussion of how medications fit into protective factors, including attention to emotional associations (e.g., "I should be able to just try harder and not need medicine.").	———
9	Provide educational booklet, contact information, and invite follow-up.	———
10	THROUGHOUT: Ensure that the interaction is a two-way dialogue with the patient in which the client is an active participant and encouraged to ask questions.	———
If relevant	If client expresses interest in discussing risk in initial contracting, explore interest again AFTER discussing vulnerability and protective factors. If still interested, provide estimates within framework of population frequencies, and context of everyone having some vulnerability, using family history to illustrate upper and lower limits of range provided.	———

Initial Contracting

As in all genetic counseling, contracting around the reasons/motivations for the visit *at a level beyond the superficial* is crucial. For example, a person attending the appointment may articulate that they are there because they want to know what the chances are for their child to develop psychiatric illness, based on the family history. However, in this case, it may be deep fear, anticipatory guilt, and/or a need for control that is motivating the need to discuss risk. The most effective counseling is achieved when these underlying factors are uncovered and addressed.

The use of scales with patients during contracting allows them to communicate their needs or identify areas in which they are experiencing distress. This practice can be helpful in preparing genetic counselors to attend more specifically to these deeper layers or emotional aspects of the session, which has been shown to lead to enhanced patient outcomes (Eijzenga et al. 2014). The GCOS is used with all patients, the IMSES is used to assess confidence to manage illness with those who live with a psychiatric diagnosis, and the SSRMI scale is used to assess internalized stigma among those with no diagnosis themselves (i.e., people who have a relative with psychiatric illness). All can be useful in establishing key issues affecting participants' lives, which are likely to need to be addressed in the genetic counseling session. For example, if a patient strongly agrees with the GCOS item concerning guilt about passing on a condition, the counselor will propose that as a topic for discussion.

In the most recent assessment, the most commonly reported indication for attending the clinic (62% of attendees) was "to understand the causes of mental illness" (Borle et al. 2018). In practice, this is characterized by existential or affective, rather than cognitive or theoretical questions. Common questions are "Why me?" and "Is there anything I can do to prevent an (other) episode of mental illness?" The second most common reported reason for attending (44% of attendees) was "to learn chance of illness recurrence" (Borle et al. 2018).

When a patient expresses during contracting that their primary—or only—motivation for attending genetic counseling is to discuss the chance for others in the family to develop psychiatric illness, or to understand options for genetic testing, the counselors do not *immediately* address either specific issue. Instead, they use the contracting process to seek to understand the deeper, emotional motivation for the patient to be seeking that information. The counselor also explains that people usually find it provides helpful context to discuss the factors that contribute both to development of illness, and to recovery first, before discussing risk and/or testing. Thus, the contracting process helps to ensure that patients' articulated needs are met as a result of probing by the counselor that facilitates insight into deeper related issues of concern. Setting expectations for a session that addresses these issues expands the likelihood of a broader benefit from the session.

Psychotherapeutically Oriented Information Exchange

Etiology

The need to establish a clear understanding of the patient's explanation of cause of the personal or familial condition is foundational for any genetic counseling encounter. The psychiatric family history can be used to reflect on perceptions of causation, and discussion of life experiences around or preceding symptom onset can be especially useful for the counselor to understand. Once a clear picture of the patients' perceptions of cause have been established, the counselor can introduce the jar model as a visual representation of the way that genetic and environmental (or experiential) vulnerability factors contribute together to the onset of psychiatric illness (see Fig. 1).

The Adapt Clinic experience has found that presenting the model first in terms of broad overall concepts, and personalizing by integrating it with concepts from the patient's initial reflections (e.g., showing how certain specific stressful life events they mentioned could have contributed environmental vulnerability into the jar, or how family history of related psychiatric disorders can translate to increased genetic vulnerability), is an effective approach.

 Cite this article as *Cold Spring Harb Perspect Med* doi: 10.1101/cshperspect.a036608

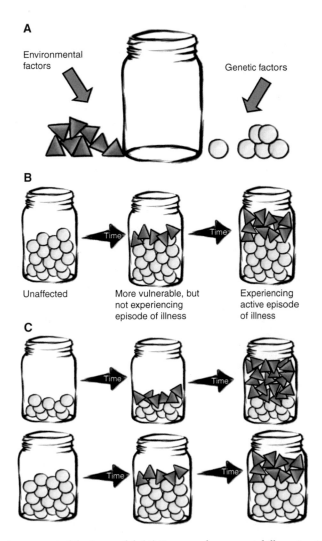

Figure 1. Introducing the concept of the jar model. (A) Everyone has a mental illness jar; it can be filled with two kinds of vulnerability factors. (B) The amount of genetic vulnerability in the jar does not change over time, unlike experiential vulnerability. To experience an active episode of illness, the jar must be full to the top. (C) Someone with a large amount of genetic vulnerability may be more likely to develop mental illness than someone with a small amount—it is more likely that the jar will fill all the way. Someone with a large amount of genetic vulnerability may develop mental illness at a younger age than someone with a small amount—it takes less time for the jar to fill up. (Figures reprinted, with permission, from Peay and Austin 2011.)

Although there are, theoretically, many possible ways of visually representing how genes and environment contribute together to the development of complex disorders, evidence supports the effectiveness of the jar model in conveying complex inheritance among psychiatric genetic counseling clients (Hippman et al. 2016; Semaka and Austin 2019).

Throughout the discussion about etiology, the counselor attends to the patient's responses and is ready, for example, to clarify how a lack of a family history of psychiatric illness does not imply the patient has no genetic vulnerability—research clearly shows that *everyone* has some genetic vulnerability to psychiatric disorders (see Fig. 2); and address issues, such as

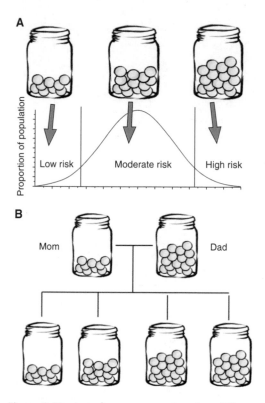

Figure 2. Everyone has some genetic vulnerability to mental illness. (*A*) Genome-wide association studies show that we all have some genetic vulnerability to mental illness—the variations are so common. We will vary in how much genetic vulnerability we each have with most people having a moderate amount. (*B*) Having no one in the family with a history of psychiatric illness does not preclude having inherited genetic vulnerability to mental illness; it simply means that other family members' jars did not fill to the top. (Figures reprinted, with permission, from Peay and Austin 2011.)

guilt, around how a history of substance use may have contributed to vulnerability, or—among parents—around how passing on "bad genes" may have contributed. The counselor's objectives are, first, to ensure that the patient understands that psychiatric illness is not their fault and is in no way a moral failing. The second objective is to establish a model of genes and environment that has direct personal relevance to the patient to provide a framework for discussing how to protect their future mental health.

Recovery

One of the components of psychiatric genetic counseling that patients find most impactful (Semaka and Austin 2019) is using the model for *cause* of illness as a framework for discussing how people can reduce risk for illness onset and/or recover from active episodes of illness. Again, the Adapt Clinic experience suggests that an effective approach to this involves first introducing the broad concept of factors that are protective to mental health (using the jar model), and then opening a two-way discussion, drawing on the patient's experience, when possible, about how some protective factors (sleep, nutrition, exercise, good social support, and finding effective ways to manage stress) are useful. The counselors find it effective to emphasize that these are examples of protective factors for mental health that can be applied to everyone, regardless of whether or not we have ever had a diagnosis of a psychiatric disorder (see Fig. 3), whereas other strategies (i.e., use of psychotropic medication) can be useful for those who have had a diagnosis, and yet other strategies (e.g., avoiding street drugs like crystal meth and cannabis) can be particularly important for those who may have elevated risk for conditions that involve psychosis.

In addition to discussing protective strategies broadly, the counselor spends time working with the patient to identify specific protective factors that they identify as important for them. Although having good social support and finding effective ways to manage stress are strategies that can be helpful for everyone, the specific ways in which people achieve this vary. Helping a patient to identify the specific activities that work for them (e.g., attending church, spending time with a pet, journaling, cooking) can be helpful. Through this discussion, the counselor is vigilant in attending to signs of emotional response (e.g., "I should just be able to try harder to get better," "I shouldn't need medicine," "it's my fault I developed [condition x] because I used too much cannabis when I was young"), which he or she then explicitly identifies and addresses (e.g., "It sounds like you feel it's your fault that you have mental illness—it's not. Many people

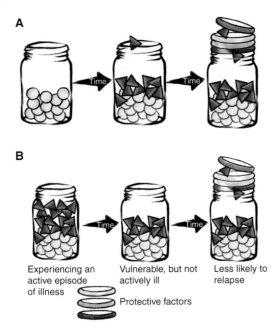

A

B

Experiencing an active episode of illness

Vulnerable, but not actively ill

Less likely to relapse

Protective factors

Figure 3. Recovery and reducing risk. (*A*) Reducing risk: Protective factors stack on top of the jar, making it taller and therefore able to accommodate more environmental factors before becoming full. Some protective factors are good for all of us (e.g., sleep, nutrition, exercise, social support, finding effective ways for managing stress), but some may be more unique to us. (*B*) Recovery from an active episode of illness: Sometimes, it may be possible to remove some of our environmental vulnerability factors (e.g., removing oneself from a particularly stressful situation, stopping use of crystal meth). But even if there are no modifiable vulnerability factors, protective factors can help increase the capacity of the jar—in addition to those noted above, medication can also be useful for those who have been diagnosed with mental illness. (Figures reprinted, with permission, from Peay and Austin 2011.)

use cannabis without getting sick, and cannabis on its own is not enough to cause [condition x]"). The counselor communicates with the patient about how use of protective factors cannot definitively *prevent* an episode of mental illness, that perfection in performing all of these self-care strategies is impossible, and the goal is not to strive for that, but to do what you can, and celebrate that with self-compassion.

Framing the discussion about strategies for recovery in terms of understanding the cause of

illness can be powerful; people intuitively want their treatment for illness to reflect what they perceive to be the cause. Clinic patients regularly experience shifts in attitude toward psychotropic medication use through genetic counseling; understanding that there is a genetic (biological) contribution helps some individuals to see how a biological treatment could be helpful, whereas for others, understanding it is not *all* biological motivates people to engage in self-care strategies to protect their mental health. In the context of psychiatric genetic counseling, a key part of facilitating meaning-making and adaptation is helping patients to apply their new or deeper understanding of cause of illness to come up with strategies to protect their mental health for the future.

Discussing Chance for Familial Recurrence

As described in Initial Contracting, data show that <50% of individuals presenting for psychiatric genetic counseling are interested in discussing chance for familial recurrence at the time of referral/initial contracting. Any discussion about risk is framed in the context of shared understanding of illness etiology, so after discussing psychiatric disorder etiology and recovery (as presented above) the counselor checks in to see if the patient would like to discuss specific numbers—essentially, this is a second contracting step that is specifically concerned with discussing chances for familial recurrence. Some patients (>25%) who indicate at initial contracting that they want to discuss chances for familial recurrence change their mind after discussing etiology, as do 11% of those who initially indicate not wanting to discuss chances for familial recurrence (Borle et al. 2018). Ultimately, after the discussions described above regarding etiology and recovery, ~40% of clinic patients opt to discuss specific chances for familial recurrence. Whereas all patients (even those who learn that the chances for familial recurrence are higher than they expected) benefit from psychiatric genetic counseling, those with the largest increases in empowerment were people who initially indicated wanting to know chances for familial recurrence, but changed their minds after discussing etiology/recovery (Borle et al. 2018). This suggests that addressing the underlying (emotional) motivation for interest in risk information, as discussed above, is sufficiently powerful.

The second contracting step focuses on chances for familial recurrence and ensures that the process of psychiatric genetic counseling is as patient-centered as possible. When specific chances are discussed, they are provided only for the patient and/or children under the patient's guardianship (i.e., specific chances for a child to develop a psychiatric disorder would not be provided to grandparents if they are *not* guardians of the child in question). Specific chances are based on analysis of a detailed, three-generation psychiatric family history using empiric data (Austin et al. 2008), provided in the form of absolute risks and frequencies with broad ranges, and always discussed in the context of population rates. As with all other components of psychiatric genetic counseling, throughout the discussion of chances for familial recurrence, the counselor works to understand and address affective responses.

Closing the Session

To close the session, the counselor ensures that the patient is aware of plans to connect by phone one month later, confirms the health-care professionals to whom consult reports should be sent, and provides resources, including referrals to other professionals or community supports, and written material to take home (an educational booklet that is available for download in several languages from the National Society of Genetic Counselors [NSGC], Canadian Association of Genetic Counselors [CAGC], and Association of Genetic Nurses and Counsellors [AGNC]websites).

Impact of Symptoms of Indicated Condition on the Appointment

In most genetic counseling specialties, patients showing symptoms of the condition for which they were referred will not directly impact the counseling interaction. This maxim holds true for the vast majority of psychiatric genetic counseling sessions as well. But one of the most common questions received by Adapt Clinic counselors from other counselors about psychiatric genetic counseling concerns the potential for symptoms of psychiatric illness to impact the counseling interaction and thereby confound our ability to meaningfully help patients—for example, what if a patient does not have insight into the fact that they have a psychiatric diagnosis? How do you counsel a patient who is experiencing psychosis? Is a patient with a psychiatric disorder too fragile to discuss the issues that constitute the focus of a genetic counseling interaction—will it trigger an illness episode?

Inherently, those who physically attend a psychiatric genetic counseling appointment are organized and functioning well enough to get there at the scheduled time. Attending an appointment after the initial phone call during which family history and the broad purpose of the session are discussed implies that the individual is ready to discuss psychiatric disorders in the context of their family, and, if they have been referred for personal history of psychiatric illness, it implies that they self-identify as having at least some component of a psychiatric diagnosis. Importantly, people can be meaningfully helped in genetic counseling even if they do not identity with the specific label they have received —for example, for someone who identifies as having experienced mania but rejects the label of bipolar disorder, the genetic counseling session can simply focus on factors that contribute to or reduce the chance for an episode of mania. Although patients can be counseled—and with positive outcomes—when experiencing active episodes of illness, it is always an option to ask a patient if they feel that rescheduling for another time may allow them to benefit maximally from the interaction. When concerned about suicidality, it is important to check in on this explicitly—many tools and checklists exist to help with this and identify critical next steps. Contrary to popular concern, asking about suicidal ideation does not trigger suicidal reactions; in fact, research shows it has the opposite effect (Dazzi et al. 2014). It is crucial to remember that people who live with psychiatric illness are often

     Cite this article as *Cold Spring Harb Perspect Med* doi: 10.1101/cshperspect.a036608

deeply resilient, and the interactions that we can offer in genetic counseling, when applied as described above, can lead to meaningful positive outcomes for individuals who are often underserved.

OUTCOMES OF PSYCHIATRIC GENETIC COUNSELING

In qualitative work, recipients of psychiatric genetic counseling report that they perceive the process described in this review as allowing them to feel heard, validated, and supported ("*[It was like] talking to someone intelligent and understanding and knowledgeable, who listened and [was] genuinely interested, you could tell she cared about the subject and helping people, helping me*") (Semaka and Austin 2019). The outcome of psychiatric genetic counseling, when using tools like the jar model as described above, was articulated beautifully by one interviewee: "*[Psychiatric genetic counseling] made me feel more empowered in the things that I can do, like it made me more empowered to be able to manage what I've been given or where I'm at now. Whatever has happened, whatever I've done or not done that has caused my jar to look*

the way it does, whether I've added the little triangles or not, I still have the power to add rings and to ensure that I'm protecting myself" (Semaka and Austin 2019).

Quantitative research shows that psychiatric genetic counseling can increase knowledge (Hippman et al. 2016), improve accuracy of risk perception (Hippman et al. 2016), and increase both empowerment and self-efficacy (Inglis et al. 2015). At least for empowerment, these effects seem to be sustained over time (see Fig. 4).

Empowerment and self-efficacy are important outcomes for two reasons. First, they are important for their own sake; in qualitative work, people talked about how the process of empowerment through psychiatric genetic counseling facilitated a deeper and fuller integration of their acceptance of their mental illness into their sense of self (Semaka and Austin 2019). Second—and crucially—empowerment and self-efficacy are necessary (although not sufficient) for people to engage in behavioral change that can help to protect their mental health. Indeed, in qualitative work, some people with mental illness who had received genetic counseling talked about how the process had

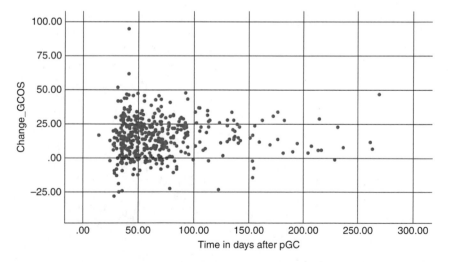

Figure 4. Empowerment scores on the genetic counseling outcomes scale (GCOS) over time after genetic counseling. Change in GCOS scores from pre- to postpsychiatric genetic counseling, shown according to length of time (in days) between pre- and postgenetic counseling data collection, showing that outcomes of psychiatric genetic counseling seem to be stable over at least 6 months postintervention. (pGC) Postgenetic counseling. (Figure provided by Emily Morris.)

empowered them to "adopt new protective strategies like good sleep, nutrition, and exercise" (Semaka and Austin 2019).

THE FUTURE

Continuing to Explore Outcomes of Psychiatric Genetic Counseling

The evidence from quantitative work showing increased empowerment and self-efficacy after psychiatric genetic counseling, and the subsequent suggestion from qualitative work that people engage in behavior changes to better protect their mental health after receiving psychiatric genetic counseling, raises the question of whether psychiatric genetic counseling can ultimately influence mental health outcomes. Ongoing work within the Translational Psychiatric Genetics Group in Vancouver is exploring the effect of psychiatric genetic counseling on behavior change, medication adherence among people with psychiatric disorders, and mental health outcomes (like frequency and duration of mental health hospitalizations). As well, we are currently exploring how variables such as specific psychiatric diagnosis, time since diagnosis, and method of service delivery may influence outcomes of psychiatric genetic counseling for those who have a psychiatric diagnosis. Among relatives of those with psychiatric diagnoses, we are examining whether genetic counseling influences internalized stigma.

Genetic Testing

Conversations continue in the psychiatric genetics community regarding whether genetic testing for copy number variations (CNVs) should routinely be offered for subgroups of patients (https://ispg.net/genetic-testing-statement/)—for example, those with schizophrenia. As well, the availability of polygenic risk scores indicating risk for complex conditions, including psychiatric disorders, has pushed this type of testing to the forefront—pressing us to consider if and how to apply or manage them in clinical settings. It is critical to remain aware that the application of any type of genetic test in the context of psychiatric genetic counseling will *not* represent a fundamental paradigm shift; these tests cannot diagnose a psychiatric disorder or definitively determine whether or not someone will develop one. Any testing applied in a psychiatric genetic counseling context will be, at best, probabilistic information about risk and must still be applied in the context of an encounter that addresses the emotional issues—guilt, blame, shame, fear, and stigma—that are so often attached to people's explanations for cause of illness. As technology evolves, we serve our patients best by remembering that the most important and impactful work of the genetic counselor is achieved by using these tools in the context of an encounter that is psychotherapeutically focused (Austin et al. 2014).

CONCLUDING REMARKS

In sum, qualitative and quantitative work point to important outcomes of psychiatric genetic counseling for those who live with psychiatric conditions and their families. These outcomes are a result of genetic counselors' use of the tools and strategies described here to successfully walk the fine line of helping people understand that mental illness is not their fault, *but* there are things that they can do to protect their mental health going forward. Given the important outcomes that genetic counseling can produce in the context of adult psychiatric disorders (even in the absence of genetic testing), it is crucial that clinical genetics services prioritize this population—who have historically been so underserved by our area of medicine—for receipt of services.

Further, the model outlined here for the evaluation of outcomes of clinical services (i.e., integration of psychometric instruments as clinical assessment tools) and exploration of a logically progressive sequence of outcomes could be more broadly applied in other genetic counseling settings.

ACKNOWLEDGMENTS

I offer sincere thanks to past and present members of the Translational Psychiatric Genetics

Group in Vancouver for their varied support, commitment, and contributions. Without all of you—past and present—the work on which this manuscript is founded would not have been possible. In particular, for this manuscript, I offer the deepest gratitude to Emily Morris, who contributed Figure 4. Finally, I was supported by the Canada Research Chairs Program and BC Mental Health and Substance Use Services.

REFERENCES

Austin JC, Honer WG. 2005. The potential impact of genetic counseling for mental illness. *Clin Genet* **67**: 134–142.

Austin JC, Honer WG. 2008. Psychiatric genetic counselling for parents of individuals affected with psychotic disorders: a pilot study. *Early Int Psychiatry* **2**: 80–89. doi:10 .1111/j.1751-7893.2008.00062.x

Austin JC, Smith GN, Honer WG. 2006. The genomic era and perceptions of psychotic disorders: genetic risk estimation, associations with reproductive decisions and views about predictive testing. *Am J Med Genet B Neuropsychiatr Genet* **141B**: 926–928. doi:10.1002/ajmg.b .30372

Austin JC, Palmer CGS, Rosen-Sheidley B, Veach PM, Gettig E, Peay HL. 2008. Psychiatric disorders in clinical genetics II: individualizing recurrence risks. *J Genet Couns* **17**: 18–29. doi:10.1007/s10897-007-9121-4

Austin J, Semaka A, Hadjipavlou G. 2014. Conceptualizing genetic counseling as psychotherapy in the era of genomic medicine. *J Genet Couns* **23**: 903–909. doi:10.1007/ s10897-014-9728-1

Bennett L, Thirlaway K, Murray AJ. 2008. The stigmatising implications of presenting schizophrenia as a genetic disease. *J Genet Couns* **17**: 550–559. doi:10.1007/s10897-008-9178-8

Bipolar Disorder and Schizophrenia Working Group of the Psychiatric Genomics Consortium, Ruderfer DM, Ripke S, McQuillin A, Boocock J, Stahl EA, Whitehead Pavlides JM, Mullins N, Charney AW, Ori APS, et al. 2018. Genomic dissection of bipolar disorder and schizophrenia, including 28 sub-phenotypes. *Cell* **173**: 1705–1715.e16. doi:10.1016/j.cell.2018.05.046

Borle K, Morris E, Inglis A, Austin J. 2018. Risk communication in genetic counseling: exploring uptake and perception of recurrence numbers, and their impact on patient outcomes. *Clin Genet* **94**: 239–245. doi:10.1111/ cge.13379

Costain G, Esplen MJ, Toner B, Hodgkinson KA, Bassett AS. 2014a. Evaluating genetic counseling for family members of individuals with schizophrenia in the molecular age. *Schizophr Bull* **40**: 88–99. doi:10.1093/schbul/ sbs124

Costain G, Esplen MJ, Toner B, Scherer SW, Meschino WS, Hodgkinson KA, Bassett AS. 2014b. Evaluating genetic counseling for individuals with schizophrenia in the molecular age. *Schizophr Bull* **40**: 78–87. doi:10.1093/schbul/ sbs138

Dazzi T, Gribble R, Wessely S, Fear NT. 2014. Does asking about suicide and related behaviours induce suicidal ideation? What is the evidence? *Psychol Med* **44**: 3361–3363. doi:10.1017/S0033291714001299

DeLisi LE, Bertisch H. 2006. A preliminary comparison of the hopes of researchers, clinicians, and families for the future ethical use of genetic findings on schizophrenia. *Am J Med Genet B Neuropsychiatr Genet* **141B**: 110–115. doi:10.1002/ajmg.b.30249

Eijzenga W, Aaronson NK, Hahn DEE, Sidharta GN, van der Kolk LE, Velthuizen ME, Ausems MGEM, Bleiker EMA. 2014. Effect of routine assessment of specific psychosocial problems on personalized communication, counselors' awareness, and distress levels in cancer genetic counseling practice: a randomized controlled trial. *J Clin Oncol* **32**: 2998–3004. doi:10.1200/JCO.2014.55.4576

Erickson JA, Kuzmich L, Ormond KE, Gordon E, Christman MF, Cho MK, Levinson DF. 2014. Genetic testing of children for predisposition to mood disorders: anticipating the clinical issues. *J Genet Couns* **23**: 566–577. doi:10 .1007/s10897-014-9710-y

Finn CT, Smoller JW. 2006. Genetic counseling in psychiatry. *Harvard Rev Psych* **14**: 109–121.

First MB, Williams JBW, Karg RS, Spitzer RL. 2015. *Structured Clinical Interview for DSM-5 Disorders, Clinical Trials Version (SCID-5-CT)*. American Psychiatric Association, Arlington, VA.

Gershon E, Alliey-Rodriguez N. 2013. New ethical issues for genetic counseling in common mental disorders. *Am J Psych* **170**: 968–976.

Hippman C, Lohn Z, Ringrose A, Inglis A, Cheek J, Austin JC. 2013. 'Nothing is absolute in life': Understanding uncertainty in the context of psychiatric genetic counseling from the perspective of those with serious mental illness. *J Genet Couns* **22**: 625–632. doi:10.1007/s10897-013-9594-2

Hippman C, Ringrose A, Inglis A, Cheek J, Albert AYK, Remick R, Honer WG, Austin JC. 2016. A pilot randomized clinical trial evaluating the impact of genetic counseling for serious mental illnesses. *J Clin Psychiatry* **77**: e190–e198. doi:10.4088/JCP.14m09710

Hodgkinson KA, Murphy J, O'Neill S, Brzustowicz L, Bassett AS. 2001 Genetic counselling for schizophrenia in the era of molecular genetics. *Can J Psych* **46**: 123–130.

Hunter MJ, Hippman C, Honer WG, Austin JC. 2009. Genetic counseling for schizophrenia: a review of referrals to a provincial medical genetics program from 1968 to 2007. *Am J Med Genet A* **152**: 147–152.

Inglis A, Koehn D, McGillivray B, Stewart SE, Austin J. 2015. Evaluating a unique, specialist psychiatric genetic counseling clinic: uptake and impact. *Clin Genet* **87**: 218–224. doi:10.1111/cge.12415

Inglis A, Morris E, Austin J. 2017. Prenatal genetic counselling for psychiatric disorders. *Prenat Diagn* **37**: 6–13. doi:10.1002/pd.4878

Kalb FM, Vincent V, Herzog T, Austin J. 2017. Genetic counseling for alcohol addiction: assessing perceptions and potential utility in individuals with lived experience and their family members. *J Genet Couns* **26**: 963–970. doi:10.1007/s10897-017-0075-x

Kay SR, Fiszbein A, Opler LA. 1987. The Positive and Negative Syndrome Scale (PANSS) for schizophrenia. *Schizophr Bull* **13:** 261–276. doi:10.1093/schbul/13.2.261

Kumar P. 1968. Genetic counselling in family planning. *Antiseptic* **65:** 831–834.

Leach E, Morris E, White HJ, Inglis A, Lehman A, Austin J. 2016. How do physicians decide to refer their patients for psychiatric genetic counseling? A qualitative study of physicians' practice. *J Genet Couns* **25:** 1235–1242. doi:10.1007/s10897-016-9961-x

Lorig K, Stewart A, Ritter P, Gonzalez V, Laurent D, Lynch J. 1996. *Outcome measures for health education and other healthcare interventions.* Sage, Thousand Oaks, CA.

Lyus VL. 2007. The importance of genetic counseling for individuals with schizophrenia and their relatives: potential clients' opinions and experiences. *Am J Med Genet B* **144B:** 1014–1021. doi:10.1002/ajmg.b.30536

Maxwell EM. 2018. *Family interview for genetic studies.* Nimhgenetics.org [accessed July 31, 2019]. https://www .nimhgenetics.org/interviews/figs/FIGS%201.0%20Manual %20-%20Aug%201992.pdf.

McAllister M, Wood AM, Dunn G, Shiloh S, Todd C. 2011. The genetic counseling outcome scale: a new patient-reported outcome measure for clinical genetics services. *Clin Genet* **79:** 413–424. doi:10.1111/j.1399-0004.2011 .01636.x

Meiser B, Mitchell PB, McGirr H, Van Herten M, Schofield PR. 2005. Implications of genetic risk information in families with a high density of bipolar disorder: an exploratory study. *Soc Sci Med* **60:** 109–118. doi:10.1016/j .socscimed.2004.04.016

Meiser B, Kasparian NA, Mitchell PB, Strong K, Simpson JM, Tabassum L, Mireskandari S, Schofield PR. 2008. Attitudes to genetic testing in families with multiple cases of bipolar disorder. *Genet Test* **12:** 233–243. doi:10.1089/ gte.2007.0100

Miller DT, Adam MP, Aradhya S, Biesecker LG, Brothman AR, Carter NP, Church DM, Crolla JA, Eichler EE, Epstein CJ, et al. 2010. Consensus statement: chromosomal microarray is a first-tier clinical diagnostic test for individuals with developmental disabilities or congenital anomalies. *Am J Hum Genet* **86:** 749–764. doi:10.1016/j .ajhg.2010.04.006

Moldovan R, Pintea S, Austin J. 2017. The efficacy of genetic counseling for psychiatric disorders: a meta-analysis. *J Genet Couns* **26:** 1341–1347. doi:10.1007/s10897-017-0113-8

Moldovan R, McGhee K, Coviello DA, Hamang A, Inglis A, Ingvoldstad Malmgren C, Johansson-Soller M, Laurino M, Meiser B, Murphy L, et al. 2019. Psychiatric genetic counseling: a mapping exercise. *Am J Med Genet B Neuropsychiatr Genet* doi:10.1002/ajmg.b.32735

Morris E, Hippman C, Murray G, Michalak EE, Boyd JE, Livingston J, Inglis A, Carrion P, Austin J. 2018. Self-stigma in relatives of people with mental illness scale: development and validation. *Br J Psychiatry* **212:** 169–174. doi:10.1192/bjp.2017.23

Peay H, Austin J. 2011. *How to talk with families about genetics and psychiatric illness.* W.W. Norton, New York.

Quaid KA, Aschen SR, Smiley CL, Nurnberger JI Jr. 2001. Perceived genetic risks for bipolar disorder in a patient population: an exploratory study. *J Genet Couns* **10:** 41–51. doi:10.1023/A:1009403329873

Quinn V, Meiser B, Wilde A, Cousins Z, Barlow-Stewart K, Mitchell PB, Schofield PR. 2014. Preferences regarding targeted education and risk assessment in people with a family history of major depressive disorder. *J Genet Couns* **23:** 785–795. doi:10.1007/s10897-013-9685-0

Resta R, Biesecker BB, Bennett RL, Blum S, Estabrooks Hahn S, Strecker MN, Williams JL. 2006. A new definition of genetic counseling: National Society of Genetic Counselors' Task Force report. *J Genet Couns* **15:** 77–83. doi:10 .1007/s10897-005-9014-3

Reveley A. 1985. Genetic counseling for schizophrenia. *Br J Psych* **147:** 107–112.

Schmitt A, Malchow B, Hasan A, Falkai P. 2014. The impact of environmental factors in severe psychiatric disorders. *Front Neurosci* **8:** 19. doi:10.3389/fnins.2014.00019

Semaka A, Austin J. 2019. Patient perspectives on the process and outcomes of psychiatric genetic counseling: an "empowering encounter". *J Genet Couns* **28:** 856–868. doi:10 .1002/jgc4.1128

Slomp C, Morris E, Inglis A, Lehman A, Austin J. 2018. Patient outcomes of genetic counseling: assessing the impact of different approaches to family history collection. *Clin Genet* **93:** 830–836.

Stancer HC, Wagener DK. 1984. Genetic counselling: its need in psychiatry and the directions it gives for future research. *Can J Psych* **29:** 289–294.

Wilhelm K, Meiser B, Mitchell PB, Finch AW, Siegel JE, Parker G, Schofield PR. 2009. Issues concerning feedback about genetic testing and risk of depression. *Br J Psychiatry* **194:** 404–410. doi:10.1192/bjp.bp.107.047514

Cite this article as *Cold Spring Harb Perspect Med* doi: 10.1101/cshperspect.a036608

Genetic Risk Assessment in Psychiatry

Holly Landrum Peay

Center for Newborn Screening, Ethics, and Disability Studies, RTI International, Research Triangle Park, North Carolina 27703, USA

Correspondence: hpeay@rti.org

Most psychiatric disorders of pediatric and adult onset are caused by a complex interplay of genetic and environmental risk factors. Risk assessment in genetic counseling is correspondingly complicated. Outside of neurodevelopmental conditions, genetic and genomic testing has not achieved clinical utility. Genetic counselors most often base risk assessment on the client's medical and family history and empiric recurrence risk data. In rare cases significant familial risk may arise from variants of large effect. New approaches such as polygenic risk scores have the potential to inform diagnosis and management of affected individuals and risk status for at-risk individuals. Research on the genetic and environmental factors that increase risk for schizophrenia and etiologically related disorders are reviewed, guidance in determining and communicating risks to families is delivered, and new opportunities and challenges that will come with translating new research findings to psychiatric risk assessment and genetic counseling are anticipated.

Psychiatric disorders are common and associated with significant functional impairment (Whiteford et al. 2013). Our knowledge about the complex etiology of psychiatric disorders is rapidly evolving. Strong evidence implicates genetic risk factors associated with psychiatric disorders of both pediatric and adult onset (Sullivan and Geschwind 2019). Genetic heritability estimates range from 40% to >80% (Polderman et al. 2015) with the heritability of bipolar disorder and schizophrenia at the high end of the range (Nöthen et al. 2010). Genome-wide association studies (GWASs) have identified common single nucleotide polymorphisms (SNPs) and rare copy-number variants (CNVs) for many psychiatric disorders of childhood and adult onset (Psychiatric GWAS Consortium Coordinating Committee et al. 2009). There is also consistent evidence for genetic risk factors that are shared across different diagnoses, including across childhood- and adolescent- or adult-onset disorders (Rasic et al. 2014; Martin et al. 2018; Sullivan and Geschwind 2019).

A wide range of environmental risk factors also contribute to etiology and may moderate genetic risk, including traumatic brain injury at birth, maternal prenatal infection or viral infection during the lifetime, environmental toxins, childhood trauma, stressful life events, head injury, and substance abuse (Peay and Austin 2011; Uher 2014; Janoutová et al. 2016; Marangoni et al. 2016; Al-Haddad et al. 2019; Bölte et al. 2019; Estrada-Prat et al. 2019). Unfortunately, with the exception of substance use, these factors are largely outside of the control of at-risk individuals. Though this is an oversimplification of the complex interplay among the risk factors, it is useful to consider genetic risk as conferring vulnerability to environmental stressors (Peay and Austin 2011).

There is some evidence that psychiatric diagnoses represent extreme ends on a continuous spectrum of related population traits (Rössler et al. 2007; Martin et al. 2018). Under this model, as genetic and/or environmental risk increases across a population, the associated traits become more extreme until a threshold is reached. After reaching the threshold, individuals would have a level of symptomatology that is sufficient to receive a psychiatric diagnosis.

And yet, even given the rapid pace of research progress in psychiatry, risk assessment is not yet meaningfully informed by genetic or genomic testing for the majority of clients. Most identified genetic risk factors play a small role in etiology (Mitchell et al. 2010; Sullivan and Geschwind 2019), and even when genetic risk factors are aggregated, they do not yet explain a clinically relevant portion of risk (Zheutlin and Ross 2018; Martin et al. 2019). Neurodevelopmental disorders are the exception (see Blesson and Cohen 2019).

The application of psychiatric genomic research to inform individual-level risk is further complicated by the current diagnostic classification, which does not define clear biological entities and thus the relationship between diagnostic categories and genomic results is probabilistic (Smoller et al. 2019). Eventual improvements in understanding the etiology are anticipated to lead to new, etiologically based diagnostic categories (Bipolar Disorder and Schizophrenia Working Group of the Psychiatric Genetics Consortium 2018; Zheutlin and Ross 2018; Sullivan and Geschwind 2019). Clinical applications related to diagnosis and management of symptomatic individuals will likely precede the ability to forecast risk in at-risk relatives.

Regardless of these limitations, genetic counseling about etiology and recurrence for psychiatric disorders is of value to clients. In most cases, risk assessment continues to be based on personal and family history. Clients have varied reasons for seeking consultation that may or may not include a desire for a quantified assessment of risk (Borle et al. 2018). This highlights the importance of contracting prior to engaging in risk assessment and customizing the counseling to client needs. Clients seek psychiatric genetic counseling for multiple indications (e.g., based on personal history, with concerns about personal or familial recurrence; based on family history with concerns about personal and/or family risk; in situations of adoption) (Peay et al. 2017; also see Austin 2019).

This review prepares genetic counselors to engage in risk assessment by providing an overview of research on the etiology of schizophrenia as an illustrative example, delivering guidance for determining and communicating risk based on family history, and anticipating benefits and challenges to translating new research findings to psychiatric genetic counseling. The focus is on risk assessment for disorders that traditionally have onset in adolescence or adulthood. Here we assume that psychiatric risk in the family is, or has become, the primary focus of the genetic counseling session, and the genetic counselor has sufficient time to engage in targeted risk assessment and counseling.

OVERVIEW OF THE ETIOLOGY OF SCHIZOPHRENIA

Schizophrenia is among the most severe of the major psychiatric disorders and is associated with considerable disability (Whiteford et al. 2013). Schizophrenia has been the focus of considerable research to elucidate specific risk factors and evidence of causative overlap with other psychiatric conditions. This research is briefly summarized below.

Common Variants with Small Effect Sizes

The mapping of genes for schizophrenia and other complex disorders is conducted using GWASs, which provides an assessment of variation across the genome through comparison of millions of polymorphisms across cases and controls. Hundreds of loci implicated in schizophrenia risk have been identified using GWASs, with enrichment of brain-expressed genes, genes associated with immune function, and genes implicated in synaptic plasticity and function (Avramopoulos 2018; Bearden and Forsyth 2018; Coelewij and Curtis 2018; Prata et al. 2019).

The variants found in GWASs can be used to develop polygenic risk scores (PRSs). PRSs provide an analytic approach to explore how genetic risk is manifested in individuals across different study populations, and thus how phenotypes are related to one another (Bipolar Disorder and Schizophrenia Working Group of the Psychiatric Genetics Consortium 2018; Mistry et al. 2018; Nakahara et al. 2018). PRSs in schizophrenia reflect a broad phenotypic spectrum, with studies finding association with factors such as cognitive ability, emotional identification, verb reasoning, and memory (Nakahara et al. 2018).

Copy-Number Variants (CNVs)

Although rare, CNVs can confer significant risk for schizophrenia (Marshall et al. 2017), and the identification of such variants may have clinical utility for medical management and genetic counseling. An example is deletion 22q11.2 (velocardiofacial syndrome [VCFS]), which confers an increased risk of psychosis. A deletion at 22q11.2 has been estimated to be present in 0.3% of those diagnosed with schizophrenia (Marshall et al. 2017). Importantly, though the genetic variations directly cause VCFS, having the 22q11.2 deletion is neither necessary nor sufficient to result in psychiatric illness.

In a large recent case-control study, 22q11.2 and seven additional CNVs achieved genome-wide significance (Marshall et al. 2017). Most of these high-risk variants are hypothesized to be de novo because of the negative selection associated with the phenotypic outcomes, which include intellectual disability and dysmorphic features as well as schizophrenia (Marshall et al. 2017; Avramopoulos 2018). Taken together, CNVs are expected to account for 2% of patients with schizophrenia (Bearden and Forsyth 2018; Coelewij and Curtis 2018). Additional research is needed to assess the lifetime risk for psychiatric outcomes associated with emerging CNV findings.

Rare Variants

There is evidence for rare, damaging variants that contribute substantially to schizophrenia risk (Avramopoulos 2018; Coelewij and Curtis 2018). Additional variants are likely to be identified through ongoing genomic sequencing studies (Bearden and Forsyth 2018).

Genetic Risk Factors Replicated across Approach

Several functional categories of genes have been implicated by studies identifying both common and rare variants. These include genes involved in central nervous system development, immunity, synaptic transmission, N-methyl-D-aspartate receptors, activity-regulated cytoskeletal complex, and fragile-X-related protein (Avramopoulos 2018; Bearden and Forsyth 2018).

Relationship of Schizophrenia to Other Psychiatric Disorders

Schizophrenia and mood disorders have a high degree of genetic overlap (e.g., ~70% for schizophrenia and bipolar disorder) (Lee et al. 2013; Zheutlin and Ross 2018), and family studies indicate that first-degree relatives of individuals with schizophrenia are at risk for psychotic and for mood disorders (Lichtenstein et al. 2009; Rasic et al. 2014). The Bipolar Disorder and Schizophrenia Working Group of the Psychiatric Genetics Consortium reported shared genetic loci across schizophrenia, bipolar disorder, major depressive disorder, autism, and attention deficit hyperactivity disorder (Ruderfer et al. 2014). A recent GWAS study of schizophrenia and bipolar disorder (Bipolar Disorder and Schizophrenia Working Group of the Psychiatric Genetics Consortium 2018) reported more than 100 loci that contributed to both schizophrenia and bipolar disorder, with enrichment for genes involved in response to potassium ions. The study also identified four loci that distinguished between the two disorders (Bipolar Disorder and Schizophrenia Working Group of the Psychiatric Genetics Consortium 2018).

Environmental and Paternal Risk Factors

There is evidence that a number of environmental factors contribute to risk for schizophrenia including prenatal and perinatal factors such as maternal malnutrition, birth trauma, and season of birth; exposure to infections; residing in an urban environment; exposure to stressful life events; and substance abuse (Mitchell et al. 2010; Peay et al. 2017; Avramopoulos 2018). Advanced paternal age has also been implicated in increased risk and earlier onset of symptoms (Wang et al. 2015; Fond et al. 2017). A recent study supported an independent role of both advanced paternal age and of very young paternal age in increasing risk for early-onset schizophrenia in offspring (Wang et al. 2019).

Summary of Schizophrenia Etiology

Schizophrenia is caused by a complex etiology wherein genetic risk originates from hundreds to thousands of genes. Although genetic risk is attributed to both common and rare variants, a large proportion of heritable risk is caused by interactions among common allele variants that each contribute small effects to one or more psychiatric conditions (Avramopoulos 2018; Bearden and Forsyth 2018; Coelewij and Curtis 2018). No evidence supports a common variant of large effect, which is likely due to the low reproductive fitness among those with schizophrenia (Power et al. 2013). Genetic risk factors likely confer a lifelong biological vulnerability (Bearden and Forsyth 2018) with the individual's mental health status ultimately determined by interactions with environmental factors.

APPLICATION OF GENOMIC RESEARCH FINDINGS TO CLIENT RISK ASSESSMENT

This overview of schizophrenia research provides background to facilitate an understanding of the complex etiology of psychiatric disorders and also a context in which to consider applying research data to individual risk assessment. There are three primary ways that genomic data can be applied to genetic counseling. The first is explanatory, in which genomic data is used to understand causation and/or confirm or reject a particular diagnosis for a person with psychiatric symptoms (Lázaro-Muñoz et al. 2018). Currently, this is limited to genes of significant effect such as well-defined CNVs. In the future, use of individual-level genomic and clinical data may be compared to others with similar PRS profiles to inform diagnostic classification and predict course of illness (Bipolar Disorder and Schizophrenia Working Group of the Psychiatric Genetics Consortium 2018).

The second is to inform the treatment and management of an affected individual. Currently available pharmacogenetic testing (Lister 2016; Routhieaux et al. 2018) aims to predict medication efficacy and risk of side effects and adverse events. Specific panels are available to inform psychotropic medication selection, such as the Assurex GeneSight Psychotropic Test (https://genesight.com/). A systematic review (Health Quality Ontario 2017) of the Assurex panel found improved patient response to depression treatment and greater response in measures of depression, but no differences in rates of complete remission (based on low to very-low quality evidence). The optimal timing of pharmacogenetic testing and extent to which results should inform treatment recommendations are still under debate (Lister 2016; Routhieaux et al. 2018). The American Psychiatric Association recently concluded that there is insufficient data to support the widespread use of pharmacogenetic testing (Zeier et al. 2018). In the future, pharmacogenetics will very likely play a role in personalized medicine through the optimization of psychotropic medication selection.

The use of polygenic risks scores also holds promise for informing treatment and management. For example, a recent study used PRSs to anticipate lithium response on people with bipolar disorder (International Consortium on Lithium Genetics et al. 2018). Research is advancing our ability to compare individual-level genetic and clinical data against others with similar profiles and thereby to inform treatment approaches that have been effective in etiologically similar patients (Bipolar Disorder and

Schizophrenia Working Group of the Psychiatric Genetics Consortium 2018).

Finally, genetic/genomic data could be applied to the prediction of risk for unaffected individual(s). Currently this risk assessment is based on personal and family history (as described in the next section) for a large majority of clients. In rare cases, the client's clinical or family history may indicate that a medical genetics evaluation and/or testing for rare variants of large effect (e.g., syndromic causes) are warranted. The identification of such variants has significant implications for psychiatric risk to unaffected relatives, although that risk may be dwarfed by other health-related or developmental risks associated the variant.

Common variants have little or no predictive value independently but could be combined with a large number of other common and/or with rare variants to result in a more stable and clinically meaningful predictive value (Mitchell et al. 2010; Bearden and Forsyth 2018). This is the concept behind PRSs that provide a sum of known risk alleles weighted by the magnitude of risk each variant is expected to confer (Mistry et al. 2018). Because PRSs result in a single value representing an individual's overall genetic risk, they have theoretical value in allowing prediction and possibly risk mitigation efforts for unaffected individuals (Zheutlin and Ross 2018) and individuals with emerging or subclinical symptoms (Calafato et al. 2018). Although PRSs could be applied to unaffected individuals at population risk, genetic counselors are likely to first use PRSs for assessing risk to individuals with a family history. Yet we are likely years from achieving this outcome in clinical psychiatric genetic counseling. PRSs are limited by insufficiently powered GWASs (e.g., in what is considered to be a very large GWAS study, only 7% of the variance in schizophrenia was explained [Schizophrenia Working Group of the Bipolar Disorder and Schizophrenia Working Group of the Psychiatric Genetics Consortium 2014]) and further limited by the lack of ancestral diversity among participants. Notably, PRSs for psychiatric disease are not yet considered to be clinically meaningful (Zheutlin and Ross 2018; Martin et al. 2019).

APPLICATION OF CLIENT FAMILY HISTORY TO RISK ASSESSMENT

Against the backdrop of rapidly advancing psychiatric genomic research that has not yet achieved clinical utility, family history–based risk assessment is still the most appropriate course for the majority of families. Individuals who seek out psychiatric genetic counseling often have multiple affected individuals in their families and have significant concerns for recurrence (Peay and Austin 2011; Peay et al. 2014; Borle et al. 2018). Risk of recurrence is often overestimated by family members (Meiser et al. 2007; Austin et al. 2012), and there is evidence that clients' perceived risk of recurrence for psychiatric disorders influences their psychological well-being and behaviors, including reproductive decision-making (Austin and Honer 2005; Meiser et al. 2007; Peay et al. 2014).

Obtaining the Family and Personal History

Because psychiatric risk assessment is rooted in the client's personal and family histories, genetic counselors have an opportunity to engage in bidirectional sharing of information to the benefit of both parties. Genetic counselors need detailed information from the client, and clients can be helped to understand psychiatric genetics through a transparent questioning process. For example, asking questions about the age of onset for a disorder provides an opportunity to discuss that increased risk may be associated with early symptom onset; and queries about relevant environmental exposures provide an opportunity to understand client perceptions of nongenetic risk factors and sets the stage for counseling about complex etiology. The implications of shared etiology across diagnoses, which include the clustering of different diagnoses in families and risk that extends beyond the diagnosis made in the proband to etiologically related conditions (e.g., if the proband has bipolar disorder, close relatives are also at increased risk for major depression; see Table 1), provide another area for shared information gathering and learning. Because the genetic counselor is reliant upon the client for an accurate and complete family and personal history, and because the client's lived

Table 1. Absolute empiric recurrence risks for selected psychiatric disorders

	Schizophrenia	Bipolar disorder	Major depressive disorder
General population	1%	0.8%–1.6%	2%–23% (female) 1%–15% (male)
Affected parent, for same diagnosis	5%–13%	6%–15%	7%–26%
Affected parent, for either schizophrenia, bipolar, or depression	27% for offspring age 20 or older	27% for offspring age 20 or older	48% for offspring age 20 or older
Affected parent, for any diagnosed mental disorder	45% for offspring age 20 or older	58% for offspring age 20 or older	65% for offspring age 20 or older
Affected sibling	9%–16%	5%–20%	5%–30%
Affected second-degree relative (pooled)	2%–8%	5%	—
Risks for additional mental disorders (not comprehensive)	Schizoaffective disorder, personality disorders, bipolar disorder, major depression, substance use	Major depression, substance use, schizophrenia, schizoaffective disorder, anxiety disorder, ADHD	Anxiety disorders, substance use, disruptive disorders, dysthymia, ADHD

Contents of table adapted from Peay et al. 2017 and Rasic et al. 2014.
(FDRR) First-degree relative recurrence risk, (ADHD) attention deficit hyperactivity disorder.

experience and perception of the illness will impact how he/she responds to the risk assessment data, the mutuality in this experience provides trust and a strong basis for client-centered counseling.

During this process, genetic counselors should obtain the client's personal history and at least three generations of psychiatric family history. Specific information to obtain includes the following (adapted from Peay and Austin 2011; Peay et al. 2017):

- Psychiatric and substance abuse diagnoses made at any time in the lifespan and the age(s) at symptom onset and at diagnosis

- Potentially important environmental exposures

- Symptomatic, undiagnosed individuals and/or undiagnosed individuals treated with a psychiatric medication or by a mental health provider

- Developmental history of relatives and achievement of normal milestones for age (i.e., independent living and employment for adults)

- History of suicide and/or self-harm

- Birth defects, mental retardation, or learning disabilities and unusual medical conditions

- The age and sex of unaffected family members

Family History–Based Risk Assessment

Family and personal histories represent both environmental and genetic risk, which cannot be reasonably separated for risk assessment. The following factors are associated with increased risk to family members when they pertain to the affected or symptomatic individual(s) in the family (adapted from Peay and Austin 2011; Peay et al. 2017):

1. Risk for recurrence in the family increases with the number of affected relatives, regardless of the type of major psychiatric disorder clustering in the family (see Table 1).

2. Individuals who have particularly severe illness, unusually early age at onset, and/or who are of the less-commonly affected sex may represent a higher load of genetic risk, and their relatives may face a higher risk.

 Cite this article as *Cold Spring Harb Perspect Med* doi: 10.1101/cshperspect.a036616

3. Assortative mating increases the risk of recurrence in subsequent generations, whether between a couple with concordant or discordant diagnoses.

4. Comorbid developmental disability, birth defects, or unusual medical history increases suspicion of rare CNVs with significant effect.

There are also important risk determinants related to the at-risk individual(s) in the family:

1. At-risk individuals have different degrees of risk based on their current age (Rasic et al. 2014). Young individuals still have all years of greatest disease risk ahead of them, whereas older individuals have "lived through" a portion of their risk.

2. Individuals with existing subthreshold psychiatric symptoms or unusual behaviors or beliefs are at greater risk (Fusar-Poli et al. 2013).

3. Exposure to known environmental triggers may increase the risk to at-risk individuals, although environmental exposures are difficult to identify and evaluate (Peay et al. 2017).

Family history–based risk assessment relies on empiric recurrence risks (see Table 1). A recent meta-analysis of family studies provides a useful set of recurrence risks for schizophrenia, bipolar disorder, and major depressive disorder and highlights the cross-disorder risk (Rasic et al. 2014). Empiric recurrence risks are limited in that they are not specific to any one individual or family but rather reflect aggregated risks for recurrence among large numbers of families that have been evaluated in research studies. Recurrence risks can be reported as relative risks or absolute risks in the literature, which is a critical distinction for disorders with a high population prevalence. Studies consistently demonstrate that recipients of health risk information understand absolute risks more accurately than relative risks (Fagerlin et al. 2011).

Even given those limitations, empiric recurrence risks provide a useful indication of average risk based on the degree of relatedness to an affected individual. Genetic counselors can evaluate the empiric risks for "fit" to any particular family (Peay et al. 2017)—for example, does the family history indicate that your at-risk client faces the same magnitude of risk as the average person who has an affected parent? The available empiric risks can be tailored, to the extent possible, based on the determinants outlined above in the family and personal history. These determinants provide evidence to raise or decrease the familial risk as determined by empiric risks. In some cases, a range of empiric risks is available and tailoring can indicate if the upper or lower end of the range may best fit the family history. In other cases, the personal and/or family history may be sufficiently compelling that the population-based empiric risks should not be seen as representing a valid measure of risk for that family. In those situations, the risk provided to the client may include the empiric risk as a lower limit while including a range that spans well above.

The primary consideration for tailoring empiric risks is the number and pattern of affected (diagnosed or symptomatic) individuals in the family in relation to unaffected individuals in the family. Empiric risks may also need to be adjusted when the proband's psychiatric history indicates an increased risk (e.g., based on early symptom onset) or when there is assortative mating, which can result in substantially elevated risks.

A final consideration is related to the status of the at-risk individual(s) for whom risk assessment is being conducted. As described above, the age of the at-risk individual should be compared to the typical age at onset for the condition. Risks of disorder manifestation are reduced considerably once at-risk individuals live beyond the expected age of onset, although, of course, late age at symptom onset is possible. Genetic counselors must be aware that risk typically extends beyond the diagnosis in the proband to etiologically related conditions (see Table 1).

The rare situations in which there are multiple affected individuals across several generations should raise red flags related to the risk of recurrence. In those situations, even without a developmental/medical history that suggests an underlying genetic syndrome, high risks of recurrence are possible. This may be due to a rare variant of significant effect or to an unusually large number of risk factors of small effect that are segregating in the family. Either situation calls into question the use of a standard empirical risk ranges and indeed suggests risks that mimic a dominant trait with reduced penetrance. Empiric risks are not available to guide assessment of such "loaded" families, and genetic counselors should use their clinical expertise and judgement to determine whether to provide a quantitative risk range and, if so, what risks to provide.

Provision of Risks

Counselors should customize their approach to risk communication to facilitate achieving the client's objectives for the genetic counseling session and making meaning of etiology and risk information. It is important to convey that genetic factors confer susceptibility through interaction with environmental factors rather than directly causing psychiatric conditions. This discussion may be facilitated through the use of an explanatory model such as the jar model (Peay and Austin 2011; also discussed in Austin 2019).

Genetic counselors should take care to explain the limitations associated with empiric recurrence risks. We simply cannot provide truly personalized risks for the majority of clients who seek psychiatric genetic counseling. In most cases, it is preferable to present a range of empiric risks that reflect the ambiguity associated with the estimates. Empiric risks should be compared to the population risk for the disorder in question to facilitate understanding and interpretation of the data. Genetic counselors should check in with the client to discuss the extent to which presented risks were in alignment with expected risks (Inglis et al. 2017).

Clients who have a psychiatric diagnosis or who are at high risk for psychiatric diagnoses may enter genetic counseling with a predisposition to respond negatively to stressful information, including predictive risk information (Peay et al. 2017). This should not be interpreted as a reason to withhold information. It is instead a

relevant contextual factor that should result in a supportive approach to risk counseling.

CONCLUDING REMARKS

Currently, genetic risk assessment for psychiatric disorders is primarily based on family history. The addition of large consortia and large study populations, methodological improvements, and advancements in approaches such as transcriptome, methylome, and neuroimaging studies will allow future research to more thoroughly explicate the relationships of protective and risk factors to phenotypes (Arslan 2018; Avramopoulos 2018; Punzi et al. 2018). These data can be integrated into PRSs and other approaches that may provide clinically meaningful information to clients with psychiatric disorders and to individuals at risk. Although PRSs that achieve clinical utility may first be used in affected individuals to inform diagnosis and treatment, it is naive to imagine that their use in at-risk family members will lag far behind.

Genetic counselors are particularly well-positioned to participate in determinations of clinical and personal utility for PRS and to educate mental health professionals. They are natural leaders for the needed research on the implementation of PRSs and related approaches first in research settings and later in clinical settings. Important research topics for the future include an assessment of the impact of PRSs on knowledge, self-concept, symptom burden, and treatment adherence for affected individuals. For at-risk individuals, studies may evaluate knowledge and risk perception, the positive and negative psychological and social impact of learning the risk information, and any resulting behavior changes for participants.

Longer-term, effective risk-reduction interventions will increase our motivation to apply PRSs to unaffected, at-risk individuals (Hirschhorn 2009) to allow targeting of early intervention approaches (Calafato et al. 2018; Martin et al. 2019). Additional research will be needed to assess health and well-being outcomes for at-risk individuals who engage in risk-reduction interventions (Martin et al. 2019). If these efforts are successful, routine use of PRSs for psychiatric disorders may extend to those at population risk. Genetic counselors should also anticipate the return of incidental findings that have implications for psychiatric outcomes if there are effective risk-reducing interventions. These outcomes would increase the relevance of psychiatric genetic counseling to counselors who work in other specialty areas.

ACKNOWLEDGMENTS

The author gratefully acknowledges her exceptional colleagues in psychiatric genetic counseling and genetic counseling research and the clients and families who taught her the most important lessons about psychiatric genetic counseling.

REFERENCES

*Reference is also in this subject collection.

Al-Haddad BJS, Jacobsson B, Chabra S, Modzelewska D, Olson EM, Bernier R, Enquobahrie DA, Hagberg H, Östling S, Rajagopal L, et al. 2019. Long-term risk of neuropsychiatric disease after exposure to infection in utero. JAMA Psychiatry 76: 594–602. doi:10.1001/jamapsychiatry.2019.0029

Arslan A. 2018. Imaging genetics of schizophrenia in the post-GWAS era. Prog Neuropsychopharmacol Biol Psychiatry 80: 155–165. doi:10.1016/j.pnpbp.2017.06.018

* Austin JC. 2019. Evidence-based genetic counseling for psychiatric disorders: a road map. Cold Spring Harb Perspect Med. doi:10.1101/cshperspect.a036608

Austin JC, Honer WG. 2005. The potential impact of genetic counseling for mental illness. Clin Genet 67: 134–142. doi:10.1111/j.1399-0004.2004.00330.x

Austin JC, Hippman C, Honer WG. 2012. Descriptive and numeric estimation of risk for psychotic disorders among affected individuals and relatives: implications for clinical practice. Psychiatry Res 196: 52–56. doi:10.1016/j.psychres.2012.02.005

Avramopoulos D. 2018. Recent advances in the genetics of schizophrenia. Mol Neuropsychiatry 4: 35–51. doi:10.1159/000488679

Bearden C E, Forsyth JK. 2018. The many roads to psychosis: recent advances in understanding risk and mechanisms. F1000Res 7: 1883. doi:10.12688/f1000research.16574.1

Bipolar Disorder and Schizophrenia Working Group of the Psychiatric Genomics Consortium. 2018. Genomic dissection of bipolar disorder and schizophrenia, including 28 subphenotypes. Cell 173: 1705–1715.e16. doi:10.1016/j.cell.2018.05.046

* Blesson A, Cohen JS. 2019. Genetic counseling in neurodevelopmental disorders. Cold Spring Harb Perspect Med doi:10.1101/cshperspect.a036533

Bölte S, Girdler S, Marschik PB. 2019. The contribution of environmental exposure to the etiology of autism spectrum disorder. Cell Mol Life Sci 76: 1275–1297. doi:10.1007/s00018-018-2988-4

Borle K, Morris E, Inglis A, Austin J. 2018. Risk communication in genetic counseling: exploring uptake and perception of recurrence numbers, and their impact on patient outcomes. Clin Genet 94: 239–245. doi:10.1111/cge.13379

Calafato MS, Thygesen JH, Ranlund S, Zartaloudi E, Cahn W, Crespo-Facorro B, Díez-Revuelta Á, Di Forti M, Hall MH, Iyegbe C, et al. 2018. Use of schizophrenia and bipolar disorder polygenic risk scores to identify psychotic disorders. Br J Psychiatry 213: 535–541. doi:10.1192/bjp.2018.89

Coelewij L, Curtis D. 2018. Mini-review: update on the genetics of schizophrenia. Ann Hum Genet 82: 239–243. doi:10.1111/ahg.12259

Estrada-Prat X, Van Meter AR, Camprodon-Rosanas E, Batlle-Vila S, Goldstein BI, Birmaher B. 2019. Childhood factors associated with increased risk for mood episode recurrences in bipolar disorder—a systematic review. Bipolar Disord doi:10.1111/bdi.12785

Fagerlin A, Zikmund-Fisher BJ, Ubel PA. 2011. Helping patients decide: ten steps to better risk communication. J Natl Cancer Inst 103: 1436–1443. doi:10.1093/jnci/djr318

Fond G, Godin O, Boyer L, Llorca PM, Andrianarisoa M, Brunel L, Aouizerate B, Berna F, Capdevielle D, D'Amato T, et al. 2017. Advanced paternal age is associated with earlier schizophrenia onset in offspring. Results from the national multicentric FACE-SZ cohort. Psychiatry Res 254: 218–223. doi:10.1016/j.psychres.2017.04.002

Fusar-Poli P, Borgwardt S, Bechdolf A, Addington J, Riecher-Rössler A, Schultze-Lutter F, Keshavan M, Wood S, Ruhrmann S, Seidman LJ, et al. 2013. The psychosis high-risk state: a comprehensive state-of-the-art review. JAMA Psychiatry 70: 107–120. doi:10.1001/jamapsychiatry.2013.269

Health Quality Ontario. 2017. Pharmacogenomic testing for psychotropic medication selection: a systematic review of the Assurex GeneSight Psychotropic Test. Ont Health Technol Assess Ser 17: 1–39.

Hirschhorn JN. 2009. Genomewide association studies—illuminating biologic pathways. N Engl J Med 360: 1699–1701. doi:10.1056/NEJMp0808934

Inglis A, Morris E, Austin J. 2017. Prenatal genetic counselling for psychiatric disorders. Prenat Diagn 37: 6–13. doi:10.1002/pd.4878

International Consortium on Lithium Genetics (ConLi+Gen), Amare AT, Schubert KO, Hou L, Clark SR, Papiol S, Heilbronner U, Degenhardt F, Tekola-Ayele F, Hsu YH, et al. 2018. Association of polygenic score for schizophrenia and HLA antigen and inflammation genes with response to lithium in bipolar affective disorder: a genome-wide association study. JAMA Psychiatry 75: 65–74.

Janoutová J, Janácková P, Serý O, Zeman T, Ambroz P, Kovalová M, Varechová K, Hosák L, Jirík V, Janout V. 2016. Epidemiology and risk factors of schizophrenia. Neuro Endocrinol Lett 37: 1–8.

Lázaro-Muñoz G, Farrell MS, Crowley JJ, Filmyer DM, Shaughnessy RA, Josiassen RC, Sullivan PF. 2018. Improved ethical guidance for the return of results from psychiatric genomics research. *Mol Psychiatry* **23:** 15–23. doi:10.1038/mp.2017.228

Lee SH, Ripke S, Neale BM, Faraone SV, Purcell SM, Perlis RH, Mowry BJ, Thapar A, Goddard ME, Witte JS, et al. 2013. Genetic relationship between five psychiatric disorders estimated from genome-wide SNPs. *Nat Genet* **45:** 984–994. doi:10.1038/ng.2711

Lichtenstein P, Yip BH, Björk C, Pawitan Y, Cannon TD, Sullivan PF, Hultman CM. 2009. Common genetic determinants of schizophrenia and bipolar disorder in Swedish families: a population-based study. *Lancet* **373:** 234–239. doi:10.1016/S0140-6736(09)60072-6

Lister JF. 2016. Pharmacogenomics: a focus on antidepressants and atypical antipsychotics. *Ment Health Clin* **6:** 48–53. doi:10.9740/mhc.2016.01.048

Marangoni C, Hernandez M, Faedda GL. 2016. The role of environmental exposures as risk factors for bipolar disorder: a systematic review of longitudinal studies. *J Affect Disord* **193:** 165–174. doi:10.1016/j.jad.2015.12.055

Marshall CR, Howrigan DP, Merico D, Thiruvahindrapuram B, Wu W, Greer DS, Antaki D, Shetty A, Holmans PA, Pinto D, et al. 2017. Contribution of copy number variants to schizophrenia from a genome-wide study of 41,321 subjects. *Nat Genet* **49:** 27–35. doi:10.1038/ng.3725

Martin J, Taylor MJ, Lichtenstein P. 2018. Assessing the evidence for shared genetic risks across psychiatric disorders and traits. *Psychol Med* **48:** 1759–1774. doi:10.1017/S0033291717003440

Martin AR, Daly MJ, Robinson EB, Hyman SE, Neale BM. 2019. Predicting polygenic risk of psychiatric disorders. *Biol Psychiatry* **86:** 97–109. doi:10.1016/j.biopsych.2018.12.015

Meiser B, Mitchell PB, Kasparian NA, Strong K, Simpson JM, Mireskandari S, Tabassum L, Schofield PR. 2007. Attitudes towards childbearing, causal attributions for bipolar disorder and psychological distress: a study of families with multiple cases of bipolar disorder. *Psychol Med* **37:** 1601–1611. doi:10.1017/S0033291707000852

Mistry S, Harrison JR, Smith DJ, Escott-Price V, Zammit S. 2018. The use of polygenic risk scores to identify phenotypes associated with genetic risk of bipolar disorder and depression: a systematic review. *J Affect Disord* **234:** 148–155. doi:10.1016/j.jad.2018.02.005

Mitchell PB, Meiser B, Wilde A, Fullerton J, Donald J, Wilhelm K, Schofield PR. 2010. Predictive and diagnostic genetic testing in psychiatry. *Clin Lab Med* **30:** 829–846. doi:10.1016/j.cll.2010.07.001

Nakahara S, Medland S, Turner JA, Calhoun VD, Lim KO, Mueller BA, Bustillo JR, O'Leary DS, Vaidya JG, McEwen S, et al. 2018. Polygenic risk score, genome-wide association, and gene set analyses of cognitive domain deficits in schizophrenia. *Schizophr Res* **201:** 393–399. doi:10.1016/j.schres.2018.05.041

Nöthen MM, Nieratschker V, Cichon S, Rietschel M. 2010. New findings in the genetics of major psychoses. *Dialogues Clin Neurosci* **12:** 85–93.

Peay H, Austin J. 2011. *How to talk to families about genetics and psychiatric illness.* W.W. Norton, New York.

Peay HL, Rosenstein DL, Biesecker BB. 2014. Parenting with bipolar disorder: coping with risk of mood disorders to children. *Soc Sci Med* **104:** 194–200. doi:10.1016/j.socscimed.2013.10.022

Peay H, Biesecker B, Austin J. 2017. Genetic counseling for psychiatric conditions. In *Kaplan & Sadock's comprehensive textbook of psychiatry* (ed. Sadock BJ, Sadock VA, Ruiz P). Wolters Kluwer, Philadelphia.

Polderman TJ, Benyamin B, de Leeuw CA, Sullivan PF, van Bochoven A, Visscher PM, Posthuma D. 2015. Meta-analysis of the heritability of human traits based on fifty years of twin studies. *Nat Genet* **47:** 702–709. doi:10.1038/ng.3285

Power RA, Kyaga S, Uher R, MacCabe JH, Långström N, Landen M, McGuffin P, Lewis CM, Lichtenstein P, Svensson AC. 2013. Fecundity of patients with schizophrenia, autism, bipolar disorder, depression, anorexia nervosa, or substance abuse vs their unaffected siblings. *JAMA Psychiatry* **70:** 22–30. doi:10.1001/jamapsychiatry.2013.268

Prata DP, Costa-Neves B, Cosme G, Vassos E. 2019. Unravelling the genetic basis of schizophrenia and bipolar disorder with GWAS: a systematic review. *J Psychiatr Res* **114:** 178–207. doi:10.1016/j.jpsychires.2019.04.007

Psychiatric GWAS Consortium Coordinating Committee, Cichon S, Craddock N, Daly M, Faraone SV, Gejman PV, Kelsoe J, Lehner T, Levinson DF, Moran A, et al. 2009. Genomewide association studies: history, rationale, and prospects for psychiatric disorders. *Am J Psychiatry* **166:** 540–556. doi:10.1176/appi.ajp.2008.08091354

Punzi G, Bharadwaj R, Ursini G. 2018. Neuroepigenetics of schizophrenia. *Prog Mol Biol Transl Sci* **158:** 195–226. doi:10.1016/bs.pmbts.2018.04.010

Rasic D, Hajek T, Alda M, Uher R. 2014. Risk of mental illness in offspring of parents with schizophrenia, bipolar disorder, and major depressive disorder: a meta-analysis of family high-risk studies. *Schizophr Bull* **40:** 28–38. doi:10.1093/schbul/sbt114

Rössler W, Riecher-Rössler A, Angst J, Murray R, Gamma A, Eich D, van Os J, Gross VA. 2007. Psychotic experiences in the general population: a twenty-year prospective community study. *Schizophr Res* **92:** 1–14. doi:10.1016/j.schres.2007.01.002

Routhieaux M, Keels J, Tillery EE. 2018. The use of pharmacogenetic testing in patients with schizophrenia or bipolar disorder: a systematic review. *Ment Health Clin* **8:** 294–302. doi:10.9740/mhc.2018.11.294

Ruderfer DM, Fanous AH, Ripke S, McQuillin A, Amdur RL, Schizophrenia Working Group of the Psychiatric Genomics Consortium, Bipolar Disorder Working Group of the Psychiatric Genomics Consortium, Cross-Disorder Working Group of the Psychiatric Genomics Consortium, Gejman PV, O'Donovan MC, et al. 2014. Polygenic dissection of diagnosis and clinical dimensions of bipolar disorder and schizophrenia. *Mol Psychiatry* **19:** 1017–1024. doi:10.1038/mp.2013.138

Schizophrenia Working Group of the Psychiatric Genomics Consortium. 2014. Biological insights from 108 schizophrenia-associated genetic loci. *Nature* **511:** 421–427. doi:10.1038/nature13595

Smoller JW, Andreassen OA, Edenberg HJ, Faraone SV, Glatt SJ, Kendler KS. 2019. Psychiatric genetics and the structure of psychopathology. *Mol Psychiatry* **24**: 409–420. doi:10.1038/s41380-017-0010-4

Sullivan PF, Geschwind DH. 2019. Defining the genetic, genomic, cellular, and diagnostic architectures of psychiatric disorders. *Cell* **177**: 162–183. doi:10.1016/j.cell.2019.01.015

Uher R. 2014. Gene-environment interactions in common mental disorders: an update and strategy for a genome-wide search. *Soc Psychiatry Psychiatr Epidemiol* **49**: 3–14. doi:10.1007/s00127-013-0801-0

Wang SH, Liu CM, Hwu HG, Hsiao CK, Chen WJ. 2015. Association of older paternal age with earlier onset among co-affected schizophrenia sib-pairs. *Psychol Med* **45**: 2205–2213. doi:10.1017/S0033291715000203

Wang SH, Hsiao PC, Yeh LL, Liu CM, Liu CC, Hwang TJ, Hsieh MH, Chien YL, Lin YT, Huang YT, et al. 2019. Advanced paternal age and early onset of schizophrenia in sporadic cases: not confounded by parental polygenic risk for schizophrenia. *Biol Psychiatry* **86**: 56–64. doi:10.1016/j.biopsych.2019.01.023

Whiteford HA, Degenhardt L, Rehm J, Baxter AJ, Ferrari AJ, Erskine HE, Charlson FJ, Norman RE, Flaxman AD, Johns N, et al. 2013. Global burden of disease attributable to mental and substance use disorders: findings from the Global Burden of Disease Study 2010. *Lancet* **382**: 1575–1586. doi:10.1016/S0140-6736(13)61611-6

Zeier Z, Carpenter LL, Kalin NH, Rodriguez CI, McDonald WM, Widge AS, Nemeroff CB. 2018. Clinical implementation of pharmacogenetic decision support tools for antidepressant drug prescribing. *Am J Psychiatry* **175**: 873–886. doi:10.1176/appi.ajp.2018.17111282

Zheutlin AB, Ross DA. 2018. Polygenic risk scores: what are they good for? *Biol Psychiatry* **83**: e51–e53. doi:10.1016/j.biopsych.2018.04.007

Psychological Issues in Managing Families with Inherited Cardiovascular Diseases

Jodie Ingles[1,2,3]

[1]Agnes Ginges Centre for Molecular Cardiology at Centenary Institute, The University of Sydney, Newtown, New South Wales NSW 2042, Australia

[2]Faculty of Medicine and Health, The University of Sydney, Sydney, New South Wales NSW 2000, Australia

[3]Department of Cardiology, Royal Prince Alfred Hospital, Camperdown, New South Wales NSW 2050, Australia

Correspondence: j.ingles@centenary.org.au

The field of cardiovascular genetic counseling has evolved dramatically in recent years largely to manage the unique psychological needs of the inherited cardiovascular disease patient population. For many, there can be difficulty in coming to terms with a diagnosis, whether it be adjusting to lifestyle recommendations such as exclusion from competitive sports or living with a small but remarkable risk of sudden cardiac death. For those considered at risk of life-threatening ventricular arrhythmias, the decision to have an implantable cardioverter defibrillator can be difficult. Living with the device, especially for those who are young and those who receive multiple shocks, can precipitate psychological distress and poor adaptation to the device. Family members who experience a sudden cardiac death of a young relative have a significant risk of poor psychological outcomes. The roles of the cardiac genetic counselor in facilitating patients' adaptation to their diagnoses and management and recognizing when additional support from a clinical psychologist is needed are key to ensuring families receive the best possible care.

Inherited cardiovascular diseases can affect all aspects of the heart. They include cardiomyopathies, arrhythmia syndromes, and metabolic and connective tissue diseases, as well as numerous other genetic syndromes that can present with cardiac manifestations (Fig. 1). As a result, the field of cardiac genetic counseling has rapidly emerged as a growing subspecialty. Sudden cardiac death is a rare but tragic outcome among those with inherited cardiovascular diseases. Managing the risk of a future sudden cardiac death can add layers of complexity to more traditional genetic counseling roles. For example, eliciting a family history from a patient newly diagnosed and told to consider an implantable cardioverter defibrillator by their cardiologist for prevention of fatal arrhythmias will be challenging. Likewise, explaining inheritance risks and undertaking pretest genetic counseling will not be straightforward for a family that have recently come to medical attention because of the sudden cardiac death of a young and otherwise healthy relative. In many cases, the pressing concern is making sense of the situation that confronts them. Providing information about their cardiovascular disease and management,

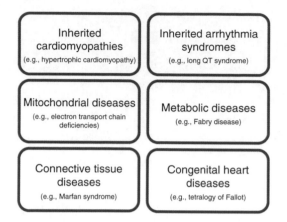

Figure 1. Inherited cardiovascular diseases.

and heritability, in combination with psychotherapeutic counseling are key skills for a cardiac genetic counselor. Counselors need to be as comfortable discussing the psychological aspects of adjusting to an implantable cardioverter defibrillator as they would be discussing inheritance risks. Counseling skills to sensitively convey important information while attending to patients' and families' psychological responses are roles that define cardiac genetic counselors as key partners in a specialized multidisciplinary team. This review aims to describe the psychological issues frequently encountered by patients and family members with inherited cardiovascular diseases, specifically inherited cardiomyopathies and arrhythmia syndromes, and to demonstrate the vital role of cardiac genetic counseling in patient care.

ASSIMILATING DIAGNOSIS OF AN INHERITED CARDIOVASCULAR DISEASE

Cardiovascular disease remains the leading cause of death in the United States (Benjamin et al. 2019), often occurring in the elderly population, with environmental factors that include lifestyle choices that contribute to disease risk and development (McGovern et al. 1996). In contrast, patients with inherited cardiovascular diseases often present with symptoms in childhood, adolescence, and early adulthood, and the cause is largely due to a gene variant that is inherited in an autosomal-dominant pattern.

Many patients with inherited cardiovascular diseases have only minimal cardiac symptoms and yet are required to make significant lifestyle adjustments to live with an elevated risk of developing the most serious outcomes: heart failure and sudden cardiac death. As such, adaptation to the diagnosis of a hereditary cardiovascular disease can be a challenging long-term process for patients and their families.

Collectively, hereditary cardiovascular diseases present similar genetic counseling issues. Adaptation to a diagnosis of any of these conditions at a young age, and specifically in cases in which there is little, if any, experience living with the disease in the family, can be challenging in many ways. There is a need to make sense of this new health threat, considering the prognostic and management information provided by the clinical team, which may require commencement of medications despite feeling well. Periodic follow-up with a cardiologist is indicated, and because of the risk of sudden cardiac death, some patients undergo prophylactic implantable cardioverter defibrillator (ICD) therapy. Other complications more frequently encountered are development of atrial fibrillation and progression to heart failure, in some instances requiring cardiac transplant. Cardiac symptoms frequently reported by patients include chest pain, shortness of breath on exertion, dizziness (presyncope), blackouts (syncope), and palpitations. For many patients with one of these conditions, there will be few if any cardiac symptoms. Clinicians' ability to predict who will experience poor clinical outcomes is limited. Even within a family, it is common to see the full phenotypic range expressed among relatives. In most cases, having a severely affected family member does not mean that other relatives will be similarly affected. For patients, direct experience with the disease will significantly affect perceptions about prognosis (Ormondroyd et al. 2014; Connors et al. 2015). As such, learning of this variability in the clinical course and uncertainty about the progression of disease can be difficult.

Hereditary arrhythmia syndromes often present in childhood and adolescence, with sudden onset of arrhythmias leading to symptoms like palpitations, dizziness, and syncope (Priori et al. 2015). They present with a greater risk of ventricular arrhythmias that can result in sudden cardiac death. Taking a family history from patients with hereditary arrhythmia syndromes will often reveal significant cardiac events among relatives, such as frequent "fainters," sudden infant death syndrome (SIDS), and epilepsy diagnoses, and, importantly, fatal ventricular arrhythmias can be misattributed to other causes if they occur while driving (motor vehicle accident) or swimming (drowning). Attention to the details of any young deaths in the family is critical to determining whether there may have been an underlying cardiac cause. Eliciting this information from the family can stir up feelings of grief and loss, and it can be evident that there is significant anxiety about the risk to other relatives, especially children (Ormondroyd et al. 2014; Burns et al. 2016). In contrast, hereditary cardiomyopathies typically present later in life, from teenage years through to adulthood, and will often result in cardiac-limiting symptoms, such as chest pain and shortness of breath at some point during life (Watkins et al. 2011). The impact on physical health-related quality of life is thus an important consideration for this group. Historical physical activity recommendations have further compounded this issue, with many patients ceasing to do any form of physical activity because of the perceived risk of sudden cardiac death (Sweeting et al. 2016).

RESEARCH FINDINGS

Numerous studies have assessed self-reported psychological outcomes or explored perceptions following diagnosis of a hereditary cardiovascular disease. Research that has generated evidence to help shape the field is presented in Table 1. In one of the largest quantitative studies, health-related quality of life was assessed in more than 400 individuals with inherited cardiovascular diseases, including both affected patients and at-risk relatives (Ingles et al. 2013b). Among those with inherited cardiomyopathies such as hypertrophic cardiomyopathy (HCM) and dilated cardiomyopathy (DCM), there was lower

physical quality of life, whereas mental quality of life was generally maintained. In those participants with long QT syndrome (LQTS), psychological quality of life was reduced compared to rates in the general population. These significant differences between hereditary cardiomyopathies and hereditary arrhythmia syndromes have been reported by others, and likely relate to the differences between the disease profile among the two groups (Table 1).

Medications are recommended for many patients with inherited cardiovascular diseases, especially for those with arrhythmia syndromes. For young patients, adherence to β-blocker medication can be strongly influenced by psychological and social factors (Ingles et al. 2015; O'Donovan et al. 2018). In the setting of LQTS, β-blocker therapy is shown to reduce the likelihood of cardiac arrhythmias and is an important first-tier treatment (Chockalingam et al. 2012). Arrhythmias while taking a β-blocker can signal the need for additional therapy such as an ICD. In the medical literature, studies of adherence to any prescribed therapy indicate that psychological factors, including illness perception, beliefs about the medication, self-efficacy (i.e., their confidence in their ability to take the medication), and psychosocial well-being, are important determinants of medication adherence (Jin et al. 2008). One study examining adherence to β-blocker therapy among patients with LQTS found 16% were not adherent (O'Donovan et al. 2018). Those who were younger and had an inherited arrhythmia syndrome reported lower self-efficacy, greater concerns, low necessity beliefs, and poorer understanding of their disease. Importantly, they were also less likely to take their prescribed β-blocker medication. Another study investigating self-reported medication adherence among patients with HCM found that 30% were nonadherent and more likely to be from a minority ethnicity with self-reported anxiety symptoms (Ingles et al. 2015). These studies provide evidence of the need for genetic counseling interventions to uphold adherence to prescribed medical therapy. This may be an achievable goal of genetic counseling. Targets of the interventions may be affirming beliefs about the cardioprotective role of the medica-

Table 1. Key studies investigating the psychological and social impact of living with an inherited cardiovascular disease

Study	Disease setting	Study design	Psychosocial measures	Key findings
Bates et al. 2019	SCD in the young	$N = 33$ first-degree relatives following a SCD of a young relative and no premorbid diagnosis	Hospital anxiety and depression scale (HADS), psychological adaptation to genetic information scale (PAGIS), impact of events scale-revised (IES-R), multidimensional scale of perceived social support (MSPSS)	More than one-third reported poor adaptation to genetic information following the postmortem molecular autopsy. Those with poor adaptation were more likely to have posttraumatic stress symptoms, depression, lower perceived social support, and support from significant others, family members, and friends.
Ingles et al. 2016	SCD in the young	$N = 103$ first-degree relatives following a SCD of a young relative and no premorbid diagnosis	Short Form Health Survey (SF-36), depression anxiety stress scale (DASS-21), IES-R, prolonged grief disorder scale	Prolonged grief was reported by 21% (36% of mothers, 42% of those who witnessed the death) and posttraumatic stress symptoms reported by 44% (59% of mothers, 67% of those who witnessed the death). Overall almost 50% of relatives reported symptoms requiring intervention with a clinical psychologist, on average 6 yr after the death.
van der Werf et al. 2014	SCD in the young	$N = 9$ first-degree relatives or spouses; semistructured interviews, analyzed thematically	Qualitative thematic analysis of interview transcripts	There was a strong need to understand the cause of the death and wanting to prevent future deaths in their family. The current multidisciplinary clinic model was appreciated, although ways to improve information processes and better support of decision-making were deemed important.

Continued

Table 1. *Continued*

Study	Disease setting	Study design	Psychosocial measures	Key findings
Ingles et al. 2013b	HCM, DCM, ARVC, LQTS, CPVT	N = 409 participants (affected and at-risk relatives); cross-sectional self-report survey	SF-36	Physical component scores were impaired for patients with HCM, DCM, and CPVT. Mental component scores were impaired in those with LQTS. In HCM, predictors of poor physical component scores were female gender, comorbidities, and higher NHYA functional class.
James et al. 2012	ARVC	N = 86 adults with ARVC and an ICD	Florida shock anxiety scale (FSAS), Florida patient acceptance survey (FPAS), Duke activity status index, HADS	Overall device acceptance was normative; however, there were marked body image concerns, particularly among younger patients. Younger patients also had greater device-specific anxiety. Those who showed poor adjustment to their device were more likely to have anxiety and depression.
Hamang et al. 2010	HCM, LQTS	N = 127 (affected and at-risk relatives); cross-sectional self-report survey	SF-36	All participants had impaired general health scores compared to the general population. Employment, higher education, and referral to genetic counseling through a family member were associated with improved health status.
Christiaans et al. 2009	HCM	N = 228 (gene-positive relatives, affected and unaffected)	SF-36, HADS, revised version of Illness Perception Questionnaire (IPQ-R)	Quality of life and distress were worse in those with manifest HCM compared to those without. Cardiac symptoms and a stronger belief in consequences of

Continued

Table 1. *Continued*

Study	Disease setting	Study design	Psychosocial measures	Key findings
				carriership were associated with worse quality of life. Illness and risk perception were major determinants of quality of life and distress.
Farnsworth et al. 2006	LQTS	N = 31 parents; qualitative phenomenological study; parents of children with LQTS	Explore fear of death and quality of life	Parents reported fear of their child dying and strategies they used to manage their fear. There was frustration about the lack of LQTS knowledge from health-care professionals.
Hendriks et al. 2005	LQTS	N = 36 parents; longitudinal surveys and one semistructured home interview; parents of gene carrier children	Impact of events scale (IES), Spielberger State Anxiety Inventory (STAI-s), and Beck Depression Inventory (BDI)	Parents of gene carrier children reported greater distress than parents of noncarriers. Up to one-third reported clinically relevant distress. More distress was reported among those with sudden death in the family.

(SCD) Sudden cardiac death, (HCM) hypertrophic cardiomyopathy, (DCM) dilated cardiomyopathy, (ARVC) arrhythmogenic right ventricular cardiomyopathy, (LQTS) long QT syndrome, (CPVT) catecholaminergic polymorphic ventricular cardiomyopathy, (NYHA) New York Heart Association, (ICD) implantable cardioverter defibrillator.

tion and their confidence to take it on a regular basis. Challenging unproductive illness and medication perceptions and acknowledging the emotional barriers to adherence have been shown to be beneficial in other settings (Petrie et al. 2012) and may be applied and studied in the care of hereditary cardiovascular syndromes.

RISK OF SUDDEN CARDIAC DEATH

Adapting to living with a risk of sudden cardiac death affects all patients. Discussing strategies to clinically manage this risk often takes precedence over all other topics in clinic visits, and being considered at increased risk of a ventricular arrhythmia is associated with heightened anxiety and poorer psychological well-being. Patients may be pragmatic about their own risk of sudden cardiac death, although in some cases, there is high anxiety and worry (Ormondroyd et al. 2014). For many, the risk to their children is forefront (Farnsworth et al. 2006). A Dutch study reported parental distress in up to one-third of parents with children who inherited a pathogenic variant for LQTS (Hendriks et al. 2005). Among those with distress, being familiar with the disease for a longer time, experiencing a sudden cardiac death in the family, having a lower formal education, and expressing dissatisfaction with the information provided were each shown to be associated with psychological distress. A qualitative study explored fear of sudden cardiac death among parents of children with LQTS (Farnsworth et al. 2006). Parents described coping mechanisms they employed to manage their fear, which included providing children with phones to allow them to check in regularly, using a baby monitor in the child's room at night, and keeping an automatic external defibrillator nearby. Many were involved in increasing awareness of LQTS and ensuring resuscitation action plans were in place in the child's community. Another qualitative study demonstrated challenges associated with a diagnosis of LQTS and risk of sudden cardiac death (Andersen et al. 2008). Those with early and gradual experience with LQTS reported less worry, but their main concern was fear for their children or grandchildren. Further, there was frustration with the limited understanding and knowledge among health professionals, in some cases providing erroneous treatment advice.

DECISION-MAKING REGARDING IMPLANTABLE CARDIOVERTER DEFIBRILLATOR THERAPY

Current knowledge about risk factors predisposing to sudden cardiac death is incomplete. Considering that the incidence of sudden cardiac death is very low, the number of patients who undergo implantation of an ICD is proportionately large. Managing risk of sudden cardiac death presents novel issues, particularly among those with heritable cardiovascular disease in which patients are often young. Given that, in most cases, prolonged ventricular arrhythmias are fatal, with little likelihood of a second chance to reconsider the management strategy, the stakes for making the correct clinical decision are high. Inherently, sudden cardiac death is an unpredictable event, often described as a moment in time. Although a patient may have been a regular runner for most of his or her life, why on one particular early morning run he or she suffered a sudden cardiac arrest is most often unknown.

The decision to undergo ICD implantation is often one the patient feels they have little control over. For example, some patients will experience an unexplained syncope and find themselves in the hospital with the decision to have an ICD seeming like an urgent and necessary next step in their management. They may later question whether the ICD was needed (Ingles et al. 2013a). Facilitating patients' efforts to weigh the benefits and potential harms of this therapy is an important component of cardiac genetic counseling (Ingles et al. 2011). Prevention of sudden cardiac death is the goal; however, being aware of the potential for low psychological well-being, inappropriate shocks, regular and ongoing device checks, and complications such as infection and device issues including lead fractures are important risks for patients to give their full consideration.

Shared decision-making as a model for deliberation over ICD recommendations has gained favor in the literature (Lewis et al. 2018). Shared decision-making in this setting includes ensuring patient values and preferences are explored and considered. One qualitative study assessed decision-making processes of patients receiving an ICD for any cardiac condition (Carroll et al. 2013). The decision-making process about the ICD was strongly influenced by the clinician's recommendation and a new awareness of the risk of sudden cardiac death. Another qualitative interview study with cardiologists and patients on their perceptions of the ICD decision-making process found that cardiologists consistently emphasized the benefits of an ICD, whereas discussion of risks was variable (Matlock et al. 2011). In contrast, many patients agreed to the ICD on the advice of their clinician without questioning the risks. Those who declined an ICD were unconvinced the ICD was necessary or believed the risks outweighed the likelihood of sudden cardiac death. These studies highlight the critical need for accurate communication of the uncertain and incremental risk of sudden cardiac death balanced with the potential risks of having an ICD.

LIVING WITH AN IMPLANTABLE CARDIOVERTER DEFIBRILLATOR

For those who choose to receive an ICD, overwhelmingly the psychological and social outcomes are favorable (Sweeting et al. 2017; Maron et al. 2018). Yet, there are certain patients who do not fare as well and would benefit from psychological counseling. In the broader cardiovascular disease literature, patients who receive an ICD at a younger age and who receive more shocks from their device are at greater risk of poor psychological outcomes (Sears and Conti 2002). The cardiac genetic counselors play an important role in this setting, in collaboration with clinical psychologists when greater support is needed (Ingles et al. 2011; Caleshu et al. 2016).

For those with inherited cardiovascular diseases, adjusting to life with an ICD can pose additional challenges. James et al. (2012) reported the experiences of patients with a diagnosis of arrhythmogenic right ventricular cardiomyopathy (ARVC) after a mean follow-up of 3 yr, finding healthy adaptation to the device overall. Despite this, body image concerns were markedly elevated and many reported device-related psychological distress—in particular, the younger patients. Among those with ARVC, having an ICD shock, younger age, poor functional capacity, and shorter time since implant all predicted greater device-specific distress, and poor adjustment was associated with greater reported anxiety and depressive symptoms (James et al. 2012). In a cohort of patients with catecholaminergic polymorphic ventricular tachycardia (CPVT) who had an ICD, those who were younger were more likely to report device-related distress and shock anxiety and even greater body image concerns than the ARVC cohort described above (Richardson et al. 2018).

Anxiety about experiencing a shock is an important consideration, as more shocks have consistently been shown to be associated with lower psychological functioning (Perini et al. 2017). In one study, one-third of patients with hereditary cardiovascular disease who received a shock from their ICD reported symptoms of posttraumatic stress (Ingles et al. 2013a). The most endorsed items on the Impact of Events Scale-Revised (IES-R) were "I tried not to think about it" and "I avoided letting myself get upset when I thought about it or was reminded of it," both of which reflect avoidance. Among the female patients, approximately half reported clinically relevant posttraumatic stress symptoms. Another study showed disparity in psychological functioning by sex, with female ARVC patients more likely to report anxiety relating to shocks (Rhodes et al. 2017).

In cardiac genetic counseling, there is an appreciation of the potential psychological consequences of an ICD that is critical to empathizing with patients and providing psychotherapeutic counseling. In a clinical setting in which busy clinicians are working to manage clinical symptoms, experienced cardiac genetic counselors who explore ICD adjustment and fears of sudden cardiac death with patients and encourage discussion of any perceived psychological distress are likely to engage in a therapeutic relationship in which psychological well-being is shared and assessed. When marked anxiety or posttraumatic stress are suspected, intervention with a clinical psychologist to manage these responses is timely and important and highlights the health benefits of established multidisciplinary networks (Caleshu et al. 2016).

SPORTS RESTRICTIONS

Physical inactivity is a major determinant of poor health outcomes, recognized as the fourth leading cause of death worldwide (Kohl et al. 2012). Among inherited cardiovascular diseases, because of the observed high incidence of cardiac events occurring during exercise (Maron et al. 2014), exclusion from high-level or competitive sports applies for essentially every patient. Two unfavorable outcomes tend to follow this guidance. The first is that the patient overinterprets this recommendation and subsequently engages in little or no physical activity, increasing risk of weight gain, risk of developing other chronic diseases, and risk of lower quality of life. Alternatively, they may disregard the recommendation and continue high-level competitive sports; this occurs most often when the patient was an avid athlete prior to diagnosis.

Managing the former situation is perhaps a more common occurrence in the inherited cardiovascular clinic and likely a reflection of fear of precipitating ventricular arrhythmias and an overall reluctance to meet recommended physical activity guidelines among the general population. Physical inactivity has been reported in >50% of patients with HCM (Sweeting et al. 2016), and HCM patients are less active overall than their general population counterparts (Reineck et al. 2013). Factors associated with physical inactivity include perceived barriers such as pain and injury/disability and older age (Sweeting et al. 2016). Most patients report purposeful reduction in physical activity due to their diagnosis of HCM, and less time in physical activity is associated with a negative impact on psychological well-being (Reineck et al. 2013). Interest in interventions to encourage safe increases in exercise capacity and increase overall time spent being physically active were motivation for two recently published trials. One randomized controlled trial investigated moderate-intensity training on peak oxygen consumption in HCM patients, showing overall improvement in exercise capacity following a 16-week training program, with no adverse events, such as arrhythmias, reported (Saberi et al. 2017). Another study used a control-theory approach to increase physical activity in HCM patients, incorporating motivational interviewing (Sweeting et al. 2018). Although the primary outcome of an increase in purposeful physical activity was not demonstrated, secondary end points relating to psychosocial outcomes such as health-related quality of life, self-efficacy, and perceived barriers to exercise were significantly improved.

For athletes who are diagnosed with an inherited cardiovascular disease, adopting the recommended lifestyle changes and subsequent adjustment to a new life can be profoundly challenging. Many individuals have built their identity around their athletic ability, and for some it has become a career. As such, the sudden loss of rigorous exercise results in grief. Just as patient acceptance and adherence to prescribed medical therapy relies on person-centered factors, such as understanding the necessity of the recommendation and perceived beliefs, discussion with athletes regarding sports participation may benefit from a similar approach. One study investigated athletes diagnosed with HCM and

reported most found it difficult to adapt to the exercise restrictions (Luiten et al. 2016). Those who reported greater negative psychological impact had a history of elite or competitive sports participation, identified as an athlete, and spent less time exercising since their diagnosis. Having friends and family participate with them in lower-intensity exercise was identified as something that helped them cope with the restrictions. A qualitative study explored perspectives of individuals with HCM and LQTS who continue to exercise against current recommendations and found many did make important modifications to their exercise program over time (Subas et al. 2019). The value of a shared decision-making model in which the athlete's preferences are considered may be beneficial to addressing exercise restrictions. Genetic counselors can be integral in facilitating decision-making and adaptation about sports participation and exercise restrictions and in supporting and motivating physically inactive patients to meet recommended exercise targets.

GRIEF AFTER SUDDEN CARDIAC DEATH

In the inherited cardiovascular clinic, the families who experience the sudden and unexpected death of a previously healthy young relative present the greatest challenges. Sudden cardiac death in the young (<35 yr) occurs at an incidence of ~1.3 per 100,000 persons, and 40% of cases will remain unexplained even after comprehensive postmortem examination (Bagnall et al. 2016). Inherited cardiomyopathies will often be identifiable at postmortem examination, because they result in structural disease. However, inherited arrhythmia syndromes, primarily diagnosed on electrocardiogram in living patients, will leave little evidence of pathophysiological changes at postmortem (Semsarian et al. 2015). In these cases, the presenting symptom can be sudden death, with no premorbid diagnosis, and those remaining unexplained at postmortem are presumed due to an inherited cardiac arrhythmia.

The grief experienced by families in this setting can be profound (Ingles and James 2017). Grief is a normal response to a loss and a process of coming to terms with a new reality. It is experienced differently among individuals, but can include disbelief, yearning, anger, sadness, and eventually acceptance (Maciejewski et al. 2007). Kübler-Ross and Kessler (2005) wrote: "The reality is that you will grieve forever. You will not 'get over' the loss of a loved one; you will learn to live with it. You will heal and you will rebuild yourself around the loss you have suffered. You will be whole again but you will not be the same. Nor should you be the same, nor would you want to." In ~7% of bereaved individuals, grief does not progress as expected, essentially resulting in the individual becoming "stuck." This is known as prolonged or complicated grief, characterized by intense yearning and an inability to imagine a new life without their loved one more than 6–12 mo after the death (Simon 2013). Prolonged grief requires intervention by a clinical psychologist or psychiatrist and poses greater risk of adverse health outcomes, suicide, and comorbid psychological conditions such as depression and posttraumatic stress disorder (Prigerson et al. 1997). Factors related to experiencing prolonged grief include being female, the nature of the relationship to the deceased, sudden and unexpected deaths, and poor understanding of the circumstances of the death, among other factors (Fujisawa et al. 2010).

The largest study to quantify psychological difficulties among first-degree relatives of a decedent in which the death was considered due to an inherited cardiovascular disease found almost 50% reported clinically significant posttraumatic stress symptoms or prolonged grief (Ingles et al. 2016). Specifically, 21% reported prolonged grief, including 36% of mothers of the decedent and 42% of those who witnessed the death. Posttraumatic stress symptoms were reported by 44% overall, including 59% of mothers of the decedent and 67% of those who witnessed the death (Fig. 2A). Witnessing the death or discovering the decedents body was associated with a four to five times greater risk of prolonged grief or posttraumatic stress symptoms. These findings are in contrast to a growing body of literature around family-witnessed resuscitation efforts by emergency medical services,

which suggest this may provide positive psychological benefits (Robinson et al. 1998; Jabre et al. 2013). Although witnessing resuscitation may be beneficial for families when an older relative suffers a cardiac arrest, when the relative is a young person, witnessing the resuscitation efforts and discovery of the body are the highest risk factors for poor psychological outcomes.

Another study investigating psychological well-being in families after a young sudden cardiac death showed that >50% reported clinically significant symptoms of anxiety, on average 4 yr after the death (Yeates et al. 2013). A qualitative analysis in the same study reported numerous factors that helped or hindered participants' ability to cope after the death (Fig. 2B). Experiences of relatives attending a specialized multidisciplinary clinic in the Netherlands identified key reasons for attending the clinic included a need to understand the cause of death and to prevent another sudden cardiac death occurring in the family (van der Werf et al. 2014). Indeed, these study outcomes reflect the overwhelming concerns expressed by relatives attending following a sudden death.

Practically, for a cardiac genetic counselor seeing families after a young sudden cardiac death, ensuring the family has the opportunity for appropriate clinical management, genetic investigations, and psychological counseling is critical. In the current clinic models for cardiovascular genetics, there can be a tendency to focus on the clinical and genetic concerns; however, based on the evidence and experiences of specialized clinics, psychological counseling should be offered to all family members (Caleshu et al. 2016; Ingles et al. 2016; Ingles and James 2017). Recognizing that in many cases these families are unlikely to encounter other health professionals following the death, the cardiologist and genetic counselor must be strong advocates for addressing psychological needs and feel comfortable discussing the family members' well-being, community support, and coping effectiveness. With growing interest in the postmortem "molecular autopsy" (genetic testing of postmortem DNA), cardiac genetic counselors must also work to ensure we limit potential harms of this technology in this vul-

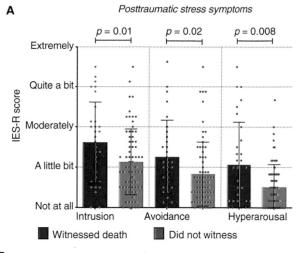

B

"Finding out that I passed on the mutation to him is probably the second hardest event of my life." [Mother of decedent]

"It is such a shock when the fittest and youngest member of your family dies suddenly. It changes your life instantly and my parents have never been the same." [Sister of decedent]

"My biggest fear is that it may happen to my other children and there doesn't seem to be a thing I can do about it." [Father of decedent]

"A's [son] passing has been the most difficult event that I have ever had to contend with in my life as I'm sure it is for any parent in the same situation." [Mother of decedent]

Figure 2. Psychological consequences for family members following the sudden cardiac death of a young relative. (*A*) There was a greater risk of posttraumatic stress symptoms among first-degree relatives who witnessed the death or discovered the decedent's body. This was true across the three subdomains of the impact of events scale (IES-R) (Ingles et al. 2016). (*B*) Key quotes from a qualitative study illustrating common difficulties reported by family members, including genetic guilt, shock, disbelief, and fear for remaining children (Yeates et al. 2013).

nerable patient group. This includes pretest genetic counseling that sets realistic expectations about the likelihood of a conclusive result and careful review of variants to avoid misclassification. A variant of uncertain significance will be identified in >40% of postmortem cases, and a causative variant identified in only ~10% (Lahrouchi et al. 2017). How a family facing profound psychological challenges can understand and adapt to the high degree of uncertainty this testing often yields is not well understood, but it has the potential to inflict harm. A small 2018 study showed that more than one-third of first-degree relatives of a young victim of sudden

cardiac death showed difficulty adapting psychologically to the postmortem genetic test information (Bates et al. 2019). Further research to establish an effective method to integrate psychotherapeutic counseling into the cardiac clinic setting is needed.

CONCLUSION

Many challenges exist for patients with inherited cardiovascular diseases and their at-risk relatives. Research-focused genetic counselors have already played an important role in understanding the issues faced by this patient population, which offers evidence to inform development of more effective management approaches. Although this review has focused on those aspects of care relating to a diagnosis, managing sudden cardiac death risk, ICD therapy, and sudden cardiac death, numerous other challenges remain, particularly related to genetic testing. For the genetic counselor, always being mindful and aware of psychological challenges that diminish our patient's quality of life is paramount and will ultimately help to improve overall management and care of families with heritable cardiovascular diseases.

ACKNOWLEDGMENTS

J.I. is the recipient of a National Health and Medical Research Council Career Development Fellowship (APP1162929). The author declares there are no conflicts of interest.

REFERENCES

Andersen J, Øyen N, Bjorvatn C, Gjengedal E. 2008. Living with long QT syndrome: a qualitative study of coping with increased risk of sudden cardiac death. *J Genet Couns* 17: 489–498. doi:10.1007/s10897-008-9167-y

Bagnall RD, Weintraub RG, Ingles J, Duflou J, Yeates L, Lam L, Davis AM, Thompson T, Connell V, Wallace J, et al. 2016. A prospective study of sudden cardiac death among children and young adults. *N Engl J Med* 374: 2441–2452. doi:10.1056/NEJMoa1510687

Bates K, Sweeting J, Yeates L, McDonald K, Semsarian C, Ingles J. 2019. Psychological adaptation to molecular autopsy findings following sudden cardiac death in the young. *Genet Med* 21: 1452–1456. doi:10.1038/s41436-018-0338-4

Benjamin EJ, Muntner P, Alonso A, Bittencourt MS, Callaway CW, Carson AP, Chamberlain AM, Chang AR, Cheng S, Das SR, et al. 2019. Heart disease and stroke statistics—2019 update: a report from the American Heart Association. *Circulation* 139: e56–e528. doi:10.1161/CIR.0000000000000659

Burns C, McGaughran J, Davis A, Semsarian C, Ingles J. 2016. Factors influencing uptake of familial long QT syndrome genetic testing. *Am J Med Genet A* 170A: 418–425. doi:10.1002/ajmg.a.37455

Caleshu C, Kasparian NA, Edwards KS, Yeates L, Semsarian C, Perez M, Ashley E, Turner CJ, Knowles JW, Ingles J. 2016. Interdisciplinary psychosocial care for families with inherited cardiovascular diseases. *Trends Cardiovasc Med* 26: 647–653. doi:10.1016/j.tcm.2016.04.010

Carroll SL, Strachan PH, de Laat S, Schwartz L, Arthur HM. 2013. Patients' decision making to accept or decline an implantable cardioverter defibrillator for primary prevention of sudden cardiac death. *Health Expect* 16: 69–79. doi:10.1111/j.1369-7625.2011.00703.x

Chockalingam P, Crotti L, Girardengo G, Johnson JN, Harris KM, van der Heijden JF, Hauer RN, Beckmann BM, Spazzolini C, Rordorf R, et al. 2012. Not all β-blockers are equal in the management of long QT syndrome types 1 and 2: higher recurrence of events under metoprolol. *J Am Coll Cardiol* 60: 2092–2099. doi:10.1016/j.jacc.2012.07.046

Christiaans I, van Langen IM, Birnie E, Bonsel GJ, Wilde AA, Smets EM. 2009. Quality of life and psychological distress in hypertrophic cardiomyopathy mutation carriers: a cross-sectional cohort study. *Am J Med Genet A* 149A: 602–612. doi:10.1002/ajmg.a.32710

Connors E, Jeremy RW, Fisher A, Sharpe L, Juraskova I. 2015. Adjustment and coping mechanisms for individuals with genetic aortic disorders. *Heart Lung Circ* 24: 1193–1202. doi:10.1016/j.hlc.2015.05.003

Farnsworth MM, Fosyth D, Haglund C, Ackerman MJ. 2006. When I go in to wake them … I wonder: parental perceptions about congenital long QT syndrome. *J Am Acad Nurse Pract* 18: 284–290. doi:10.1111/j.1745-7599.2006.00132.x

Fujisawa D, Miyashita M, Nakajima S, Ito M, Kato M, Kim Y. 2010. Prevalence and determinants of complicated grief in general population. *J Affect Disord* 127: 352–358. doi:10.1016/j.jad.2010.06.008

Hamang A, Eide GE, Nordin K, Rokne B, Bjorvatn C, Øyen N. 2010. Health status in patients at risk of inherited arrhythmias and sudden unexpected death compared to the general population. *BMC Med Genet* 11: 27. doi:10.1186/1471-2350-11-27

Hendriks KS, Grosfeld FJ, van Tintelen JP, van Langen IM, Wilde AA, van den Bout J, ten Kroode HF. 2005. Can parents adjust to the idea that their child is at risk for a sudden death?: psychological impact of risk for long QT syndrome. *Am J Med Genet A* 138A: 107–112. doi:10.1002/ajmg.a.30861

Ingles J, James C. 2017. Psychosocial care and cardiac genetic counseling following sudden cardiac death in the young. *Prog Pediatr Cardiol* 45: 31–36. doi:10.1016/j.ppedcard.2017.03.001

Ingles J, Yeates L, Semsarian C. 2011. The emerging role of the cardiac genetic counselor. *Heart Rhythm* 8: 1958–1962. doi:10.1016/j.hrthm.2011.07.017

Ingles J, Sarina T, Kasparian N, Semsarian C. 2013a. Psychological wellbeing and posttraumatic stress associated with implantable cardioverter defibrillator therapy in young adults with genetic heart disease. *Int J Cardiol* 168: 3779–3784. doi:10.1016/j.ijcard.2013.06.006

Ingles J, Yeates L, Hunt L, McGaughran J, Scuffham PA, Atherton J, Semsarian C. 2013b. Health status of cardiac genetic disease patients and their at-risk relatives. *Int J Cardiol* 165: 448–453. doi:10.1016/j.ijcard.2011.08.083

Ingles J, Johnson R, Sarina T, Yeates L, Burns C, Gray B, Ball K, Semsarian C. 2015. Social determinants of health in the setting of hypertrophic cardiomyopathy. *Int J Cardiol* 184: 743–749. doi:10.1016/j.ijcard.2015.03.070

Ingles J, Spinks C, Yeates L, McGeechan K, Kasparian N, Semsarian C. 2016. Posttraumatic stress and prolonged grief after the sudden cardiac death of a young relative.

JAMA Intern Med **176:** 402–405. doi:10.1001/jamai
nternmed.2015.7808

Jabre P, Belpomme V, Azoulay E, Jacob L, Bertrand L, La-
postolle F, Tazarourte K, Bouilleau G, Pinaud V, Broche
C, et al. 2013. Family presence during cardiopulmonary
resuscitation. *N Engl J Med* **368:** 1008–1018. doi:10.1056/
NEJMoa1203366

James CA, Tichnell C, Murray B, Daly A, Sears SF, Calkins
H. 2012. General and disease-specific psychosocial
adjustment in patients with arrhythmogenic right ventricular dysplasia/cardiomyopathy with implantable cardioverter defibrillators: a large cohort study. *Circ Cardiovasc Genet* **5:** 18–24. doi:10.1161/CIRCGENETICS.111
.960898

Jin J, Sklar GE, Min Sen Oh V, Chuen Li S. 2008. Factors
affecting therapeutic compliance: a review from the patient's perspective. *Ther Clin Risk Manag* **4:** 269–286.
doi:10.2147/TCRM.S1458

Kohl HW III, Craig CL, Lambert EV, Inoue S, Alkandari JR,
Leetongin G, Kahlmeier S, Lancet Physical Activity Series
Working Group. 2012. The pandemic of physical inactivity: global action for public health. *Lancet* **380:** 294–305.
doi:10.1016/S0140-6736(12)60898-8

Kübler-Ross E, Kessler D. 2005. *On grief and grieving. Finding the meaning of grief through the five stages of loss.*
Simon & Schuster, New York.

Lahrouchi N, Raju H, Lodder EM, Papatheodorou E, Ware
JS, Papadakis M, Tadros R, Cole D, Skinner JR, Crawford
J, et al. 2017. Utility of post-mortem genetic testing in
cases of Sudden Arrhythmic Death Syndrome. *J Am
Coll Cardiol* **69:** 2134–2145. doi:10.1016/j.jacc.2017.02
.046

Lewis KB, Carroll SL, Birnie D, Stacey D, Matlock DD. 2018.
Incorporating patients' preference diagnosis in implantable cardioverter defibrillator decision-making: a review
of recent literature. *Curr Opin Cardiol* **33:** 42–49. doi:10
.1097/HCO.0000000000000464

Luiten RC, Ormond K, Post L, Asif IM, Wheeler MT, Caleshu C. 2016. Exercise restrictions trigger psychological
difficulty in active and athletic adults with hypertrophic
cardiomyopathy. *Open Heart* **3:** e000488. doi:10.1136/
openhrt-2016-000488

Maciejewski PK, Zhang B, Block SD, Prigerson HG. 2007.
An empirical examination of the stage theory of grief. *J
Am Med Assoc* **297:** 716–723. doi:10.1001/jama.297.7.716

Maron BJ, Haas TS, Murphy CJ, Ahluwalia A, Rutten-
Ramos S. 2014. Incidence and causes of sudden death
in U.S. college athletes. *J Am Coll Cardiol* **63:** 1636–
1643. doi:10.1016/j.jacc.2014.01.041

Maron BJ, Casey SA, Olivotto I, Sherrid MV, Semsarian
C, Autore C, Ahmed A, Boriani G, Francia P, Winters
SL, et al. 2018. Clinical course and quality of life in
high-risk patients with hypertrophic cardiomyopathy
and implantable cardioverter-defibrillators. *Circ Ar-
rhythm Electrophysiol* **11:** e005820. doi:10.1161/CIRCEP
.117.005820

Matlock DD, Nowels CT, Masoudi FA, Sauer WH, Bekelman
DB, Main DS, Kutner JS. 2011. Patient and cardiologist
perceptions on decision making for implantable cardioverter-defibrillators: a qualitative study. *Pacing Clin Electrophysiol* **34:** 1634–1644. doi:10.1111/j.1540-8159.2011
.03237.x

McGovern PG, Pankow JS, Shahar E, Doliszny KM, Folsom
AR, Blackburn H, Luepker RV. 1996. Recent trends in
acute coronary heart disease—mortality, morbidity, med-
ical care, and risk factors. The Minnesota Heart Survey
Investigators. *N Engl J Med* **334:** 884–890. doi:10.1056/
NEJM199604043341403

O'Donovan CE, Waddell-Smith KE, Skinner JR, Broadbent
E. 2018. Predictors of β-blocker adherence in cardiac inherited disease. *Open Heart* **5:** e000877. doi:10.1136/
openhrt-2018-000877

Ormondroyd E, Oates S, Parker M, Blair E, Watkins H. 2014.
Pre-symptomatic genetic testing for inherited cardiac
conditions: a qualitative exploration of psychosocial and
ethical implications. *Eur J Hum Genet* **22:** 88–93. doi:10
.1038/ejhg.2013.81

Perini AP, Kutyifa V, Veazie P, Daubert JP, Schuger C, Za-
reba W, McNitt S, Rosero S, Tompkins C, Padeletti L, et al.
2017. Effects of implantable cardioverter/defibrillator
shock and antitachycardia pacing on anxiety and quality
of life: a MADIT-RIT substudy. *Am Heart J* **189:** 75–84.
doi:10.1016/j.ahj.2017.03.009

Petrie KJ, Perry K, Broadbent E, Weinman J. 2012. A text
message programme designed to modify patients' illness
and treatment beliefs improves self-reported adherence to
asthma preventer medication. *Br J Health Psychol* **17:** 74–
84. doi:10.1111/j.2044-8287.2011.02033.x

Prigerson HG, Bierhals AJ, Kasl SV, Reynolds CF III, Shear
MK, Day N, Beery LC, Newsom JT, Jacobs S. 1997. Traumatic grief as a risk factor for mental and physical morbidity. *Am J Psychiatry* **154:** 616–623. doi:10.1176/ajp.154
.5.616

Priori SG, Blomström-Lundqvist C, Mazzanti A, Blom N,
Borggrefe M, Camm J, Elliott PM, Fitzsimons D, Hatala
R, Hindricks G, et al. 2015. 2015 ESC Guidelines for the
management of patients with ventricular arrhythmias
and the prevention of sudden cardiac death: the Task
Force for the Management of Patients with Ventricular
Arrhythmias and the Prevention of Sudden Cardiac
Death of the European Society of Cardiology (ESC). Endorsed by Association for European Paediatric and Congenital Cardiology (AEPC). *Eur Heart J* **36:** 2793–2867.
doi:10.1093/eurheartj/ehv316

Reineck E, Rolston B, Bragg-Gresham JL, Salberg L, Baty L,
Kumar S, Wheeler MT, Ashley E, Saberi S, Day SM. 2013.
Physical activity and other health behaviors in adults with
hypertrophic cardiomyopathy. *Am J Cardiol* **111:** 1034–
1039. doi:10.1016/j.amjcard.2012.12.018

Rhodes AC, Murray B, Tichnell C, James CA, Calkins H,
Sears SF. 2017. Quality of life metrics in arrhythmogenic
right ventricular cardiomyopathy patients: the impact of
age, shock and sex. *Int J Cardiol* **248:** 216–220. doi:10
.1016/j.ijcard.2017.08.026

Richardson E, Spinks C, Davis A, Turner C, Atherton J,
McGaughran J, Semsarian C, Ingles J. 2018. Psychosocial
implications of living with catecholaminergic polymorphic ventricular tachycardia in adulthood. *J Genet Couns*
27: 549–557. doi:10.1007/s10897-017-0152-1

Robinson SM, Mackenzie-Ross S, Campbell Hewson GL,
Egleston CV, Prevost AT. 1998. Psychological effect of
witnessed resuscitation on bereaved relatives. *Lancet*
352: 614–617. doi:10.1016/S0140-6736(97)12179-1

Saberi S, Wheeler M, Bragg-Gresham J, Hornsby W, Agarwal PP, Attili A, Concannon M, Dries AM, Shmargad Y, Salisbury H, et al. 2017. Effect of moderate-intensity exercise training on peak oxygen consumption in patients with hypertrophic cardiomyopathy: a randomized clinical trial. *J Am Med Assoc* **317:** 1349–1357. doi:10.1001/jama.2017.2503

Sears SF Jr, Conti JB. 2002. Quality of life and psychological functioning of ICD patients. *Heart* **87:** 488–493. doi:10.1136/heart.87.5.488

Semsarian C, Ingles J, Wilde AA. 2015. Sudden cardiac death in the young: the molecular autopsy and a practical approach to surviving relatives. *Eur Heart J* **36:** 1290–1296. doi:10.1093/eurheartj/ehv063

Simon NM. 2013. Treating complicated grief. *J Am Med Assoc* **310:** 416–423. doi:10.1001/jama.2013.8614

Subas T, Luiten R, Hanson-Kahn A, Wheeler M, Caleshu C. 2019. Evolving decisions: perspectives of active and athletic individuals with inherited heart disease who exercise against recommendations. *J Genet Couns* **28:** 119–129. doi: 10.1007/s10897-018-0297-6.

Sweeting J, Ingles J, Timperio A, Patterson J, Ball K, Semsarian C. 2016. Physical activity in hypertrophic cardiomyopathy: prevalence of inactivity and perceived barriers. *Open Heart* **3:** e000484. doi:10.1136/openhrt-2016-000484

Sweeting J, Ball K, McGaughran J, Atherton J, Semsarian C, Ingles J. 2017. Impact of the implantable cardioverter defibrillator on confidence to undertake physical activity in inherited heart disease: a cross-sectional study. *Eur J Cardiovasc Nurs* **16:** 742–752. doi:10.1177/1474515117715760

Sweeting J, Ingles J, Ball K, Semsarian C. 2018. A control theory-based pilot intervention to increase physical activity in patients with hypertrophic cardiomyopathy. *Am J Cardiol* **122:** 866–871. doi:10.1016/j.amjcard.2018.05.023

van der Werf C, Onderwater AT, van Langen IM, Smets EM. 2014. Experiences, considerations and emotions relating to cardiogenetic evaluation in relatives of young sudden cardiac death victims. *Eur J Hum Genet* **22:** 192–196. doi:10.1038/ejhg.2013.126

Watkins H, Ashrafian H, Redwood C. 2011. Inherited cardiomyopathies. *N Engl J Med* **364:** 1643–1656. doi:10.1056/NEJMra0902923

Yeates L, Hunt L, Saleh M, Semsarian C, Ingles J. 2013. Poor psychological wellbeing particularly in mothers following sudden cardiac death in the young. *Eur J Cardiovasc Nurs* **12:** 484–491. doi:10.1177/1474515113485510

Cite this article as *Cold Spring Harb Perspect Med* doi: 10.1101/cshperspect.a036558

A Person-Centered Approach to Cardiovascular Genetic Testing

Julia Platt

Stanford Center for Inherited Cardiovascular Disease, Falk Cardiovascular Research Center, Stanford, California 94305, USA

Correspondence: jplatt@stanfordhealthcare.org

Cardiovascular genetic counselors provide guidance to people facing the reality or prospect of inherited cardiovascular conditions. Key activities in this role include discussing clinical cardiac screening for at-risk family members and offering genetic testing. Psychological factors often influence whether patients choose to have genetic testing and how they understand and communicate the results to at-risk relatives, so psychological counseling increases the impact of genetic education and medical recommendations. This work reviews the literature on the factors that influence patient decisions about cardiovascular genetic testing and the psychological impact of results on people who opt to test. It also models use of a psychological framework to apply themes from the literature to routine cardiovascular genetic counseling practice. Modifications of the framework are provided to show how it can be adapted to serve the needs of both new and experienced genetic counselors.

Genetic counselors have a unique role that spans medical, genetic, and psychological domains (Accreditation Council for Genetic Counseling 2015). In the context of the complex psychological challenges faced by their patients, genetic counselors need to convey accurate medical information in a way that patients can understand, retain, and put to meaningful use (Veach et al. 2007). In a cardiology setting, the core medical activities of a genetic counselor include (a) family history risk assessment, (b) selecting the most informative relative(s) for genetic testing, (c) selecting the appropriate genetic test, (d) educating the patient and family members about the purpose of the test, (e) ensuring accurate variant interpretation, and (f)

communicating the implications of the results effectively to the patient and family (Somers et al. 2014; Cirino et al. 2017).

Cardiovascular genetic counselors interact with patients who often have multiple disease-related stressors, including the physical burden of cardiac symptoms and difficulties adjusting to a treatment plan or medically indicated lifestyle changes. Patients may also feel uncertainty about their or their family's health, fear the risk of sudden death, and exhibit psychological distress related to their own condition or witnessing the decline and/or sudden death of a relative (Aatre and Day 2011; Caleshu et al. 2016). Psychological distress adversely affects a person's ability to retain and process information with downstream effects on factors affecting key medical outcomes, such as decision-making, adherence to the medical plan, and the desire and ability to communicate risk information to others (Borders et al. 2006). Therefore, effective psychological counseling skills are imperative for cardiovascular genetic counselors to meet their medical goals of education and management (see Ingles 2019).

UTILITY OF GENETIC TESTING FOR CARDIOVASCULAR DISEASES

Inherited cardiovascular diseases are a heterogeneous group of conditions with a wide range of clinical findings and causative genes (see Table 1; Cox 2007; Ackerman et al. 2011; MacCarrick et al. 2014; Pugh et al. 2014; Burke et al. 2016; Garcia et al. 2016; Maurer et al. 2016; Goyal et al. 2017; Muchtar et al. 2017; Arbustini et al. 2018; Geddes and Earing 2018; Giudicessi et al. 2018; Pierpont et al. 2018; Ingles et al. 2019; Stout et al. 2019). Conditions that cause isolated cardiovascular disease include familial hyperlipidemias, cardiomyopathies, arrhythmias, disorders of the aorta, and congenital heart disease (Cirino et al. 2017). Most follow an autosomal-dominant inheritance pattern, meaning that first-degree relatives of an affected person are at 50% risk of inheriting the genetic predisposition for the condition. X-linked, autosomal-recessive, or mitochondrial inheritance patterns occur less often and, rarely, multiple pathogenic Mendelian variants may be present in the same family (Cirino et al. 2017).

Diagnostic and predictive genetic testing may be helpful for the care of people with known or suspected inherited cardiovascular disease. Diagnostic testing may confirm a genetic etiology for index patients who have a borderline phenotype or environmental risk factors that might contribute to the phenotype. For example, a clinical diagnosis of hypertrophic cardiomyopathy is sometimes unclear if a person with left ventricular hypertrophy also has a history of hypertension, in which case a genetic diagnosis clarifies the need to recommend cardiac screening for their relatives. Diagnostic testing can also guide treatment in cases in which a causative

gene is associated with an increased likelihood of adverse events, such as life-threatening arrhythmias. For example, identifying a disease-causing variant in the *LMNA* gene lowers the threshold for recommending placement of an implantable cardiac defibrillator. In other cases, a positive genetic result diagnoses a phenocopy with a treatment that might not otherwise be considered, such as enzyme replacement therapy for people diagnosed with Fabry disease. Most often, however, genetic test results do not change the care of the index patient because the clinical diagnosis is sufficient to guide treatment (Aatre and Day 2011).

The primary goal of genetic testing is to identify a gene variant that facilitates predictive testing of at-risk family members. Relatives with positive results from predictive testing become eligible for appropriate follow-up care, and those with negative results can forgo ongoing cardiac screening. It is important, therefore, that genetic test results be as accurate as possible, because the classification of a variant may have far-reaching implications for the patient and their family. It is also very important for the genetic counselor to communicate the results and associated clinical recommendations in a way that maximizes the person's ability to understand and transmit that information to their family, particularly against a backdrop of multiple concurrent psychological stressors.

ISSUES IN DECISION-MAKING

There are several topics that cardiovascular genetic counselors need to discuss with patients and families considering genetic testing, including (a) the purpose of the test, (b) the likelihood of identifying a causative variant, (c) implications of positive, negative, and uncertain results for the patient, (d) implications for family members, and (e) the possibility of variant reclassification (Cirino et al. 2017). Eliciting details about a patient's life allows personalized education so the person can better conceptualize and remember how medical information applies to their specific situation. For example, education about family screening recommendations and the possibility of predictive testing can be per-

Table 1. Inherited cardiovascular conditions and the most common causative genes and/or other genetic causes

Inherited cardiovascular conditions	Common causative genes and cellular functions
Cardiomyopathies	
Hypertrophic cardiomyopathy	Sarcomere and associated genes *ACTC1, FHL1, MYBPC3, MYH7, MYL2, MYL3, TNNI3, TNNI2, TPM1*

Dilated cardiomyopathy
  Phenotypic and genetic overlap with
  arrhythmogenic cardiomyopathy
	Sarcomere and nonsarcomere structure, electrolyte homeostasis,
	 and other genes
	*BAG3, DES, FLNC, LMNA, MYH7, PLN, RBM20, SCN5A,
	TNNC1, TNNI3, TNNT2, TPM1, TTN*

Arrhythmogenic cardiomyopathy
  Includes arrhythmogenic right ventricular
  cardiomyopathy
	Desmosome, electrolyte homeostasis, and other genes
	*ACTC1, DES, DSC2, DSG2, DSP, FLNC, JUP, LMNA, PKP2,
	PLN, RBM20, RYR2, SCN5A, TMEM43*

Left ventricular noncompaction	Sarcomere gene *MYH7*
Restrictive cardiomyopathy	Sarcomere genes *MYH7, TNNI3, TNNT2*
Arrhythmias	
Long QT syndrome	Transmembrane ion channel genes *KCNQ1, KCNH2, SCN5A, KCNE1, KCNE2, KCNJ2*
Catecholaminergic polymorphic ventricular tachycardia	Calcium-handling genes *CALM1, CALM2, CALM3, CASQ2, RYR2, TRDN*
Brugada syndrome	Transmembrane sodium channel gene *SCN5A*
Aortopathies	
Marfan syndrome	Extracellular matrix protein gene *FBN1*
Vascular Ehlers–Danlos syndrome	Type III procollagen gene *COL3A1*
Loeys–Dietz syndrome	Transforming growth factor-β signaling pathway genes *SMAD3, TGFB2, TGFB3, TGFBR1, TGFBR2*
Familial thoracic aortic aneurysms and dissections	Smooth muscle contraction and metabolism *ACTA2, MYH11, MYLK, PRKG1, TGFBR2*
Hyperlipidemias	
Familial hypercholesterolemia	LDL receptor and associated proteins *LDLR, APOB, PCSK9*
Congenital heart disease	
Anomalous pulmonary venous connections	Varied etiologies include:
Atrioventricular septal defects	Multifactorial
Complex lesions	Aneuploidies
Conotruncal lesions	Chromosomal microdeletions or duplications
Heterotaxy	Single genes vary according to type of structural difference
Left ventricular outflow tract	
Right ventricular outflow tract	
Other septal lesions	
Inherited conditions with extracardiac involvement	
Hereditary amyloidosis	Transthyretin gene *TTR*
Hereditary muscle disorders	Genes vary according to clinical diagnosis
Biochemical genetic disorders	Genes vary according to clinical diagnosis
Other syndromic conditions	
RASopathies and other syndromic conditions	Genes vary according to clinical diagnosis

sonalized by using the names of specific family members.

The credibility of medical information is increased when it is delivered by a trusted provider, and trust is established by identifying and exploring patient concerns (Werner-Lin et al. 2016). Assessing the patient's motivations, expectations, and hopes may also reveal critical information that enhances the effectiveness of a counseling session. For example, two people with the same well-established, stable cardiovascular diagnosis may have very different hopes for the outcome of testing. One person may hope for a positive result because they want predictive testing to be an option for their family, but another may harbor hope that their diagnosis is not truly genetic because they do not want to believe their family is at risk. The first person would be motivated to have genetic testing and consider a positive test result to be desirable news, whereas the second person might downplay the validity of the test and decide not to pursue it because they fear a positive result. Understanding the second person's perspective allows the genetic counselor to guide the session toward exploring their fear of a heritable condition, which likely influences their ability to talk about their condition as well as the choice to have or forego genetic testing.

Literature on Decision-Making and Uptake of Genetic Testing

In literature spanning genetics specialties, qualitative studies report that people have fairly consistent motivations and concerns about genetic testing, but the consistency of the predictors of behavior are not borne out in quantitative studies (Sweeny et al. 2014). Of note, perceived test-related factors (benefits of testing, few risks or barriers, and positive attitudes toward testing) seem to more consistently predict uptake of genetic testing than condition-related factors (perceptions of risk, control, severity of the condition) or demographic factors (Sweeny et al. 2014).

In cardiovascular genetics, most studies have focused on the uptake of predictive testing, whereas studies investigating diagnostic testing

have looked at mixed populations of affected and at-risk individuals (Smart 2010; Khouzam et al. 2015; Burns et al. 2016). Reasons for testing include external factors (provider recommendations or family pressure) and internal factors (reducing uncertainty, risk clarification for relatives, an explanation for a cardiac event, or an unexplained sudden death in the family) (van Maarle et al. 2001; Smart 2010; Erskine et al. 2014). One study found the strongest predictor of uptake to be cues to action like a doctor's recommendation, an offer of testing, or other variables that precede genetic counseling (Khouzam et al. 2015). Among those who chose to undergo testing, perceived negative aspects of testing include the potential emotional impact of a positive result and the possibility of continued uncertainty (Smart 2010; Erskine et al. 2014).

Several studies have focused on the behavior of at-risk relatives. One study found that family screening uptake rates were higher among people at risk for arrhythmias than cardiomyopathies (80% vs. 45%) (van der Roest et al. 2009). This finding is consistent with a predictive testing uptake rate of 60% in a study of a Long QT population and several studies that have reported an uptake rate of 39% for predictive testing in hypertrophic cardiomyopathy (HCM) and dilated cardiomyopathy (DCM) populations (Charron et al. 2002; Christiaans et al. 2008; Miller et al. 2013; Burns et al. 2016). One study reported higher education and household income levels among people who got genetic testing, which led the authors to consider whether others may lack access to genetic services (Burns et al. 2016).

Of interest, a family history of sudden death does not significantly affect genetic testing uptake rates (Christiaans et al. 2008; Khouzam et al. 2015). There is some qualitative evidence that a family history of sudden death may play a strong role in some individual choices, either reinforcing the need for testing or leading people to avoid learning their genetic status (Smart 2010; Khouzam et al. 2015; Burns et al. 2016). In other cases, people use defense strategies to minimize their risk, such as hypothesizing nongenetic causes for a death or emphasizing that deaths occurred in another branch of the family (Ormondroyd et al. 2014). The varied ways that people navigate sudden death or other disease-related factors reinforces the need for personalized genetic counseling with a strong psychological emphasis in the cardiovascular setting.

Themes to explore during a pretest discussion are highlighted by two qualitative studies of people at risk to inherit highly penetrant variants causing arrhythmogenic right ventricular cardiomyopathy. Among people who had predictive genetic testing, the decision-making process either progressed gradually and included consideration of a variety of factors or happened so quickly that it was not necessarily viewed as a decision at all (Manuel and Brunger 2014; Etchegary et al. 2015). The interlocking factors that people weighed included (a) the availability and reliability of the test, (b) awareness of numerous losses or deaths in the family, (c) physical signs and symptoms of disease, (d) their own gender, (e) sense of relational responsibility or moral obligation to other family members, and (f) family support or pressure (Manuel and Brunger 2014; Etchegary et al. 2015).

Literature on the Psychological Impact of Genetic Test Results

On the whole, studies indicate that the outcome of genetic testing itself is less predictive of long-term emotional distress than other factors such as the person's baseline emotional state, stressors related to the effects of the disease itself, and relational factors within the family. The results may act as a trigger of an underlying psychological dynamic or they may provide a sense of clarity that promotes adaptive coping. There are certain factors that correlate with distress among subgroups of people that genetic counselors can use to identify people who are most likely to need guidance and support.

Studies of the impact of diagnostic testing on affected people have not shown negative effects on long-term well-being beyond those caused by the diagnosis itself (Vansenne et al. 2009; Ingles et al. 2012; Hickey et al. 2014). People who had diagnostic testing reported that they expected clear answers from genetic results and had difficulty recalling the implications of uncertain re-

sults, highlighting the need for genetic counselors to clarify expectations and facilitate clear communication of results (Burns et al. 2017). People with anxiety and depression are more likely to perceive barriers to communicating results with their families (Burns et al. 2016).

Studies on the impact of predictive genetic testing among at-risk family members have also indicated that, beyond some initial distress, for most people it does not cause serious or prolonged distress, regardless of the result (Hoedemaekers et al. 2007; Christiaans et al. 2009; Ingles et al. 2012; Oliveri et al. 2018). Over the long term, most people report a relative decrease in distress after testing, with this response being understandably larger and more rapid in people who receive negative results from predictive testing (Broadstock et al. 2000; Hendriks et al. 2008; Christiaans et al. 2009; Oliveri et al. 2018). These findings are consistent with the literature on genetic testing for treatable conditions as a whole (Broadstock et al. 2000; Oliveri et al. 2018). There is also evidence that some of this lack of perceived effect is driven by threat minimization or denial, which are coping strategies that people use to maintain a sense of control (Broadstock et al. 2000). Careful assessment and counseling can help people identify ways to adapt to the situation in a more realistic and sustainable way. It should be noted that ascertainment is an issue for most of these studies, and there is some evidence that people who seek predictive genetic testing may be more resilient than others (Broadstock et al. 2000).

Other factors such as the presence of symptoms, medically indicated lifestyle changes, and pretest emotional or mental health status correlate better with measures of distress than the outcome of genetic testing (Broadstock et al. 2000; Hendriks et al. 2008; Christiaans et al. 2009; Bonner et al. 2018; Oliveri et al. 2018). However, there appears to be a subset of people who experience significant distress from predictive test results (Heshka et al. 2008; Bonner et al. 2018). Correlates of distress include having felt pressured to get the test, not being emotionally prepared, perceiving oneself to be at high risk for a severe condition, and lack of social support (Christiaans et al. 2009; Bonner et al. 2018).

Across all ages, findings support the importance of feeling certain and in control for adaptation (Meulenkamp et al. 2008; MacLeod et al. 2014). This is further explored below.

Studies indicate that different considerations are prompted by tests done during different life stages. In adolescents and young people, as seen in other populations, test results do not directly correlate with emotional outcomes (Godino et al. 2016). Those who have had predictive testing report that, even if it caused short-term distress, it was empowering if they felt it increased their sense of control over the course of future events (MacLeod et al. 2014). Young people report that lack of emotional experience made it difficult for them to foresee how the news might impact them, reinforcing the need for ongoing support as their reactions unfold, as they begin to disclose the results, and at new life stages when more implications of the result become clear (MacLeod et al. 2014). Especially for those who live with their parents, family dynamics may also play a larger role for young people in the forms of family pressure, worries about how their result will impact others, or because living with an affected parent contributes to their perception of the condition (Geelen et al. 2011; Godino et al. 2016).

Genetic testing of minors can be considered for inherited cardiovascular diseases, given the possibility for childhood onset and the availability of treatment. Most studies indicate that children cope effectively with the knowledge of being at-risk for a genetic condition and report similar quality of life to peers or their baseline emotional state (Meulenkamp et al. 2008; Wakefield et al. 2016). Children may worry more if they identify with a family member who has died or experienced serious consequences from the condition or worry more about the health of a parent who has had positive genetic testing even if they do not perceive themselves to be at risk (Meulenkamp et al. 2008; Wakefield et al. 2016). Other sources of distress include forgetting to take medication, parental worry, and feeling different from peers or being bullied (Meulenkamp et al. 2008). Worry may be expressed in subtle or avoidant ways, such as looking away, changing the subject, or making jokes (Meulenkamp et al.

2008). During a time of life in which much is out of their control already, children seem to rely on markers of vicarious control by expressing confidence in medication or the medical profession or by responding to the differences between their situation and those of an affected relative (Meulenkamp et al. 2008). Of note, positive predictive genetic test results may have more impact on parents, with a longitudinal study showing that 30% of parents of children who have tested positive for Long QT had clinically significant levels of long term distress (Hendriks et al. 2005, 2008). Parents also report more worry for their children than themselves and may have inflated perceptions of adverse psychological impact on their children (Smets et al. 2008; Burns et al. 2017).

INTERPRETING THE TEST RESULT

The uncertainty that sometimes accompanies genetic test results can contribute to the confusion and distress that patients may experience. The American College of Medical Genetics and Genomics (ACMG) and Association for Molecular Pathology (AMP) guidelines for variant interpretation provide a scoring system to structure and standardize the process, but application of the criteria requires clinical judgement (Richards et al. 2015). The inherent difficulty of variant classification is reflected in discrepant classifications between laboratories that persist despite efforts to resolve discordance (Rehm et al. 2015; Amendola et al. 2016; Balmaña et al. 2016; Garber et al. 2016; Furqan et al. 2017; Harrison et al. 2017; Yang et al. 2017). Expert panels continue to refine the ACMG-AMP guidelines for various disease areas through ClinGen expert panels funded by the National Institutes of Health (NIH), but much work remains (Rehm et al. 2015; Kelly et al. 2018; Renard et al. 2018).

There is some evidence that cardiovascular genetics is an area of particularly high discordance (Balmaña et al. 2016; Harrison et al. 2017; Yang et al. 2017). This has led the majority of cardiovascular genetic counselors to perform independent variant assessments, most frequently by using resources such as PubMed, ClinVar, gene or disease-specific databases,

Cite this article as *Cold Spring Harb Perspect Med* doi: 10.1101/cshperspect.a036624

allele frequency databases, or searching for additional case data by other means (Reuter et al. 2018). Clinicians tend to require a higher degree of certainty than laboratories to classify a result pathogenic or likely pathogenic, the categories suitable for use in predictive testing (Bland et al. 2018).

Genetic and condition-specific expertise are critical at all stages of interpreting results, including assessing the genes being tested, because large panels frequently include genes with limited clinical validity (Hosseini et al. 2018; Ingles et al. 2019). At the variant level, interpretation challenges include the lack of condition-specific population frequency thresholds to establish when a variant is insufficiently rare to be considered pathogenic. Accurate assessment of population frequency is also compounded by the lack of diverse ancestral representation in population data sets (Manrai et al. 2016; Popejoy et al. 2018). Variant interpretation processes and resources are rapidly evolving, so patients should be made aware of the need for periodic follow-up with a genetic counselor for reassessment of their test results (Das K et al. 2014).

MAXIMIZING THE VALUE OF CARDIOVASCULAR GENETIC COUNSELING

As reflected in coping theory and the genetic testing literature, a person's sense of certainty and control is an important influence on their willingness to engage with the implications of testing and their level of distress about the results (McConkie-Rosell and Sullivan 1999). Genetic counselors can increase perceptions of control in numerous ways throughout the visit, starting with a psychological assessment that gives the patient a chance to voice their needs and provides information to the counselor about the person's baseline emotional state, family dynamics, and the significance they attribute possible test results. This valuable context enables genetic counselors to use their clinical judgement to tailor information to the patient's needs by deferring, shortening, or omitting less relevant information. A sense of control is also created by engaging people in a two-way educational process and prompting people to apply

biomedical information to their own lives rather than presenting information as a lecture.

The educational goals during genetic test result disclosure include communicating (a) the impact of a genetic test result on medical management, if applicable, (b) family screening recommendations, (c) whether predictive testing is available, and (d) if further steps are needed to clarify the results (Cirino et al. 2017). However, even if these topics are explained clearly, many of the desired outcomes depend on how well the practitioner understands and engages the patient.

Cascade screening depends on people effectively relaying the implications of the result to their relatives. As noted in the literature, it is common for people to have unrealistic expectations about testing and to misconstrue the meaning of results, particularly uncertain results, in a way that underrepresents familial risk (Burns et al. 2017). The influence of family dynamics on screening and testing behavior also means that it is important to help people process the results in a way that will lead them to positively influence others. Studies of genetic counselor communication patterns have found that almost 60% of genetic counselors communicate in "teaching" styles characterized by high verbal dominance and that only ~4% of counselor speech is targeted at helping the patient to apply the information to their own lives (Roter et al. 2006; Ellington et al. 2011). It has also been reported that the majority of counselors' empathic statements tend to be slanted toward negative emotions even when the patient expresses mostly positive emotion (Ellington et al. 2011). In contrast, communication styles that draw out more psychological content and more patient speech lead to higher satisfaction and promote both cognitive and emotional processing (Roter et al. 2006; Ellington et al. 2011). Further, adaptive behavior is facilitated by eliciting patient dialogue that builds motivation for change (Ash 2017).

Promoting Adaptive Coping and Behavior

A patient-centered approach requires that the genetic counselor focus on understanding and working with the state of the patient. These goals include (a) eliciting the person's ideas and expectations about testing, (b) understanding their motivations and concerns, (c) assessing factors likely to influence perceptions of the results, (d) identifying emotional states that the test results may alleviate or exacerbate, (e) guiding decision-making about whether to do the test and timing of testing, and (f) helping the person make meaning of the result, and take appropriate action.

As genetic counselors develop their psychological counseling skills, common concerns include discomfort with asking about personal matters, uncertainty about how to phrase questions, fear of a negative patient response, or not knowing how to respond if the person does choose to share difficult experiences or emotions (Borders et al. 2006; Shugar 2017). As counselors progress in their careers, there is also the concern that these conversations might be too time consuming. These concerns are addressed by approaching psychological counseling in a systematic way, such as using a counseling framework to guide the session.

Psychological counseling can be learned by the same structured methods used for developing other clinical skills. For example, genetic counseling trainees learn to take a family history by first memorizing specific questions to ask, and then gradually progress to a more fluid, tailored approach as they learn more about genetic conditions. Developmentally, it has been noted that trainees tend to conceptualize psychological counseling as something they can "do" through rote application of a learned technique (Shugar 2017). Starting with a framework supports people as they gain the clinical experience that leads to sound judgement over time so they can progress from "doing" psychological behaviors to a state of "being" psychologically engaged with their patient. Below, the BATHE method is used as an example framework to model one approach of structuring a psychological assessment.

Using a Counseling Framework: an Example of the BATHE Psychological Assessment

An optimal approach to the psychological aspects of a cardiovascular counseling session is flexible enough to allow a practitioner to conduct a brief assessment or spend a longer time building a deeper relationship. It provides a balance between phrasing that is not invasive but also does not gloss over or shy away from pertinent issues. It also provides a structure that is concrete enough for a beginner to grasp easily and fluid enough for a seasoned practitioner to use without feeling constrained.

Originally developed as a brief screen for primary care physicians, the BATHE method is a framework that summarizes the core elements of a psychological assessment (Lieberman and Stuart 1999). The BATHE acronym refers to the following elements: Background, Affect, Trouble, Handling, Empathy (Table 2). It has been shown to improve patient satisfaction with physician encounters without significantly increasing a 15-min visit time (Leiblum et al. 2008; De Maria et al. 2010, 2011; Kim et al. 2012; Pace et al. 2017).

Although it was designed as a psychological screen, BATHE can easily be adapted to specific situations. A key feature is that the method does not require psychological hypothesis testing by the clinician. Rather, it is a patient-led method that can be used to assess the person's concerns

Table 2. The BATHE method of psychological assessment

Background
Elicit context of the person's life outside of the indication for the visit.
"Outside of this, what is going on in your life?"
Affect
Elicit psychosocial impact of the condition, visit, or other topic.
"How does that affect your quality of life?"
Trouble
Elicit patient's concerns.
"What troubles you the most about this?"
Handling
Assess how the patient is coping with the topic.
"How are you handling that?"
Empathy
Give an empathetic response, preferably in the form of a reflective statement.
"You're worried." "It's difficult." etc.

Data adapted from Lieberman and Stuart 1999.

Cite this article as *Cold Spring Harb Perspect Med* doi: 10.1101/cshperspect.a036624

Table 3. An empathic interviewing style of BATHE-guided psychological assessment

GC: How has thinking about testing been affecting your quality of life?	Affect
P: Well, not that much. I think I want to know. I'm not sure. I mean, I'd want to know if it's good news…	Affect
GC: You have mixed feelings.	Reflection
P: Yes, I have been going back and forth, but I decided that now is the time.	Affect
GC: What bothers you the most when you think about the test?	Trouble
P: I've been a big part of my father's care and it's been hard seeing what he's gone through. I don't want to think that I might have to face something similar.	Trouble
GC: It's been hard.	Reflection
P: Yeah, I've spent a lot of time in this hospital. It's a little stressful to be here now.	Affect
GC: How do you handle your stress?	Handling
P: I spend a lot of time at the gym. It really grounds me. And my family, we're really close, so that helps.	
GC: And what else is going on in your life that might influence how this is for you?	Background
P: Well, I'm here now because I'm starting to think about having kids, so now I have a reason.	Background
GC: You're thinking about the next generation.	Reflection
P: Yeah, it's not just about me now, you know.	

in a way that builds the clinical experience and judgement of the practitioner, while it is supplemented by other counseling skills over time. If used without modification, as a sequence of four questions and an empathic response, it can feel somewhat stilted. This can be counteracted by interspersing the questions with reflective and empathic responses to encourage elaboration. By adding a few basic counseling skills, such as reflection statements, the BATHE questions can be expanded into a concise, information-rich, and empathic brief interview (Table 3).

When there is more time, or with people in psychological distress, it can also be used as a framework to guide an in-depth discussion by using an empathic conversational style (Table 4).

For example, an experienced genetic counselor can use it as a mental checklist of points that can be reordered and expanded in a more fluid and natural discussion of the patient's concerns. An empathic conversational style intersperses questions with reflection and empathic statements. This style is more patient-led than the interviewing styles and uses more reflection statements to demonstrate empathy and draw out patient speech.

The BATHE framework can also be supplemented with an assessment tailored to the clinical context. For example, for a person considering predictive testing, appropriate questions might assess motivations for testing, their experience with the condition, family dynamics

Table 4. An empathic conversational style of BATHE-guided psychological assessment

Patient: I've been so anxious waiting for this appointment.	Affect
GC: You've been anxious.	Reflection
P: Yes, I've been having symptoms, but I don't know if they're serious or not.	Trouble
GC: It's felt uncertain.	Reflection
P: I don't deal well with uncertainty. I know I'm at risk because of my brother, and 1 just want to know what's going on.	Trouble
GC: The uncertainty really bothers you.	Reflection
P: Yes, it does! I like to know the facts so I can make a plan.	Affect, handling
GC: Knowing what you're dealing with would feel better.	Reflection
P: Yes.	
GC: And what else is going on in your life that might influence how this is for you?	Background
P: Well, I'm thinking about getting a new job, but I don't want my insurance to lapse if this ends up being serious.	Background

and pressure, baseline emotional state, and hopes and expectations for the results. As long as the clinician varies the wording of the questions slightly each time, the BATHE elements of affect, trouble, and handling can be used repeatedly to understand each subtopic in greater detail.

The structure can also be used to guide disclosure of a genetic result. Eliciting information about any changes in the patient's life (reassessing the background element) alerts a genetic counselor to new life stressors or changes in the person's thoughts or emotions. This is generally quick; however, at times new context greatly influences how the disclosure is framed. For example, a serious medical event or news about a new job or pregnancy can drastically change a person's state of mind between sessions.

During the disclosure session there are a number of ways to encourage someone to apply information and recommendations to their own life. One generalizable method is to make a high-level statement and prompt feedback by asking "What do you think when I say that?" or "So when you hear these recommendations, how do you think it might work in your family?" Further education then becomes interactive. It also makes discussions about interpersonal or logistical barriers feel more like shared problem-solving rather than unsolicited advice.

When disclosing genetic testing results that are perceived as good news by the patient, clinical judgment dictates use of the BATHE elements. For example, if the person has a purely positive reaction to negative predictive test results, assessing what troubles them may be unnecessary and awkward. It can still be helpful, however, to address the Handling topic in a conversational way by asking them whom they plan to tell first or how they think others will react. If there are indications of survival guilt or other mixed feelings, a softened version of the Trouble question can be used, such as, "Is there any part of you that has mixed feelings about the result?"

When giving unwanted news, the trouble and handling elements are clearly relevant. Eliciting a person's thoughts and feelings about what troubles them allows the genetic counselor to bear witness to their reaction and facilitate emotional processing. Transitioning to the topic of Handling, when appropriate, leads to the opportunity to reinforce existing coping methods or to elicit unrecognized strengths and resilience factors. Rather than using the retrospective, "How have you handled situations in the past?" or the present tense "How are you handling this?" recommended by the standard BATHE method, it can also be phrased as "How will you handle this?" Use of future tense language engages the person in visualizing concrete future behavior.

It is important to tailor the emphasis and time spent on these topics to the needs of the patient. For example, it is often appropriate to allow a distressed person to focus more on the troubling aspects of the situation before gently introducing the topic of handling. Aligning with the patient's state of mind avoids seeming dismissive of their reaction and builds trust so that the genetic counselor can act as a guide toward adaptive thoughts and behaviors over time.

CONCLUDING REMARKS

Although this work references literature and uses examples from inherited cardiovascular conditions, these methods can be generalized to other settings. Despite the challenges of communicating technical information, using a counseling framework such as the BATHE method provides a structure that can be adapted to the needs of each genetic counseling session. Psychological counseling increases the impact of education and allows genetic counselors to better help people understand information, guide them to make decisions about their care that fit their needs and values, and promote their well-being as they move forward in their lives.

ACKNOWLEDGMENTS

Many thanks to Colleen Caleshu and Tia Moscarello for their work adapting the BATHE method for use in genetic counseling settings.

REFERENCES

*Reference is also in this collection.

Aatre RD, Day SM. 2011. Psychological issues in genetic testing for inherited cardiovascular diseases. *Circ Cardiovasc Genet* **4:** 81–90. doi:10.1161/CIRCGENETICS.110.957365

Accreditation Council for Genetic Counseling. 2015. *Practice-Based Competencies for Genetic Counselors.* Accreditation Council for Genetic Counseling. www.gceducation.org/wp-content/uploads/2019/02/ACGC-Core-Competencies-Brochure_15_Web.pdf

Ackerman MJ, Priori SG, Willems S, Berul C, Brugada R, Calkins H, Camm AJ, Ellinor PT, Gollob M, Hamilton R, et al. 2011. HRS/EHRA expert consensus statement on the state of genetic testing for the channelopathies and cardiomyopathies: this document was developed as a partnership between the Heart Rhythm Society (HRS) and the European Heart Rhythm Association (EHRA). *Europace* **13:** 1077–1109.

Amendola LM, Jarvik GP, Leo MC, McLaughlin HM, Akkari Y, Amaral MD, Berg JS, Biswas S, Bowling KM, Conlin LK, et al. 2016. Performance of ACMG-AMP variant-interpretation guidelines among nine laboratories in the Clinical Sequencing Exploratory Research Consortium. *Am J Hum Genet* **98:** 1067–1076. doi:10.1016/j.ajhg.2016.03.024

Arbustini E, Di Toro A, Giuliani L, Favalli V, Narula N, Grasso M. 2018. Cardiac phenotypes in hereditary muscle disorders: JACC state-of-the-art review. *J Am Coll Cardiol* **72:** 2485–2506.

Ash E. 2017. Motivational interviewing in the reciprocal engagement model of genetic counseling: a method overview and case illustration. *J Genet Couns* **26:** 300–311. doi:10.1007/s10897-016-0053-8

Balmaña J, Digiovanni L, Gaddam P, Walsh MF, Joseph V, Stadler ZK, Nathanson KL, Garber JE, Couch FJ, Offit K, et al. 2016. Conflicting interpretation of genetic variants and cancer risk by commercial laboratories as assessed by the prospective registry of multiplex testing. *J Clin Oncol* **34:** 4071–4078. doi:10.1200/JCO.2016.68.4316

Bland A, Harrington EA, Dunn K, Pariani M, Platt JCK, Grove ME, Caleshu C. 2018. Clinically impactful differences in variant interpretation between clinicians and testing laboratories: a single-center experience. *Genet Med* **20:** 369–373. doi:10.1038/gim.2017.212

Bonner C, Spinks C, Semsarian C, Barratt A, Ingles J, McCaffery K. 2018. Psychosocial impact of a positive gene result for asymptomatic relatives at risk of hypertrophic cardiomyopathy. *J Genet Couns* **27:** 1040–1048. doi:10.1007/s10897-018-0218-8

Borders LDA, Eubanks S, Callanan N. 2006. Supervision of psychosocial skills in genetic counseling. *J Genet Couns* **15:** 211–223. doi:10.1007/s10897-006-9024-9

Broadstock M, Michie S, Marteau T. 2000. Psychological consequences of predictive genetic testing: a systematic review. *Eur J Hum Genet* **8:** 731–738. doi:10.1038/sj.ejhg.5200532

Burke MA, Cook SA, Seidman JG, Seidman CE. 2016. Clinical and mechanistic insights Into the genetics of cardiomyopathy. *J Am Coll Cardiol* **68:** 2871–2886.

Burns C, McGaughran J, Davis A, Semsarian C, Ingles J. 2016. Factors influencing uptake of familial long QT syndrome genetic testing. *Am J Med Genet A* **170:** 418–425. doi:10.1002/ajmg.a.37455

Burns C, Yeates L, Spinks C, Semsarian C, Ingles J. 2017. Attitudes, knowledge and consequences of uncertain genetic findings in hypertrophic cardiomyopathy. *Eur J Hum Genet* **25:** 809–815. doi:10.1038/ejhg.2017.66

Caleshu C, Kasparian NA, Edwards KS, Yeates L, Semsarian C, Perez M, Ashley E, Turner CJ, Knowles JW, Ingles J. 2016. Interdisciplinary psychosocial care for families with inherited cardiovascular diseases. *Trends Cardiovasc Med* **26:** 647–653. doi:10.1016/j.tcm.2016.04.010

Charron P, Héron D, Gargiulo M, Richard P, Dubourg O, Desnos M, Bouhour JB, Feingold J, Carrier L, Hainque B, et al. 2002. Genetic testing and genetic counselling in hypertrophic cardiomyopathy: the French experience. *J Med Genet* **39:** 741–746. doi:10.1136/jmg.39.10.741

Christiaans I, Birnie E, Bonsel GJ, Wilde AA, van Langen IM. 2008. Uptake of genetic counselling and predictive DNA testing in hypertrophic cardiomyopathy. *Eur J Hum Genet* **16:** 1201–1207. doi:10.1038/ejhg.2008.92

Christiaans I, van Langen IM, Birnie E, Bonsel GJ, Wilde AAM, Smets EMA. 2009. Quality of life and psychological distress in hypertrophic cardiomyopathy mutation carriers: a cross-sectional cohort study. *Am J Med Genet A* **149A:** 602–612. doi:10.1002/ajmg.a.32710

Cirino AL, Harris S, Lakdawala NK, Michels M, Olivotto I, Day SM, Abrams DJ, Charron P, Caleshu C, Semsarian C, et al. 2017. Role of genetic testing in inherited cardiovascular disease: a review. *JAMA Cardiol* **2:** 1153–1160. doi:10.1001/jamacardio.2017.2352

Cox GF. 2007. Diagnostic approaches to pediatric cardiomyopathy of metabolic genetic etiologies and their relation to therapy. *Prog Pediatr Cardiol* **24:** 15–25.

Das K J, Ingles J, Bagnall RD, Semsarian C. 2014. Determining pathogenicity of genetic variants in hypertrophic cardiomyopathy: importance of periodic reassessment. *Genet Med* **16:** 286–293. doi:10.1038/gim.2013.138

De Maria S, De Maria AP, Weiner M, Silvay G. 2010. Use of the BATHE method to increase satisfaction amongst patients undergoing cardiac and major vascular operations. *Cleve Clin J Med* **77:** eS25.

DeMaria S Jr, DeMaria AP, Silvay G, Flynn BC. 2011. Use of the BATHE method in the preanesthetic clinic visit. *Anesth Analg* **113:** 1020–1026. doi:10.1213/ANE.0b013e318229497b

Ellington L, Kelly KM, Reblin M, Latimer S, Roter D. 2011. Communication in genetic counseling: cognitive and emotional processing. *Health Commun* **26:** 667–675. doi:10.1080/10410236.2011.561921

Erskine KE, Hidayatallah NZ, Walsh CA, McDonald TV, Cohen L, Marion RW, Dolan SM. 2014. Motivation to pursue genetic testing in individuals with a personal or family history of cardiac events or sudden cardiac death. *J Genet Couns* **23:** 849–859. doi:10.1007/s10897-014-9707-6

Etchegary H, Pullman D, Simmonds C, Young TL, Hodgkinson K. 2015. "It had to be done": genetic testing decisions for arrhythmogenic right ventricular cardiomyopathy. *Clin Genet* **88:** 344–351. doi:10.1111/cge.12513

Furqan A, Arscott P, Girolami F, Cirino AL, Michels M, Day SM, Olivotto I, Ho CY, Ashley E, Green EM, et al. 2017. Care in specialized centers and data sharing increase agreement in hypertrophic cardiomyopathy genetic test interpretation. *Circ Cardiovasc Genet* **10**: 1–9. doi:10.1161/CIRCGENETICS.116.001700

Garber KB, Vincent LM, Alexander JJ, Bean LJH, Bale S, Hegde M. 2016. Reassessment of genomic sequence variation to harmonize interpretation for personalized medicine. *Am J Hum Genet* **99**: 1140–1149. doi:10.1016/j.ajhg.2016.09.015

Garcia J, Tahiliani J, Johnson NM, Aguilar S, Beltran D, Daly A, Decker E, Haverfield E, Herrera B, Murillo L, et al. 2016. Clinical genetic testing for the cardiomyopathies and arrhythmias: a systematic framework for establishing clinical validity and addressing genotypic and phenotypic heterogeneity. *Front Cardiovasc Med* **3**: 20.

Geddes GC, Earing MG. 2018. Genetic evaluation of patients with congenital heart disease. *Curr Opin Pediatr* **30**: 707–713.

Geelen E, Van Hoyweghen I, Doevendans PA, Marcelis CLM, Horstman K. 2011. Constructing "best interests": genetic testing of children in families with hypertrophic cardiomyopathy. *Am J Med Genet A* **155**: 1930–1938. doi:10.1002/ajmg.a.34107

Giudicessi JR, Wilde AAM, Ackerman MJ. 2018. The genetic architecture of long QT syndrome: a critical reappraisal. *Trends Cardiovasc Med* **28**: 453–464.

Godino L, Turchetti D, Jackson L, Hennessy C, Skirton H. 2016. Impact of presymptomatic genetic testing on young adults: a systematic review. *Eur J Hum Genet* **24**: 496–503. doi:10.1038/ejhg.2015.153

Goyal A, Keramati AR, Czarny MJ, Resar JR, Mani A. 2017. The genetics of aortopathies in clinical cardiology. *Clin Med Insights Cardiol* **11**: 1179546817709787.

Harrison SM, Dolinsky JS, Knight Johnson AE, Pesaran T, Azzariti DR, Bale S, Chao EC, Das S, Vincent L, Rehm HL. 2017. Clinical laboratories collaborate to resolve differences in variant interpretations submitted to ClinVar. *Genet Med* **19**: 1096–1104. doi:10.1038/gim.2017.14

Hendriks KSWH, Grosfeld FJM, Wilde AAM, van den Bout J, van Langen IM, van Tintelen JP, ten Kroode HFJ. 2005. High distress in parents whose children undergo predictive testing for long QT syndrome. *Community Genet* **8**: 103–113. doi:10.1159/000084778

Hendriks KSWH, Hendriks MMWB, Birnie E, Grosfeld FJM, Wilde AAM, van den Bout J, Smets EMA, van Tintelen JP, ten Kroode HFJ, van Langen IM. 2008. Familial disease with a risk of sudden death: a longitudinal study of the psychological consequences of predictive testing for long QT syndrome. *Heart Rhythm* **5**: 719–724. doi:10.1016/j.hrthm.2008.01.032

Heshka JT, Palleschi C, Howley H, Wilson B, Wells PS. 2008. A systematic review of perceived risks, psychological and behavioral impacts of genetic testing. *Genet Med* **10**: 19–32. doi:10.1097/GIM.0b013e31815f524f

Hickey KT, Sciacca RR, Biviano AB, Whang W, Dizon JM, Garan H, Chung WK. 2014. The effect of cardiac genetic testing on psychological well-being and illness perceptions. *Heart Lung* **43**: 127–132. doi:10.1016/j.hrtlng.2014.01.006

Hoedemaekers E, Jaspers JPC, Van Tintelen JP. 2007. The influence of coping styles and perceived control on emotional distress in persons at risk for a hereditary heart disease. *Am J Med Genet A* **143A**: 1997–2005. doi:10.1002/ajmg.a.31871

Hosseini SM, Kim R, Udupa S, Costain G, Jobling R, Liston E, Jamal SM, Szybowska M, Morel CF, Bowdin S, et al. 2018. Reappraisal of reported genes for sudden arrhythmic death. *Circulation* **138**: 1195–1205. doi:10.1161/CIRCULATIONAHA.118.035070

* Ingles J. 2019. Psychological issues in managing families with inherited cardiovascular diseases. *Cold Spring Harb Perpect Med* doi:10.1101/cshperspect.a036558

Ingles J, Yeates L, O'Brien L, McGaughran J, Scuffham PA, Atherton J, Semsarian C. 2012. Genetic testing for inherited heart diseases: longitudinal impact on health-related quality of life. *Genet Med* **14**: 749–752. doi:10.1038/gim.2012.47

Ingles J, Goldstein J, Thaxton C, Caleshu C, Corty EW, Crowley SB, Dougherty K, Harrison SM, McGlaughon J, Milko LV, et al. 2019. Evaluating the clinical validity of hypertrophic cardiomyopathy genes. *Circ Genom Precis Med* **12**: e002460.

Kelly MA, Caleshu C, Morales A, Buchan J, Wolf Z, Harrison SM, Cook S, Dillon MW, Garcia J, Haverfield E, et al. 2018. Adaptation and validation of the ACMG/AMP variant classification framework for *MYH7*-associated inherited cardiomyopathies: recommendations by ClinGen's Inherited Cardiomyopathy Expert Panel. *Genet Med* **20**: 351–359. doi:10.1038/gim.2017.218

Khouzam A, Kwan A, Baxter S, Bernstein JA. 2015. Factors associated with uptake of genetics services for hypertrophic cardiomyopathy. *J Genet Couns* **24**: 797–809. doi:10.1007/s10897-014-9810-8

Kim JH, Park YN, Park EW, Cheong YS, Choi EY. 2012. Effects of BATHE interview protocol on patient satisfaction. *Korean J Fam Med* **33**: 366–371. doi:10.4082/kjfm.2012.33.6.366

Leiblum SR, Schnall E, Seehuus M, DeMaria A. 2008. To BATHE or not to BATHE: patient satisfaction with visits to their family physician. *Fam Med* **40**: 407–411.

Lieberman JA III, Stuart MR. 1999. The BATHE method: incorporating counseling and psychotherapy into the everyday management of patients. *Prim Care Companion J Clin Psychiatry* **1**: 35–38. doi:10.4088/PCC.v01n0202

MacCarrick G, Black JH 3rd, Bowdin S, El-Hamamsy I, Frischmeyer-Guerrerio PA, Guerrerio AL, Sponseller PD, Loeys B, Dietz HC 3rd. 2014. Loeys–Dietz syndrome: a primer for diagnosis and management. *Genet Med* **16**: 576–587.

MacLeod R, Beach A, Henriques S, Knopp J, Nelson K, Kerzin-Storrar L. 2014. Experiences of predictive testing in young people at risk of Huntington's disease, familial cardiomyopathy or hereditary breast and ovarian cancer. *Eur J Hum Genet* **22**: 396–401. doi:10.1038/ejhg.2013.143

Manrai AK, Funke BH, Rehm HL, Olesen MS, Maron BA, Szolovits P, Margulies DM, Loscalzo J, Kohane IS. 2016. Genetic misdiagnoses and the potential for health disparities. *N Engl J Med* **375**: 655–665. doi:10.1056/NEJMsa1507092

Manuel A, Brunger F. 2014. Making the decision to participate in predictive genetic testing for arrhythmogenic right ventricular cardiomyopathy. *J Genet Couns* **23**: 1045–1055. doi:10.1007/s10897-014-9733-4

Maurer MS, Hanna M, Grogan M, Dispenzieri A, Witteles R, Drachman B, Judge DP, Lenihan DJ, Gottlieb SS, Shah SJ, et al. 2016. Genotype and phenotype of transthyretin cardiac amyloidosis: THAOS (Transthyretin Amyloid Outcome Survey). *J Am Coll Cardiol* **68**: 161–172.

McConkie-Rosell A, Sullivan JA. 1999. Genetic counseling-stress, coping, and the empowerment perspective. *J Genet Couns* **8**: 345–357. doi:10.1023/A:1022919325772

Meulenkamp TM, Tibben A, Mollema ED, van Langen IM, Wiegman A, de Wert GM, de Beaufort ID, Wilde AAM, Smets EMA. 2008. Predictive genetic testing for cardiovascular diseases: impact on carrier children. *Am J Med Genet A* **146A**: 3136–3146. doi:10.1002/ajmg.a.32592

Miller EM, Wang Y, Ware SM. 2013. Uptake of cardiac screening and genetic testing among hypertrophic and dilated cardiomyopathy families. *J Genet Couns* **22**: 258–267. doi:10.1007/s10897-012-9544-4

Muchtar E, Blauwet LA, Gertz MA. 2017. Restrictive cardiomyopathy: genetics, pathogenesis, clinical manifestations, diagnosis, and therapy. *Circ Res* **121**: 819–837.

Oliveri S, Ferrari F, Manfrinati A, Pravettoni G. 2018. A systematic review of the psychological implications of genetic testing: a comparative analysis among cardiovascular, neurodegenerative and cancer diseases. *Front Genet* **9**: 624. doi:10.3389/fgene.2018.00624

Ormondroyd E, Oates S, Parker M, Blair E, Watkins H. 2014. Pre-symptomatic genetic testing for inherited cardiac conditions: a qualitative exploration of psychosocial and ethical implications. *Eur J Hum Genet* **22**: 88–93. doi:10.1038/ejhg.2013.81

Pace EJ, Somerville NJ, Enyioha C, Allen JP, Lemon LC, Allen CW. 2017. Effects of a brief psychosocial intervention on inpatient satisfaction: a randomized controlled trial. *Fam Med* **49**: 675–678.

Pierpont ME, Brueckner M, Chung WK, Garg V, Lacro RV, McGuire AL, Mital S, Priest JR, Pu WT, Roberts A, et al. 2018. Genetic basis for congenital heart disease: revisited: a scientific statement from the American Heart Association. *Circulation* **138**: e653–e711.

Popejoy AB, Ritter DI, Crooks K, Currey E, Fullerton SM, Hindorff LA, Koenig B, Ramos EM, Sorokin EP, Wand H, et al. 2018. The clinical imperative for inclusivity: race, ethnicity, and ancestry (REA) in genomics. *Hum Mutat* **39**: 1713–1720. doi:10.1002/humu.23644

Pugh TJ, Kelly MA, Gowrisankar S, Hynes E, Seidman MA, Baxter SM, Bowser M, Harrison B, Aaron D, Mahanta LM, et al. 2014. The landscape of genetic variation in dilated cardiomyopathy as surveyed by clinical DNA sequencing. *Genet Med* **16**: 601–608.

Rehm HL, Berg JS, Brooks LD, Bustamante CD, Evans JP, Landrum MJ, Ledbetter DH, Maglott DR, Martin CL, Nussbaum RL, et al. 2015. ClinGen—the Clinical Genome Resource. *N Engl J Med* **372**: 2235–2242. doi:10.1056/NEJMsr1406261

Renard M, Francis C, Ghosh R, Scott AF, Witmer PD, Adès LC, Andelfinger GU, Arnaud P, Boileau C, Callewaert BL, et al. 2018. Clinical validity of genes for heritable thoracic aortic aneurysm and dissection. *J Am Coll Cardiol* **72**: 605–615. doi:10.1016/j.jacc.2018.04.089

Reuter C, Grove ME, Orland K, Spoonamore K, Caleshu C. 2018. Clinical cardiovascular genetic counselors take a leading role in team-based variant classification. *J Genet Couns* **27**: 751–760. doi:10.1007/s10897-017-0175-7

Richards S, Aziz N, Bale S, Bick D, Das S, Gastier-Foster J, Grody WW, Hegde M, Lyon E, Spector E, et al. 2015. Standards and guidelines for the interpretation of sequence variants: a Joint Consensus Recommendation of the American College of Medical Genetics and Genomics and the Association for Molecular Pathology. *Genet Med* **17**: 405–423. doi:10.1038/gim.2015.30

Roter D, Ellington L, Erby LH, Larson S, Dudley W. 2006. The Genetic Counseling Video Project (GCVP): models of practice. *Am J Med Genet C Semin Med Genet* **142C**: 209–220. doi:10.1002/ajmg.c.30094

Shugar A. 2017. Teaching genetic counseling skills: incorporating a genetic counseling adaptation continuum model to address psychosocial complexity. *J Genet Couns* **26**: 215–223. doi:10.1007/s10897-016-0042-y

Smart A. 2010. Impediments to DNA testing and cascade screening for hypertrophic cardiomyopathy and Long QT syndrome: a qualitative study of patient experiences. *J Genet Couns* **19**: 630–639. doi:10.1007/s10897-010-9314-0

Smets EMA, Stam MMH, Meulenkamp TM, van Langen IM, Wilde AAM, Wiegman A, de Wert GM, Tibben A. 2008. Health-related quality of life of children with a positive carrier status for inherited cardiovascular diseases. *Am J Med Genet A* **146A**: 700–707. doi:10.1002/ajmg.a.32218

Somers AE, Ware SM, Collins K, Jefferies JL, He H, Miller EM. 2014. Provision of cardiovascular genetic counseling services: current practice and future directions. *J Genet Couns* **23**: 976–983. doi:10.1007/s10897-014-9719-2

Stout KK, Daniels CJ, Aboulhosn JA, Bozkurt B, Broberg CS, Colman JM, Crumb SR, Dearani JA, Fuller S, Gurvitz M, et al. 2019. 2018 AHA/ACC guideline for the management of adults with congenital heart disease: a report of the American College of Cardiology/American Heart Association Task Force on Clinical Practice Guidelines. *Circulation* **139**: e698–e800.

Sweeny K, Ghane A, Legg AM, Huynh HP, Andrews SE. 2014. Predictors of genetic testing decisions: a systematic review and critique of the literature. *J Genet Couns* **23**: 263–288. doi:10.1007/s10897-014-9712-9

van der Roest WP, Pennings JM, Bakker M, van den Berg MP, van Tintelen JP. 2009. Family letters are an effective way to inform relatives about inherited cardiac disease. *Am J Med Genet A* **149A**: 357–363. doi:10.1002/ajmg.a.32672

van Maarle MC, Stouthard MEA, Marang-van de Mheen PJ, Klazinga NS, Bonsel GJ. 2001. How disturbing is it to be approached for a genetic cascade screening programme for familial hypercholesterolaemia? Psychological impact and screenees' views. *Community Genet* **4**: 244–252.

Vansenne F, Bossuyt PMM, de Borgie CAJM. 2009. Evaluating the psychological effects of genetic testing in symptomatic patients: a systematic review. *Genet Test Mol Biomarkers* **13**: 555–563. doi:10.1089/gtmb.2009.0029

Veach PM, Bartels DM, LeRoy BS. 2007. Coming full circle: a reciprocal-engagement model of genetic counseling practice. *J Genet Couns* **16:** 713–728. doi:10.1007/s10897-007-9113-4

Wakefield CE, Hanlon LV, Tucker KM, Patenaude AF, Signorelli C, McLoone JK, Cohn RJ. 2016. The psychological impact of genetic information on children: a systematic review. *Genet Med* **18:** 755–762. doi:10.1038/gim.2015.181

Werner-Lin A, McCoyd JLM, Bernhardt BA. 2016. Balancing genetics (science) and counseling (art) in prenatal chromosomal microarray testing. *J Genet Couns* **25:** 855–867. doi:10.1007/s10897-016-9966-5

Yang S, Lincoln SE, Kobayashi Y, Nykamp K, Nussbaum RL, Topper S. 2017. Sources of discordance among germ-line variant classifications in ClinVar. *Genet Med* **19:** 1118–1126. doi:10.1038/gim.2017.60

Genetic Counseling and Genome Sequencing in Pediatric Rare Disease

Alison M. Elliott

Department of Medical Genetics, University of British Columbia Investigator, BC Children's Hospital Research Institute and BC Women's Health Research Institute, and Provincial Medical Genetics Program, Vancouver, British Columbia V6H 3N1, Canada

Correspondence: aelliott@bcchr.ca

Both genome sequencing (GS) and exome sequencing (ES) have proven to be revolutionary in the diagnosis of pediatric rare disease. The diagnostic potential and increasing affordability make GS and ES more accessible as a routine clinical test in some centers. Herein, I review aspects of rare disease in pediatrics associated with the use of genomic technologies with an emphasis on the benefits and limitations of both ES and GS, complexities of variant classification, and the importance of genetic counseling. Indications for testing, the role of genetic counselors in genomic test selection, and the diagnostic potential of ES and GS in various pediatric multisystem disorders are discussed. The neonatal population represents an important cohort in pediatric rare disease. Rapid ES and GS in critically ill neonates can have an immediate impact on medical management and present unique genetic counseling challenges. This work includes reviews of recommendations for genetic counseling for families considering genome-wide sequencing, and issues of access to genetic counseling that affect clinical use and will necessitate implementation of innovative methods such as online decision aids. Finally, this work will also review the challenges of having a child with a rare disease, the impact of results from ES and GS on these families, and the role of various support agencies.

RARE DISEASE—DEFINITION AND FREQUENCY

In Europe, a rare disease is defined as one that affects less than one in 2000 individuals (European Organisation for Rare Diseases 2005). In the United States, the Rare Diseases Act defines a rare disease to be one that affects fewer than one in 200,000 individuals (H.R.4013 — 107th Congress 2002). Others consider a frequency of less than one in 2,000,000 to represent an "ultra-rare" disease (Hennekam 2011). Although, individually, each disease is characterized as "rare," collectively, rare diseases affect 4%–8% of the general population (Baird et al. 1988; Boycott et al. 2017). There are ∼7000 rare diseases, and ∼80% are thought to have a genetic basis (Amberger et al. 2009, 2015, 2019). The majority affect the pediatric population, and it is estimated that 30% of affected children do not survive beyond 5 years of age (European Organisation for Rare Diseases 2005). Rare diseases are generally multisystem and complex in nature. Because of their rarity, these disorders are generally not familiar to most clinicians. As a result, they represent significant diagnostic challenges.

Genetic heterogeneity (allelic and locus), variable clinical presentation (expressivity), incomplete penetrance, genetic modifiers, and environmental factors can further complicate the ability to obtain a specific diagnosis in individuals with rare disorders (as reviewed in Wright et al. 2018a). Furthermore, extremely rare or ultra-rare diagnoses have been even less well-characterized from a genetic and phenotypic standpoint.

Genetic conditions and malformations are the leading causes of infant deaths in the neonatal intensive care unit (NICU) (Weiner et al. 2011). Identification of rare disorders in this setting is uniquely challenging for a number of reasons: The complete clinical phenotype may not have evolved, clinical signs can be compounded by prematurity, and underlying genetic heterogeneity may delay diagnosis through a candidate gene approach or serial multiple gene panel testing. As a result, many patients with genetic disorders are discharged or die before the diagnosis has been established (Elliott et al. 2019).

THE CHANGING LANDSCAPE OF GENETIC TESTING

Serial investigations of single genes are time-consuming, expensive, and not always available. In the case of intellectual disability, more than 700 genes have been implicated (Grozeva et al. 2015; Kochinke et al. 2016), making serial single-gene investigations impractical. Consequently multigene panels are often used, but typically lack a complete comprehensive inclusion of all potentially relevant genes and variants. Chromosomal microarray analysis (CMA) allows for interrogation of unbalanced structural changes and copy number variants and, with greater resolution than cytogenetic analysis (50–100 kb vs. 5–7 Mb), it is instrumental in the detection of chromosomal microdeletion syndromes. However, it has been estimated that only ~10% of patients with a rare pediatric disorder are diagnosed when using array-based techniques (Sagoo et al. 2009; Clark et al. 2018). Thus, technologies like genome sequencing/exome sequencing (GS/ES) are increasingly being used (as reviewed in Wright et al. 2018). These are more comprehensive evaluations than panel-based tests or CMA as a result of their unbiased approach to disease gene identification.

GS/ES has allowed for the discovery and further characterization of thousands of variants in Mendelian disorders—many of which represent rare disease in the pediatric population. Approximately 250 new gene-disease and 9200 variant-disease associations are reported each year (Wenger et al. 2017). This constantly evolving landscape not only contributes to rare disease discovery, but also underlies the importance of reanalyzing patients for whom a disorder was not identified on initial GS/ES analysis. Large-scale studies involving individuals with intellectual disability and suspected genetic disease have shown the diagnostic potential of this technology. The diagnostic rate of certain disorders of particular organ systems with ES is listed in Table 1 (Wright et al. 2018).

Many individuals enlisted in GS/ES studies can have more than one diagnosis. At least 4% of patients who enroll in GS/ES testing cohorts have at least two distinct genetic diagnoses (Balci et al. 2017). The disorders can be blended with overlapping phenotypes (e.g., variants resulting in both CHARGE and Kabuki syndromes) or distinct phenotypes (e.g., variants resulting in both neurofibromatosis and epileptic encephalopathy). The rate of dual diagnoses is even greater (14%) in cohorts of selected phenotypes (e.g., inclusion requirements of two or more conditions such as metabolic disease and intellectual disability) (Tarailo-Graovac et al. 2016).

From a genetic counseling perspective, the introduction of this technology is accompanied by a need for a shift from traditional single-gene counseling (e.g., for cystic fibrosis) to "genomic counseling" with its unique, inherent issues, which include variants of uncertain significance (VUSs) and incidental (secondary) findings (Ormond 2013; Patch and Middleton 2018).

Table 1. Diagnostic rate of certain disease classes with exome sequencing (ES)

Osteogenesis imperfecta	100%	Primary immunodeficiency	40%
Ciliary dyskinesia	76%	Limb girdle muscular dystrophy	37%
Epileptic encephalopathy	70%	Early-onset generalized dystonia	37%
Neurometabolic disorders	70%	Severe short stature	36%
Nonsyndromic hearing loss	56%	Inherited bone marrow failure	27%
Retinopathy	56%	Nephrolithiasis and/or nephrocalcinosis	17%
Suspected inborn errors of metabolism	50%	Congenital diaphragmatic hernia	<12%
Inherited thrombocytopenia	46%	Childhood solid tumors	10%
Intellectual disability	42%	Syndromic congenital heart disease	9.7%
Sporadic infantile spasms	40%	Autism spectrum disorder	8.4%

Adapted from Wright et al. 2018.

EXOME VERSUS GENOME APPROACHES

Although GS and ES use overlapping advances in DNA-sequencing technology, there are distinct differences. ES only captures and reports on the exonic or protein-coding sequences, which represent <2% of the genome. GS attempts to capture and report on the entirety of the genome, but because of the challenges in sequencing certain regions, GS examines ~90% of the genome.

By focusing on only the exonic, more well-characterized regions of the genome, ES offers pragmatic advantages over GS in overall cost savings in reagents and data storage; greater depth of coverage (100× vs. 30×) of reported sequences; and quicker, cheaper, and easier data interpretation (Warr et al. 2015).

However, as a result of the enrichment steps that are necessary for ES, the coverage of the final DNA sequences evaluated is not uniform. This results in reads with "hotspots" represented by an overabundance of coverage and other regions with too little coverage to provide reliable calls. Because GS does not require this upfront enrichment step, it generates much more uniform coverage of the genome, has the advantage of producing longer reads, and is better at capturing the high GC-rich content within exonic sequences. In addition to reporting on intronic sequences, the longer reads also allow for better detection and characterization of copy number variations, rearrangements, and other structural variations. Also, although ES has the ability to provide data for the mitochondrial genome, these results are not reported by many reference laboratories (Posey 2019).

LIMITATIONS OF EXOME AND GENOME SEQUENCING

Both ES and GS have limitations that affect their effectiveness as a diagnostic tool. Neither technology is capable of detecting changes in gene expression because of disorders of imprinting or providing reliable detection of variants in GC-rich regions such as centromeres and telomeres, trinucleotide repeat expansion, regions of highly homologous sequences, and low-level or tissue-specific mosaicism. It is also worth noting that with either ES or GS, the bioinformatic interpretation may rely on familiarity with the causative gene or variant to recognize its significance (Salgado et al. 2016).

INDICATIONS FOR AND APPROACHES TO EXOME AND GENOME SEQUENCING

Different groups have published indications for GS/ES, and a general summary is included here (adapted from Elliott et al. 2018).

The patient has a suspected genetic disorder plus one or more of the following:

- Previous genetic investigations, including CMA, appropriate single-gene or available panel testing, and first-tier biochemical testing for intellectual disability (Van Karnebeek et al. 2014) have not identified the genetic cause.

- The condition shows extensive genetic heterogeneity.

- The family history suggests a Mendelian single-gene disorder (e.g., affected parent and child, unaffected parents and affected child, parental consanguinity, multiple affected siblings).

As with older pediatric patients undergoing GS/ES, suspected neurologic disorders and multiple congenital anomalies are frequent indications for ES/GS studies in the neonatal intensive care setting (Willig et al. 2015; Petrikin et al. 2018; Stark et al. 2018; Elliott et al. 2019). Identifying appropriate patients for GS/ES and other genomic testing is an important clinical and economic consideration for providers. The integration of genetic counselors in the triage process of genomic tests has shown advantages that include a reduction in the misordering of tests, decreased time to diagnosis, cost efficiencies, education of ordering clinicians, and an increased diagnostic rate in ES/GS studies (Miller et al. 2014; Suarez et al. 2017; Dragojlovic et al. 2018; Elliott et al. 2018; Wakefield et al. 2018).

Trio-based (proband plus both biologic parents) approaches to GS/ES in rare pediatric disorders have diagnostic and efficiency advantages, particularly in neonates given the importance of rapid turnaround time for critically ill babies to immediately inform medical management. For example, de novo variants that present only in the child are easily identified, and the phase of variants in imprinted or recessive disorders can be detected. There is an ~10-fold reduction in the number of candidate variants, as well as a 50% increase in diagnostic yield with the trio approach, and a result can be obtained earlier for families (Wright et al. 2015). For individuals with severe developmental disabilities, a diagnostic rate of 40% can be achieved using trios (Wright et al. 2018b).

GS/ES IN THE NEONATAL POPULATION AND RARE DISEASE

Rapid GS/ES have shown impressive diagnostic capability in the NICU setting, immediately influencing clinical management by enabling pre-cision treatment, including therapeutic intervention or customized palliative care, as well as accurate genetic counseling (Saunders et al. 2012; Willig et al. 2015; Bowdin 2016; Berg et al. 2017; Meng et al. 2017; van Diemen et al. 2017; Petrikin et al. 2018; Stark et al. 2018; Elliott et al. 2019). Rapid GS/ES in the NICU setting can identify the genomic diagnosis in as little as 26 hours (Miller et al. 2015). A groundbreaking 2015 study showed the diagnostic effectiveness of rapid GS in 35 critically ill neonates: 57% diagnosed compared to 9% with conventional genetic testing, with results available in 5 days for some patients (Willig et al. 2015). Other studies have shown diagnostic rates of ~30%–60% (Meng et al. 2017; Stark et al. 2018; Elliott et al. 2019). Combining ES results with multi-gene panels and CMA outcomes can result in diagnostic yields of >70% with significant impact on immediate medical management (e.g., customizing anticonvulsant medication) in 83% of patients given a diagnosis (Elliott et al. 2019).

VARIANT CLASSIFICATION AND RECLASSIFICATION

The diagnostic potential of GS/ES is realized only when interpretation of GS/ES data and subsequent classification of the variant(s) are appropriate. Each of us has 4,000,000–5,000,000 nucleotides in our genome that differ from those of the human reference genome, but in most cases only one or two of these variants are responsible for the disease in an individual with Mendelian disorder (Elliott and Friedman 2018). Identifying most of these nonreference variants as benign and classifying most of the remainder as unlikely to be causal with respect to disease is an important component of GS/ES interpretation.

Previously, terms such as "mutation" and "polymorphism" were used to describe nucleotide changes, with the latter describing a change in >1% of the population and, therefore, presumed to be benign. As we accrue data through GS/ES, we have found that this simplified form of classification is not optimally relevant or useful. Furthermore, genetic changes can have different implications and frequencies among different ethnic groups. Underrepresentation of certain populations in reference databases results in suboptimal variant classification and the generation of an increased number of variants of unknown significance.

In 2015, the American College of Medical Genetics and Genomics (ACMG) recommended the replacement of the terms mutation and polymorphism with the term "variant" (Richards et al. 2015) and published guidelines regarding the classifications of the spectrum of pathogenicity. There are five resulting categories: "pathogenic," "likely pathogenic," "uncertain significance," "likely benign," and "benign." Corresponding details relevant to scoring, classification, and prediction algorithms (e.g., criteria including segregation with disease, evidence of the variant being de novo [i.e., not inherited], and its effect or predicted effect on the protein) were included.

The ACMG guidelines serve to standardize nomenclature for variant interpretation across laboratories, and outline criteria for classification of variants based on various lines of evidence. Classifying variants according to the guidelines takes into account different lines of evidence of pathogenicity. One of the supporting lines of evidence used is the agreement of multiple in silico tools (representing computational evidence), which predict a deleterious effect of the specific variant on the gene or gene product. An example of a common and preferred annotation tool that assists in the interpretation of a sequence variant is the Combined Annotation Dependent Depletion Score (aka "CADD score"). The CADD score integrates multiple annotations into a single metric and generates a score that is a measure of deleteriousness (pathogenicity of the variant) (Kircher et al. 2014). Typically, a CADD score of 20 means that a variant is among the top 1% of deleterious variants in the human genome, and a CADD score of 30 indicates the variant is in the top 0.1%. In spite of these attempts at standardization, interpretations of pathogenicity can vary among laboratories.

A number of databases are available to assess the pathogenicity of a variant, and include those associated with (1) sequence variation, (2) disease, and (3) population. Population databases may include presumably unaffected individuals (e.g., gnomAD) or genomic data from affected and unaffected individuals (e.g., dbSNP). Examples are shown in Tables 2–4 (Richards et al. 2015).

LIMITATIONS OF VARIANT CLASSIFICATION

GS/ES provide a higher diagnostic rate if performed in a hospital setting rather than a reference laboratory remote from the patient (Clark et al. 2018). This is likely because the hospital setting provides access to patient health records and consultation with clinicians who have deep knowledge of the patient's family history, clinical presentation, and laboratory and imaging results. This is distinct from the clinical laboratories that classify genomic variants according to the guidelines mentioned, through an algorithmic assessment and limited phenotypic correlation (Elliott and Friedman 2018). Given their access to additional information, clinicians may interpret a variant as causal for the patient's phenotype in spite of the variant being reported as a VUS (Shashi et al. 2016). These discrepancies can result in challenges for the patient and health-care team. Standardization of phenotypic information provided to laboratories can help to reduce these discrepancies (Bowdin et al. 2016).

One of the challenges to interpretation of variants is that certain groups—such as Indigenous populations—are underrepresented in ref-

Table 2. Sequence databases

NCBI Genome, www.ncbi.nlm.nih.gov/genome	Contains full human genome reference sequences
RefSeqGene, www.ncbi.nlm.nih.gov/refseq/rsg	Contains medically relevant gene reference sequences
MitoMap, www.mitomap.org/MITOMAP/ HumanMitoSeq	Revised Cambridge Reference Sequence (rCRS) for human mitochondrial DNA

Adapted from Richards et al. 2015.

Table 3. Disease databases

The Human Gene Mutation Database (HGMD), www.hgmd.org	Contains genetic variants associated with disease in the literature, not including somatic variants and variants in the mitochondrial genome
The Online Mendelian Inheritance in Man (OMIM), www.ncbi.nlm.nih.gov/omim	Contains content from the published scientific literature. Many of the genes and associated variants in OMIM are considered pathogenic and associated with human disease (the "Clinome"), but some genes not currently associated with human disease are also included (the "NonClinome"), and can serve as potential research candidates
ClinVar, www.ncbi.nlm.nih.gov/clinvar	Unlike HGMD and OMIM, variants do not have to be reported in the literature to be entered into ClinVar
The Database of Genomic Variation and Phenotype in Humans Using Ensembl Resources (DECIPHER/ DDD), decipher.sanger.ac.uk/ddd#research-variants	Contains sequence variants and copy number variants with corresponding patient data and phenotypic information

Adapted from Richards et al. 2015.

erence databases. These shortcomings (genomic inequities) of reference databases have been acknowledged as critical areas to address in reducing challenges associated with variant interpretation (Morgan et al. 2019).

Another challenge to variant interpretation is that it is a dynamic process and, as new data emerge, variants can be reclassified (e.g., VUS to benign). This has significant implications for medical management and family planning. But who is responsible for informing the family? Recent European guidelines indicate that the duty to recontact family exists for findings with clinical or personal utility, and the best interest of the family must be kept in mind. The process should be a joint effort involving

Table 4. Population databases

gnomAD, gnomad.broadinstitute.org	Contains 125,748 exomes and 15,708 genomes from unrelated individuals sequenced as part of various disease-specific and population genetic studies, totaling 141,456 individuals. Although individuals with severe pediatric disease have been reported to be removed, some remain
Exome Variant Server, evs.gs.washington.edu/ EVS	Database of variants (>6000 samples) found during exome sequencing (ES) of several large cohorts of individuals of European and African American ancestry
1000 Genomes, browser.1000genomes.org; grch37.ensembl.org/index.html	The 1000 Genomes Project ran between 2008 and 2015, creating the largest public catalog of human variation and genotype data. As the project ended, the Data Coordination Centre at the European Molecular Biology Laboratory-European Bioinformatics Institute (EMBL-EBI) received continued funding from the Wellcome Trust to maintain and expand the resource
dbSNP, www.ncbi.nlm.nih.gov/snp	Database of short genetic variations (typically 50 base pairs or less) submitted from many sources. May lack details of originating study and may contain pathogenic variants
dbVar, www.ncbi.nlm.nih.gov/dbvar	Database of structural variation (typically >50 base pairs) submitted from many sources

Adapted from Richards et al. 2015.

 Cite this article as *Cold Spring Harb Perspect Med* doi: 10.1101/cshperspect.a036632

the clinical team, the family, and the laboratory and needs to be sustainable for the health-care system (Carrieri et al. 2019). The shared responsibility indicated in the European guidelines is echoed in the recently published American College of Genetics and Genomics position statement on recontacting patients after revision of genomic results and should be discussed as part of pretest genetic counseling (David et al. 2019). Many individuals undergo GS and ES through research protocols, in part, because genome-wide sequencing is not routinely available clinically. Bombard et al. (2019) recently published a position statement on behalf of the American Society of Human Genetics, which includes 12 recommendations for recontacting research participants after reinterpretation of genetic and genomic testing.

GENETIC COUNSELING AND GS/ES

Genetic counseling is a communication process that is both educational and supportive; it involves helping people understand and adapt to the medical, psychological, and familial aspects of the genetic contribution to disease (Resta et al. 2006).

Informed consent is not the same as genetic counseling. Issues specific to GS/ES and pre- and posttest genetic counseling have been recently addressed (Elliott and Friedman 2018). The complex issues associated with genetic counseling and GS/ES include the challenges connected with a rare disease diagnosis, and the possibility that some patients will be diagnosed with more than one genetic disorder. Pretest genetic counseling should include taking a detailed family history, explanation of the method of testing used, the associated risks and benefits, and the possibility that uncertain, undefined, and difficult to interpret results can be generated. In addition, it is necessary to communicate that results may only explain a portion of the child's phenotype. Incidental findings and potential implications for relatives need to be discussed, including the chance that other individuals may need to be tested to determine whether or not a particular variant is carried by all family members with a disease or condi-

tion. Privacy issues and concerns regarding insurance need to be addressed. The pretest counseling process should include emotional support for families and help to guide them to make informed decisions that are consistent with their values. Families need to be prepared for the uncertainties that can accompany GS/ES results and that a diagnosis may not be reached. Importantly, families should understand that they can decline GS/ES before providing "informed consent."

Canadian clinical practice guidelines state that before diagnostic GS/ES, genetic counseling should be provided for the patient/family and documented in the medical record by a qualified individual with a thorough understanding of clinical GS/ES (Boycott et al. 2015). An explanation of what will happen with data, including how long they will be stored and if and when additional analysis or reanalysis will be performed in the future, should be provided. Patients/families should be given the option of having coded or anonymized genome-wide and phenotypic data deposited and stored in an international database to assist in interpretation of genome-wide studies of themselves and other patients and the opportunity to enroll in current or future research studies (Boycott et al. 2015).

European guidelines also indicate that pretest counseling is necessary for families considering GS/ES and should include a discussion on both expected results and the potential for unsolicited and secondary findings (Matthijs et al. 2016). The ACMG recommends GS/ES be accompanied by consultation with a genetics professional and adequate genetic counseling (Green et al. 2013). International recommendations require adequate pretest counseling, interpretation results, and provision of posttest counseling when GS/ES is being considered clinically (Bowdin et al. 2016).

GUIDELINES REGARDING INCIDENTAL AND SECONDARY FINDINGS

Determining the patient's preference for receiving incidental and secondary findings (IFs/SFs) is an important component of pretest genetic counseling for GS/ES and a key recommendation of multiple guidelines (Knoppers et al. 2015; Williams et al. 2015). American diagnostic laboratories provide patients undergoing GS/ES with the choice to opt in or out of the reporting of SFs. For individuals who opt in, the patient's sample is interrogated independently for the 59 genes identified by the working group (Green et al. 2013). Furthermore, it is recommended that the seeking and reporting of SFs to ordering clinicians not be limited by the age of the person being sequenced.

Unlike diagnostic laboratories in the United States, the handling of SFs in the Canadian context follows the Canadian College of Medical Genetics guidelines for research studies "…until the benefits of reporting incidental findings are established, we do not endorse the intentional clinical analysis of disease-associated genes other than those linked to the primary indication; clinicians should provide genetic counseling and obtain informed consent prior to undertaking clinical genome-wide sequencing. Counseling should include discussion of the limitations of testing, likelihood and implications of diagnosis and incidental findings…" (Boycott et al. 2015). The Canadian approach is similar to European guidelines (Matthijs et al. 2016). For adults who opt in, pathogenic, medically actionable adult-onset variants are disclosed. For children, pathogenic, pediatric-onset medically actionable variants are automatically disclosed and adult-onset variants are not disclosed (Zawati et al. 2014; Boycott et al. 2015). SFs/IFs are an inevitable consequence of GS/ES and an important genetic counseling issue to discuss with families considering GS/ES.

Access to genetic counseling is a necessity for families considering GS/ES, and this requirement puts increased pressure on an already strained resource. This will be compounded once GS/ES becomes routinely available clinically. Consequently, innovative methods (e.g., videoconferencing, telehealth, group counseling, telephone counseling, and online decision aids) for the provision of genetic counseling services for GS/ES must be considered.

Given their inherent scalability, online decision aids are a logical solution to support the increasing demand for GS/ES. Decision aids have been shown to be effective when users are faced with complex and difficult treatment or screening decisions (Sheehan and Sherman 2012), including genetic screening and single-gene testing (Wakefield et al. 2008; Kuppermann et al. 2009; Yee et al. 2014). Decision aids have been proposed, but not yet thoroughly evaluated, in applications as complex as prenatal GS/ES and newborn screening (Lewis et al. 2016; Chen and Wasserman 2017). Compared to usual care, decision aid users are better informed, and have shown improved knowledge and more accurate understanding of the pros and cons of their options. Several studies suggest that users typically participate to a greater extent in the decision, feel clearer about what matters to them, and make choices that are more congruent with their values and respectful of patient autonomy (Johnston et al. 2017; Stacey et al. 2017).

The effectiveness and cost efficiency of decision aids compared to in-person genetic counseling for GS/ES have yet to be established from the perspective of either the patient-user or the health-care system. Decision aids for GS/ES have been introduced in some clinical settings. "DECIDE" (decision aid and e-counseling for inherited disorder evaluation) is designed to streamline, enhance, and improve the accessibility of genetic counseling for clinical GS/ES (Birch et al. 2016). DECIDE provides educational material in a variety of formats, tailored to users' needs, and presents pros and cons of GS/ES and IFs for users to evaluate in the context of their own values. In a recent study of families undergoing GS/ES for pediatric intractable epilepsy, a noninferiority trial comparing DECIDE to conventional in-person genetic counseling revealed a significant increase in parental knowledge in both groups, with no difference between the two groups (Adam et al. 2019).

The genomics "ADvISER" is a patient-centered support tool that is specific to the selection of incidental sequencing results. Usability testing showed this tool to be effective for delivering genomic information, was acceptable to patients, and was sufficient for them to make an informed hypothetical decision (Bombard et al. 2018).

GS/ES AND THE DIAGNOSTIC ODYSSEY

Genetic counselors represent the front line of genomic medicine and are instrumental in guiding and supporting patients and their families throughout their clinical journey, often well beyond the delivery of diagnostic results.

Pediatric patients with rare diseases often have complex, multisystem involvement and require the care of numerous health-care providers from diverse subspecialties. Adherence to medical management recommendations is an important component of the care trajectory for these families. The inclusion of a genetic counselor in initial pediatric visits in genetics has been shown to significantly increase patient adherence and can be considered a metric by which genetic counseling is assessed. A chart review of 198 pediatric patients seen for their initial appointment in clinical genetics revealed that appointments including a genetic counselor were associated with significantly increased adherence across three domains (follow-up with genetics, referral to specialist, and testing) as compared to appointments in which there was no genetic counselor involved in the initial visit (Rutherford et al. 2014).

The increased uncertainty experienced by parents of a child with an undiagnosed disorder has been associated with lower levels of optimism and feelings of control, in addition to increased perception of severity (O'Daniel et al. 2010). Isolation, frustration, and hopelessness have been reported in families with rare disorders (Zurynski et al. 2008; Helm 2015; Baumbusch et al. 2019). Uncertainty, fear, and loss of control were shown in another study of parents of children with undiagnosed disorders (Spillmann et al. 2017). Families dealing with rare diseases experience challenges that affect their psychological well-being and include financial burden, lack of information, lack of access to appropriate care, delays in diagnosis, and increased social isolation and uncertainty (Zurynski et al. 2008; Kole and Faurisson 2009; Baumbusch et al. 2019), whereas, for some parents of undiagnosed children, the uncertainty associated with their child's disorder can be associated with parental social integration, a component of adaptation (Yanes et al. 2017).

ES and GS have helped to end the "diagnostic odyssey" for thousands of pediatric patients with suspected genetic disease. A diagnosis has clinical, social, and economic benefits. These include optimization of patient management (and in some cases, customizing treatment), an improved understanding of prognosis and surveillance, and a reduction in unnecessary testing. For families, an answer can provide access to social programs, allied health-care services, specialized educational programs, and relevant support groups (Strande and Berg 2016; Boycott et al. 2017). From a genetic counseling perspective, a diagnosis allows for informed family planning, more accurate recurrence risks, and the option of prenatal diagnosis or preimplantation diagnosis for interested parents. Additionally, it allows genetic counselors to characterize genetic risk for other family members.

Diagnostic sessions have been shown to be correlated with a positive parental experience when a genetic counselor is present, likely attributable to the specialized training, emotional support, and resources provided (Waxler et al. 2013). Using semistructured in-person interviews with families whose children received a diagnosis, emotional support, counseling, providing hope, and perspective and explaining follow-up were associated with more positive experiences (Ashtiani et al. 2014). The importance of families being "prepared" was emphasized, which is particularly relevant in the space of genetic counseling for ES/GS. Genetic counselors taking an active and defined role in diagnosis sessions can result in a more positive experience for families (Waxler et al. 2013; Ashtiani et al. 2014).

Education is an important component of pre- and posttest genetic counseling, and ES/GS is a complex topic to navigate. Parents of pediatric patients who underwent clinical ES were surveyed to assess perceived and actual understanding (Tolusso et al. 2017). Parents, in general, had a good understanding (actual and perceived), but areas for improved understanding included how genes are analyzed and the lack of protection with respect to life insurance discrimination. There was also low actual understanding related to certain aspects of SFs. The importance of providers to explain to families at the pretest counseling session that ES may not find a diagnosis was a common theme emphasized by multiple respondents independently.

In a study surveying 192 parents whose children had diagnostic ES, the parents' interpretation of the child's result agreed with the clinicians' interpretation in 79% of cases. There was more frequent discordance when the clinician's interpretation was uncertain. Most parents (79%) reported no regret with respect to their decision to pursue ES, and most (65%) reported complete satisfaction with the genetic counseling encounter. Satisfaction was positively correlated with their genetic counselor's number of years of clinical experience (Wynn et al. 2018).

Rosell and colleagues examined parental perceptions of ES in pediatric disorders that were previously diagnosed. Some families experienced frustration and isolation because of the limited information available about the rare disorder. Parents wanted more information and hoped to identify with other families with the disorder (Rosell et al. 2016). A recent meta-analysis of psychological outcomes for individuals who underwent GS/ES for a variety of indications (including a pediatric cohort) found that there were no significant psychological harms from the return of GS/ES results across multiple clinical settings (Robinson et al. 2019).

GS/ES has the capacity to identify ultra-rare disorders and is a constant source of gene discovery in the pediatric rare disease space. A recent study of parents whose children had been diagnosed with "new" genetic conditions revealed that, in spite of limited information about the child's disorder, most parents experienced relief and perceived value in the diagnosis (having an explanation for the cause of the condition). Some families reported a fear that their child would develop cancer after parental internet searches turned up literature mentioning changes in the gene in cancer cell lines (e.g., somatic, not germline) (Inglese et al. 2019). Other investigators have indicated that a parent's desire to obtain an answer is a strong motivator for pursuing GS/ES (Sapp et al. 2014; Rosell et al. 2016; Smith et al. 2019).

Informational and emotional support are essential benefits of peer support for children with a rare disease (Baumbusch et al. 2019). Various investigators have described the benefits to families of connecting with other families who have children with the same disorder (Krabbenborg et al. 2016; Rosell et al. 2016; Inglese et al. 2019). Social networking sites are helpful when a support group specific to a child's disorder has not been established. Some resources for families who have a child with a rare genetic disorder are listed in Table 5.

Parents who have children with a suspected genetic disease considering GS/ES should be informed during pretest genetic counseling that, even if a diagnosis is found, there may not be many resources or information specific to the disorder. It is important that genetic counselors and other health-care providers identify relevant support groups for families once a diagnosis is established. If a relevant support group does not exist, encouraging linkages through

Table 5. Resources for families with rare disorders

Organization name	Website
Rare Diseases Europe (EURODIS)	www.eurordis.org
Orphanet	www.orpha.net
National Organization for Rare Disorders (NORD)	rarediseases.org
Canadian Organization for Rare Disorders (CORD)	www.raredisorders.ca
Rare Disease Foundation (RDF)	rarediseasefoundation.org
Unique	www.rarechromo.org
RareShare	rareshare.org
MyGene2	mygene2.org/MyGene2
Facebook	www.facebook.com

portals such as MyGene2, or establishing an independent group through Facebook are options for interested families.

GENETIC COUNSELING ISSUES AND GS/ES IN THE NEONATAL INTENSIVE CARE UNIT

The importance of availability of genetic counseling for rapid GS/ES has previously been addressed (Stark et al. 2018; Smith et al. 2019). There are challenges to enrollment of parents in genomic sequencing studies of newborns. Parents declining to participate have cited such reasons as feeling overwhelmed, logistical issues, lack of interest in research involving genetic testing, and unavailability of both parents for trio-based testing (Willig et al. 2015; Petrikin et al. 2018; Genetti et al. 2019).

Maximizing communication between the clinical and study teams, in addition to flexibility in the genetic counselor's availability, can ensure appropriate pretest genetic counseling occurs. This may require counseling parents separately to accommodate family schedules (Elliott et al. 2019).

Parents of patients in the NICU experience increased stress and depression (Chourasia et al. 2013; Alkozei et al. 2014; Turner et al. 2015). Lack of enrollment in neonatal GS/ES studies indicates different concerns and issues in this group than in families of older children with suspected genetic disorders. A recent study comparing parents of neonates and parents of older pediatric children (average age of 10 years) undergoing GS/ES revealed that after pretest genetic counseling, parents of the NICU patients were significantly less likely to opt in to find out incidental findings for themselves, significantly more likely to identify "diagnosis" as their primary motivation for pursuing GS/ES, and significantly less likely to identify "no concerns" than the parents of older children. Parents in both cohorts had increased anxiety and depression relative to the general population (Smith et al. 2019).

CONCLUSION

Families of children with rare disease can benefit from the tremendous diagnostic potential of GS/ES. However, GS/ES should be delivered as a holistic service that includes pre- and posttest genetic counseling. The eventual clinical implementation of GS/ES results in the need for innovative methods to deliver genetic counseling services. Genetic counselors represent the front line of genomic medicine and are integral to ensuring families receive appropriate support as new genetic disorders continue to be discovered and characterized.

ACKNOWLEDGMENTS

The author is grateful to Dr. Alan Rope, Dr. Jill Mwenifumbo, and Courtney B. Cook for assistance.

REFERENCES

Adam S, Birch PH, Coe RR, Bansback N, Jones AL, Connolly MB, Demos MK, Toyota EB, Farrer MJ, Friedman JM. 2019. Assessing an interactive online tool to support parents' genomic testing decisions. *J Genet Couns* **28:** 10–17. doi:10.1007/s10897-018-0281-1

Alkozei A, McMahon E, Lahav A. 2014. Stress levels and depressive symptoms in NICU mothers in the early postpartum period. *J Matern Neonatal Med* **27:** 1738–1743. doi:10.3109/14767058.2014.942626

Amberger J, Bocchini CA, Scott AF, Hamosh A. 2009. Online Mendelian Inheritance in Man (OMIM), an online catalog of human genes and genetic disorders. *Nucleic Acids Res* **37:** D793–D796. doi:10.1093/nar/gkn665

Amberger JS, Bocchini CA, Schiettecatte F, Scott AF, Hamosh A. 2015. OMIM.org: Online Mendelian Inheritance in Man (OMIM®), an online catalog of human genes and genetic disorders. *Nucleic Acids Res* **43:** D789–D798. doi:10.1093/nar/gku1205

Amberger JS, Bocchini CA, Scott AF, Hamosh A. 2019. OMIM.org: leveraging knowledge across phenotype–gene relationships. *Nucleic Acids Res* **47:** D1038–D1043. doi:10.1093/nar/gky1151

Ashtiani S, Makela N, Carrion P, Austin J. 2014. Parents' experiences of receiving their child's genetic diagnosis: a qualitative study to inform clinical genetics practice. *Am J Med Genet Part A* **164:** 1496–1502. doi:10.1002/ajmg.a.36525

Baird PA, Anderson TW, Newcombe HB, Lowry RB. 1988. Genetic disorders in children and young adults: a population study. *Am J Hum Genet* **42:** 677–693.

Balci TB, Hartley T, Xi Y, Dyment DA, Beaulieu CL, Bernier FP, Dupuis L, Horvath GA, Mendoza-Londono R, Prasad C, et al. 2017. Debunking Occam's razor: diagnosing multiple genetic diseases in families by whole-exome sequencing. *Clin Genet* **92:** 281–289. doi:10.1111/cge.12987

Baumbusch J, Mayer S, Sloan-Yip I. 2019. Alone in a crowd? Parents of children with rare diseases' experiences of navigating the healthcare system. *J Genet Couns* **28:** 80–90. doi:10.1007/s10897-018-0294-9

Berg JS, Agrawal PB, Bailey DB, Beggs AH, Brenner SE, Brower AM, Cakici JA, Ceyhan-Birsoy O, Chan K, Chen F, et al. 2017. Newborn sequencing in genomic medicine and public health. *Pediatrics* **139:** e20162252. doi:10.1542/peds.2016-2252

Birch P, Adam S, Bansback N, Coe RR, Hicklin J, Lehman A, Li KC, Friedman JM. 2016. DECIDE: a decision support tool to facilitate parents' choices regarding genome-wide sequencing. *J Genet Couns* **25:** 1298–1308. doi:10.1007/s10897-016-9971-8

Bombard Y, Clausen M, Mighton C, Carlsson L, Casalino S, Glogowski E, Schrader K, Evans M, Scheer A, Baxter N, et al. 2018. The Genomics ADvISER: Development and usability testing of a decision aid for the selection of incidental sequencing results. *Eur J Hum Genet* **26:** 984–995. doi:10.1038/s41431-018-0144-0

Bombard Y, Brothers KB, Fitzgerald-Butt S, Garrison NA, Jamal L, James CA, Jarvik GP, McCormick JB, Nelson TN, Ormond KE, et al. 2019. The responsibility to recontact research participants after reinterpretation of genetic and genomic research results. *Am J Hum Genet* **104:** 578–595. doi:10.1016/J.AJHG.2019.02.025

Bowdin SC. 2016. The clinical utility of next-generation sequencing in the neonatal intensive care unit. *CMAJ* **188:** 786–787. doi:10.1503/cmaj.160490

Bowdin S, Gilbert A, Bedoukian E, Carew C, Adam MP, Belmont J, Bernhardt B, Biesecker L, Bjornsson HT, Blitzer M, et al. 2016. Recommendations for the integration of genomics into clinical practice. *Genet Med* **18:** 1075–1084. doi:10.1038/gim.2016.17

Boycott K, Hartley T, Adam S, Bernier F, Chong K, Fernandez BA, Friedman JM, Geraghty MT, Hume S, Knoppers BM, et al. 2015. The clinical application of genome-wide sequencing for monogenic diseases in Canada: Position Statement of the Canadian College of Medical Geneticists. *J Med Genet* **52:** 431–437. doi:10.1136/jmedgenet-2015-103144

Boycott KM, Rath A, Chong JX, Hartley T, Alkuraya FS, Baynam G, Brookes AJ, Brudno M, Carracedo A, den Dunnen JT, et al. 2017. International cooperation to enable the diagnosis of all rare genetic diseases. *Am J Hum Genet* **100:** 695–705. doi:10.1016/J.AJHG.2017.04.003

Carrieri D, Howard HC, Benjamin C, Clarke AJ, Dheensa S, Doheny S, Hawkins N, Halbersma-Konings TF, Jackson L, Kayserili H, et al. 2019. Recontacting patients in clinical genetics services: recommendations of the European Society of Human Genetics. *Eur J Hum Genet* **27:** 169–182. doi:10.1038/s41431-018-0285-1

Chen SC, Wasserman DT. 2017. A framework for unrestricted prenatal whole-genome sequencing: respecting and enhancing the autonomy of prospective parents. *Am J Bioeth* **17:** 3–18. doi:10.1080/15265161.2016.1251632

Chourasia N, Surianarayanan P, Adhisivam B, Vishnu Bhat B. 2013. NICU admissions and maternal stress levels. *Indian J Pediatr* **80:** 380–384. doi:10.1007/s12098-012-0921-7

Clark MM, Stark Z, Farnaes L, Tan TY, White SM, Dimmock D, Kingsmore SF. 2018. Meta-analysis of the diagnostic and clinical utility of genome and exome sequencing and chromosomal microarray in children with suspected ge-

netic diseases. *NPJ Genom Med* **3:** 16. doi:10.1038/s41525-018-0053-8

David KL, Best RG, Brenman LM, Bush L, Deignan JL, Flannery D, Hoffman JD, Holm I, Miller DT, O'Leary J, et al. 2019. Patient re-contact after revision of genomic test results: points to consider—a statement of the American College of Medical Genetics and Genomics (ACMG). *Genet Med* **21:** 769–771. doi:10.1038/s41436-018-0391-z

Dragojlovic N, Elliott AM, Adam S, van Karnebeek C, Lehman A, Mwenifumbo JC, Nelson TN, du Souich C, Friedman JM, Lynd LD. 2018. The cost and diagnostic yield of exome sequencing for children with suspected genetic disorders: a benchmarking study. *Genet Med* **20:** 1013–1021. doi:10.1038/gim.2017.226

Elliott AM, Friedman JM. 2018. The importance of genetic counselling in genome-wide sequencing. *Nat Rev Genet* **19:** 735–736. doi:10.1038/s41576-018-0057-3

Elliott AM, du Souich C, Adam S, Dragojlovic N, van Karnebeek C, Nelson TN, Lehman A, The CAUSES Study, Lynd LD, Friedman JM. 2018. The Genomic Consultation Service: a clinical service designed to improve patient selection for genome-wide sequencing in British Columbia. *Mol Genet Genomic Med* **6:** 592–600. doi:10.1002/mgg3.410

Elliott AM, du Souich C, Lehman A, Guella I, Evans DM, Candido T, Tooman L, Armstrong L, Clarke L, Gibson W, et al. 2019. RAPIDOMICS: Rapid genome-wide sequencing in a neonatal intensive care unit—successes and challenges. *Eur J Pediatr* **178:** 1207–1218. doi:10.1007/s00431-019-03399-4

European Organisation for Rare Diseases. 2005. Rare diseases: Understanding this public health priority. www.eurordis.org/publication/rare-diseases-understanding-public-health-priority

Genetti CA, Schwartz TS, Robinson JO, VanNoy GE, Petersen D, Pereira S, Fayer S, Peoples HA, Agrawal PB, Betting WN, et al. 2019. Parental interest in genomic sequencing of newborns: enrollment experience from the BabySeq Project. *Genet Med* **21:** 622–630. doi:10.1038/s41436-018-0105-6

Green RC, Berg JS, Grody WW, Kalia SS, Korf BR, Martin CL, McGuire AL, Nussbaum RL, O'Daniel JM, Ormond KE, et al. 2013. ACMG recommendations for reporting of incidental findings in clinical exome and genome sequencing. *Genet Med* **15:** 565–574. doi:10.1038/gim.2013.73

Grozeva D, Carss K, Spasic-Boskovic O, Tejada MI, Gecz J, Shaw M, Corbett M, Haan E, Thompson E, Friend K, et al. 2015. Targeted next-generation sequencing analysis of 1,000 individuals with intellectual disability. *Hum Mutat* **36:** 1197–1204. doi:10.1002/humu.22901

Helm BM. 2015. Exploring the genetic counselor's role in facilitating meaning-making: rare disease diagnoses. *J Genet Couns* **24:** 205–212. doi:10.1007/s10897-014-9812-6

Hennekam RCM. 2011. Care for patients with ultra-rare disorders. *Eur J Med Genet* **54:** 220–224. doi:10.1016/j.ejmg.2010.12.001

H.R.4013—107th Congress (2002). An Act to amend the Public Health Service Act to establish an Office of Rare Diseases at the National Institutes of Health, and for other purposes. Government Printing Office, Washington, DC.

Cite this article as *Cold Spring Harb Perspect Med* doi: 10.1101/cshperspect.a036632

https://www.govinfo.gov/content/pkg/STATUTE-116/pdf/STATUTE-116-Pg1988.pdf

Inglese CN, Elliott AM, Lehman A, Lehman A. 2019. New developmental syndromes: understanding the family experience. *J Genet Couns* 28: 202–212. doi:10.1002/jgc4.1121

Johnston J, Farrell RM, Parens E. 2017. Supporting women's autonomy in prenatal testing. *N Engl J Med* 377: 505–507. doi:10.1056/NEJMp1703425

Kircher M, Witten DM, Jain P, O'Roak BJ, Cooper GM, Shendure J. 2014. A general framework for estimating the relative pathogenicity of human genetic variants. *Nat Genet* 46: 310–315. doi:10.1038/ng.2892

Knoppers BM, Zawati MH, Sénécal K. 2015. Return of genetic testing results in the era of whole-genome sequencing. *Nat Rev Genet* 16: 553–559. doi:10.1038/nrg3960

Kochinke K, Zweier C, Nijhof B, Fenckova M, Cizek P, Honti F, Keerthikumar S, Oortveld MAW, Kleefstra T, Kramer JM, et al. 2016. Systematic phenomics analysis deconvolutes genes mutated in intellectual disability into biologically coherent modules. *Am J Hum Genet* 98: 149–164. doi:10.1016/J.AJHG.2015.11.024

Kole A, Faurisson F. 2009. The voice of 12,000 patients—Experiences and expectations of rare disease patients on diagnosis and care in Europe. www.eurordis.org/publication/voice-12000-patients

Krabbenborg L, Vissers LELM, Schieving J, Kleefstra T, Kamsteeg EJ, Veltman JA, Willemsen MA, Van der Burg S. 2016. Understanding the psychosocial effects of WES test results on parents of children with rare diseases. *J Genet Couns* 25: 1207–1214. doi:10.1007/s10897-016-9958-5

Kuppermann M, Norton ME, Gates E, Gregorich SE, Learman LA, Nakagawa S, Feldstein VA, Lewis J, Washington AE, Nease RF. 2009. Computerized prenatal genetic testing decision-assisting tool: a randomized controlled trial. *Obstet Gynecol* 113: 53–63. doi:10.1097/AOG.0b013e31818e7ec4

Lewis MA, Paquin RS, Roche MI, Furberg RD, Rini C, Berg JS, Powell CM, Bailey DB Jr. 2016. Supporting parental decisions about genomic sequencing for newborn screening: the NC NEXUS Decision Aid. *Pediatrics* 137: S16–S23. doi:10.1542/peds.2015-3731E

Matthijs G, Souche E, Alders M, Corveleyn A, Eck S, Feenstra I, Race V, Sistermans E, Sturm M, Weiss M, et al. 2016. Erratum: Guidelines for diagnostic next-generation sequencing. *Eur J Hum Genet* 24: 1515. doi:10.1038/ejhg.2016.63

Meng L, Pammi M, Saronwala A, Magoulas P, Ghazi AR, Vetrini F, Zhang J, He W, Dharmadhikari AV, Qu C, et al. 2017. Use of exome sequencing for infants in intensive care units: ascertainment of severe single-gene disorders and effect on medical management. *JAMA Pediatr* 171: e173438. doi:10.1001/jamapediatrics.2017.3438

Miller CE, Krautscheid P, Baldwin EE, Tvrdik T, Openshaw AS, Hart K, Lagrave D. 2014. Genetic counselor review of genetic test orders in a reference laboratory reduces unnecessary testing. *Am J Med Genet A* 164: 1094–1101. doi:10.1002/ajmg.a.36453

Miller NA, Farrow EG, Gibson M, Willig LK, Twist G, Yoo B, Marrs T, Corder S, Krivohlavek L, Walter A, et al. 2015. A 26-hour system of highly sensitive whole genome sequencing for emergency management of genetic diseases. *Genome Med* 7: 100. doi:10.1186/s13073-015-0221-8

Morgan J, Coe RR, Lesueur R, Kenny R, Price R, Makela N, Birch PH. 2019. Indigenous peoples and genomics: starting a conversation. *J Genet Couns* 28: 407–418. doi:10.1002/jgc4.1073

O'Daniel JM, Haga SB, Willard HF. 2010. Considerations for the impact of personal genome information: a study of genomic profiling among genetics and genomics professionals. *J Genet Couns* 19: 387–401. doi:10.1007/s10897-010-9297-x

Ormond KE. 2013. From genetic counseling to "genomic counseling." *Mol Genet Genomic Med* 1: 189–193. doi:10.1002/mgg3.45

Patch C, Middleton A. 2018. Genetic counselling in the era of genomic medicine. *Br Med Bull* 126: 27–36. doi:10.1093/bmb/ldy008

Petrikin JE, Cakici JA, Clark MM, Willig LK, Sweeney NM, Farrow EG, Saunders CJ, Thiffault I, Miller NA, Zellmer L, et al. 2018. The NSIGHT1-randomized controlled trial: rapid whole-genome sequencing for accelerated etiologic diagnosis in critically ill infants. *NPJ Genom Med* 3: 6. doi:10.1038/s41525-018-0045-8

Posey JE. 2019. Genome sequencing and implications for rare disorders. *Orphanet J Rare Dis* 14: 153. doi:10.1186/s13023-019-1127-0

Resta R, Biesecker BB, Bennett RL, Blum S, Hahn SE, Strecker MN, Williams JL. 2006. A new definition of genetic counseling: National Society of Genetic Counselors' Task Force report. *J Genet Couns* 15: 77–83. doi:10.1007/s10897-005-9014-3

Richards S, Aziz N, Bale S, Bick D, Das S, Gastier-Foster J, Grody WW, Hegde M, Lyon E, Spector E, et al. 2015. Standards and guidelines for the interpretation of sequence variants: a joint consensus recommendation of the American College of Medical Genetics and Genomics and the Association for Molecular Pathology. *Genet Med* 17: 405–423. doi:10.1038/gim.2015.30

Robinson JO, Wynn J, Biesecker B, Biesecker LG, Bernhardt B, Brothers KB, Chung WK, Christensen KD, Green RC, McGuire AL, et al. 2019. Psychological outcomes related to exome and genome sequencing result disclosure: a meta-analysis of seven Clinical Sequencing Exploratory Research (CSER) Consortium studies. *Genet Med* doi:10.1038/s41436-019-0565-3

Rosell AMC, Pena LDM, Schoch K, Spillmann R, Sullivan J, Hooper SR, Jiang YH, Mathey-Andrews N, Goldstein DB, Shashi V. 2016. Not the end of the odyssey: parental perceptions of whole exome sequencing (WES) in pediatric undiagnosed disorders. *J Genet Couns* 25: 1019–1031. doi:10.1007/s10897-016-9933-1

Rutherford S, Zhang X, Atzinger C, Ruschman J, Myers MF. 2014. Medical management adherence as an outcome of genetic counseling in a pediatric setting. *Genet Med* 16: 157–163. doi:10.1038/gim.2013.90

Sagoo GS, Butterworth AS, Sanderson S, Shaw-Smith C, Higgins JPT, Burton H. 2009. Array CGH in patients with learning disability (mental retardation) and congenital anomalies: updated systematic review and meta-analysis of 19 studies and 13,926 subjects. *Genet Med* 11: 139–146. doi:10.1097/GIM.0b013e318194ee8f

Salgado D, Bellgard MI, Desvignes JP, Béroud C. 2016. How to identify pathogenic mutations among all those variations: variant annotation and filtration in the genome sequencing era. *Hum Mutat* **37:** 1272–1282. doi:10.1002/humu.23110

Sapp JC, Dong D, Stark C, Ivey LE, Hooker G, Biesecker LG, Biesecker BB. 2014. Parental attitudes, values, and beliefs toward the return of results from exome sequencing in children. *Clin Genet* **85:** 120–126. doi:10.1111/cge.12254

Saunders CJ, Miller NA, Soden SE, Dinwiddie DL, Noll A, Alnadi NA, Andraws N, Patterson ML, Krivohlavek LA, Fellis J, et al. 2012. Rapid whole-genome sequencing for genetic disease diagnosis in neonatal intensive care units. *Sci Transl Med* **4:** 154ra135. doi:10.1126/scitranslmed.3004041

Shashi V, McConkie-Rosell A, Schoch K, Kasturi V, Rehder C, Jiang YH, Goldstein DB, McDonald MT. 2016. Practical considerations in the clinical application of whole-exome sequencing. *Clin Genet* **89:** 173–181. doi:10.1111/cge.12569

Sheehan J, Sherman KA. 2012. Computerised decision aids: a systematic review of their effectiveness in facilitating high-quality decision-making in various health-related contexts. *Patient Educ Couns* **88:** 69–86. doi:10.1016/j.pec.2011.11.006

Smith EE, du Souich C, Dragojlovic N, Elliott AM, Elliott AM. 2019. Genetic counseling considerations with rapid genome-wide sequencing in a neonatal intensive care unit. *J Genet Couns* **28:** 263–272. doi:10.1002/jgc4.1074

Spillmann RC, McConkie-Rosell A, Pena L, Jiang YH, Schoch K, Walley N, Sanders C, Sullivan J, Hooper SR, Shashi V. 2017. A window into living with an undiagnosed disease: illness narratives from the Undiagnosed Diseases Network. *Orphanet J Rare Dis* **12:** 71. doi:10.1186/s13023-017-0623-3

Stacey D, Légaré F, Lewis K, Barry MJ, Bennett CL, Eden KB, Holmes-Rovner M, Llewellyn-Thomas H, Lyddiatt A, Thomson R, et al. 2017. Decision aids for people facing health treatment or screening decisions. *Cochrane Database Syst Rev* **4:** CD001431. doi:10.1002/14651858.CD001431.pub5

Stark Z, Lunke S, Brett GR, Tan NB, Stapleton R, Kumble S, Yeung A, Phelan DG, Chong B, Fanjul-Fernandez M, et al. 2018. Meeting the challenges of implementing rapid genomic testing in acute pediatric care. *Genet Med* **20:** 1554–1563. doi:10.1038/gim.2018.37

Strande NT, Berg JS. 2016. Defining the clinical value of a genomic diagnosis in the era of next-generation sequencing. *Annu Rev Genomics Hum Genet* **17:** 303–332. doi:10.1146/annurev-genom-083115-022348

Suarez CJ, Yu L, Downs N, Costa HA, Stevenson DA. 2017. Promoting appropriate genetic testing: the impact of a combined test review and consultative service. *Genet Med* **19:** 1049–1054. doi:10.1038/gim.2016.219

Tarailo-Graovac M, Shyr C, Ross CJ, Horvath GA, Salvarinova R, Ye XC, Zhang LH, Bhavsar AP, Lee JJY, Drögemöller BI, et al. 2016. Exome sequencing and the management of neurometabolic disorders. *N Engl J Med* **374:** 2246–2255. doi:10.1056/NEJMoa1515792

Tolusso LK, Collins K, Zhang X, Holle JR, Valencia CA, Myers MF. 2017. Pediatric whole exome sequencing: an assessment of parents' perceived and actual understanding. *J Genet Couns* **26:** 792–805. doi:10.1007/s10897-016-0052-9

Turner M, Chur-Hansen A, Winefield H, Stanners M. 2015. The assessment of parental stress and support in the neonatal intensive care unit using the Parent Stress Scale—Neonatal Intensive Care Unit. *Women Birth* **28:** 252–258. doi:10.1016/J.WOMBI.2015.04.001

van Diemen CC, Kerstjens-Frederikse WS, Bergman KA, de Koning TJ, Sikkema -Raddatz B, van der Velde JK, Abbott KM, Herkert JC, Löhner K, Rump P, et al. 2017. Rapid targeted genomics in critically ill newborns. *Pediatrics* **140:** e20162854. doi:10.1542/peds.2016-2854

Van Karnebeek CDM, Shevell M, Zschocke J, Moeschler JB, Stockler S. 2014. The metabolic evaluation of the child with an intellectual developmental disorder: diagnostic algorithm for identification of treatable causes and new digital resource. *Mol Genet Metab* **111:** 428–438. doi:10.1016/j.ymgme.2014.01.011

Wakefield CE, Meiser B, Homewood J, Ward R, O'Donnell S, Kirk J; Australian GENetic testing Decision Aid Collaborative Group. 2008. Randomized trial of a decision aid for individuals considering genetic testing for hereditary nonpolyposis colorectal cancer risk. *Cancer* **113:** 956–965. doi:10.1002/cncr.23681

Wakefield E, Keller H, Mianzo H, Nagaraj CB, Tawde S, Ulm E. 2018. Reduction of health care costs and improved appropriateness of incoming test orders: the impact of genetic counselor review in an academic genetic testing laboratory. *J Genet Couns* **27:** 1067–1073. doi:10.1007/s10897-018-0226-8

Warr A, Robert C, Hume D, Archibald A, Deeb N, Watson M. 2015. Exome sequencing: Current and future perspectives. *G3 Genes, Genomes, Genet* **5:** 1543–1550. doi:10.1534/G3.115.018564

Waxler JL, Cherniske EM, Dieter K, Herd P, Pober BR. 2013. Hearing from parents: the impact of receiving the diagnosis of Williams syndrome in their child. *Am J Med Genet* **161:** 534–541. doi:10.1002/ajmg.a.35789

Weiner J, Sharma J, Lantos J, Kilbride H. 2011. How infants die in the neonatal intensive care unit: trends from 1999 through 2008. *Arch Pediatr Adolesc Med* **165:** 630–634. doi:10.1001/archpediatrics.2011.102

Wenger AM, Guturu H, Bernstein JA, Bejerano G. 2017. Systematic reanalysis of clinical exome data yields additional diagnoses: Implications for providers. *Genet Med* **19:** 209–214. doi:10.1038/gim.2016.88

Williams JK, Cashion AK, Brooks PJ. 2015. Return of anticipated and incidental results from next-generation sequencing: implications for providers and patients. Discussion Paper, Institute of Medicine, Washington, DC.

Willig LK, Petrikin JE, Smith LD, Saunders CJ, Thiffault I, Miller NA, Soden SE, Cakici JA, Herd SM, Twist G, et al. 2015. Whole-genome sequencing for identification of Mendelian disorders in critically ill infants: a retrospective analysis of diagnostic and clinical findings. *Lancet Respir Med* **3:** 377–387. doi:10.1016/S2213-2600(15)00139-3

Wright CF, Fitzgerald TW, Jones WD, Clayton S, McRae JF, van Kogelenberg M, King DA, Ambridge K, Barrett DM, Bayzetinova T, et al. 2015. Genetic diagnosis of developmental disorders in the DDD study: a scalable analysis of genome-wide research data. *Lancet* **385:** 1305–1314. doi:10.1016/S0140-6736(14)61705-0

Wright CF, FitzPatrick DR, Firth HV. 2018a. Paediatric genomics: diagnosing rare disease in children. *Nat Rev Genet* **19:** 253–268. doi:10.1038/nrg.2017.116

Wright CF, McRae JF, Clayton S, Gallone G, Aitken S, FitzGerald TW, Jones P, Prigmore E, Rajan D, Lord J, et al. 2018b. Making new genetic diagnoses with old data: Iterative reanalysis and reporting from genome-wide data in 1,133 families with developmental disorders. *Genet Med* **20:** 1216–1223. doi:10.1038/gim.2017.246

Wynn J, Ottman R, Duong J, Wilson AL, Ahimaz P, Martinez J, Rabin R, Rosen E, Webster R, Au C, et al. 2018. Diagnostic exome sequencing in children: a survey of parental understanding, experience and psychological impact. *Clin Genet* **93:** 1039–1048. doi:10.1111/cge.13200

Yanes T, Humphreys L, McInerney-Leo A, Biesecker B. 2017. Factors associated with parental adaptation to children with an undiagnosed medical condition. *J Genet Couns* **26:** 829–840. doi:10.1007/s10897-016-0060-9

Yee LM, Wolf M, Mullen R, Bergeron AR, Cooper Bailey S, Levine R, Grobman WA. 2014. A randomized trial of a prenatal genetic testing interactive computerized information aid. *Prenat Diagn* **34:** 552–557. doi:10.1002/pd.4347

Zawati MH, Parry D, Knoppers BM. 2014. The best interests of the child and the return of results in genetic research: International comparative perspectives. *BMC Med Ethics* **15:** 72. doi:10.1186/1472-6939-15-72

Zurynski Y, Frith K, Leonard H, Elliott E. 2008. Rare childhood diseases: how should we respond? *Arch Dis Child* **93:** 1071–1074. doi:10.1136/ADC.2007.134940

Bridging the Gap between Scientific Advancement and Real-World Application: Pediatric Genetic Counseling for Common Syndromes and Single-Gene Disorders

Julie A. McGlynn[1] and Elinor Langfelder-Schwind[2]

[1]Department of Obstetrics, Gynecology, and Reproductive Sciences, Yale School of Medicine, New Haven, Connecticut 06510, USA

[2]The Cystic Fibrosis Center, Mount Sinai Beth Israel, Icahn School of Medicine at Mount Sinai, New York, New York 10003, USA

Correspondence: julie.mcglynn@yale.edu

Screening and diagnostic testing for single-gene disorders and common syndromes in the pediatric setting frequently generate data that are challenging to interpret, and the ability to diagnose genetic conditions has outpaced the development of successful treatments or cures. Genetic testing is now integrated purposefully into a variety of primary and specialty care clinics, creating an increased requirement for genetic literacy among providers and patients, as well as a growing need to incorporate genetic counseling services into mainstream clinical practice. The practice of pediatric genetic counseling encompasses a unique combination of skills and training designed to address the evolving psychological, social, educational, medical, and reproductive concerns of patients and their families, which complements the multidisciplinary services of physicians, nurses, and other allied health professionals caring for patients with pediatric-onset genetic conditions. The potential range of genetic counseling needs in the pediatric setting transcends the diagnostic period. The sustained nature of pediatric care presents opportunities for development of trusting and longstanding professional relationships that permit the evolving genetic counseling needs of patients and families to be met. A discussion of cystic fibrosis, a common autosomal recessive single-gene disorder with an increasingly broad clinical spectrum and genotype–phenotype variability, serves as a useful case study to illustrate the current and emerging genetic counseling practices, goals, and challenges impacting patients and their families.

Genetic testing and diagnostic services once delivered exclusively by genetic specialists are now likely to be offered by nongeneticists in various subspecialties (Rigter et al. 2014). Concurrently, advances in diagnostic technologies are driving the need for providers to address the increasing complexity in interpretation and application of genetic testing (Helm et al. 2018; Stoll et al. 2018). Through the facilitation of informed decision-making and provision of anticipatory guidance, genetic counseling aims to empower patients and families to make knowledgeable choices about whether or not to pursue testing and gain preemptive awareness about potential genetic test results. There is no doubt that genetic counseling has informational and educational aspects, but to be effective, there must be a deeply human dimension as well (Witt and Jankowska 2018). Genetic counseling requires active use of communication skills to establish rapport, to identify, assess, and address immediate and evolving concerns and to provide targeted psychological counseling to promote establishment of a therapeutic relationship between a genetic counseling provider and patient that, from its start in the pediatric setting, may span the patient's lifetime.

The clinical, educational, psychological, and social needs of pediatric patients and their families who are referred for genetic counseling regarding common single-gene disorders and chromosomal syndromes vary according to the indication for referral and the timing and nature of the diagnosis. Common indications for genetic counseling in the pediatric setting are summarized in Figure 1. As illustrated in this figure, the potential genetic counseling requirements for children with genetic conditions and/or their parents are not confined to the initial diagnostic period. Genetic counseling may be relevant prior to delivery, when a fetus is identified to be at risk for, or conclusively diagnosed with, a genetic condition for which expectant parents would like information about prognosis and treatment. After delivery and throughout childhood and adulthood, genetic counseling needs and priorities may shift, as individuals require information, guidance, and psychological support to promote adaptation and facilitate decision-making later in life.

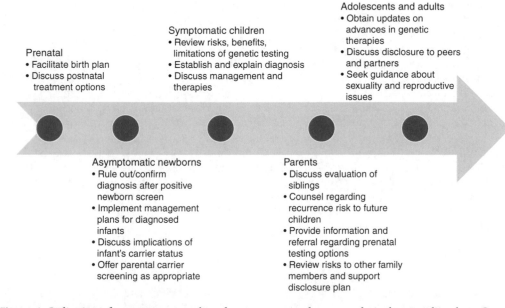

Figure 1. Indications for genetic counseling for common syndromes and single-gene disorders. Common indications for genetic counseling in the pediatric setting to discuss single-gene or chromosomal syndromes are (1) in the prenatal period, to prepare for the birth of a child with a genetic condition when positive parental carrier screening results convey increased risk for a heritable condition, suggestive ultrasound findings are identified, or prenatal diagnostic testing is confirmatory; (2) in the neonatal period following positive newborn screening (NBS) results to confirm or rule out the diagnosis, implement a management plan, and discuss implications of carrier status as appropriate; (3) during the process of seeking and confirming a diagnosis for a child with suggestive features, developmental delay, or intellectual disability; (4) for parents to discuss recurrence risk and reproductive options as well as obtain guidance regarding disclosure of diagnosis and familial implications; and (5) for a child with a genetic condition or parents to obtain updated information about the condition, learn about advances in therapeutic options, and receive ongoing support. As the child grows up, their own genetic counseling needs manifest as well.

Because patients and families most often access genetic counseling services through referral by a pediatric provider, patient access to these services depends on a provider's ability to assess, recognize, and respond to a range of potential genetic counseling needs.

The original genetic counseling practice model involved counselors working alongside or under the supervision of medical geneticists as part of a tertiary care genetics team. Today, however, genetic counselors are frequently embedded in medical subspecialty clinics, working directly with neurologists, cardiologists, oncologists, pulmonologists, ophthalmologists, and others in clinical and/or clinical-research roles (National Society of Genetic Counselors 2019). Including genetic counseling services within specialty clinics such as metabolic or cystic fibrosis (CF) centers is correlated with improved cost efficiency, decreased parental anxiety, increased retention of genetic information, and improved adherence to medical management (Cavanagh et al. 2010; Lang et al. 2011; Rutherford et al. 2014).

Pediatric genetic counseling may involve multiple sessions or be confined to one visit. Although a single meeting may be sufficient to resolve an individual's or family's immediate needs, multiple sessions may permit the opportunity to address multifaceted and developing needs at the point in time when issues are most pertinent to the patient or parents. An advantage of having a series of sessions over a child's early years is the longstanding therapeutic relationship that develops between the genetic counselor, child, and parents. When built upon a foundation of mutual trust and respect, this ongoing partnership enables illness-related challenges and the concerns of those with genetic conditions and their families to be discussed as they arise. Additional beneficial outcomes can emerge from this type of genetic counseling beyond helping individuals and families to understand the diagnostic process. These can include developing and sustaining effective coping strategies to manage the medical and emotional burdens of the condition, making difficult but informed choices about treatment, and obtaining anticipatory guidance about the testing of other family members and future reproductive concerns.

Genetic counseling for CF, a classical Mendelian disorder inherited in an autosomal recessive pattern, provides a useful example to illustrate current practices, goals, and challenges in pediatric genetic counseling and will be used as a case study throughout this review. Cystic fibrosis in its classic form is a progressive multisystem disease resulting in chronic pulmonary infections, pancreatic insufficiency, and male infertility, requiring multidisciplinary specialty care. Seen with the greatest frequency among persons of northern European ancestry, CF occurs in nearly all populations throughout the world. Researchers have systematically categorized the pathogenicity and penetrance of more than 400 of the more than 2000 known variants in the cystic fibrosis transmembrane conductance regulator (CFTR) gene on the path to understanding genotype–phenotype correlations and identifying pharmacogenomic approaches to correction at the cellular level (Sosnay et al. 2013) (www.cftr2.org). Advances in early detection and treatment have transformed CF from a predominantly fatal disease of childhood to one in which survival into adulthood is now expected (Ronan et al. 2017). In addition, it is now recognized that patients with CFTR genetic variants may exhibit only a subset of the full clinical syndrome, and/or remain asymptomatic until adulthood. The phenotypic range and enduring impact of this condition suggests that individuals with CF and their parents may have different genetic counseling needs at various stages throughout their lifetimes.

FAMILY-CENTERED COUNSELING

Genetic counseling involves eliciting the psychological, social, and medical concerns of patients and their families, and tailoring each session with the goal of meeting their expressed and implied needs. This is accomplished in large part by explaining the purpose of the session, asking open-ended questions about the patient and family, acknowledging and validating their thoughts and feelings, checking and restating concerns, and providing information and guidance to facilitate planning and decision-making (Redlinger-Grosse et al. 2017).

During these conversations, potentially complex subjects are broached, including diagnostic testing for the proband (primary patient) and at-risk relatives, current management and treatment options, carrier testing for unaffected siblings, and reproductive recurrence risks. Sometimes the lines between the patient's needs and the family's needs become blurred. The genetic counselor endeavors to recognize, acknowledge, and respond to conflicting priorities, being mindful of the need to protect the patient's autonomy while supporting the wishes of the family, as appropriate. The rapport established during this process is vital for the development of successful professional relationships between the counselor, patient, family, and other members of the health-care team (Bernhardt et al. 2000). These relationships will facilitate the effective delivery of individualized and family-centered care over time and across a patient's life span.

THE DIAGNOSTIC WORKUP

A mainstay of pediatric medicine involves providing consultation for symptomatic individuals referred for assessment and diagnosis. With many common single-gene and chromosome disorders, a child will exhibit physical, medical, and/or developmental features suggestive of a condition; symptoms may initially present in utero or postdelivery. Therefore, the genetic counseling process necessitates meticulous collection of baseline data about the child and family, gathering information about medical, developmental, and multigenerational family histories that facilitate accurate diagnosis and/or determination of appropriate testing.

Prior to the diagnostic consultation, parents may have conducted online searches about their child's signs and symptoms (Nicholl et al. 2017), reached out to a primary care pediatrician or other trusted medical providers for information, and/or asked relatives and friends about their knowledge and experiences. Thus, the parents often arrive at a pediatric genetic counseling session with some familiarity with, and expectation of, a diagnosis. This familiarity neither predicts

accurate understanding nor obviates the need for psychological counseling. Genetic conditions have variable clinical presentations and/or degrees of severity; this can hold true even within families and among those carrying the same genetic pathogenic variant(s) or chromosomal aneuploidy. Consequently, the name recognition of a condition may mask a limited or inaccurate understanding of its true nature and phenotypic range. Media accounts, including social media, may also provide a skewed or one-dimensional view of the phenotype. Through the process of genetic counseling, patient's and family's perceptions, needs, and experiences prior to diagnosis are integrated to facilitate posttest counseling, diagnostic destigmatization, and adaptation.

Test Selection and Interpretation

Historically, diagnostic genetic testing has involved the selection of targeted technology with a high degree of sensitivity and specificity based on a physical examination by a medical geneticist, review of records about previous diagnostic tests and imaging, and documentation of compatible clinical features. For example, a blood draw for chromosome analysis may be ordered to diagnose trisomy 21 (Down syndrome) in a newborn with suggestive facial features and a congenital heart defect, or targeted molecular testing of the *FGFR3* gene may be ordered in a child with short stature and skeletal findings consistent with achondroplasia. Although genetic testing has always had the potential to unveil unrelated or poorly understood findings, changes in the types of testing used to detect and diagnose these children with suspected genetic disease has increased the frequency of tests that reveal unexpected diagnoses or results whose significance is unclear, and magnified the resulting genetic counseling challenges.

Chromosome microarray analysis (CMA) is now designated as a first-line test to evaluate a child with intellectual disabilities (Moeschler and Shevell 2014). Exome sequencing has become standard for infants with congenital anomalies in the neonatal intensive care unit (NICU) (Meng et al. 2017), prior to referral to a genetic specialist (see Elliott 2019). As genetic

Cite this article as *Cold Spring Harb Perspect Med* doi: 10.1101/cshperspect.a036640

tests are increasingly utilized in pediatric care settings, concerns have been raised that nongenetic professionals may not fully appreciate nor have adequately educated patients and their families about the scope of testing and possible implications of results (Arora et al. 2017). When genetic counseling services are incorporated into the diagnostic workup, testing protocols for common diagnostic scenarios can be optimized and patients and families can be appropriately educated about the benefits, limitations, and interpretive complexities of various tests (Arscott et al. 2016).

The utilization of broad-based testing platforms in the initial workup can generate unrelated and unanticipated findings even while confirming the possible or presumptive diagnosis. For example, clinical diagnostic laboratories performing exome or genome sequencing will report the detection of pathogenic or likely pathogenic variants in genes that are analyzed as part of the diagnostic process, as well as those in other genes unrelated to the primary medical reason for testing, termed secondary findings (Kalia et al. 2017; Ormond et al. 2019). Some, but not all, of the information generated by broad-based testing will be actionable or of immediate benefit to the child and/or family. The prevalence of secondary findings in medically actionable genes likely differs based on the ethnic and racial background of the population, the choice of technology and reporting practices, and the indication for testing. However, multiple studies have reported frequencies of actionable secondary findings in the range of 1%–9% (Chen et al. 2018), a diagnostic reality that considerably magnifies the scope of pre- and posttest counseling.

Other Counseling Issues

Genetic testing may also increase rather than resolve uncertainty. Outside of applying well-characterized genetic testing methodologies for specific etiologies (e.g., investigating a panel of known pathogenic variants in a single gene or testing for aneuploidy via karyotype), there can be difficulty in distinguishing between pathogenic findings and rare polymorphisms resulting in the identification of variants of uncertain

significance (VUSs). Disease-specific clinical and genetic databases, bioinformatic prediction tools, and publicly available allele frequency databases facilitate the assessment of the likelihood of pathogenicity (Richards et al. 2015). Nevertheless, there are many instances in which the data generated by testing leads to increased diagnostic uncertainty or complexity.

Additionally, as a result of using broad-based technology, genetic information critical to the provision of a complete and accurate diagnosis may be missed (Kalia et al. 2017). For example, microarray technology cannot distinguish unbalanced Robertsonian translocations from acrocentric trisomies. Thus, when a child with Down syndrome is diagnosed via microarray analysis rather than karyotyping, parents may be unaware of the possibility of a balanced parental translocation with reproductive implications unless a knowledgeable provider engages the parents in posttest genetic counseling to discuss appropriate reflex testing.

Pretest genetic counseling and the informed consent process are designed to ensure patients and families understand the limitations of genetic testing and have opportunities to ask questions and seek clarification as they make decisions about whether or not genetic testing is appropriate for their child. In addition to educating parents about why testing has been recommended and instilling confidence that most positive findings will be conclusive in confirming a genetic diagnosis, there is also a need to create a foundation for potential future discussions about clinical uncertainty or secondary finding(s). Providing anticipatory guidance about the possible clinical and psychological impact of detecting unexpected or uninterpretable findings is an important part of building rapport. In the event such a finding is identified, posttest counseling can begin from a place of trust and is more likely to meet parental expectations.

GENETIC COUNSELING FOR PARENTS OF ASYMPTOMATIC OR "AT-RISK" CHILDREN

Newborn Screening (NBS) Referrals

A notable exception to pretest informed consent as the standard of care occurs in NBS, for which practices vary across the United States (Ross 2010; Blout et al. 2014). What began in the early 1960s as a simple blood-spot test used for mass screening of infants for phenylketonuria has expanded to include a wide variety of inherited conditions, both common and rare. Universal NBS has been available in the majority of states since 1990 and in all 50 states since 2006 (Benson and Therrell 2010). Although the total number of disorders screened for differs by state, the federally commissioned Recommended Uniform Screening Panel has defined a core for which all states offer screening, as well as secondary conditions that are offered at the state's discretion (https://www.hrsa.gov/advisory-committees/heritable-disorders/rusp/index.html). Early conditions included in NBS panels met a standard based on modified criteria formulated by Wilson and Junger (Wilson et al. 1968), which were reaffirmed and updated by the American College of Medical Genetics in 2006: The disorders are not readily detectable in neonates through clinical observation, the test has appropriate sensitivity and specificity, there are demonstrable benefits with early detection and timely intervention, and efficacious treatment is available (American College of Medical Genetics Newborn Screening Expert Group 2006).

Advances in molecular technology with corresponding relative decreases in the cost of adding new diseases to NBS panels have resulted in the expansion of NBS panels in ways which have challenged some of the fundamental principles. Additions to NBS panels have resulted in improvements in early and accurate diagnosis and treatment of affected individuals, but it has also led to the identification of individuals for whom NBS was not originally designed. This includes affected individuals for whom no significant treatment is available, mildly affected individuals who may never require treatment or remain largely asymptomatic until adulthood, those with a positive diagnosis for a disorder that lacks a clear point of onset or treatment, individuals with a poorly understood genotype who may never become symptomatic, and unaffected carriers (Salinas et al. 2016). Reports on early efforts to introduce exome sequencing to NBS programs suggest that these challenges will increase in the near future (Holm et al. 2018).

As an example, newborn screening for cystic fibrosis (CFNBS) has been available in all 50 states since 2010 (Parker-McGill et al. 2016). A tailored, family-centered approach to CFNBS genetic counseling that anticipates and responds to parents' emotional distress and adjusts the type and amount of information based on elicited parental preferences has been delineated (Tluczek et al. 2011). The technical approach to CFNBS identifies not only patients with classic CF disease requiring immediate medical treatment, but also CF carriers, and infants who fall into an intermediate diagnostic category of CFTR-related metabolic syndrome/CF screen–positive, inconclusive diagnosis (CRMS/CFSPID). CRMS/CFSPID does not meet the typical NBS standard of requiring immediate and essential treatment in the newborn period. Infants in the CRMS/CFSPID diagnostic category are asymptomatic but have a positive CFNBS (Farrell et al. 2017), with subsequent determination of normal or intermediate sweat chloride levels and one or more CFTR variants of unknown or variable clinical consequence (VCC) (Ren et al. 2017). Both the prognosis and the clinical benefit of obtaining this information in the newborn period remain unclear. Some CFTR VCCs may confer an increased risk for later onset, CFTR-related disorders including the congenital bilateral absence of the vas deferens, chronic pancreatitis, or sinusitis. However, a subset of children with a positive CFNBS and one or more VCCs will eventually meet diagnostic criteria for CF (Ren et al. 2015).

Some parents of children with positive CFNBS results express disbelief at the possibility of a CF diagnosis given that one or both members of the couple have had CF carrier screening with negative results. Yet this can occur, as current carrier screening methodologies do not have the sensitivity and specificity to detect all pathogenic variants (Punj et al. 2018) in all genes included on the test panel, including CF. Parents may not appreciate that there is a residual risk to have a child with a genetic disorder after "negative" carrier screening (Ioannou et al. 2014). Moreover, providers who order genetic carrier screening may be unaware of differences

Cite this article as *Cold Spring Harb Perspect Med* doi: 10.1101/cshperspect.a036640

in laboratory practices regarding the inconsistent reporting of VUSs and the classification of variants in the screening versus diagnostic settings (Westerfield et al. 2014; Vears et al. 2015). VUS results are not disclosed routinely or consistently on prenatal carrier screening reports. With a growing trend to use sequencing technology for prenatal carrier screening and the resulting apparent reassurance of a negative carrier screening result, it can be even more disconcerting and confusing for parents when their baby has a positive NBS result.

Variability in the general public's and providers' baseline understanding of the process, scope, and limitations of NBS and its interface with prenatal screening practices present significant challenges for genetic counseling in the neonatal period. As NBS for more conditions becomes standard practice, with genetic sequencing as the method of choice, ensuring that parents of newborns understand the implications of complex NBS results will increase the need for follow up pediatric genetic counseling. Additionally, parental preferences for pre-screening information and delivery of results, including diagnoses of nonmedically actionable conditions in children and genetic information with unclear prognoses may differ based on educational, cultural, and socioeconomic factors (Joseph et al. 2016). Because of the educational, clinical, and diagnostic challenges associated with NBS, it is critical that genetic counseling services be provided to bridge the gaps between genetic technology, scientific advancement, and real-world application with the goal of improving patient care. Inclusion of genetic counseling in NBS follow-up protocols and establishing professional relationships between genetic counselors and designated NBS treatment facilities will be key to ensuring genetic counseling access to patients and families of children with positive NBS results.

Referrals Generated in Response to Positive Parental Carrier Screening Results

Genetic carrier screening has been an established part of preconception and prenatal care for decades. Initially, carrier screening was limited to a small panel of well-characterized pathogenic variants and targeted to ethnic and racial groups with increased carrier frequencies for specific conditions and individuals with positive family histories. Over time, it has become clear that self-reported ethnicity is not always a reliable predictor of genetic risk (Shraga et al. 2017). Concurrently, new developments in laboratory technology have led to the availability of numerous expanded carrier screening panels capable of detecting hundreds of pathogenic variants associated with genetic disease that are designed for universal use and are not tied to an ethnic predilection or family history (Nazareth et al. 2015). Use of expanded reproductive screening methodologies intended to improve the detection of at-risk individuals/couples results in downstream opportunities for genetic counseling, including addressing psychological responses and helping to navigate diagnostic challenges in the pediatric setting. Ensuring that parents have the opportunity to ask questions of a knowledgeable pediatric care provider prior to conceiving future pregnancies will allow them time to process information and facilitate informed reproductive decision-making in keeping with their values and priorities. Some couples will decide not to alter their reproductive planning or decision-making based on carrier status. However, knowledge of positive carrier status and the medical and developmental concerns associated with a particular condition can impart benefits in the perinatal and pediatric settings, including management of a potentially high-risk pregnancy, preparation for possible birth complications, and facilitation of testing and/or early intervention in the newborn period (Ciarleglio et al. 2003; Langfelder-Schwind et al. 2014; Nazareth et al. 2015).

DIAGNOSTIC CONFIRMATION AND RESULTS DISCLOSURE

The genetic counseling needs of pediatric patients and their families vary depending on the timing, nature, and implications of the diagnosis. In family sessions, information is shared and the families' understanding is assessed, expanded, refined, and/or corrected. Simultaneously, genetic counseling encourages, validates, and addresses affective responses that may interfere with comprehension and coping strategies. The finality of a diagnosis can cause tremendous emotional distress, triggering a cascade of feelings: shock, confusion, helplessness, anger, and grief (Ashtiani et al. 2014). Parents may also experience positive feelings including relief, validation, or empowerment upon receiving their child's diagnosis (Carmichael et al. 1999; Makela et al. 2009). A genetic counselor's ability to appropriately gauge and respond to the need for information specific to the diagnosis, including phenotype, prognosis, mode of inheritance, and recurrence risk, while effectively attending to a family's primary concerns and emotional needs is a product of training that addresses knowledge, rapport, and strong communication skills.

Treatment of Genetic Conditions

One of the primary goals of establishing a genetic diagnosis is to create a medical management and treatment plan. With rare exceptions, the ability to diagnose has far outpaced the development of effective treatments or preventative measures for most "simple" single-gene and chromosome disorders. Although genetic counselors may not have hands-on involvement in the provision of treatment for genetic conditions, it is within their scope of practice to discuss the existence of current treatment and management protocols. Pediatric genetic counseling routinely includes descriptions of the current state of therapeutic options and the research pipeline, as well as information about how to access current and future clinical trials.

CF is one of a handful of conditions at the forefront of genotype-driven approaches to therapy for genetic disease, and the experience gained will likely serve as a primer for other diseases to follow. Ensuring that patients are well-informed about the current state of research and the likelihood of new treatments becoming available without creating unrealistic expectations is an important and difficult balance to maintain. Concerns have been raised that introducing the topic of novel and experi-mental therapies in a genetic counseling session may instill false hope in parents and potentially impact reproductive decision-making (Elsas et al. 2017). The unfulfilled initial promise of gene therapy as a cure for genetic conditions such as CF and the more recent success of genetic modulator therapies for CF illustrate the challenge of balancing appropriate levels of hope and caution.

Shortly after *CFTR* was discovered (Riordan et al. 1989), clinical trials began for gene replacement therapy (Crystal et al. 1994). It was widely anticipated that CF would be the first genetic disease successfully treated and cured by gene therapy (Lindee and Mueller 2011); however, efforts to develop a gene therapy–based cure for CF have not been successful to date. Greater success has been achieved for spinal muscular atrophy (SMA), a debilitating neurodegenerative condition, for which a life-extending and symptom-mediating gene therapy is now clinically available (Groen et al. 2018). In large part because of leadership, strategic planning, and funding from a patient advocacy organization, The Cystic Fibrosis Foundation (CFF) (Heltshe et al. 2017), several effective, FDA-approved CFTR modulators are now clinically available to patients with specific genotypes beginning as early as infancy (Taylor-Cousar et al. 2017; Rosenfeld et al. 2018). The cost of novel genetic-based therapies remains a potential barrier to their widespread use (Friedmann 2017; Sharma et al. 2018). As more FDA-approved genetic modulators come to market, ensuring awareness of and facilitating access to cutting-edge therapies is an increasingly essential component of genetic counseling, particularly in a pediatric specialty care setting.

Carrier Testing for Siblings

The rights of parents to make decisions about what testing is performed on their child may come into conflict with the genetic counselor's duty to uphold standards of practice (see Hercher 2019). It is common for parents to be curious or invested in learning the carrier status of their children. Although there is little debate about the use of genetic testing to diagnose a pediatric-onset genetic condition in the siblings of an affected child, particularly when there are medical benefits attributed to early diagnosis and intervention, carrier testing in unaffected minor siblings is generally not recommended (Botkin et al. 2015). When working with a family that is seeking pediatric testing in the absence of medical necessity, it is challenging to maintain a trusting relationship and balance conflicting desires. Some clinicians agree to facilitate carrier testing for unaffected siblings, taking into account the maturity and wishes of the minor, parental concerns, and the health and reproductive implications for the child, depending on the genetic condition (Vears et al. 2015). Exploring the parents' motivation for testing, explaining the reasoning behind professional recommendations, and facilitating a plan to offer carrier testing to siblings when they are able to make an informed decision about the test and understand the implications of the result may serve to reassure parents that information about their carrier status will be accessible to their child when they need it.

REPRODUCTIVE PLANNING FOR PARENTS

Pediatric genetic counseling encompasses a discussion of recurrence risks and options for prenatal/preconception testing with the parents of children with genetic conditions. Initiating the conversation in pediatric genetic counseling provides parents with the opportunity to consider all available reproductive options, including preimplantation genetic testing (PGT), targeted prenatal diagnosis, conception with gamete donation, adoption, and natural conception without prenatal testing, and sets a foundation for future reproductive planning and decision-making. For parents who plan to pursue assisted reproductive technology with PGT or targeted prenatal testing, proper documentation of familial genotypes and parental carrier test results is needed, prior to a subsequent conception. Although comprehensive discussions regarding the process, benefits, limitations, and risks of the preimplantation and prenatal procedures preferably include both a reproductive endocrinologist and a prenatal genetic counselor, assisting parents in the pediatric setting can facilitate future reproductive planning and implementation.

In providing education about reproductive test options and reproductive rights, genetic counselors risk alienating or distressing their clients. For example, some parents of children with genetic conditions may have feelings of guilt and remorse about either considering or planning to terminate a subsequent affected pregnancy. Alternatively, parents may express feelings of disappointment, disbelief, or anger when trusted clinicians previously perceived as staunch advocates for their children seem to be biased against them having additional children with the same condition. It may appear as if the lives of those affected with a genetic condition are being judged as inadequate and that technology is contributing to a world less tolerant of disability (Madeo et al. 2011). The tension that exists between the disability and genetic counseling communities has inspired many genetic counselors to explore how their own and societal attitudes and beliefs concerning disability color their practice and to pursue opportunities to collaborate with and learn from the disability community (Madeo et al. 2011). It has also motivated individuals involved in the development of genetic counseling training curricula to include disability studies and provide additional opportunities for trainees to have experiences with people with disabilities outside of a medical setting (Sanborn and Patterson 2014). It is clear that genetic counseling must elicit patient values and concerns while offering accurate and balanced portrayals of both biomedical and human aspects of having and raising a child with a genetic condition. Ideally, this will help to ensure that individuals learning about their reproductive test options and rights feel fully supported and validated by their professional care team to make autonomous decisions.

GENETIC COUNSELING NEEDS ACROSS THE LIFE SPAN

Much of genetic counseling in the pediatric setting focuses on providing information and support to parents of children with genetic conditions about their child's condition, prognosis, and expectations for adulthood. As children grow up, their genetic counseling needs may simultaneously overlap and become distinct from those of their parents (Hufnagel et al. 2016). Genetic counselors who are connected to families throughout the life span of the child can provide support to parents and also meet directly with children and adolescents to provide developmentally appropriate information (Hartley et al. 2011). In this way, genetic counseling services complement the lifelong advocacy provided by disease-specific support groups and services that often provide patients and families with invaluable assistance and guidance in both the short term and long term.

Some adolescent needs for genetic information and counseling are invariably linked to their own reproductive behaviors and goals. Genetic counselors' role in providing sex education is variable, in part because of perceptions of the scope of practice (Murphy et al. 2016). In addition, patients and parents may be embarrassed or uncomfortable introducing questions about sexual health and reproduction in a health-care setting (Havermans et al. 2011). Positive perceptions of genetic counseling by adolescents are strengthened by understanding the genetic counselor's role in their care and feeling empowered to maintain personal control of the session, such as by being asked whether they would prefer to have their parents present. Adolescents report that normalizing the concept of genetic variation among all people fosters adaptation to their condition (Pichini et al. 2016).

Today, more than half of all people living with CF in the United States are over the age of 18 (Cystic Fibrosis Foundation 2018). A genetic condition once firmly in the wheelhouse of pediatric providers, with parents expecting to outlive their children, CF treatment now requires adult care providers and a willingness on the part of pediatricians and parents to transition patients to a new health-care team and support structure that may involve a significant other assuming the role of caregiver. Reproductive genetic counseling sessions once limited to discussions with parents are now a point of care offered to adults with CF. Men and women with CF are having children of their own and benefit from genetic counseling to provide anticipatory guidance, identify resources, and set expectations (Tsang et al. 2010). Recognition of the evolving reproductive needs of individuals with CF and other genetic conditions associated with increased longevity is an essential component of genetic counseling.

CONCLUSION

As seen in the example of genetic counseling for CF, genetic counseling services in the pediatric setting can transcend the diagnostic period and are strengthened through establishing rapport and developing therapeutic relationships over time. Rooted in the areas of health-care communication and education, risk analysis, facilitated decision-making, patient advocacy, and short-term psychological counseling, genetic counseling complements ongoing multidisciplinary care and aims to address evolving concerns across the life span. Increasing demands for genetic expertise in various fields of practice will continue to drive the expansion of genetic counseling services into general and specialized multidisciplinary care settings, whereas ongoing advances in technology and personalized medicine will intensify the demand for knowledgeable providers to meet the genetic educational, social, and psychological needs of patients and families.

The core principles of establishing rapport and maintaining a trusting patient–provider relationship are fundamental components of genetic counseling for children and their families undergoing genetic evaluation, and as their genetic counseling needs evolve over time. Diagnostic testing for common single-gene disorders and syndromes frequently generates interpretive complexities and outpaces the existence of successful treatments. Moreover, broad-based genetic testing platforms are expanding the capabilities of modern medicine to assign genetic contributions to thousands of diseases, driving mainstream medicine to better recognize, diagnose, and manage genetic aspects of rare and common conditions. To effectively bridge the gap between laboratory findings and clinical application for the benefit of patients and families, the field of genetic counseling must endeavor to retain and expand the paramount human element of genetic counseling care.

REFERENCES

Reference is also in this subject collection.

American College of Medical Genetics Newborn Screening Expert Group. 2006. Newborn screening: toward a uniform screening panel and system—executive summary. *Pediatrics* **117:** S296–S307. doi:10.1542/peds.2005-2633I

Arora NS, Davis JK, Kirby C, McGuire AL, Green RC, Blumenthal-Barby JS, Ubel PA. 2017. Communication challenges for nongeneticist physicians relaying clinical genomic results. *Per Med* **14:** 423–431. doi:10.2217/pme-2017-0008

Arscott P, Caleshu C, Kotzer K, Kreykes S, Kruisselbrink T, Orland K, Cherny S. 2016. A case for inclusion of genetic counselors in cardiac care. *Cardiol Rev* **24:** 49–55. doi:10.1097/CRD.0000000000000081

Ashtiani S, Makela N, Carrion P, Austin J. 2014. Parents' experiences of receiving their child's genetic diagnosis: a qualitative study to inform clinical genetics practice. *Am J Med Genet A* **164:** 1496–1502. doi:10.1002/ajmg.a.36525

Benson JM, Therrell BL Jr. 2010. History and current status of newborn screening for hemoglobinopathies. *Semin Perinatol* **34:** 134–144.

Bernhardt BA, Biesecker BB, Mastromarino CL. 2000. Goals, benefits, and outcomes of genetic counseling: client and genetic counselor assessment. *Am J Med Genet* **94:** 189–197.

Blout C, Walsh Vockley C, Gaviglio A, Fox M, Croke B, Williamson Dean L. 2014. Newborn screening: education, consent, and the residual blood spot. The Position of the National Society of Genetic Counselors. *J Genet Couns* **23:** 16–19. doi:10.1007/s10897-013-9631-1

Botkin JR, Belmont JW, Berg JS, Berkman BE, Bombard Y, Holm IA, Levy HP, Ormond KE, Saal HM, Spinner NB, et al. 2015. Points to consider: ethical, legal, and psychosocial implications of genetic testing in children and adolescents. *Am J Hum Genet* **97:** 6–21. doi:10.1016/j.ajhg.2015.05.022

Carmichael B, Pembrey M, Turner G, Barnicoat A. 1999. Diagnosis of fragile-X syndrome: the experiences of parents. *J Intellect Disabil Res* **43:** 47–53. doi:10.1046/j.1365-2788.1999.43120157.x

Cavanagh L, Compton CJ, Tluczek A, Brown RL, Farrell PM. 2010. Long-term evaluation of genetic counseling following false-positive newborn screen for cystic fibrosis. *J Genet Couns* **19:** 199–210. doi:10.1007/s10897-009-9274-4

Chen W, Li W, Ma Y, Zhang Y, Han B, Liu X, Zhao K, Zhang M, Mi J, Fu Y, et al. 2018. Secondary findings in 421 whole exome-sequenced Chinese children. *Hum Genomics* **12:** 42. doi:10.1186/s40246-018-0174-2

Ciarleglio LJ, Bennett RL, Williamson J, Mandell JB, Marks JH. 2003. Genetic counseling throughout the life cycle. *J Clin Invest* **112:** 1280–1286. doi:10.1172/JCI200320170

Crystal RG, McElvaney NG, Rosenfeld MA, Chu CS, Mastrangeli A, Hay JG, Brody SL, Jaffe HA, Eissa NT, Danel C. 1994. Administration of an adenovirus containing the human *CFTR* cDNA to the respiratory tract of individuals with cystic fibrosis. *Nat Genet* **8:** 42–51. doi:10.1038/ng0994-42

Cystic Fibrosis Foundation. 2018. *2017 Patient registry*. Annual data report. Bethesda, MD.

* Elliott AM. 2019. Genetic counseling and genome sequencing in pediatric rare disease. *Cold Spring Harb Perspect Med* doi:10.1101/cshperspect.a036632

Elsas CR, Schwind EL, Hercher L, Smith MJ, Young KG. 2017. Attitudes toward discussing approved and investigational treatments for cystic fibrosis in prenatal genetic counseling practice. *J Genet Couns* **26:** 63–71. doi:10.1007/s10897-016-9978-1

Farrell PM, White TB, Howenstine MS, Munck A, Parad RB, Rosenfeld M, Sommerburg O, Accurso FJ, Davies JC, Rock MJ, et al. 2017. Diagnosis of cystic fibrosis in screened populations. *J Pediatr* **181:** S33–S44.e2. doi:10.1016/j.jpeds.2016.09.065

Friedmann T. 2017. Gene therapy for spinomuscular atrophy: a biomedical advance, a missed opportunity for more equitable drug pricing. *Gene Ther* **24:** 503–505. doi:10.1038/gt.2017.48

Groen EJN, Talbot K, Gillingwater TH. 2018. Advances in therapy for spinal muscular atrophy: promises and challenges. *Nat Rev Neurol* **14:** 214–224. doi:10.1038/nrneurol.2018.4

Hartley JN, Greenberg CR, Mhanni AA. 2011. Genetic counseling in a busy pediatric metabolic practice. *J Genet Couns* **20:** 20–22. doi:10.1007/s10897-010-9324-y

Havermans T, Abbott J, Colpaert K, De Boeck K. 2011. Communication of information about reproductive and sexual health in cystic fibrosis. Patients, parents and caregivers' experience. *J Cyst Fibros* **10:** 221–227. doi:10.1016/j.jcf.2011.04.001

Helm BM, Freze SL, Spoonamore SM, Ware MD, Dean AC. 2018. The genetic counselor in the pediatric arrhythmia clinic: review and assessment of services. *J Genet Couns* **27:** 558–564. doi:10.1007/s10897-017-0169-5

Heltshe SL, Cogen J, Ramos KJ, Goss CH. 2017. Cystic fibrosis: the dawn of a new therapeutic era. *Am J Respir Crit Care Med* **195:** 979–984. doi:10.1164/rccm.201606-1250PP

* Hercher L. 2019. Discouraging elective genetic testing of minors: a norm under siege in a new era of genomic medicine. *Cold Spring Harb Perspect Med* doi:10.1101/cshperspect.a036657

Holm IA, Agrawal PB, Ceyhan-Birsoy O, Christensen KD, Fayer S, Frankel LA, Genetti CA, Krier JB, LaMay RC, Levy HL, et al. 2018. The BabySeq project: implementing genomic sequencing in newborns. *BMC Pediatr* **18:** 225. doi:10.1186/s12887-018-1200-1

Hufnagel SB, Martin LJ, Cassedy A, Hopkin RJ, Antommaria AHM. 2016. Adolescents' preferences regarding disclosure of incidental findings in genomic sequencing that are not medically actionable in childhood. *Am J Med Genet A* **170:** 2083–2088. doi:10.1002/ajmg.a.37730

Ioannou L, Mcclaren BJ, Massie J, Lewis S, Metcalfe SA, Forrest L, Delatycki MB. 2014. Population-based carrier screening for cystic fibrosis: a systematic review of 23 years of research. *Genet Med* **16:** 207–216. doi:10.1038/gim.2013.125

Joseph G, Chen F, Harris-Wai J, Puck J, Young C, Koenig B. 2016. Parental views on expanded newborn screening using whole-genome sequencing. *Pediatrics* **137**(Suppl.): S36–S46. doi:10.1542/peds.2015-3731h

Kalia SS, Adelman K, Bale SJ, Chung WK, Eng C, Evans JP, Miller DT. 2017. Recommendations for reporting of secondary findings in clinical exome and genome sequencing, 2016 update (ACMG SF v2.0): a policy statement of the American College of Medical Genetics and Genomics. *Genet Med* **19:** 249–255. doi:10.1038/gim.2016.190

Lang CW, McColley SA, Lester LA, Ross LF. 2011. Parental understanding of newborn screening for cystic fibrosis after a negative sweat-test. *Pediatrics* **127:** 276–283. doi:10.1542/peds.2010-2284

Langfelder-Schwind E, Karczeski B, Strecker MN, Redman J, Sugarman EA, Zaleski C, Darrah R. 2014. Molecular testing for cystic fibrosis carrier status practice guidelines: recommendations of the National Society of Genetic Counselors. *J Genet Couns* **23:** 5–15. doi:10.1007/s10897-013-9636-9

Lindee S, Mueller R. 2011. Is cystic fibrosis genetic medicine's canary? *Perspect Biol Med* **54:** 316–331. doi:10.1353/pbm.2011.0035

Madeo AC, Biesecker BB, Brasington C, Erby LH, Peters KF. 2011. The relationship between the genetic counseling profession and the disability community: a commentary. *Am J Med Genet A* **155:** 1777–1785 doi:10.1002/ajmg.a.34054

Makela NL, Birch PH, Friedman JM, Marra CA. 2009. Parental perceived value of a diagnosis for intellectual disability (ID): a qualitative comparison of families with and without a diagnosis for their child's ID. *Am J Med Genet A* **149A:** 2393–2402. doi: 10.1002/ajmg.a.33050

Meng L, Pammi M, Saronwala A, Magoulas P, Ghazi AR, Vetrini F, Lalani SR. 2017. Use of exome sequencing for infants in intensive care units. *JAMA Pediatr* **171:** e173438. doi:10.1001/jamapediatrics.2017.3438

Moeschler JB, Shevell M, Committee on Genetics. 2014. Comprehensive evaluation of the child with intellectual disability or global developmental delays. *Pediatrics* **134:** e903–e918. doi:10.1542/peds.2014-1839

Murphy C, Lincoln S, Meredith S, Cross EM, Rintell D. 2016. Sex education and intellectual disability: practices and insight from pediatric genetic counselors. *J Genet Couns* **25:** 552–560. doi:10.1007/s10897-015-9909-6

National Society of Genetic Counselors. 2019. *National Society of Genetic Counselors professional status survey 2019: work environment*. Retrieved from https://www.nsgc.org/p/do/sd/sid=8405&fid=9097&req=direct

Nazareth SB, Lazarin GA, Goldberg JD. 2015. Changing trends in carrier screening for genetic disease in the United States. *Prenat Diagn* **35:** 931–935. doi:10.1002/pd.4647

Nicholl H, Tracey C, Begley T, King C, Lynch AM. 2017. Internet use by parents of children with rare conditions: findings from a study on parents' Web information needs. *J Med Internet Res* **19:** e51. doi:10.2196/jmir.5834

Ormond KE, O'Daniel JM, Kalia SS. 2019. Secondary findings: how did we get here, and where are we going? *J Genet Couns* **28:** 326–333. doi:10.1002/jgc4.1098

Parker-McGill L, Rosenberg M, Farrell P. 2016. Access to primary care and subspecialty care after positive cystic fibrosis newborn screening. *WMJ* **115:** 295–299.

Pichini A, Shuman C, Sappleton K, Kaufman M, Chitayat D, Babul-Hirji R. 2016. Experience with genetic counseling: the adolescent perspective. *J Genet Couns* **25:** 583–595. doi:10.1007/s10897-015-9912-y

Punj S, Akkari Y, Huang J, Yang F, Creason A, Pak C, Richards CS. 2018. Preconception carrier screening by genome sequencing: results from the clinical laboratory. *Am J Hum Genet* **102:** 1078–1089. doi:10.1016/j.ajhg.2018.04.00

Redlinger-Grosse K, Veach P, LeRoy B, Zierhut H. 2017. Elaboration of the reciprocal-engagement model of genetic counseling practice: a qualitative investigation of goals and strategies. *J Genet Couns* **26:** 1372–1387. doi:10.1007/s10897-017-0114-7

Ren CL, Fink AK, Petren K, Borowitz DS, McColley SA, Sanders DB, Rosenfeld M, Marshall BC. 2015. Outcomes of infants with indeterminate diagnosis detected by cystic fibrosis newborn screening. *Pediatrics* **135:** e1386–e1392. doi:10.1542/peds.2014-3698

Ren CL, Borowitz DS, Gonska T, Howenstine MS, Levy H, Massie J, Milla C, Munck A, Southern KW. 2017. Cystic fibrosis transmembrane conductance regulator-related metabolic syndrome and cystic fibrosis screen positive, inconclusive diagnosis. *J Pediatr* **181:** S45–S51.e1. doi:10.1016/j.jpeds.2016.09.066

Richards S, Aziz N, Bale S, Bick D, Das S, Gastier-Foster J, Grody WW, Hedge M, Lyon E, Spector E, et al. 2015. Standards and guidelines for the interpretation of sequence variants: a joint consensus recommendation of the American College of Medical Genetics and Genomics and the Association for Molecular Pathology. *Genet Med* **17:** 405–424. doi:10.1038/gim.2015.30

Rigter T, Henneman L, Broerse JE, Shepherd M, Blanco I, Kristoffersson U, Cornel MC. 2014. Developing a framework for implementation of genetic services: learning from examples of testing for monogenic forms of common diseases. *J Community Genet* **5:** 337–347. doi:10.1007/s12687-014-0189-x

Riordan J, Rommens J, Kerem B, Alon N, Rozmahel R, Grzelczak Z, Zielenski J, Lok S, Plavsic N, Chou J, et al. 1989. Identification of the cystic fibrosis gene: cloning and characterization of complementary DNA. *Science* **245:** 1066–1073. doi:10.1126/science.2475911

Ronan NJ, Elborn JS, Plant BJ. 2017. Current and emerging comorbidities in cystic fibrosis. *Presse Med* **46:** e125–e138. doi:10.1016/j.lpm.2017.05.011

Rosenfeld M, Wainwright CE, Higgins M, Wang LT, McKee C, Campbell D, Tian S, Schneider J, Cunningham S, Davies JC, et al. 2018. Ivacaftor treatment of cystic fibrosis in children aged 12 to <24 months and with a CFTR gating mutation (ARRIVAL): a phase 3 single-arm study. *Lancet Respir Med* **6:** 545–553. doi:10.1016/S2213-2600(18)30202-9

Ross LF. 2010. Mandatory versus voluntary consent for newborn screening? *Kennedy Inst Ethics J* **20:** 299–328.

Rutherford S, Zhang X, Atzinger C, Ruschman J, Myers MF. 2014. Medical management adherence as an outcome of genetic counseling in a pediatric setting. *Genet Med* **16:** 157–163. doi:10.1038/gim.2013.90

Salinas DB, Sosnay PR, Azen C, Young S, Raraigh KS, Keens TG, Kharrazi M. 2016. Benign and deleterious cystic fibrosis transmembrane conductance regulator mutations identified by sequencing in positive cystic fibrosis newborn screen children from California. *PLoS One* **11:** e0155624. doi:10.1371/journal.pone.0155624

Sanborn E, Patterson AR. 2014. Disability training in the genetic counseling curricula: bridging the gap between genetic counselors and the disability community. *Am J Med Genet A* **164:** 1909–1915. doi:10.1002/ajmg.a.36613

Sharma D, Xing S, Hung YT, Caskey RN, Dowell ML, Touchette DR. 2018. Cost-effectiveness analysis of lumacaftor and ivacaftor combination for the treatment of patients with cystic fibrosis in the United States. *Orphanet J Rare Dis* **13:** 172. doi:10.1186/s13023-018-0914-3

Shraga R, Yarnall S, Elango S, Manoharan A, Rodriguez SA, Bristow SL, Puig O. 2017. Evaluating genetic ancestry and self-reported ethnicity in the context of carrier screening. *BMC Genet* **18:** 99. doi:10.1186/s12863-017-0570-y

Sosnay PR, Siklosi KR, van Goor F, Kaniecki K, Yu H, Sharma N, Ramalho AS, Amaral MD, Dorfman R, Zielenski J, et al. 2013. Defining the disease liability of variants in the cystic fibrosis transmembrane conductance regulator gene. *Nat Genet* **45:** 1160–1167. doi:10.1038/ng.2745

Stoll K, Kubendran S, Cohen SA. 2018. The past, present and future of service delivery in genetic counseling: keeping up in the era of precision medicine. *Am J Med Genet C Semin Med Genet* **178:** 24–37. doi: 10.1002/ajmg.c.31602

Taylor-Cousar JL, Munck A, McKone EF, van der Ent CK, Moeller A, Simard C, Wang LT, Ingenito EP, McKee C, Lu Y, et al. 2017. Tezacaftor–Ivacaftor in patients with cystic fibrosis homozygous for Phe508del. *N Engl J Med* **377:** 2013–2023. doi:10.1056/NEJMoa1709846

Tluczek A, Zaleski C, Stachiw-Hietpas D, Modaff P, Adamski CR, Nelson MR, Reiser CA, Ghate S, Josephson KD. 2011. A tailored approach to family-centered genetic counseling for cystic fibrosis newborn screening: the Wisconsin Model. *J Genet Couns* **20:** 115–128. doi:10.1007/s10897-010-9332-y

Tsang A, Moriarty C, Towns S. 2010. Contraception, communication and counseling for sexuality and reproductive health in adolescents and young adults with CF. *Paediatr Respir Rev* **11:** 84–89. doi:10.1016/j.prrv.2010.01.002

Vears DF, Delany C, Gillam L. 2015. Carrier testing in children: exploration of genetic health professionals' practices in Australia. *Genet Med* **17:** 380–385. doi:10.1038/gim.2014.116

Westerfield L, Darilek S, van den Veyver I. 2014. Counseling challenges with variants of uncertain significance and incidental findings in prenatal genetic screening and diagnosis. *J Clin Med* **3:** 1018–1032. doi:10.3390/jcm3031018

Wilson JMG, Jungner G. 1968. *Principles of screening for disease*, pp. 26–39. World Health Organization, Geneva.

Witt MM, Jankowska KA. 2018. Breaking bad news in genetic counseling—problems and communication tools. *J Appl Genet* **59:** 449–452. doi:10.1007/s13353-018-0469-y

Genetic Counseling and Assisted Reproductive Technologies

Debra Lilienthal[1] and Michelle Cahr[2]

[1]The Ronald O. Perelman and Claudia Cohen Center for Reproductive Medicine, Weill Cornell Medicine, New York, New York 10021, USA

[2]California Cryobank Life Sciences, Los Angeles, California 90025, USA

Correspondence: dpl9001@med.cornell.edu

Despite the ever-increasing number of patients undergoing fertility treatments and the expanded use of genetic testing in this context, there has been limited focus in the literature on the involvement of genetics professionals in the assisted reproductive technology (ART) setting. Here we discuss the importance of genetic counseling within reproductive medicine. We review how genetic testing of embryos is performed, the process of gamete donation, the challenges associated with genetic testing, and the complexities of genetic test result interpretation.

In 2018, Louise Brown, the first child born from in vitro fertilization (IVF), turned 40 yr old (Steptoe and Edwards 1978; Fishel 2018). For people who were previously unable to have a biological child, the ability to conceive via IVF meant that their dream of parenthood had a greater chance of becoming a reality. IVF has come a long way over the last four decades, and the number of children born following the use of assisted reproductive technology (ART) continues to grow rapidly. In 2016 alone, data from the Centers for Disease Control and Prevention reported 263,577 IVF cycles, resulting in 76,930 total infants born; the number of IVF cycles documented between January 9, 1969 and August 1, 1978 was 457, resulting in two live births (Centers for Disease Control and Prevention 2018; Fishel 2018). Given the societal trends of increasing parental age and delayed child-

bearing, both of which correlate with reduced fertility, these numbers are expected to continue to rise (Botting and Dunnell 2000; Armstrong and Akande 2013).

Advances in genetic testing have become increasingly intertwined with the world of IVF, and are critical components of reproductive medicine. Given the complexity of these advances, the inclusion of trained genetics professionals within the field of ART is vital.

BACKGROUND

There are many reasons for individuals and couples to seek out ART, including female infertility, male factor infertility, unexplained infertility, secondary infertility (infertility after a previous successful pregnancy), and the use of a gestational carrier. Some who choose to use

ART do not meet the traditional definition of infertility (i.e., not conceiving after 6 or 12 months of unprotected intercourse, depending on age); this includes individuals wishing to prevent genetic disease, same-sex couples, or single patients who require donor gametes, those pursuing sex selection, and those interested in fertility preservation. Fertility preservation may be used when health-related issues or medical treatments affect future fertility, such as gonadotoxic treatments for cancer or delayed childbearing. The reasons for seeking treatment will likely continue to change over time.

Before initiating any form of treatment, American Society for Reproductive Medicine (ASRM) guidelines suggest a thorough evaluation including a physical exam and laboratory testing (Practice Committee of the American Society for Reproductive 2015a,b). This evaluation should elicit personal and family history of genetic disease and review any prior genetic test results (i.e., chromosome analysis, carrier screening, etc.) that could affect the course of treatment. Patients should also be counseled about additional genetic testing that may be indicated before starting treatment. When issues arise as a result of this evaluation, referral to a genetics specialist is recommended.

ART is not synonymous with IVF and less-invasive treatments may be an appropriate first step in many instances. For some individuals and couples, initial approaches may include timed intercourse or intrauterine insemination (IUI). Reasons to consider IVF as a first-line therapy include the need for any of the following: preimplantation genetic testing (PGT), a gestational carrier, donor oocytes, or intracytoplasmic sperm injection (ICSI) for patients with significant male factor infertility. Regardless of the indication, it is important to note that the IVF process can be physically, emotionally, and financially demanding.

IVF BASICS

The process of IVF begins with ovarian stimulation. Injectable gonadotropin medications are given to stimulate the ovaries to produce more mature oocytes than could naturally be produced in one cycle. During this stage, near-daily monitoring by blood and ultrasound occurs in order for the physician to assess the response to treatment and determine when the oocytes are mature and ready for retrieval. When the oocytes appear to be potentially mature, the patient is given medication to trigger ovulation and undergoes surgery to remove the oocytes (egg retrieval). Following retrieval, mature oocytes are fertilized with the designated sperm sample in the laboratory and embryo development is monitored. Embryo(s) are usually transferred into the uterus on the third or fifth day of development. Excess embryos may be cryopreserved for later use.

Oocytes are fertilized either by insemination, in which they are mixed with sperm in a Petri dish, or by ICSI, in which a single sperm is injected into the oocyte. The sperm source may be a fresh ejaculate from the patient's partner, frozen sperm from the patient's partner, or sperm obtained via microsurgical epididymal sperm aspiration (MESA) or testicular sperm extraction (TESE). Alternatively, patients may use frozen sperm from an anonymous donor or a directed donor, who may be a friend or a family member.

ICSI was first used to treat male factor infertility for cases in which IVF by insemination did not lead to fertilization, or when a man had few sperm (Lanzendorf et al. 1988). ICSI is now also used in situations in which there is no male factor infertility, such as with PGT, to reduce the potential paternal contamination of the DNA tested from extraneous sperm that are not fertilizing the oocyte. ICSI is also used to fertilize frozen oocytes or to maximize fertilization if only few or poor-quality oocytes are retrieved.

As mentioned above, following fertilization embryos are monitored for cell division. For patients trying to achieve pregnancy within the cycle, embryo(s) are selected for transfer to the uterus. Experienced embryologists carefully assess embryo quantity and quality to select the best embryo(s) and also determine the day of embryo transfer, which varies from center to center.

Vitrification, a method of rapid freezing, is currently the most widely used technique for cryopreservation, the process of freezing reproductive tissue (Practice Committees of American

Society for Reproductive and Society for Assisted Reproductive 2013). Some patients choose to cryopreserve all of their suitable embryos without transferring any fresh embryos during the cycle. Patients who require up-front cryopreservation include those who do not have medical clearance to achieve pregnancy, those who are using a gestational carrier, and those who are undergoing PGT or fertility preservation. A frozen embryo transfer (FET) is the process by which a frozen embryo is thawed and transferred into a woman's uterus in a subsequent cycle.

The potential risks to offspring conceived by ART have been debated since its early days. Any potential risks of ART treatments must be carefully weighed against their benefits for individuals and couples who may not otherwise be able to conceive.

In any pregnancy, there is a baseline risk of 3%–4% for a child to have a birth defect or developmental abnormality (Bodurtha and Strauss 2012). Although most babies conceived via ART treatments are healthy, some studies suggest that children born following ART have an increased risk of birth defects compared to those born from spontaneous conceptions (Kurinczuk et al. 2004; Hansen et al. 2005). ICSI, in particular, has been reported in some studies to be associated with an increased incidence of hypospadias, sex chromosome abnormalities, and imprinting defects (i.e., Angelman syndrome, Prader–Willi syndrome, and Beckwith–Wiedemann syndrome) (Alukal and Lamb 2008; Practice Committee of American Society for Reproductive and Practice Committee of Society for Assisted Reproductive 2008). It is unclear if these risks are due to the use of ART itself, possible epigenetic factors, or the use of technologies to overcome infertility that may have a genetic basis, whereby abnormal genetic traits may thus be transmitted to offspring (Alukal and Lamb 2008).

Some patients may delay or decline ART treatment because of their concerns about the possible risks. Additionally, those who are worried about the risks of ICSI may forgo its use, even though it may improve their chances of getting pregnant. For some patients, concerns about the risks following ART may persuade them to request special prenatal diagnostic testing for chromosome abnormalities and/or imprinting defects.

GENETIC TESTING OF EMBRYOS

Preimplantation Genetic Testing Overview

PGT involves the removal of one or more nuclei from an oocyte (polar body biopsy) or an embryo (blastomere or trophectoderm cells) to test for mutations or aneuploidy prior to embryo transfer (Practice Committee of the Society for Assisted Reproductive and Practice Committee of the American Society for Reproductive 2008). The purpose of PGT is to reduce the risk of having a child with a genetic disorder or a chromosome abnormality.

There are three types of PGT. PGT for aneuploidy (PGT-A; previously called PGS) is the testing of embryos for chromosome abnormalities. PGT for structural rearrangements (PGT-SR) is the testing of embryos for structural chromosome rearrangements (i.e., translocations and inversions). PGT for monogenic disorders (PGT-M; previously called PGD) is the testing of embryos for single gene defects. PGT requires the use of IVF, oocyte or embryo biopsy, and the transfer of embryo(s) into the uterus based on genetic test results.

A biopsy must be performed to obtain DNA for analysis. Historically, embryo biopsies were performed at the cleavage stage, on day 3, when embryos have approximately six to eight cells. A single blastomere was removed and analyzed, allowing for a fresh embryo transfer on day 5 post–egg retrieval. After the fresh transfer, excess embryos were considered for cryopreservation for possible future use. Current practice has shifted, and most centers now perform biopsies on day-5 or day-6 blastocyst embryos which have more than 100 cells. This type of biopsy allows for more cells to be analyzed. For this biopsy, a sample of cells is taken from the trophectoderm layer of the embryo, which will ultimately become the placenta. In most instances, the embryo is frozen after the biopsy because there is not enough time to obtain PGT results to allow for a fresh embryo transfer. Once results are obtained, a woman will return for a FET, by which the desired embryos(s) will be transferred into the uterus. Because they are performed at a later developmental stage, and the number of cells removed is smaller on a percentage basis, blastocyst biopsies are considered less disruptive to the developing embryo than those done at the cleavage stage, while providing more DNA for testing (Practice Committees of the American Society for Reproductive et al. 2018).

Although less commonly performed, a polar body biopsy involves removing the first and second polar bodies from an oocyte. This method can be used to detect maternally inherited mutations or meiotic errors during oocyte development. The results obtained from the polar bodies indirectly reflect the genetic status of the oocyte, as the polar body contains the reciprocal result of what is in the oocyte that would be fertilized to create an embryo. With this method of testing, the paternal contribution cannot be assessed.

The ASRM strongly recommends genetic counseling before starting an IVF cycle for patients considering PGT (Practice Committee of the Society for Assisted Reproductive and Practice Committee of the American Society for Reproductive 2008). The initial genetic counseling consultation should include a discussion of the risk for the individual or couple to have an affected child with a single gene defect and/or a chromosome abnormality, as well as the natural history of the disease. All diseases are not created equal, and not all patients at risk of having a child with a genetic disorder or chromosome abnormality elect to pursue PGT or prenatal testing. All testing options should be discussed, including IVF with PGT, as well as prenatal testing via chorionic villus sampling (CVS) or amniocentesis. The risks, benefits, and limitations of PGT technologies, including the risk for false-positive and false-negative results, should also be reviewed, and recommendations made for follow-up diagnostic testing during pregnancy.

There are some inherent risks associated with PGT, which is an invasive procedure. One risk is possible damage to the embryo from the biopsy, whereby the embryo arrests and is no longer usable. This risk is generally thought to be low. Extended culture may pose an additional risk; the embryos may not survive in the laboratory until day 5 or 6 for testing or transfer. The biopsy itself may also potentially adversely affect an embryo's ability to implant (Scott et al. 2013). Although initial reports on children born following day-3 biopsies did not show an increased risk for physical or mental disabilities, large, long-term studies, specific to the use of day-5/6 biopsy (currently the mostly widely used biopsy technique), have not yet been performed (Liebaers et al. 2010; Heijligers et al. 2018).

Although the goal of PGT is to select normal embryos suitable for transfer, patients should be aware of the issues that may arise with its use. For instance, it is possible that embryos may not reach the appropriate developmental stage to be biopsied. Embryos tested on day 3 may not develop until day 5 for transfer. There also may not be any PGT-normal embryos suitable for transfer. When patients are having a FET of PGT-normal embryos, the embryos may not survive the thaw to be transferred. Moreover, patients should be counseled that PGT-normal embryos do not guarantee pregnancy or a healthy live birth, and that it may take multiple cycles to obtain enough PGT-normal embryos to achieve their family. Because of the complex nature of PGT, patients need to be prepared for the potential outcomes of their test results and have access to a genetics specialist to help interpret them.

Preimplantation Genetic Testing— Aneuploidy

The purpose of PGT-A is to assess the presence or absence of aneuploidy in order to reduce the risk of transferring chromosomally abnormal embryos and to improve the probability of a successful pregnancy. Aneuploidy is the most common cause of miscarriage and implantation failure (Hassold et al. 1996; Kang et al. 2016; Friedenthal et al. 2018). Currently there are no standard indications for PGT-A (Practice Committee of the Society for Assisted Reproductive and Practice Committee of the American Society for Reproductive 2008). Common reasons for considering its use include, but are not limited to, the following: advanced maternal age

(AMA), recurrent pregnancy loss (RPL), and recurrent IVF/implantation failure. Additional indications may include patients who have had a prior child or pregnancy with aneuploidy or individuals with a chromosome abnormality who are at an increased risk for having chromosomally abnormal offspring. PGT-A is used for sex selection when there is a risk for sex-linked conditions, sex-specific issues without a clear Mendelian pattern of inheritance (i.e., autism), or family balancing/personal choice. Patients who are undergoing fertility preservation or are using a gestational carrier may elect to undergo PGT-A to maximize the success of a future transfer.

PGT-A has been available as an adjunct to IVF since the early 1990s (Verlinsky et al. 1995). Since then, testing techniques have improved, and utilization has expanded significantly. Some studies have shown an increased live birth rate following PGT-A and single embryo transfer (SET) (Dahdouh et al. 2015; Simon et al. 2018), whereas others have only found improvements for certain patient populations, such as those who are of advanced maternal age (AMA) (Kang et al. 2016). Despite this, debates regarding its use remain (Rosenwaks et al. 2018).

PGT-A has evolved significantly since its first use. Initially, PGT-A was performed by fluorescent in situ hybridization (FISH), and it screened for a limited number of chromosome abnormalities via polar body or day-3 biopsy (Delhanty et al. 1993; Munné et al. 1993; Verlinsky et al. 1995). FISH continued to have limitations, such as overlapping fluorescent signals which could lead to misinterpretation of the embryo's chromosome status. Additionally, many chromosome abnormalities were missed because of the small number of chromosomes analyzed and higher incidence of mosaicism in day-3 embryos (Treff et al. 2010; Mertzanidou et al. 2013). These issues resulted in false-positive and false-negative results. Many studies also did not demonstrate an improved pregnancy rate with its use (Mastenbroek et al. 2007, 2011).

Because of these drawbacks, there was a desire from experts in the field to move toward molecular technologies, which would eliminate some of the technical limitations of FISH and would allow screening of all 24 chromosomes.

These techniques include array comparative genomic hybridization (aCGH), quantitative polymerase chain reaction (qPCR), single-nucleotide polymorphism (SNP) microarray, and next-generation sequencing (NGS). Although all of these assess aneuploidy, they vary in resolution and depth of read. Newer technologies, such as SNP and NGS, can detect segmental gains and losses of chromosome material. SNP can also detect polyploidies, parental origin of aneuploidy, and uniparental disomy (UPD).

Although they all provide similar results, there may be some circumstances in which a particular technology may prove more beneficial. For instance, triploidy and tetraploidy most frequently result in early miscarriage and are generally considered sporadic with a low risk for recurrence. However, there have been several case reports of women with multiple triploid pregnancies (Pergament et al. 2000; Brancati et al. 2003). In some of these cases, this may be due to an inherited predisposition to meiotic errors (Huang et al. 2004; Filges et al. 2015). Thus, although unlikely, patients with histories of polyploid pregnancies may benefit from the use of SNP microarray for PGT-A. Additionally, for patients with multiple failed fertility treatments and/or miscarriages who are considering the use of donor gametes, determining parental origin of aneuploidy in embryos via SNP microarray may aid in their decision.

Because SNP analysis requires DNA samples from both biological parents, there may be cases in which it is not recommended and/or another PGT-A technology may be more appropriate. Obtaining DNA samples from both biological parents may be challenging when donor gametes are used. However, in some cases, cryobanks may have DNA available from the gamete donor for such purposes. Additionally, consanguinity may complicate SNP analysis because of the high level of shared genetic markers increasing the risk for uninformative results.

The presence of mosaicism may further complicate the interpretation of PGT-A results. Embryonic mosaicism is the presence of two or more genetically different cell types within the same embryo. Mosaicism is very common in human embryos, particularly at the cleavage stage (Mertzanidou et al. 2013; Harton et al. 2017). Although less common, mosaicism is still often present in the blastocyst (Fragouli et al. 2011; Harton et al. 2017). Despite trophecto-derm biopsies containing several cells, these cells may not be representative of other cells within the trophectoderm and/or inner cell mass. This is the likely explanation for many false-positive and false-negative PGT-A results.

As PGT-A technology has improved, the detection of mosaicism within biopsy samples has become far more common. Previously, re-sults were classified as normal or abnormal. However, a third category has now emerged with the reporting of mosaic results. Although any PGT-A-normal or -abnormal result may theoretically not represent the eventual chromo-some makeup of a developing embryo, an initial mosaic result presents a higher level of uncer-tainty as to an embryo's ultimate fitness.

The decision of whether or not to transfer mosaic embryos can be challenging. Several stud-ies have demonstrated healthy live births follow-ing the transfer of mosaic embryos (Greco et al. 2015; Fragouli et al. 2017; Munné et al. 2017; Spinella et al. 2018). Some centers are comfort-able transferring these embryos as long as pa-tients have had genetic counseling to discuss the results. Prenatal diagnostic testing is strongly recommended in these cases, and patients should be counseled on the possible outcomes of trans-fer, including implantation failure, miscarriage, abnormal or mosaic pregnancy detected via pre-natal screening/diagnosis, live birth of a child with abnormal or mosaic chromosomes, or a healthy live birth (Harton et al. 2017; Sachdev et al. 2017). A review of mosaic results should distinguish between embryos with abnormalities that have been reported at the time of prenatal diagnosis and live birth versus those that are less likely to be compatible with life. Also, embryos with mosaic chromosome abnormalities associ-ated with an increased risk for UPD need be given special consideration. The uncertainties associ-ated with mosaic results may be viewed by some as a reason to not transfer these embryos.

Because of the limitations of PGT-A tech-nology and the issue of mosaicism, it is impor-tant for patients to be counseled that PGT-A is a screening test. Because all screening tests have the potential to give false-positive and false-negative results, prenatal diagnostic testing should be performed to confirm the chromo-some status of a pregnancy following PGT-A (Brezina et al. 2016; Kimelman et al. 2018). In addition, prenatal testing can look for smaller deletions and duplications than PGT-A can de-tect, while also studying more cells, thus further reducing the risk of undetected mosaicism, al-though this risk cannot be completely eliminat-ed. Although patients who have undergone IVF may be understandably wary of having invasive diagnostic testing that carries a small risk of miscarriage, they need to be counseled about the differences between screening and diagnos-tic testing so that they can make a well-informed decision (Devers et al. 2013).

Preimplantation Genetic Testing—Structural Rearrangement

It is estimated that ∼1 in 560 to 1 in 1100 indi-viduals in the general population carries an apparently balanced structural chromosome re-arrangement (Forabosco et al. 2009; Tobler et al. 2014; Simioni et al. 2017). Given that such rearrangements increase the risk for infertility and RPL, it is not uncommon for there to be a higher frequency of these rearrangements found among patients referred for fertility treatment. Some patients may also be identified with struc-tural rearrangements following prenatal diagno-sis, analysis of products of conception (POC), birth of a child with an abnormality, or detection in another family member. In others, their rear-rangement may have been known since birth, if their mothers underwent CVS or amniocentesis.

For patients who are at an increased risk for unbalanced chromosome products because of a structural rearrangement, PGT-SR allows them to select embryos that have an apparently ba-lanced chromosome complement. Currently, most PGT-SR testing uses the same technology as PGT-A (i.e., NGS or SNP). To assess whether or not PGT-SR is possible, the testing laboratory first needs to review karyotypes. Effectiveness of PGT-SR will depend on the ability of the chosen PGT technology to detect the unbalanced segments of the rearrangement. In some cases, such as small deletions or inversions not detectable by NGS or SNP testing, linkage analysis may be necessary, requiring the participation of family members for the development of customized probes.

Although in most cases NGS and SNP are equally effective to screen for unbalanced translocation products, SNP is the preferred technology for carriers of Robertsonian translocations involving chromosomes 14 or 15 because it can detect UPD, which is a concern in these cases.

Because none of the technologies currently available can distinguish between a normal karyotype and a balanced rearrangement, PGT-SR cannot eliminate the risk that a balanced rearrangement is transmitted to the fetus. Research is ongoing to perfect technologies capable of distinguishing between normal and balanced karyotypes (Treff et al. 2011).

Preimplantation Genetic Testing— Monogenic

PGT-M tests embryos for a specific genetic disorder. There are many reasons why patients may consider this type of testing. Individuals who are affected by and/or have a family history of an autosomal dominant disorder, as well as those who are at risk of having offspring with an autosomal recessive or X-linked condition, are candidates for PGT-M. Others may choose PGT-M for HLA-matching for a child in need of a stem cell transplant, whereas those with antigen sensitivity from a prior pregnancy may use PGT-M to select embryos that are not at risk for hemolytic disease of the newborn.

PGT-M is unique to each patient and requires a custom setup that can take several weeks or months to create. Unlike other genetic tests performed pre- or postnatally from samples in which there are hundreds of cells, PGT-M involves the use of a very small amount of DNA. The DNA is amplified with PCR, and then targeted analysis is performed. Because of the small amount of DNA obtained in an embryo biopsy, PGT-M is prone to allele dropout and amplification failure. To increase the likelihood of getting clear results, linkage analysis is used, which

can be done with short tandem repeats (STRs) or karyomapping/SNP analysis.

Linkage analysis is based on the concept that genetic markers closer together on a chromosome are more likely to be inherited together. This type of analysis involves participation of family members, usually from a generation above or below the individual or couple undergoing treatment. This can be challenging for those who are trying to keep their fertility treatments private, who have family members who may not be supportive of their decision to do PGT, whose parents are deceased, who are estranged from their families, or who are adopted. For conditions that are autosomal dominant or recessive, DNA from an affected parent or child can be used. For autosomal recessive conditions that are detected by general population carrier screening, testing of the couple's parents is needed to determine which parent carries the trait in order to establish linkage.

For de novo cases in which there is no family history of the disorder, PGT-M may or may not be a viable option. For males with an autosomal dominant de novo disorder, a semen sample can be used to establish linkage because sperm contain a single haplotype. Unfortunately, there is no equivalent way to obtain single cells from a female to establish linkage. Some laboratories will be able to work backward to complete the interpretation and establish linkage by testing the patient's embryos. This method may not permit diagnosis of an embryo as PGT-normal or PGT-abnormal if the embryo yield is low or if there is not a mixture of PGT-affected and PGT-unaffected embryos.

Although PGT-M may be performed for multiple genetic disorders, patients must be counseled regarding the likelihood of finding embryos free of these diseases.

SPECIAL TYPES OF PREIMPLANTATION GENETIC TESTING

Nondisclosing PGT-M

Nondisclosing PGT-M is used when patients at risk for a serious genetic disease do not want to know their mutation status but wish to ensure that no mutation-containing embryos are transferred (Stern et al. 2002). Although it is commonly used for patients at risk for Huntington's disease, nondisclosing PGT-M can be used for any disease for which a prospective parent is not comfortable knowing his or her affected status.

There are two approaches to nondisclosing PGT-M: indirect testing and direct testing. Both methods require the use of PGT-A in conjunction with PGT-M. Because chromosome abnormalities are common in embryos, if a patient were to see a report showing that there were embryos not recommended for transfer, they would not be able to know if those embryos were PGT-M-abnormal or PGT-A-abnormal. The purpose of this is make it less likely for patients to inadvertently learn their disease status.

Indirect testing excludes embryos that inherited the disease-associated allele from the affected grandparent of the future offspring, without determining whether it was the grandparent's normal or disease-carrying allele. Only embryos that have the allele inherited from the unaffected grandparent would be recommended for transfer. With indirect testing, typically 50% of embryos would be expected to be rejected as potentially disease-carrying. Patients need to be prepared for the possibility that all tested embryos inherited the allele from the affected grandparent and would not be recommended for use, even though some or all of these embryos may carry the normal allele.

Direct testing is performed by the PGT lab, which tests the patient for the disease under an alias through a clinical laboratory. If the laboratory determines that the patient is affected with the disease, it will proceed with PGT-M and PGT-A to make sure that only disease-free embryos are recommended for use. If the patient is not affected with the disease, only PGT-A will be performed, and embryos will be recommended for transfer based on these results. With this method, someone who is not affected with the disease would not have embryos unnecessarily eliminated from use.

Once the results of nondisclosing PGT-M are complete, the IVF center is informed and the individual or couple undergoes a FET cycle when they are ready to get pregnant. If there are

PGT-normal embryos available, the embryo(s) will be transferred followed by a pregnancy test, which will be given to determine if the FET was successful. If the FET was unsuccessful, and there are additional frozen embryos, they will proceed with FET cycles until they achieve a pregnancy. If there are no additional frozen embryos available, they would need to proceed with a new IVF cycle. In cases in which there are no embryos recommended for use, because of PGT-M or PGT-A results, the patient will have a mock embryo transfer, a FET of no embryos, and a negative pregnancy test. Because a FET can fail to result in a pregnancy even when embryos are transferred, patients should not assume they are affected with the disease in question based on a negative pregnancy test. In this case, the patient will be told that there are no additional frozen embryos and that they will need to proceed with a new IVF cycle.

Although nondisclosing PGT-M is done regularly, it is not without controversy. In giving patients the right not to know their disease status, fertility centers may be performing unnecessary treatments that are costly and carry physical risks. The diseases for which nondisclosing PGT-M is typically performed have significant morbidity with no treatments or cure. Patients at risk for these diseases may choose not to know their status to avoid the stress and other psychological sequelae of knowing they may develop the condition. They may also be concerned about their ability to get health or life insurance should their affected status be known. Regardless of their reasoning, all patients must weigh the risks and benefits to decide the best course of treatment.

PGT for HLA Matching

PGT for human leukocyte antigen (HLA) typing is used to conceive a child who could donate cord blood or stem cells for transplantation to save an ill sibling (Tur-Kaspa and Jeelani 2015). PGT for this purpose is performed by linkage analysis only. The chance of having an HLA-matched embryo to a full sibling is 25%, assuming that recombination has not occurred. If recombination occurred in an ill child, it would be nearly impossible to find a suitable match, as the same recombination event would need to occur again. If the affected child has an autosomal recessive condition, the likelihood of an embryo being PGT-M-unaffected, HLA matched is 18.8%. If the affected child has an X-linked recessive disorder, the likelihood of finding an embryo that is PGT-M-unaffected, HLA-matched is 12.5%.

Those considering PGT for HLA matching need to be counseled regarding the amount of time required for the PGT test to be developed and applied and for an HLA-matched sibling to be born (Tur-Kaspa and Jeelani 2015). They also should be counseled about the possibility of not having any HLA-matched embryos that are PGT-M-normal. Additionally, the transfer of an HLA-matched embryo may not result in pregnancy and live birth. When no HLA-matched embryos are found, patients who want to have more children must decide if they will proceed with additional IVF treatments to try to find a matched embryo, do a FET of an unmatched embryo, or try to conceive naturally.

PGT-M for Mitochondrial Disease

Mitochondrial diseases can be caused by nuclear DNA mutations or mutations in the mitochondrial DNA (mtDNA) itself. The former can follow any traditional Mendelian inheritance pattern (autosomal recessive, autosomal dominant, or X-linked) and is readily amenable to standard PGT-M. The latter is most commonly maternally inherited, as typically only the oocyte transmits mitochondria and its DNA.

There are multiple biological factors that make mitochondrial inheritance complicated. Heteroplasmy is the presence of both normal and abnormal mtDNA in the same individual. A higher level of abnormal mtDNA increases the chance of an individual being affected with a mitochondrial disorder. However, individuals, including transmitting females, may or may not have clinical symptoms even in the presence of low levels of abnormal mtDNA, depending on tissue distribution. The threshold effect is the percentage of abnormal mtDNA necessary in a given tissue to lead to symptoms of a particular mitochondrial disorder. The bottleneck effect may lead to significant and unpredictable changes in the amount of abnormal mtDNA inherited from one generation to the next.

Options for women at an increased risk of having children with mitochondrial disorders because of mtDNA mutations include prenatal diagnostic testing, PGT-M for mtDNA mutations, and use of an unrelated ovum donor. Another option, although controversial for ethical reasons, and currently not permitted in the United States, is the use of mitochondrial transfer therapy in which a woman has her nuclear DNA transferred into an enucleated oocyte of an unrelated donor, containing the cytoplasm of the ovum donor and thus the donor's mtDNA (Wolf et al. 2015).

PGT-M for mtDNA mutations has significant limitations, as there may be fluctuations in the mtDNA mutation load during embryonic development (Mitalipov et al. 2014; Diot et al. 2016; Craven et al. 2017). Although absence or very low levels of abnormal mtDNA may indicate a reduced likelihood of an embryo being affected with a mitochondrial disorder, the risk cannot be entirely eliminated. Additionally, multiple IVF cycles may be required to find embryos suitable for transfer.

PGT-M for Variants of Uncertain Significance

With advances in sequencing technology, detection of genetic variants of uncertain significance (VUSs) has increased. When no known pathogenic variant is identified, it is not uncommon for patients to hope that a VUS could explain the disease in their family. PGT-M for a VUS necessitates laboratory approval, as there is liability in testing for a variant that may not be causative of the disease within a given family. Patients need to be cautioned that PGT-M results will only reduce the risk for disease if the VUS is truly causative of the disease in question. If the VUS is not causative, PGT-M will not reduce the risk of that disease in offspring.

GAMETE DONATION

Gamete donation involves the donation of either oocytes or sperm to assist another individual or couple in achieving pregnancy. Oocyte donors are usually between the ages of 21 and 34. Indications for the use of an oocyte donor include women with the following: hypergonadotropic hypogonadism, advanced maternal age, diminished ovarian reserve, poor oocyte/embryo quality, or multiple failed attempts to conceive after fertility treatments. Additional indications include women known to be affected with or carriers of a significant genetic disorder or those who have a family history of a heritable condition for which testing is not available. Oocyte donation is also necessary for male same-sex couples and single men.

Sperm donors are usually under the age of 40. These donors are most frequently obtained from a sperm bank that has a large pool of donors from which recipients may choose. Indications for the use of a sperm donor include the following: azoospermia, severe oligospermia, or other significant sperm or seminal fluid abnormality; ejaculatory dysfunction; or other significant male factor infertility that has not responded to previous attempts at ART. A male may also be affected with or a carrier of a significant genetic disorder or have a family history of a heritable condition for which testing is not available. Additional reasons for use of a sperm donor include an STD that cannot be eradicated or an Rh-positive male with a female partner who is Rh-negative and severely Rh-isoimmunized. Sperm donors are also necessary for women who are trying to conceive without a male partner or female same-sex couples.

Gamete Donor Screening

The ASRM recommends that donors be in good health with no known genetic abnormalities (Practice Committee of American Society for Reproductive and Practice Committee of Society for Assisted Reproductive 2013). According to guidelines, a complete medical, family, and sexual history should be obtained. Donors need to meet FDA regulations and applicable state tissue banking requirements to reduce the risk of transmitting infectious diseases, and they must also have a complete physical and psychological evaluation.

In addition, the ASRM recommends that a genetic counselor, or other medical professional trained in genetics, perform an assessment of hereditary risk factors for donors (Practice Committee of American Society for Reproductive and Practice Committee of Society for Assisted Reproductive 2013). A complete three-generation pedigree will help determine if there are any known or suspected major Mendelian disorders or conditions with variable expressivity and/or incomplete penetrance, which may put donor-conceived offspring at an increased risk of being affected. If such a condition were identified, the recipient should be informed of the risks so that he/she can make an informed decision regarding use of that donor, or the applicant may be excluded from participating in the donor program. Donors and their first-degree relatives should be free from the following: major malformation with multifactorial or polygenic inheritance, known chromosome abnormalities, and mental retardation of unknown etiology. The ASRM defines major malformation as one that carries serious functional or cosmetic handicap.

According to ASRM guidelines, genetic screening should be performed on all donors, including carrier screening for cystic fibrosis (CF) and other genetic testing as indicated by the donor's ethnic background in accordance with current ASRM recommendations. Chromosome analysis is not required for all donors; it may be performed on all donors as a routine part of a program's screening process to rule out donors with chromosome abnormalities that could affect the success of fertility treatments or may be performed on a donor only if indicated by the donor's medical history.

Because carrier screening will most likely continue to evolve, it should be performed based on professional society recommendations at the time of donation. With the current use of sequencing technology, expanded carrier screening has brought both ease and challenges to the donor selection process.

Prior to the advent of expanded carrier screening, there were a limited number of diseases for which donors could be tested. Given the specific ASRM recommendation for CF screening, it was rare that a donor would not have been screened for the disorder; however, the extent and methodology of screening was variable across donor programs. Some centers would routinely test for spinal muscular atrophy and hemoglobinopathies for all donors, and fragile-X syndrome for oocyte donors. Additional testing may have been based on the donor's ethnicity, such as Tay–Sachs disease testing if the donor were of French Canadian or Cajun ancestry, or as a part of a panel of diseases that were more common in the Ashkenazi Jewish population. Historically, testing was performed by genotyping and, in the case of Tay–Sachs disease, by enzyme analysis. Centers varied in how they managed donors who were identified as carriers for autosomal recessive diseases, with some allowing the use of a donor who was a carrier as long as the recipient partner tested negative for the condition in question. Other centers would disqualify donors who were carriers because they believed that inadequate detection rates of genotyping resulted in unacceptable residual risk in some populations. Gene sequencing and deletion/duplication testing, which would have resulted in lower residual risk, were time-consuming and expensive.

Although not standard, it is now common for many donors to have expanded genetic screening via sequencing methodologies, with or without duplication/deletion analysis. Because it is estimated that an average individual may carry eight to 10 recessive gene mutations (Nussbaum et al. 2007), it is common that people who do expanded testing will learn that they are a carrier of at least one mutation for a disease on a panel. Unless a donor is a carrier of a recessive disease that might have clinical consequences in the heterozygous state (i.e., carriers of ataxia-telangiectasia are at an increased risk of developing cancer) or is an oocyte donor who carries an X-linked recessive disease mutation, most donors are no longer disqualified from donation based solely on carrier status, as gene sequencing now provides a high level of sensitivity to achieve the lowest residual risk for the reproductive partner. If a donor or recipient is a carrier of an autosomal recessive trait, it is important that both parties consider appropriate genetic screen-

ing and genetic counseling. Given the significant variability of disease panels within and between laboratories that perform expanded carrier screening, screening results must be carefully reviewed to make sure that appropriate testing has been performed on both the donor and recipient (and/or sperm/egg source).

Types of Gamete Donors

The majority of gamete donors are anonymous by intent, meaning that the recipient does not know the identity of the donor at the time of donor selection. Although anonymous at the time of donation, donors may agree to varying degrees of contact with donor-conceived individuals once they reach the age of majority. Despite a preference for anonymity by either the donor or the recipient, it is plausible that persons conceived from donor gametes will be able to identify their donors or biologically related individuals and learn that they were conceived via gamete donation. This means that a donor, who believed he or she could remain anonymous, could be identified. Because of the complex nature of gamete donation and advances in genetic medicine, donors and recipients must be aware of the genetic ties that bind them to each other.

Ovum and sperm donors may also be nonanonymous or directed donors. Nonanonymous ovum donors are women found through a donor agency that allows the recipient to know the donor's identity. Directed donors could be the recipient's family member (i.e., sister, brother, aunt, uncle, cousin, niece, or nephew) or friend.

Selecting which type of donor to use is a very personal decision. Use of a relative as a directed donor allows the recipient to have a genetic connection to the child (De Jonge and Barratt 2006). In some cases, an individual selected as a directed donor may be older than the recommended age limit for donation or may harbor a genetic risk that is typically excluded in an anonymous donor program. Recipients must be counseled on the increased risks, such as for aneuploidy in oocytes from women who are of advanced maternal age, and for the risk of new dominant mutations, autism, and schizophrenia in off-spring from men who are over the age of 40 (Brandt et al. 2019).

Genetic Issues following Gamete Donation

Changes to the health of donor-conceived off-spring, donors themselves, and donor family members can have implications for the health of biologically related individuals. Donors are often asked to notify their gamete donation facilities of changes in their personal or family medical histories. Likewise, when medical issues occur in pregnancies or individuals conceived from gamete donors, it is important that recipient families notify their fertility center and donor facility.

If any new information is believed to have significant implications for the health of related individuals and is clinically relevant or actionable, the gamete donor facility may contact those individuals to whom they believe the information is relevant. This allows the donors and recipients of the donor's gamete to discuss the information with their and their children's personal health-care providers, and to consider pursuing genetic counseling services or other follow-up as applicable to their personal needs and concerns.

SUMMARY AND FUTURE DIRECTIONS

The landscape of genetics is changing rapidly, particularly in the field of ART. Today, the aim of PGT is to select out embryos that are abnormal. Research is underway using gene editing technologies, such as CRISPR/Cas9, which in the future may allow embryos currently deemed abnormal to be used rather than being discarded or donated to research. Given the multitude of factors necessary for a successful IVF cycle, this will significantly impact the outcome for many patients.

As we identify more candidate genes associated with conditions with complex, multifactorial, and polygenic inheritance, the question of what is appropriate to test for will be a crucial one for IVF centers and patients. Thus, there is a greater need for pre- and posttest counseling in

order to fully understand the implications and interpretation of these results. Given the number of genetic issues that may arise in a fertility setting, the expertise of a genetic counselor may prove particularly beneficial for ART clinics.

ACKNOWLEDGMENTS

The authors thank Justine Witzke, PhD, MPH, CCRP, Alexandra MacWade, MFA, and Pamela Callum, MS, for their editorial comments.

REFERENCES

Alukal JP, Lamb DJ. 2008. Intracytoplasmic sperm injection (ICSI)—what are the risks? *Urol Clin North Am* **35:** 277–288, ix–x. doi:10.1016/j.ucl.2008.01.004

Armstrong S, Akande V. 2013. What is the best treatment option for infertile women aged 40 and over? *J Assist Reprod Genet* **30:** 667–671. doi:10.1007/s10815-013-9980-6

Bodurtha J, Strauss JF III. 2012. Genomics and perinatal care. *N Engl J Med* **366:** 64–73. doi:10.1056/NEJMra1105043

Botting B, Dunnell K. 2000. Trends in fertility and contraception in the last quarter of the 20th century. *Popul Trends* 32–39.

Brancati F, Mingarelli R, Dallapiccola B. 2003. Recurrent triploidy of maternal origin. *Eur J Hum Genet* **11:** 972–974. doi:10.1038/sj.ejhg.5201076

Brandt JS, Cruz Ithier MA, Rosen T, Ashkinadze E. 2019. Advanced paternal age, infertility, and reproductive risks: a review of the literature. *Prenat Diagn* **39:** 81–87. doi:10.1002/pd.5402

Brezina PR, Kutteh WH, Bailey AP, Ke RW. 2016. Preimplantation genetic screening (PGS) is an excellent tool, but not perfect: a guide to counseling patients considering PGS. *Fertil Steril* **105:** 49–50. doi:10.1016/j.fertnstert.2015.09.042

Centers for Disease Control and Prevention ASfRM, Society for Assisted Reproductive Technology. 2018. *2016 Assisted reproductive technology fertility clinic success rates report.* U.S. Department of Health and Human Services, Atlanta.

Craven L, Tang MX, Gorman GS, De Sutter P, Heindryckx B. 2017. Novel reproductive technologies to prevent mitochondrial disease. *Hum Reprod Update* **23:** 501–519. doi:10.1093/humupd/dmx018

Dahdouh EM, Balayla J, García-Velasco JA. 2015. Comprehensive chromosome screening improves embryo selection: a meta-analysis. *Fertil Steril* **104:** 1503–1512. doi:10.1016/j.fertnstert.2015.08.038

De Jonge C, Barratt CL. 2006. Gamete donation: a question of anonymity. *Fertil Steril* **85:** 500–501. doi:10.1016/j.fertnstert.2005.07.1304

Delhanty JD, Griffin DK, Handyside AH, Harper J, Atkinson GH, Pieters MH, Winston RM. 1993. Detection of aneuploidy and chromosomal mosaicism in human embryos during preimplantation sex determination by fluorescent *in situ* hybridisation (FISH). *Hum Mol Genet* **2**: 1183–1185. doi:10.1093/hmg/2.8.1183

Devers PL, Cronister A, Ormond KE, Facio F, Brasington CK, Flodman P. 2013. Noninvasive prenatal testing/noninvasive prenatal diagnosis: the position of the National Society of Genetic Counselors. *J Genet Couns* **22**: 291–295. doi:10.1007/s10897-012-9564-0

Diot A, Dombi E, Lodge T, Liao C, Morten K, Carver J, Wells D, Child T, Johnston IG, Williams S, et al. 2016. Modulating mitochondrial quality in disease transmission: towards enabling mitochondrial DNA disease carriers to have healthy children. *Biochem Soc Trans* **44**: 1091–1100. doi:10.1042/BST20160095

Filges I, Manokhina I, Peñaherrera MS, McFadden DE, Louie K, Nosova E, Friedman JM, Robinson WP. 2015. Recurrent triploidy due to a failure to complete maternal meiosis II: whole-exome sequencing reveals candidate variants. *Mol Hum Reprod* **21**: 339–346. doi:10.1093/molehr/gau112

Fishel S. 2018. First in vitro fertilization baby—this is how it happened. *Fertil Steril* **110**: 5–11. doi:10.1016/j.fertnstert.2018.03.008

Forabosco A, Percesepe A, Santucci S. 2009. Incidence of non-age-dependent chromosomal abnormalities: a population-based study on 88965 amniocenteses. *Eur J Hum Genet* **17**: 897–903. doi:10.1038/ejhg.2008.265

Fragouli E, Alfarawati S, Daphnis DD, Goodall NN, Mania A, Griffiths T, Gordon A, Wells D. 2011. Cytogenetic analysis of human blastocysts with the use of FISH, CGH and aCGH: scientific data and technical evaluation. *Hum Reprod* **26**: 480–490. doi:10.1093/humrep/deq344

Fragouli E, Alfarawati S, Spath K, Babariya D, Tarozzi N, Borini A, Wells D. 2017. Analysis of implantation and ongoing pregnancy rates following the transfer of mosaic diploid-aneuploid blastocysts. *Hum Genet* **136**: 805–819. doi:10.1007/s00439-017-1797-4

Friedenthal J, Maxwell SM, Munné S, Kramer Y, McCulloh DH, McCaffrey C, Grifo JA. 2018. Next generation sequencing for preimplantation genetic screening improves pregnancy outcomes compared with array comparative genomic hybridization in single thawed euploid embryo transfer cycles. *Fertil Steril* **109**: 627–632. doi:10.1016/j.fertnstert.2017.12.017

Greco E, Minasi MG, Fiorentino F. 2015. Healthy babies after intrauterine transfer of mosaic aneuploid blastocysts. *N Engl J Med* **373**: 2089–2090. doi:10.1056/NEJMc1500421

Hansen M, Bower C, Milne E, de Klerk N, Kurinczuk JJ. 2005. Assisted reproductive technologies and the risk of birth defects—a systematic review. *Hum Reprod* **20**: 328–338. doi:10.1093/humrep/deh593

Harton GL, Cinnioglu C, Fiorentino F. 2017. Current experience concerning mosaic embryos diagnosed during preimplantation genetic screening. *Fertil Steril* **107**: 1113–1119. doi:10.1016/j.fertnstert.2017.03.016

Hassold T, Abruzzo M, Adkins K, Griffin D, Merrill M, Millie E, Saker D, Shen J, Zaragoza M. 1996. Human aneuploidy: incidence, origin, and etiology. *Environ Mol Mutagen* **28**: 167–175. doi:10.1002/(SICI)1098-2280(1996)28:3<167::AID-EM2>3.0.CO;2-B

Heijligers M, van Montfoort A, Meijer-Hoogeveen M, Broekmans F, Bouman K, Homminga I, Dreesen J, Paulussen A, Engelen J, Coonen E, et al. 2018. Perinatal follow-up of children born after preimplantation genetic diagnosis between 1995 and 2014. *J Assist Reprod Genet* **35**: 1995–2002. doi:10.1007/s10815-018-1286-2

Huang B, Prensky L, Thangavelu M, Main D, Wang S. 2004. Three consecutive triploidy pregnancies in a woman: genetic predisposition? *Eur J Hum Genet* **12**: 985–986. doi:10.1038/sj.ejhg.5201274

Kang HJ, Melnick AP, Stewart JD, Xu K, Rosenwaks Z. 2016. Preimplantation genetic screening: who benefits? *Fertil Steril* **106**: 597–602. doi:10.1016/j.fertnstert.2016.04.027

Kimelman D, Confino R, Confino E, Shulman LP, Zhang JX, Pavone ME. 2018. Do patients who achieve pregnancy using IVF-PGS do the recommended genetic diagnostic testing in pregnancy? *J Assist Reprod Genet* **35**: 1881–1885. doi:10.1007/s10815-018-1289-z

Kurinczuk JJ, Hansen M, Bower C. 2004. The risk of birth defects in children born after assisted reproductive technologies. *Curr Opin Obstet Gynecol* **16**: 201–209. doi:10.1097/00001703-200406000-00002

Lanzendorf SE, Maloney MK, Veeck LL, Slusser J, Hodgen GD, Rosenwaks Z. 1988. A preclinical evaluation of pronuclear formation by microinjection of human spermatozoa into human oocytes. *Fertil Steril* **49**: 835–842. doi:10.1016/S0015-0282(16)59893-8

Liebaers I, Desmyttere S, Verpoest W, De Rycke M, Staessen C, Sermon K, Devroey P, Haentjens P, Bonduelle M. 2010. Report on a consecutive series of 581 children born after blastomere biopsy for preimplantation genetic diagnosis. *Hum Reprod* **25**: 275–282. doi:10.1093/humrep/dep298

Mastenbroek S, Twisk M, van Echten-Arends J, Sikkema-Raddatz B, Korevaar JC, Verhoeve HR, Vogel NE, Arts EG, de Vries JW, Bossuyt PM, et al. 2007. In vitro fertilization with preimplantation genetic screening. *N Engl J Med* **357**: 9–17. doi:10.1056/NEJMoa067744

Mastenbroek S, Twisk M, van der Veen F, Repping S. 2011. Preimplantation genetic screening: a systematic review and meta-analysis of RCTs. *Hum Reprod Update* **17**: 454–466. doi:10.1093/humupd/dmr003

Mertzanidou A, Wilton L, Cheng J, Spits C, Vanneste E, Moreau Y, Vermeesch JR, Sermon K. 2013. Microarray analysis reveals abnormal chromosomal complements in over 70% of 14 normally developing human embryos. *Hum Reprod* **28**: 256–264. doi:10.1093/humrep/des362

Mitalipov S, Amato P, Parry S, Falk MJ. 2014. Limitations of preimplantation genetic diagnosis for mitochondrial DNA diseases. *Cell Rep* **7**: 935–937. doi:10.1016/j.celrep.2014.05.004

Munné S, Lee A, Rosenwaks Z, Grifo J, Cohen J. 1993. Fertilization and early embryology: diagnosis of major chromosome aneuploidies in human preimplantation embryos. *Hum Reprod* **8**: 2185–2191. doi:10.1093/oxfordjournals.humrep.a138001

Munné S, Blazek J, Large M, Martinez-Ortiz PA, Nisson H, Liu E, Tarozzi N, Borini A, Becker A, Zhang J, et al. 2017. Detailed investigation into the cytogenetic constitution and pregnancy outcome of replacing mosaic blastocysts detected with the use of high-resolution next-generation sequencing. *Fertil Steril* **108**: 62–71.e8. doi:10.1016/j.fertnstert.2017.05.002

Nussbaum R, McInnes R, Williard H. 2007. *Thompson & Thompson genetics in medicine.* Saunders Elsevier, Philadelphia.

Pergament E, Confino E, Zhang JX, Roscetti L, Xien Chen P, Wellman D. 2000. Recurrent triploidy of maternal origin. *Prenat Diagn* **20:** 561–563. doi:10.1002/1097-0223(200007)20:7<561::AID-PD875>3.0.CO;2-1

Practice Committee of American Society for Reproductive Medicine; Practice Committee of Society for Assisted Reproductive Technology. 2008. Genetic considerations related to intracytoplasmic sperm injection (ICSI). *Fertil Steril* **90:** S182–S184.

Practice Committee of American Society for Reproductive Medicine; Practice Committee of Society for Assisted Reproductive Technology. 2013. Recommendations for gamete and embryo donation: a committee opinion. *Fertil Steril* **99:** 47–62.e1. doi:10.1016/j.fertnstert.2012.09.037

Practice Committees of American Society for Reproductive Medicine; Society for Assisted Reproductive Technology. 2013. Mature oocyte cryopreservation: a guideline. *Fertil Steril* **99:** 37–43. doi:10.1016/j.fertnstert.2012.09.028

Practice Committee of the American Society for Reproductive Medicine. 2015a. Diagnostic evaluation of the infertile female: a committee opinion. *Fertil Steril* **103:** e44–e50. doi:10.1016/j.fertnstert.2015.03.019

Practice Committee of the American Society for Reproductive Medicine. 2015b. Diagnostic evaluation of the infertile male: a committee opinion. *Fertil Steril* **103:** e18–e25.

Practice Committees of the American Society for Reproductive Medicine and the Society for Assisted Reproductive Technology. 2018. The use of preimplantation genetic testing for aneuploidy (PGT-A): a committee opinion. *Fertil Steril* **109:** 429–436. doi:10.1016/j.fertnstert.2018.01.002

Practice Committee of the Society for Assisted Reproductive Technology; Practice Committee of the American Society for Reproductive Medicine. 2008. Preimplantation genetic testing: a Practice Committee opinion. *Fertil Steril* **90:** S136–S143.

Rosenwaks Z, Handyside AH, Fiorentino F, Gleicher N, Paulson RJ, Schattman GL, Scott RT Jr, Summers MC, Treff NR, Xu K. 2018. The pros and cons of preimplantation genetic testing for aneuploidy: clinical and laboratory perspectives. *Fertil Steril* **110:** 353–361. doi:10.1016/j.fertnstert.2018.06.002

Sachdev NM, Maxwell SM, Besser AG, Grifo JA. 2017. Diagnosis and clinical management of embryonic mosaicism. *Fertil Steril* **107:** 6–11. doi:10.1016/j.fertnstert.2016.10.006

Scott RT Jr, Upham KM, Forman EJ, Zhao T, Treff NR. 2013. Cleavage-stage biopsy significantly impairs human embryonic implantation potential while blastocyst biopsy does not: a randomized and paired clinical trial. *Fertil Steril* **100:** 624–630. doi:10.1016/j.fertnstert.2013.04.039

Simioni M, Artiguenave F, Meyer V, Sgardioli IC, Viguetti-Campos NL, Lopes Monlleó I, Maciel-Guerra AT, Steiner CE, Gil-da-Silva-Lopes VL. 2017. Genomic investigation of balanced chromosomal rearrangements in patients with abnormal phenotypes. *Mol Syndromol* **8:** 187–194. doi:10.1159/000477084

Simon AL, Kiehl M, Fischer E, Proctor JG, Bush MR, Givens C, Rabinowitz M, Demko ZP. 2018. Pregnancy outcomes from more than 1,800 in vitro fertilization cycles with the use of 24-chromosome single-nucleotide polymorphism-based preimplantation genetic testing for aneuploidy. *Fertil Steril* **110:** 113–121. doi:10.1016/j.fertnstert.2018.03.026

Spinella F, Fiorentino F, Biricik A, Bono S, Ruberti A, Cotroneo E, Baldi M, Cursio E, Minasi MG, Greco E. 2018. Extent of chromosomal mosaicism influences the clinical outcome of in vitro fertilization treatments. *Fertil Steril* **109:** 77–83. doi:10.1016/j.fertnstert.2017.09.025

Steptoe PC, Edwards RG. 1978. Birth after the reimplantation of a human embryo. *Lancet* **312:** 366. doi:10.1016/S0140-6736(78)92957-4

Stern HJ, Harton GL, Sisson ME, Jones SL, Fallon LA, Thorsell LP, Getlinger ME, Black SH, Schulman JD. 2002. Non-disclosing preimplantation genetic diagnosis for Huntington disease. *Prenat Diagn* **22:** 503–507. doi:10.1002/pd.359

Tobler KJ, Brezina PR, Benner AT, Du L, Xu X, Kearns WG. 2014. Two different microarray technologies for preimplantation genetic diagnosis and screening, due to reciprocal translocation imbalances, demonstrate equivalent euploidy and clinical pregnancy rates. *J Assist Reprod Genet* **31:** 843–850. doi:10.1007/s10815-014-0230-3

Treff NR, Levy B, Su J, Northrop LE, Tao X, Scott RT Jr. 2010. SNP microarray-based 24 chromosome aneuploidy screening is significantly more consistent than FISH. *Mol Hum Reprod* **16:** 583–589. doi:10.1093/molehr/gaq039

Treff NR, Tao X, Schillings WJ, Bergh PA, Scott RT Jr, Levy B. 2011. Use of single nucleotide polymorphism microarrays to distinguish between balanced and normal chromosomes in embryos from a translocation carrier. *Fertil Steril* **96:** e58–e65. doi:10.1016/j.fertnstert.2011.04.038

Tur-Kaspa I, Jeelani R. 2015. Clinical guidelines for IVF with PGD for HLA matching. *Reprod Biomed Online* **30:** 115–119. doi:10.1016/j.rbmo.2014.10.007

Verlinsky Y, Cieslak J, Freidine M, Ivakhnenko V, Wolf G, Kovalinskaya L, White M, Lifchez A, Kaplan B, Moise J, et al. 1995. Pregnancies following pre-conception diagnosis of common aneuploidies by fluorescent in-situ hybridization. *Hum Reprod* **10:** 1923–1927. doi:10.1093/oxfordjournals.humrep.a136207

Wolf DP, Mitalipov N, Mitalipov S. 2015. Mitochondrial replacement therapy in reproductive medicine. *Trends Mol Med* **21:** 68–76. doi:10.1016/j.molmed.2014.12.001

Discouraging Elective Genetic Testing of Minors: A Norm under Siege in a New Era of Genomic Medicine

Laura Hercher

Sarah Lawrence College, Joan H. Marks Graduate Program in Human Genetics, Bronxville, New York 10708, USA

Correspondence: lhercher@sarahlawrence.edu

Consistently, the field of genetic counseling has advocated that parents be advised to defer elective genetic testing of minors until adulthood to prevent a range of potential harms, including stigma, discrimination, and the loss of the child's ability to decide for him- or herself as an adult. However, consensus around the policy of "defer-when-possible" obscures the extent to which this norm is currently under siege. Increasingly, routine use of full or partial genome sequencing challenges our ability to control what is discovered in childhood or, when applied in a prenatal context, even before birth. The expansion of consumer-initiated genetic testing services challenges our ability to restrict what is available to minors. As the barriers to access crumble, medical professionals should proceed with caution, bearing in mind potential risks and continuing to assess the impact of genetic testing on this vulnerable population.

For decades, the field of genetic counseling has articulated a single, consistent position on the genetic testing of minors: When possible, counselors should advise parents to defer genetic testing until adulthood. The phrase "when possible" functions as a caveat for medical necessity; genetic testing with an immediate medical purpose is noncontroversial. The rationale for discouraging elective genetic testing of minors was first described in the early 1990s and has been reaffirmed in more than 25 subsequent guidelines and position statements (Borry et al. 2006). In the year 1995 alone, consensus statements to that effect were adopted by six organizations of medical professionals, including the American Medical Association (AMA Council on Ethical and Judicial Affairs 1995), the American Society of Human Genetics (ASHG), and the American College of Medical Genetics (ACMG) (The American Society of Human Genetics Board of Directors and The American College of Medical Genetics Board of Directors Rapid 1995). The 2017 position statement of the National Society of Genetic Counselors (NSGC) echoes the language of its predecessors (the Public Policy Committee of NSGC 2017):

> The National Society of Genetic Counselors ... encourages deferring predictive genetic testing of minors for adult-onset conditions when results will not impact childhood medical management

or significantly benefit the child. Predictive testing should optimally be deferred until the individual has the capacity to weigh the associated risks, benefits, and limitations of this information, taking his/her circumstances, preferences, and beliefs into account to preserve his/her autonomy and right to an open future.

Today's genetic counselors are likely to be taught that the field discourages elective testing of minors as students. Bonnie Jeanne Baty, writing in *A Guide to Genetic Counseling*, strikes a note of concern about the motivation of parents who seek elective genetic testing for their children: "adults may coerce children into testing because of adult needs rather than appropriateness of testing for the child" (Uhlmann et al. 2009). Baty summarizes a representative ethical debate on the subject thusly: "they still recommend that the outcome (of the counseling process) preserve future autonomy by denying childhood testing… ." Similarly, Dawn Allain, writing in Ethical Dilemmas in Genetics and Genetic Counseling, describes an historical consensus that "unless medical intervention could benefit the minor, testing should be deferred until he or she reaches adult status" (Berliner 2015).

However, the consistent invocation of the "defer-when-possible" mantra masks the way in which this norm is under siege. To an extent, this reflects greater experience: A handful of early studies have tempered some of our worst fears about people's ability to handle the results of predictive genetic testing, although few of these studies examined children or adolescents (Green et al. 2009; Roberts et al. 2011). At the same time, the increasingly routine use of sequencing, which looks at great swaths of the genome rather than specific targets, challenges our ability to control what is discovered in childhood. Cell-free fetal DNA testing and other new developments in prenatal testing challenge our ability to control what is known about a child before he or she is born. Finally, the explosion of companies marketing genetic testing services outside of traditional clinical settings challenges our ability to draw an enforceable distinction between what is available to minors and what is available to adults.

THE ARGUMENT AGAINST ELECTIVE GENETIC TESTING OF MINORS

The ACMG/ASHG report "Points to Consider: Ethical, Legal and Psychosocial Implications of Genetic Testing in Children and Adolescents," published in 1995, lays out a medical and ethical basis for the recommendation to defer testing (The American Society of Human Genetics Board of Directors and The American College of Medical Genetics Board of Directors Rapid 1995). The standard used is the best interests of the child: "the primary goal of genetic testing should be to promote the well-being of the child." For the most part, "well-being" is defined as medical benefit, but the authors note that psychological benefits may be compelling as well, particularly in the case of a mature adolescent. And that benefit must be timely: "If the medical or psychosocial benefits of a genetic test will not accrue until adulthood, as in the case of carrier status or adult-onset diseases, genetic testing generally should be deferred."

The ACMG/ASHG paper identifies a number of potential harms that could result from the receipt of predictive genetic information at a young age. There is the possibility of discrimination by insurers, employers, teachers, or others. There is the question of stigma; the authors speculate that genetic information could alter a person's self-image or the expectations of those close to them. The authors suggest that genetic testing of minors may compromise their rights of autonomy and privacy. If tested as children, individuals lose the opportunity to decide for themselves as adults what they wish to know. They also lose the ability to control who has access to that information. Genetic testing is a lifelong commitment because what is known cannot be unknown; it is equally true that what is divulged cannot be undivulged.

Although the arguments against genetic testing of minors typically focus on adult-onset conditions, the emphasis on deferral of testing extends to childhood-onset conditions as well, although with less vehemence and more room for discretion. Pediatric cancer specialist Eric Kodish proposed the influential "rule of earliest onset" in 1999: Genetic testing should not occur

Cite this article as *Cold Spring Harb Perspect Med* doi: 10.1101/cshperspect.a036657

before the age of earliest onset for the condition in question (Kodish 1999). This emphasis on immediate benefit has remained a touchstone and influenced practice guidelines (Saelaert et al. 2018).

Although Kodish is more definitive in his rule-making than some of his contemporaries (Wertz et al. 1994; Ross 2001), there is general agreement that parents who request genetic testing for their at-risk children before the age of earliest onset may well be acting on their own behalf and not in the best interests of their child (Wertz et al. 1994; Kodish 1999; Botkin et al. 2015; Lim et al. 2017). Wertz states tartly, "If no clear benefits exist, parents should restrain their desire to know" (Wertz et al. 1994).

Inherently, the push to defer decisions about testing to adulthood encompasses an unusual level of skepticism about the parents' ability or even their willingness to act in their child's best interests. Notably, the 1995 ACMG/ASHG guidelines place the burden of determining the minor's best interest on the providers and not the parents, directing providers to defer the choice to families or competent adolescents only "when the balance of benefits and harms is uncertain." This recommendation, a striking departure from the routine assumption that parents are the best arbiters of their child's interests, goes beyond the suggestion that providers discourage requests for elective genetic testing and advocates for them to refuse these requests if they disapprove: "A health care provider has no obligation to provide a medical service for a child or adolescent that is not in the best interest of the child or adolescent" (The American Society of Human Genetics Board of Directors and The American College of Medical Genetics Board of Directors Rapid 1995).

Seen in this light, the emphasis on deferring genetic testing of minors when possible is not simply a cautious default position—although it is that—but an assertion of a right so fundamental that it trumps the routine prerogatives of parents and guardians. The human right on which this argument rests is autonomy—specifically, the right of individuals to determine for themselves when and what genetic information they choose to receive: the right to know, but

also (as it is called) "the right not to know." This claim in turn engenders what the 2017 NSGC position statement refers to as "the right to an open future." As described in 2019 by Garrett et al., philosopher Joel Feinberg first used the phrase "open future" to define a category of rights that children have but cannot exercise as minors. Parents and guardians, Feinberg argues, should not be allowed to act in such a way that their children's ability to make these choices is lost before it is gained (Garrett et al. 2019a). In genetics, the "right to an open future" guarantees that a child's right to make decisions about genetic testing for him or herself is not forestalled.

THE RIGHT NOT TO KNOW

The right to know and the right not to know are two sides of the same autonomy coin, so it is interesting and significant and perhaps unexpected that bioethicists and genetic clinicians have, historically speaking, chosen to emphasize the latter (Laurie 2014). Why? Why did UNESCO, in its 1998 Declaration on the Human Genome and Human Rights, feel it was necessary to stipulate "the right of every individual to decide whether or not to be informed of the results of genetic examination…" (UNESCO 1998)? Philosophically speaking, we may agree that people are free to remain uninformed as to the results of their own medical tests, but it hardly bears mentioning as a right. We do not debate whether or not people are adequately consented before being confronted with the results of their cholesterol tests or bone density scans or other common prognosticators of morbidity and mortality.

The significance of the right not to know in genetics hinges on the idea that because of the unique nature of genetic information, there are things a reasonable person might not wish to know. This was anticipated by ethicists and moral philosophers before genetic testing was widely available (Laurie 1999; Takala 1999; Wertz et al. 2003; Andorno and Laurie 2004), and the concern has been borne out in certain real-life scenarios in which the demand for testing has failed to meet expectations. An oft-cited example of the phenomenon is predictive testing for Huntington's disease, in which surveys taken before a predictive test was available showed a high degree of interest among at-risk individuals, with much lower uptake once it was an existing rather than hypothetical alternative (Quaid and Morris 1993; Meiser and Dunn 2000). That said, Huntington's disease may not be the best model for predictive genetic testing overall, because as a lethal, adult-onset condition with essentially 100% penetrance and (to date) no way to treat or prevent it, it is the exception and not the rule.

But even in more typical scenarios, when genetics is a contributing and not determinative factor, our experience to date suggests that many people, adequately advised, choose not to know. They may be concerned that the results of testing will influence how other people view them or even how they view themselves. The theory of "genetic essentialism" posits that people will overvalue information on genetic attribution and fail to contextualize the genetic contribution to traits or risks even when environmental or other nongenetic factors are paramount (Dar-Nimrod and Heine 2011). A number of studies appear to bear this out (Dar-Nimrod et al. 2013; Lineweaver et al. 2014), including a recent report demonstrating that participants given sham genetic test results showed physiological changes in exercise capacity and appetite in line with what they believed to be their genetic predisposition (Turnwald et al. 2019).

Studies show that some people choose not to test based on concerns about privacy and/or genetic discrimination (Harris et al. 2005; Rothstein 2008). Although to date evidence for genetic discrimination among employers and insurers is thin on the ground (Rothstein and Anderlik 2001; Green et al. 2015), it does not change the fact that significant numbers of people have determined that genetic testing is not worth the perceived risk. Because circumstances change but genetic information does not, it is impossible to prove that their concerns for the future are unfounded.

What is true with regard to adults is even truer with regard to children. Adults may rationally decide they do not want to know about

unavoidable bad things that will happen in the future; for a child, those same risks are that much farther away and harder to comprehend. As adults, they may have access to new information that makes testing more or less desirable. In addition, concerns have been voiced about whether or not minors are particularly vulnerable in terms of how the information affects their developing self-image, and whether or not they will be treated differently by their parents and others.

PRENATAL TESTING

Any concerns about genetic testing of minors—stigma, discrimination, loss of autonomy—are equally relevant to the question of prenatal genetic testing, because there is no mechanism by which information disclosed during pregnancy can be expunged at birth. However, practice in the prenatal setting is complicated by the possibility of elective termination.

Although it is possible that the best interests of the future child may be impinged upon by tests done before birth, it is unequivocally true that restricting prenatal testing denies parents the right to make decisions about whether or not to continue a pregnancy based on the best available information. Guidelines and positions statements from NSGC, ACMG, ASHG, and the American College of Obstetricians and Gynecologists (ACOG) all strongly advise that genetic counseling for parents considering prenatal genetic testing for adult disease include the potential negative implications of testing for themselves and their future child, should they choose to continue the pregnancy (American College of Obstetricians and Gynecologists 2008; Botkin et al. 2015; the Public Policy Committee of NSGC 2017). However, none of the organizations suggest that parents be denied the opportunity to test. The NSGC position statement published in 2017 states, "NSGC does not recommend prenatal genetic testing for known adult-onset conditions *if pregnancy or childhood management will not be affected*" (emphasis added here, not in original) (NSGC 2018).

In fact, there is always the possibility that pregnancy management will be affected, and it is not possible, or even rational, to suggest that expectant parents may know in advance how they will react to a positive test. Genetic counselors may discourage expectant parents from testing if they do not plan to abort an affected fetus, but restrictions on testing in the prenatal setting compromise the parents' reproductive rights. Therefore, tests that would not be offered to parents in the pediatric setting are available prenatally.

THE RECOMMENDATION TO DEFER: OTHER CAVEATS AND EXCEPTIONS

Despite widespread support, in principle, for discouraging elective genetic testing of minors, there are many agreed-upon exceptions to the rule. A quick overview of guidelines from 1995 to the present shows a steady growth of context-dependent caveats and exemptions. The ASHG/ACMG guidelines written back in 1995 essentially divide genetic tests into two monolithic categories: those that do and those that do not provide a benefit during childhood. Carrier testing and adult-onset testing are mentioned as examples of the latter, but without any substantive exploration of the circumstances that make each of these settings unique (The American Society of Human Genetics Board of Directors and The American College of Medical Genetics Board of Directors Rapid 1995).

By contrast, a 2001 position paper on genetic testing from the American Association of Pediatrics (AAP 2001) goes into considerable detail on the ethical issues associated with carrier screening and predictive testing for adult-onset disorders, as well as newborn screening programs. Both carrier screening and predictive testing for adult-onset disease are identified as tests without immediate medical value for the child, and testing for these purposes is not recommended. In the case of adult-onset disease testing, pediatricians are advised to decline requests for testing from parents or guardians; whereas with carrier screening, there is a more nuanced recommendation that results obtained incidentally (e.g., through newborn screening) should be conveyed to parents, but that screening should not be instituted for the sole purpose of determining carrier status. In this distinction

the subtleties are decisive; choosing not to get the information is quite different from choosing not to disclose the information.

Since those practice guidelines were published in 2001, genetic testing has grown rapidly, with a greater variety of tests being used in a greater variety of settings. The 2013 paper published jointly by the AAP and the ACMG reviews not only newborn screening, carrier screening, and predictive testing, but pharmacogenetic testing, histocompatibility testing, direct-to-consumer testing, and testing during the adoption process (Ross et al. 2013). In each case, the principles are the same, but context shapes the recommendations. With histocompatibility testing, for example, the AAP/ACMG guidelines suggest that the benefit to immediate family members creates such a conflict of interest that "a donor advocate or similar mechanism should be in place from the outset to avoid coercion and safeguard the interests of the child." In the case of adoption, the authors speculate that the best interests of an adoptable child with a known genetic risk may require that he or she is placed with "a family capable of and willing to accept the child's potential medical and developmental challenges… ." For this reason, the guidelines stipulate that predictive testing of a minor up for adoption may be ethically permissible. In one circumstance, the parents themselves cannot order testing without the oversight of a third party, whereas in the other, testing can be requested on behalf of prospective parents with no legal relationship at all.

Beyond what is reflected in guidelines and position statements, an increasing number of voices have resisted the idea that deferring elective genetic testing of minors is the preferable outcome and that medical providers are the best arbiters of the question rather than the families themselves (Wakefield et al. 2016; Lim et al. 2017; Garrett et al. 2019). In a 2016 commentary, Barbara Biesecker argues that families should be empowered to make their own decision about predictive genetic testing of minor children through a dynamic genetic counseling process that encourages them to consider the full range of potential outcomes in the context of their individual values, beliefs, and circumstances. "Families should be encouraged to discuss the predictive testing of minors in genetic counseling and, after collaborative deliberation, be offered testing if that is the preferred outcome" (Biesecker 2016).

What this evolution of recommendations demonstrates is the importance of context in determining best practices in the genetic testing of minors. It is not just a question of who is being tested, but of how, when, where, and why the tests are being used. In this context-dependent analysis, the caveat of "when possible" in "defer testing when possible" has come to encompass a wider definition of value.

ASSESSING VALUE

Medical value—the ability to diagnose, treat, or prevent illness—is the most straightforward to assess and the least controversial, but there is also potential psychological value to performing genetic testing in minors. Testing may alleviate anxiety or allow a presymptomatic or at-risk individual (and their family) the opportunity to prepare. It can also inform life choices, such as financial planning or career decisions. Having more information about likely or potential future medical needs may be a factor in where a family lives or what home they choose. Exposure over time to the myriad ways in which genetic risk affects families has expanded our awareness of the complex dynamics involved in determining whether or not a test is useful.

Another way in which value can be assessed narrowly or widely is in how closely we hew to the idea that the sole beneficiary must be the minor child. Parental anxiety is a variable that merits consideration. Although genetic counselors have been advised not to view parental anxiety as a justification for testing (NSGC 2018), it may not help the child to ignore the needs of the parent. As Lainie Ross Friedman has pointed out, children's well-being is to some extent indistinguishable from that of their family (Ross 2002). However, if parental needs alone are the deciding factor in their desire to test, it is a weak case for testing the child and a strong argument for genetic counseling to explore their motivation and expectations.

It is an inherent complication of genomic medicine that the boundaries between patients and their families are porous. Genetic test results in minors may have implications for their relatives. Genetic testing of a child may inform reproductive decision-making for the parents or other family members. Parents may well argue that the loss of autonomy for a minor in these circumstances is less important than the value added for the family as a whole. Genetic tests of minors for late-onset disease may have immediate medical utility not for minors themselves but for their relatives. Whether or not this is a benefit depends on how the relative in question feels about receiving the information. With the introduction of exome sequencing into clinical practice, this once rare hypothetical situation has become an everyday question for genetic clinicians and is best discussed in the larger context of how exome and genome sequencing of minors alters the dynamics of practice.

EXOME AND GENOME SEQUENCING OF MINORS

One of the major drivers in the extraordinary growth in clinical genetic sequencing since 2009 is its use in the pediatric setting, in which exome testing has become a first-line approach for children with suspected genetic disease and/or developmental delay (Meng et al. 2017; Stark et al. 2017; Clark et al. 2018). Full or partial genome sequencing generates unprecedented amounts of data, raising issues of what to examine and what to report. Although some have argued that we are not compelled to analyze the excess information embedded in a clinical exome, just as we are not compelled to run every available blood test every time we phlebotomize a patient (Gliwa and Berkman 2013; Klitzman et al. 2013), 10 years of fractious debate over what to report suggests there is a perceived difference between tests not ordered and test results left unexamined in a file.

If the patient is a minor, the situation is particularly fraught. Though default settings for what is examined may differ, untargeted testing reduces our ability to pick and choose between tests with immediate benefit and those without.

Cite this article as *Cold Spring Harb Perspect Med* doi: 10.1101/cshperspect.a036657

By design, clinical exome flags all variants with known medical implications. To opt out, we must actively seek to avoid getting the information, either by not interrogating those parts of the genome where it lies (if this is compatible with the medical purposes of the test) or by withholding the results, which, as we have seen in the earlier discussion of carrier screening, is an option that makes ethicists and clinicians uneasy.

Because having information in hand creates such a powerful incentive to disclose it, laboratories offering exome sequencing have chosen at times to eliminate the option. A 2012 version of Ambry's consent form for clinical exome sequencing gives adult patients the choice of whether or not to receive results in four categories: carrier disease status, later-onset disease risk, predisposition to increased risk for cancer, and early-onset disease risk not related to the reason for testing. For probands under 18, all four categories were blinded proactively. The test was configured not only to avoid passing along information to a clinician that normatively would not be shared with the parent or guardian, but even to avoid obtaining that information in the first place. This same assumption of the need for disclosure of any results motivated the AAP to advocate for return of carrier screening results to minors in cases in which the information was obtained incidentally, while discouraging its use as a rule (AAP 2001). Apparently, the risks and benefits of giving out secondary findings to minors are so finely balanced that ordering exome sequencing does not compel disclosure, and the labs and practitioners can safely ignore results that sit unexamined in a file, but having results—that is, having data that has been read by an informatics program—tips the scales.

In 2013, the ACMG published the first guidelines for the return of secondary or incidental findings in which, for a limited number of genes in which the health risk was well-established, the medical implications were serious and an intervention was available, the group stipulated that results should be both sought and disclosed in all instances of clinical sequencing regardless of patient preferences (Green et al.

2013). Although recognizing that their recommendations violated the norm of the "right not to know," the authors stated that "we felt clinicians and laboratory personnel have a fiduciary duty to prevent harm by warning patients and their families about certain incidental findings and that this principle supersedes concerns about autonomy... ."

Controversially, the ACMG guidelines explicitly included tests done on minors, despite the fact that the diseases associated with the genes on the curated panel were largely adult-onset. There were three parts to the rationale for this decision. First, that information on their future health could prove useful to minors. Second, that it would be unwieldy for laboratories to selectively reconfigure genetic sequencing reports for minors as the volume of testing increased. Third, that information found via testing of minors might be of immediate benefit to their adult family members.

The authors of the ACMG paper identify the potential to benefit relatives as an early-days effect, likely to disappear as genetic testing becomes more routine and widespread. Today, when clinical use of genetic testing is contingent on family or personal medical history, many carriers of risk-related variants will never be offered testing, and serendipity may be their only chance at finding out that they are at risk.

Despite these arguments, the recommendation to treat minors no differently than adults and the absence of a chance to opt out were identified as areas of concern by the ACMG membership, and in 2015 the college revised its guidelines to permit patients to opt out after counseling (Kalia et al. 2017), specifically noting that parents should be offered the opportunity to refuse analysis of genes unrelated to the indication for testing as a part of the informed consent process (Watson 2015).

This modification should not obscure the fact that the ACMG guidelines represent a significant shift in policy as regards genetic testing of minors. Although no longer mandatory, under these guidelines the option of testing minors for this limited set of adult-onset diseases is not only permissible but promoted as the default position. The 2013 ACMG working group acknowledged and even embraced this as a break with prior practice. Earlier recommendations, the ACMG working group writes, were "inconsistent with the general practice of respecting parental decision-making about their child's health …" (Green et al. 2013).

Although the ACMG guidelines refer to a limited, curated fraction of available information, the arguments they make about clinical utility and parental rights could be used to support broader disclosure of test results.

There are additional reasons why genetic counselors and laboratories might not wish to be the custodians of genetic information that does not get released to the family or the patient. There is liability associated with withholding information, if the results may have clinical value to family members now or the patient in the future. A plan that involves giving out the results at some later date assumes recontact, and neither the genetic counselor nor the family can reliably ensure that recontact occurs at the appropriate time, especially if that time is well in the future.

Further, patients do not have a legal right to demand all tests, but they do have a legal right to demand all test results. The Clinical Laboratory Improvement Amendments (CLIA) act of 1988 guaranteed patients (and their legal guardians) access to their medical records (1997), and an added rule passed in 2014 required laboratories to provide each person with their test results (HHS 2012). Under the "CLIA Program and Health Insurance Portability and Accountability (HIPAA) Privacy Rule: Patient Access to Test Reports," a motivated parent or guardian can obtain copies of any test report, even if it was not returned to the clinician. Does data need to be interpreted to be considered a result? The courts have yet to clarify that question (Guerrini et al. 2017), but what is clear is that exclusion violates the spirit of the rule change, which was intended to "provide individuals with a greater ability to access their health information, empowering them to take a more active role in managing their health and health care" (HHS 2012).

This push to "empower" people via their test results reflects a growing emphasis on data-driven and preventative care. These hopes lean heavily on genetics and genomics as a vehicle for predicting disease risk and personalizing care. Naturally, this vision is at odds with norms against genetic testing of minors. Although earlier positions presupposed that clinical utility was limited and circumscribed except when proven otherwise, the push to get data to people reflects a more optimistic view of the power of genomic information. At the same time, the logic of preventive medicine leads us to think about testing earlier, before conditions manifest, expanding the borderlines of "immediate" clinical utility. For medical problems like heart disease, diabetes, and obesity, it is hard to conceive of any programmatic intervention based on genetic risk assessment that is not optimized by starting in childhood. As a result, although it remains controversial, it is now commonplace for people to discuss sequencing infants at birth or even testing prenatally, and a number of NIH-funded research projects are examining the risks and benefits of each of these potentialities (Bianchi and Chiu 2018; Holm et al. 2018).

The temptation to test minors is likely to grow as we understand more about the health implications of genome-scale data. Clinical utility may increase rapidly or modestly, but the more we understand, the more we will have reason to use exome or genome sequencing, and the more likely it is that predictive information with and without clinical utility will be delivered as a package deal.

EXOME AND GENOME SEQUENCING IN THE PRENATAL SETTING

Prenatal screening today principally focuses on chromosome anomalies, although some forms of cell-free DNA testing provide information on microdeletions and copy number variants. However, proof-of-principle experiments have demonstrated that the technique can be used for anything up to and including genome sequencing (Klitzman et al. 2012; Mao et al. 2018). For a number of reasons, the move from targeted testing to untargeted testing would likely increase the number of children born with preexisting information on their genetic status.

Targeted testing is both specific to a known disease or disease risk and rare, both of which make it amenable to counseling. If a man carries a known *BRCA1* mutation and the couple wishes to test the fetus, the counselor can discuss the pros and cons with some degree of specificity. In many cases, the individual is already familiar with the condition because it runs in his or her family. But if the couple tests the fetus proactively for a catalog of pathogenic variants, practically, they cannot assess each one in advance to decide whether or not it would affect pregnancy management. One of the stated goals of genetic counseling for prenatal testing for adult-onset conditions is to steer parents away from testing if it would not change pregnancy management (NSGC 2018). Expanding the scope of what tests include will make this process less effective and push much of the decision-making to posttest counseling, when choosing not to have the information is no longer an option.

Untargeted tests can also be offered to a much wider audience than targeted testing, which was reserved for those at higher risk, and for that reason individual test takers have a lower prior probability of a positive result. Because there are more patients getting tested, it is harder to provide universal pretest counseling, and at the same time there is less motivation for it, because the average patient is more likely to come back with negative results.

Relying on genetic counseling to avoid return of results from prenatal testing that would not affect pregnancy management is a flawed strategy because pretest counseling is likely to be both less effective and more perfunctory as testing becomes increasingly routine. Realistically, if broader genetic testing becomes a standard part of prenatal care, for an increasing number of children, genetic information will be part of their medical record at birth.

DIRECT-TO-CONSUMER GENETIC TESTING AND MINORS

The use of direct-to-consumer genetic testing (DTC-GT) for minors is discouraged by organizations for genetic professionals (Caulfield et al. 2016). In fact, the potential to test minors is often cited as a part of the case against DTC-GT in general (Botkin et al. 2015). However, no laws or regulations limit what DTC-GT is allowed for minors. Companies can choose to put a minimum age in their terms of service, but practically speaking that is hard to enforce for an online business model. In fact, some DTC-GT, such as tests claiming to identify athleticism or other "innate" abilities, are marketed specifically to parents with children in mind.

The menu of what is available via DTC without involvement of a medical professional changes frequently, but multiple tests include genes for susceptibility to Alzheimer's disease, breast cancer, Parkinson's disease, and other late-onset conditions. This includes the health risk test from 23andMe, in which customer care offers directions for registering children under 13 (customercare.23andme.com/hc/en-us/articles/202904710-Account-options-for-families), and the company's founder has spoken publicly about her decision to test her own children (Rochman and Wojcicki 2012). Ethicists and clinicians may have their doubts about whether or not this information should be available for minors, but the DTC-GT industry clearly believes that this is a question the market will answer.

One potential regulator of DTC-GT is the Food and Drug Administration (FDA), which draws a distinction between tests for susceptibility, which it calls genetic health risk tests, and those that may be diagnostic, which are subject to greater scrutiny and restriction (2017). In 2017, the FDA articulated a policy of distinguishing between the two, labeling those without immediate medical value as genetic health risk tests, and those with immediate medical value as diagnostic, although FDA Commissioner Scott Gottlieb conceded that striking the right balance was a challenge (FDA News Release 2017). There is a clear logic in restricting the use of DTC-GT for tests with immediate medical implications, in which misapplied or inaccurate results might misinform medical care. But ironically, this policy permits companies to market with no review or oversight, exactly the sort of predictive genetic testing that genetics professionals identified in the past as particularly inappropriate for minors.

CONCLUSIONS

The genetic counseling community has taken a clear stand against elective genetic testing of minors, although the position has long been subject to qualifications. Even the definition of elective testing has been up for debate, as it is not always easy to draw a clean line between tests with and without immediate medical value. Genes are pleiotropic and do not divide neatly into categories, meaning that testing for one reason can produce secondary findings. Medical value can be defined narrowly to include only information that changes clinical management in a direct and literal sense, or it can be defined broadly to include other types of well-being. Prohibiting testing is meant to protect the autonomy of the child, but it can also limit the autonomy of the parent or guardian, who is by law and tradition the principal decision-maker when it comes to the best interests of their child. However, caveats and all, the idea that testing should be deferred when possible remains the norm in the field of clinical genetics.

However, although it may violate the beliefs and instincts of many genetic counselors, convergent trends suggest that genetic testing of minors may soon be commonplace. Rapid growth in the use of genome sequencing and better understanding of the role genetic variation plays in health and disease increase the number of health-related insights genetic testing might provide, making it harder to restrict genetic information as inappropriate for minors. Broad-based movements in support of preventive and personalized medicine favor the development of gene-based prevention strategies that are likely to lead us toward ever-earlier frontiers in genome sequencing, which not only includes but focuses on predictive information. Finally, with increased use of prenatal and direct-to-consumer testing, parents may relocate the question of what testing is permitted to a realm in which their authority is unchallenged.

There has never been a metric by which we could measure the amount of harm or benefit averted by deferring genetic testing to adulthood. Although it is clear the barrier to access is crumbling, it is not clear how much will be lost or gained as genetic testing of minors becomes commonplace. Medical professionals should proceed with some caution, bearing in mind potential risks and continuing to assess the impact of testing on this vulnerable population.

REFERENCES

AAP. 2001. Ethical issues with genetic testing in pediatrics. *Pediatrics* **107**: 1451–1455. doi:10.1542/peds.107.6.1451

AMA Council on Ethical and Judicial Affairs. 1995. Testing children for genetic status. In *Medical Ethics Reports*, Vol. 6. AMA, Chicago.

American College of Obstetricians and Gynecologists. 2008. ACOG committee opinion No. 410: ethical issues in genetic testing. *Obstet Gynecol* **111**: 1495–1502. doi:10.1097/AOG.0b013e31817d252f

Andorno R, Laurie G. 2004. The right not to know: an autonomy based approach. *J Med Ethics* **30**: 435–439.

Berliner J, ed. 2015. *Ethical dilemmas in genetics and genetic counseling.* Oxford University Press, New York

Bianchi DW, Chiu RWK. 2018. Sequencing of circulating cell-free DNA during pregnancy. *N Engl J Med* **379**: 464–473. doi:10.1056/NEJMra1705345

Biesecker BB. 2016. Predictive genetic testing of minors: evidence and experience with families. *Genet Med* **18**: 763–764. doi:10.1038/gim.2015.191

Borry P, Fryns JP, Schotsmans P, Dierickx K. 2006. Carrier testing in minors: a systematic review of guidelines and position papers. *Eur J Hum Genet* **14**: 133–138.

Botkin JR, Belmont JW, Berg JS, Berkman BE, Bombard Y, Holm IA, Levy HP, Ormond KE, Saal HM, Spinner NB, et al. 2015. Points to consider: ethical, legal, and psychosocial implications of genetic testing in children and adolescents. *Am J Hum Genet* **14**: 133–138.

Caulfield T, Borry P, Toews M, Elger BS, Greely HT, McGuire A. 2016. Marginally scientific? Genetic testing of children and adolescents for lifestyle and health promotion. *J Law Biosci* **2**: 627–644.

Clark MM, Stark Z, Farnaes L, Tan TY, White SM, Dimmock D, Kingsmore SF. 2018. Meta-analysis of the diagnostic and clinical utility of genome and exome sequencing and chromosomal microarray in children with suspected genetic diseases. *NPH Genom Med* **3**: 16. doi:10.1038/s41525-018-0053-8

Dar-Nimrod I, Heine SJ. 2011. Genetic essentialism: on the deceptive determinism of DNA. *Psychol Bull* **137**: 800–818. doi:10.1037/a0021860

Dar-Nimrod I, Zuckerman M, Duberstein PR. 2013. The effects of learning about one's own genetic susceptibility to alcoholism: a randomized experiment. *Genet Med* **15**: 132–138. doi:10.1038/gim.2012.111

FDA News Release. 2017. Press Announcements—Statement from FDA Commissioner Scott Gottlieb, M.D., on implementation of agency's streamlined development and review pathway for consumer tests that evaluate genetic health risks. www.fda.gov/NewsEvents/Newsroom/PressAnnouncements/ucm583885.htm

Garrett JR, Lantos JD, Biesecker LG, Childerhose JE, Chung WK, Holm IA, Koenig BA, McEwen JE, Wilfond BS, Brothers K, et al. 2019. Rethinking the "open future" argument against predictive genetic testing of children. *Genet Med* doi:10.1038/s41436-019-0483-4

Gliwa C, Berkman BE. 2013. Response to open peer commentaries on "Do researchers have an obligation to actively look for genetic incidental findings?" *Am J Bioeth* **13:** W10–W11. doi:10.1080/15265161.2013.781470

Green RC, Roberts JS, Cupples LA, Relkin NR, Whitehouse PJ, Brown T, Eckert SL, Butson M, Sadovnick AD, Quaid KA, et al. 2009. Disclosure of *APOE* genotype for risk of Alzheimer's disease. *N Engl J Med* **361:** 245–254. doi:10.1056/NEJMoa0809578

Green RC, Berg JS, Grody WW, Kalia SS, Korf BR, Martin CL, Mcguire AL, Nussbaum RL, Daniel JMO, Ormond KE, et al. 2013. American College of Medical Genetics and Genomics ACMG recommendations for reporting of incidental findings in clinical exome and genome sequencing. *Genet Med* **15:** 565–574. doi:10.1038/gim.2013.73

Green RC, Lautenbach D, McGuire AL. 2015. GINA, genetic discrimination, and genomic medicine. *N Engl J Med* **372:** 397–399. doi:10.1056/NEJMp1404776

Guerrini CJ, McGuire AL, Majumder MA. 2017. Myriad take two: can genomic databases remain secret? *Science* **356:** 586–587. doi:10.1126/science.aal3224

Harris M, Winship I, Spriggs M. 2005. Controversies and ethical issues in cancer-genetics clinics. *Lancet Oncol* **6:** 301–310. doi:10.1016/S1470-2045(05)70166-2

HHS. 2012. Part II Department of Health and Human Services SUPPLEMENTARY INFORMATION. *Fed Regist* **77:** 1–58.

Holm IA, Agrawal PB, Ceyhan-Birsoy O, Christensen KD, Fayer S, Frankel LA, Genetti CA, Krier JB, LaMay RC, Levy HL, et al. 2018. The BabySeq project: implementing genomic sequencing in newborns. *BMC Pediatr* **18:** 225. doi:10.1186/s12887-018-1200-1

Kalia SS, Adelman K, Bale SJ, Chung WK, Eng C, Evans JP, Herman GE, Hufnagel SB, Klein TE, Korf BR, et al. 2017. Recommendations for reporting of secondary findings in clinical exome and genome sequencing, 2016 update (ACMG SF v2.0): a policy statement of the American College of Medical Genetics and Genomics. *Genet Med* **19:** 249–255. doi:10.1038/gim.2016.190

Klitzman JO, Snyder MW, Ventura M, Lewis AP, Qiu R, Simmons LE, Gammill HS, Rubens CE, Santillan DA, Murray JC, et al. 2012. Noninvasive whole-genome sequencing of a human fetus. *Sci Transl Med* **4:** 137ra76.

Klitzman R, Appelbaum PS, Chung W. 2013. Return of secondary genomic findings vs patient autonomy: implications for medical care. *J Am Med Assoc* **310:** 369–370. doi:10.1001/jama.2013.41709

Kodish ED. 1999. Testing children for cancer genes: the rule of earliest onset. *J Pediatr* **135:** 390–395. doi:10.1016/S0022-3476(99)70142-3

Laurie G. 1999. In defence of ignorance: genetic information and the right not to know. *Eur J Heal Law* **6:** 119–132. doi:10.1163/15718099920522730

Laurie G. 2014. Recognizing the right not to know: conceptual, professional, and legal implications. *J Law Med Ethics* **42:** 53–63. doi:10.1111/jlme.12118

Lim Q, McGill BC, Quinn VF, Tucker KM, Mizrahi D, Patenaude AF, Warby M, Cohn RJ, Wakefield CE. 2017. Parents' attitudes toward genetic testing of children for health conditions: a systematic review. *Clin Genet* **92:** 569–578. doi:10.1111/cge.12989

Lineweaver TT, Bondi MW, Galasko D, Salmon DP. 2014. Effect of knowledge of APOE genotype on subjective and objective memory performance in healthy older adults. *Am J Psychiatry* **171:** 201–208. doi:10.1176/appi.ajp.2013.12121590

Mao Q, Chin R, Xie W, Deng Y, Zhang W, Xu H, Zhang RY, Shi Q, Peters EE, Gulbahce N, et al. 2018. Advanced whole-genome sequencing and analysis of fetal genomes from amniotic fluid. *Clin Chem* **64:** 715–725. doi:10.1373/clinchem.2017.281220

Meiser B, Dunn S. 2000. Psychological impact of genetic testing for Huntingtons disease: an update of the literature. *J Neurol Neurosurg Psychiatry* **69:** 574–578. doi:10.1136/jnnp.69.5.574

Meng L, Pammi M, Saronwala A, Magoulas P, Ghazi AR, Vetrini F, Zhang J, He W, Dharmadhikari AV, Qu C, et al. 2017. Use of exome sequencing for infants in intensive care units ascertainment of severe single-gene disorders and effect on medical management. *JAMA Pediatr* **171:** e173438. doi:10.1001/jamapediatrics.2017.3438

National Society of Genetic Counselors (NSGC). 2017. Genetic counseling of minors for adult-onset conditions. https://www.nsgc.org/p/bl/et/blogaid=860

National Society of Genetic Counelors (NSGC). 2018. Prenatal testing for adult-onset conditions. https://www.nsgc.org/p/bl/et/blogaid=1066

Quaid KA, Morris M. 1993. Reluctance to undergo predictive testing: the case of Huntington disease. *Am J Med Genet* **45:** 41–45. doi:10.1002/ajmg.1320450112

Roberts JS, Christensen KD, Green RC. 2011. Using Alzheimer's disease as a model for genetic risk disclosure: implications for personal genomics. *Clin Genet* **80:** 407–414. doi:10.1111/j.1399-0004.2011.01739.x

Rochman BB, Wojcicki A. 2012. *Test your DNA for diseases—no doctor required*, pp. 10–13. Center for Genetics and Society, Berkeley, CA.

Ross LF. 2001. Ethical and policy issues in genetic testing. *Pancreatology* **1:** 576–580. doi:10.1159/000055866

Ross LF. 2002. Predictive genetic testing for conditions that present in childhood. *Kennedy Inst Ethics J* **12:** 225–244. doi:10.1353/ken.2002.0019

Ross LF, Saal HM, David KL, Anderson RR. 2013. Technical report: ethical and policy issues in genetic testing and screening of children. *Genet Med* **15:** 234–245. doi:10.1038/gim.2012.176

Rothstein MA. 2008. Currents in contemporary ethics: GINA, the ADA, and genetic discrimination in employment. *J Law Med Ethics* **36:** 837–840. doi:10.1111/j.1748-720X.2008.00341.x

Rothstein MA, Anderlik MR. 2001. What is genetic discrimination, and when and how can it be prevented? *Genet Med* **3:** 354–358. doi:10.1097/00125817-200109000-00005

Saelaert M, Mertes H, De Baere E, Devisch I. 2018. Incidental or secondary findings: an integrative and patient-inclusive approach to the current debate. *Eur J Hum Genet* **26:** 1424–1431. doi:10.1038/s41431-018-0200-9

Stark Z, Schofield D, Alam K, Wilson W, Mupfeki N, Macciocca I, Shrestha R, White SM, Gaff C. 2017. Prospective comparison of the cost-effectiveness of clinical whole-exome sequencing with that of usual care overwhelmingly supports early use and reimbursement. *Genet Med* **19:** 867–874. doi:10.1038/gim.2016.221

Takala T. 1999. The right to genetic ignorance confirmed. *Bioethics* **13:** 288–293. doi:10.1111/1467-8519.00157

The American Society of Human Genetics Board of Directors and The American College of Medical Genetics Board of Directors Rapid. 1995. Points to consider: ethical, legal, and psychosocial implications of genetic testing in children and adolescents. *Am J Hum Genet* 57: 1233–1241.

Turnwald BP, Goyer JP, Boles DZ, Silder A, Delp SL, Crum AJ. 2019. Learning one's genetic risk changes physiology independent of actual genetic risk. *Nat Hum Behav* **3:** 48–56. doi:10.1038/s41562-018-0483-4

Uhlmann WR, Schuette J, Yashar B. 2009. *A guide to genetic counseling.* John Wiley & Sons, Hoboken, NJ

UNESCO. 1998. UNESCO: Universal Declaration on the Human Genome and Human Rights. *J Med Philos* **23:** 234–246.

Wakefield CE, Hanlon LV, Tucker KM, Patenaude AF, Signorelli C, McLoone JK, Cohn RJ. 2016. The psycholog-

ical impact of genetic information on children: a systematic review. *Genet Med* **18:** 755–762. doi:10.1038/gim.2015.181

Watson MS. 2015. ACMG policy statement: updated recommendations regarding analysis and reporting of secondary findings in clinical genome-scale sequencing. *Genet Med* **17:** 68–69. doi:10.1038/gim.2014.151

Wertz DC, Fanos JH, Reilly PR. 1994. Genetic testing for children and adolescents: who decides? *J Am Med Assoc* **272:** 875–881. doi:10.1001/jama.1994.03520110055029

Wertz DC, Fletcher JC, Berg K. 2003. *Review of ethical issues in medical genetics. Report of Consultants to WHO.* WHO, Geneva. World Health Organization Human Genetetics Program: www.who.int

1997. Clinical Laboratory Improvement Amendments of 1988. Note that this document is provided as a convenience and includes excerpts from the full CLIA '88 Law to demonstrate the changes. This document does not represent the full context or intent of the law. **123:** 2324.

2017. Federal Register: Medical Devices; Immunology and Microbiology Devices; Classification of the Genetic Health Risk Assessment System. *Fed Regist* federalregister.gov/d/2017-24159. www.federalregister.gov/documents/2017/11/07/2017-24159/medical-devices-immunology-and-microbiology-devices-classification-of-the-genetic-health-risk

Legal Challenges in Genetics, Including Duty to Warn and Genetic Discrimination

Sonia Suter

George Washington University, Washington, D.C. 20052, USA

Correspondence: ssuter@law.gwu.edu

This review will explore two legal issues in genetic counseling: genetic discrimination and the duty to warn. It emphasizes the complexity and variability of federal and state genetic non-discrimination protections in the United States and how the many gaps in such protections may affect people pursuing genetic testing. The limited law addressing legal obligations genetic counselors owe at-risk relatives likely does not require counselors to warn relatives directly about genetic risks. Whether it permits them to make such disclosures, however, is more uncertain and may depend on the jurisdiction.

LEGAL ISSUES IN GENETIC COUNSELING

Central to genetic counseling is the collection and sharing of medical and genetic information with clients. Genetic counselors, of course, do more than obtain and share information—they work with clients to help them make informed decisions based on their personal values and life plans. This involves discussions about clients' goals, medical and personal circumstances, family situation, reproductive plans, and, of course, the reason for genetic counseling in the first place. The genetic or genomic information at the center of genetic counseling, therefore, has as much psychological and social as medical value.

Genetic information has social significance in other ways. It can affect how third parties view and treat us. Entities like employers, insurers, mortgage companies, schools, and future part-ners may find our genetic information relevant, raising issues about who can access our genetic information and how they can use it. In addition, because we share many genetic variants with family members, genetic information that is personally and medically significant to us may be equally consequential to relatives. That raises questions about the legal obligations genetic counselors have toward a client's family members to warn them about genetic risks.

This review explores these two different legal issues, with a focus on the United States. The first section describes genetic nondiscrimination and privacy laws, which mirror laws in many other countries in both their reach and variety. It highlights the scope and limitation of these laws so that genetic counselors can understand how variable these protections are across jurisdictions. Although genetic counselors cannot become experts as to every legal nuance, they should understand the general legal landscape to help clients consider the implications of obtaining genetic information, as well as the optimal timing and context.

The second section explores cases, statutes, and regulatory law to consider what legal and ethical duties genetic counselors owe a client's relatives. Although the focus is on U.S. law, many of the same principles apply internationally, particularly in Europe. Although a genetic counselor's primary obligation is to her client, she owes certain ethical and legal obligations to the client's relatives. Although there is limited law in this area and some uncertainty about its reach, this section suggests genetic counselors likely have no duty to warn relatives directly. It might, however, be permissible for them to do so in very rare instances.

Genetic Discrimination—What Is Protected and What Is Not

This section lays out the areas where genetic nondiscrimination and privacy protections exist and where they do not. The goal is to give genetic counselors tools for helping clients understand the implications of undergoing genetic testing or counseling in light of the complex range of U.S. federal and state genetic nondiscrimination and privacy protections. Helping patients navigate this legal landscape is not easy. These legal protections vary considerably from state to state and depend on the context in which genetic information is obtained. Furthermore, there is a tension between the counselor's ethical obligation to ensure that clients understand the limits of privacy and nondiscrimination protections and the danger that overemphasizing these risks may prevent the client from undergoing genetic testing that could be beneficial to their health and well-being.

Worries about genetic discrimination arose not long after the Human Genome Project began when ethics and legal scholars sounded the alarm that employers and insurers might use genetic information to discriminate. Although the threat of genetic discrimination was mostly theoretical in the 1990s, state legislatures began to craft laws to prohibit genetic discrimination,

primarily in employment and health insurance. Combining antidiscrimination features with some genetic privacy features, the state laws varied in many respects: who was proscribed from certain uses of genetic information (antidiscrimination protections), the restrictions on third-party access to genetic information (privacy protections), and the definition of genetic information (Suter 2001).

For over a decade, Congress unsuccessfully attempted to pass legislation to address this patchwork of genetic nondiscrimination and privacy protections. After 13 years of failed attempts, it finally passed the Genetic Information Nondiscrimination Act ("GINA"),[1] which President Bush signed into law in 2008. GINA was not the first federal law prohibiting genetic discrimination. The Health Insurance Portability and Accountability Act had already prohibited genetic discrimination by employer-sponsored group health plans,[2] although these protections were not fully comprehensive and left the individual insurance market untouched.

GINA differs from other antidiscrimination laws in that it was a preemptive measure to prevent future discrimination, rather than a response to historical and pervasive discrimination (Roberts 2010). Congress was also motivated by a concern about the public health and research implications of the public's fear of genetic discrimination, regardless of the prevalence of such discrimination. Specifically, members worried these fears would prevent people from undergoing genetic testing or participating in research (Roberts 2010; Suter 2019). In addition, Congress wanted to remedy the "patchwork of State and Federal laws," which were "confusing and inadequate to protect [the public] from discrimination."[3] For all of these reasons, Congress enacted GINA to "fully protect the public from discrimination."[4] To that end, it defined genetic information more broadly than

---

[1]Public Law No. 110-233, 122 Stat. 881 (2008) (codified in scattered sections of 26, 29, and 42 U.S. Congress).
[2]29 U.S. Congress §§ 1181-82 (2006), 42 U.S. Congress §§ 3000 gg-41 (2006).
[3]Public Law No. 110-223, § 2(5).
[4]Id.

most state laws.[5] Congress also wanted to ensure a minimum level of protections across the United States by making GINA's provisions a floor below which no state could go (Roberts 2010).

GINA achieves its goal of nondiscrimination in health insurance and employment in two ways. First, it prohibits discriminatory *uses* of genetic information. Health insurers cannot make determinations of eligibility and premium rates or create preexisting condition exclusions based on genetic information with respect to group health plans and individual insurance markets.[6] Similarly, employers may not use genetic information to make employment decisions with respect to such matters as hiring, compensation, or conditions of employment.[7] Second, GINA bolsters the nondiscrimination features by prohibiting health insurers[8] (Sarata et al. 2011) and employers (with some exceptions)[9] from *accessing* genetic information through requests or inquiries (Roberts 2015). GINA, therefore, provides an element of federal privacy protection, at least with respect to health insurers and employers.

Although GINA represented progress in proscribing genetic discrimination, in several ways, GINA does not "fully protect the public from discrimination." First, although its definition of genetic information is broad in including family history (and an expansive conception of family history at that) (Suter 2019), its definition is narrow in another sense. GINA does not cover discriminatory provisions "based on the mani-

festation of a disease or disorder of an individual."[10] This is consistent with the general consensus that genetic discrimination is problematic when based on a predisposition to, as opposed to manifestation of, an illness (Hall 1999). Yet it is not clear why such a distinction is morally sound. Surely a genetic counseling client's need for health insurance is even greater once a genetic risk develops into disease (Suter 2001). Genetic counseling clients, and perhaps even some genetic counselors, however, may not understand that GINA provides no protections against discrimination based on illness.

Any federal protection against health insurance or employment discrimination based on manifested genetic conditions comes from other laws. For example, the significant health insurance reforms of the Patient Protection and Affordable Care Act of 2010 ("ACA")[11] protects against health insurance discrimination based on most health factors, including predisposition to disease, a protection that largely overlaps with GINA, and manifestation of a genetic condition, which goes beyond GINA (Sarata et al. 2011).

Although growing public support of the ACA quelled Congressional efforts to repeal the ACA (Everett 2018), the law is once again under attack in the courts. In 2018, 20 state governors and attorneys general challenged the statute's constitutionality, and a federal judge ruled that the ACA's individual mandate was unconstitutional, rendering the entire statute invalid.[12] Legal scholars strongly question the decision, which has been appealed to the Fifth Circuit. In a "highly unusual move," President Trump's Department of Defense not only failed to defend the federal law (Keith 2018), it asked the federal appeals court to invalidate it (Pear 2019). Unfortunately, recent oral arguments suggest the three-judge panel of the Fifth Circuit is skeptical about the ACA's constitutionality (Goodnough 2019). The case may ultimately end up before

---

[5]Public Law No. 110-233, §§ 101(d), 102(a)(1)(B), 103(d), 104(b), 201(4)(A)(i)–(iii); 42 U.S. Congress §2000ff(A)(i)–(iii) (defining family history as "the manifestation of a disease or disorder in family members"); GINA could have modified the definition as "the manifestation of an *inheritable* disease or disorder in family members," but it did not (Suter 2019).

[6]Public Law No. 110-223 § 102. GINA "fortified" existing protections from HIPAA with respect to group plans, but its "main value" was with respect to protections against discrimination in the individual insurance market (Rothstein 2009).

[7]Public Law No. 110-233 § 202(b).

[8]Public Law No. 110-233, §§101-106.

[9]Section 202(b); Genetic Information Nondiscrimination Act, Public Law No. 110-233, § 101(d), 112 Stat. 881, 884-5; § 102 (d)(2)(A), 112 Stat at 896; § 103(d), 122 Stat. at 898-00; § 104(b)(2), 122 Stat. at 901.

[10]Public Law No. 110-233, § 101(1)(3); 29 USCA. § 1182.

[11]Public Law No. 111-48, as modified by the Health Care and Education Reconciliation Act, Public Law No. 111-52.

[12]*Texas v. United States* (340 F. Supp. 579) (E.D. Ark. 2018).

the Supreme Court. Thus, the fate of the ACA, including its protections against preexisting condition exclusions, once again hangs in the balance. Such litigation creates uncertainty about the scope of protections against discrimination based on genetic illness.

In the employment context, The American with Disabilities Act ("ADA")[13] offers protections for people with manifested genetic conditions if they can show the employer's allegedly discriminatory action was based on the ADA's definition of a disability.[14] Unfortunately, some individuals will lie in the uncovered gap between GINA's and the ADA's employment nondiscrimination protections. For example, someone in the early stages of a disease would not be protected by GINA, but their mild symptoms or biomarkers might not meet the criteria for ADA protections (Rothstein 2018a). This presents genetic counseling challenges because it is difficult to determine whether someone has demonstrated enough signs of illness to be denied GINA's protections against employment discrimination, but not enough for the ADA's protections.

An additional vulnerability with respect to employment discrimination protections involves the regulation of employee wellness programs, which employers offer to encourage healthy behavior in employees to reduce health-care costs (Wieczner 2013). Although GINA prohibits employers from asking employees for genetic information, it allows them to collect such information as part of a wellness program if certain conditions are met. For example, the employee must give "knowing, voluntary, and written authorization," and the employer can only receive information in an aggregated and unidentifiable form (Sarata et al. 2011). In 2017, some Congressional Representatives introduced a bill to reduce employers' burdens in implementing wellness programs. One provision

would have allowed employers to collect identifiable medical history (including family history) and genetic tests,[15] which would have undone many of the privacy protections of GINA and the ADA. News of these efforts worried individuals considering genetic testing or counseling. Although the legislative attempts failed, they highlight GINA's vulnerability to political whims, which can impact public attitudes and fears regarding genetic testing.

Another significant inadequacy of GINA is that, in focusing only on health insurance and employment, it ignores areas in which genetic discrimination may be even more likely. For example, GINA does not address life, long-term care, and disability insurance. Similarly, GINA fails to protect against discriminatory uses of genetic information for mortgages or commercial transactions (Rothstein 2009), admission to schools, etc. Some states have enacted laws prohibiting discrimination in some of these areas, but their protections vary dramatically. The patchwork of legislation here is even sparser than the health insurance and employment laws that motivated GINA's passage.

To give an example, some state laws prohibit disability and life insurers from discriminating based on genetic information that is not actuarially justified. That means such insurers can underwrite insurance based on genetic information if evidence demonstrates it truly indicates an increased risk for a relevant disability or serious medical condition.[16] In other words, they can charge higher premiums or deny coverage based on genetic predispositions, precisely the type of discrimination people feared with advances in genetics.

Other state laws prohibit certain types of insurers from *requiring* genetic tests, but they do not prohibit the use of *existing* genetic information, for underwriting.[17] This does not pre-

---

[13]Public Law 101-336.

[14]42 U.S. Congress § 12101 (1) (defining a disability as an "impairment that substantially limits one or more major life activities," having a "record of such an impairment," or being "regarded as having such an impairment.").

---

[15]H.R. 1313, 115th Congress (2017).

[16]*See generally*, Genome Statute and Legislation Database Search. *National Human Genome Research Institute*. https:// www.genome.gov/policyethics/legdatabase/pubsearchresult .cfm?content_type=1&content_type_id=1&topic=4&topic_ id=1&source_id=1&keyword= &search=Search.

[17]*Id.*

Cite this article as *Cold Spring Harb Perspect Med* doi: 10.1101/cshperspect.a036665

vent genetic discrimination against clients who have undergone genetic testing, but it avoids their being coerced into testing simply to obtain insurance. With respect to other areas of potential discrimination, the protections are extremely limited or nonexistent. Genetic counseling clients should be aware of such genetic discrimination risks. The challenge, however, is that helping them understand when genetic discrimination is still possible may lead to fears that discourage their pursuit of genetic testing for medical purposes. Genetic counselors can help clients navigate these issues by encouraging them to consider the ideal time for testing. For example, some clients may benefit from seeking such insurance before pursuing genetic testing.

GINA and state genetic nondiscrimination laws also fail to protect against future areas of discrimination that may arise as technologies refine our ability to predict disease risks through indicators other than genetic variants. For example, these statutes do not address, let alone consider, the possibility of third parties using information about epigenetic changes or microbiomes to discriminate. As Professor Rothstein notes, "GINA has been frozen in time for at least ten years" (Rothstein 2009). These limitations will be increasingly relevant in genetic counseling as the scope of information patients obtain expands beyond analysis of genetic variants.

Finally, GINA provides only limited privacy protections. It restricts to some extent the information that health insurers and employers may request or require, but it does not address privacy protections in other realms. The most comprehensive law addressing genetic privacy is the Health Insurance Portability and Privacy Act (HIPAA) Privacy Rule, which protects "identifiable health information"[18] by prohibiting its disclosure without written authorization if such disclosure is not for treatment, payment, or health-care operations.[19] After GINA was enacted, the Privacy Rule was modified to expressly include genetic information within its

protected health information.[20] HIPAA's federal privacy protections, however, only apply to health-care providers, health plans, health clearing houses, and business associates of these entities.[21] Consequently, genetic privacy protections outside the health-care context depend on the patchwork of state legislation.

Not all states have genetic privacy laws, and those that do vary considerably. Some protect genetic privacy by requiring an individual's (written and/or informed) consent to perform a genetic test or DNA analysis, to obtain or retain an individual's genetic information, or to release genetic information to third parties. A few states define genetic information as the individual's property, although it is not precisely clear what added protections that label offers (Suter 2004). Sometimes these statutes are limited to specific entities such as insurers, employers, and/or schools. Most of these laws create exceptions for activities like diagnosis and treatment, newborn screening, forensics, etc.[22] Although genetic counselors cannot be expected to know the particular privacy protections of every jurisdiction, they can assure patients that information obtained through genetic counseling clinics or other health-care facilities is protected by HIPAA.

As more people seek genetic testing outside the medical context, however, the protections over genetic information will be more limited. GINA, HIPAA, and many state genetic privacy or nondiscrimination statutes simply do not apply in this context. For example, direct-to-con-

---

[18]45 C.F.R. § 160.103 (2014).
[19]45 C.F.R. § 164.508 (2013).

[20]Modifications to the HIPAA Privacy, Security, Enforcement, and Breach Notification Rules Under the Health Information Technology for Economic and Clinical Health Act and the Genetic Information Nondiscrimination Act; Other Modifications to the HIPAA rules, 78 Fed. Reg. 5566-01, 5661-62 (Jan. 25, 2013) (noting that prior to enactment of GINA, the Department of Health and Human Services had issued guidance that genetic information was considered protected health information under the Privacy Rule).

[21]45 C.F.R. § 160.103 (2018).

[22]See generally, Genome Statute and Legislation Database Search. National Human Genome Research Institute. https://www.genome.gov/policyethics/legdatabase/pubsearchresult.cfm?content_type=1&content_type_id=1&topic=4&topic_id=1&source_id=1&keyword=&search=Search.

sumer ("DTC") genetic testing companies like 23andMe are not among the four listed entities to which HIPAA applies. GINA is only relevant to DTC genetic testing to the extent it prohibits employers or health insurers from seeking genetic information, including information generated by DTC companies. One's control over genetic data generated through DTC companies largely depends on their varied practices and policies (Hazel and Sogbogan 2018). Some state laws might grant consumers limited control over this information, but most would not. Moreover, some elements of genetic privacy are virtually impossible to protect as the expansion of DTC genetic testing allows individuals to identify the presence or absence of genetic links in families (Holger 2018). Although such information may not reveal disease propensity, it highlights the limits of our control over our genetic information.

This brief survey of genetic privacy and nondiscrimination laws suggests that people contemplating genetic testing should understand the complexity, range, and limits of state and federal protections. Genetic counselors can play a role in emphasizing the variability in protections against discrimination outside of health insurance and employment. They can also indicate that, in many jurisdictions, protections against genetic discrimination in areas like long-term care, life, and disability insurance are limited or nonexistent. Genetic counseling clients should also understand that, once their genetic information is generated, their ability to control third-party access to it might be restricted. Although the greatest protections exist for information generated in the health-care context, they are minimal in the realm of DTC genetic testing.

Genetic counseling clients should also understand that entities other than health insurers and employers who might have an interest in one's genetic information—life insurers, long-term care insurers, schools, mortgage brokers, etc.—are legally entitled in most jurisdictions to ask individuals whether they or family members have received genetic testing or counseling, and if so what the results are. A failure to answer such questions truthfully would invalidate the terms of any insurance contract they obtain.[23] Clients should therefore consider not only the personal and medical value of obtaining genetic information, but also the optimal timing based on their plans and needs for other lines of insurance. In some jurisdictions, however, even if one planned to undergo genetic testing after seeking life, disability, or long-term care insurance, these insurers may be legally entitled to make genetic testing a precondition for obtaining insurance.

Asking genetic counselors to address, in general terms, the complicated nature of legal nondiscrimination protections demonstrates how elusive GINA's goal to eliminate public fears of genetic discrimination may have been. Congress hoped GINA would inspire more people to seek genetic counseling and testing and to participate in genetic research, but helping clients understand its limitations may not offer the sense of security GINA was intended to instill. Nevertheless, such information is necessary to assist clients in making decisions consistent with their values, goals, and life plans.

Even though it is appropriate for genetic counselors to highlight the general gaps and variation in federal and state genetic nondiscrimination and privacy laws, they cannot be expected to become experts about the nuanced variations in protections (or lack thereof) across jurisdictions and the range of contexts in which discrimination might occur. Genetic counselors are increasingly burdened with broadening disclosure obligations as genetic/genomic testing becomes more complex and generates ever larger amounts of information. Furthermore, as more people seek genetic information through DTC companies, we cannot rely solely on genetic counselors to educate the public about the inadequacies of genetic nondiscrimination and privacy protections. Instead, we must develop mechanisms to provide greater awareness of these issues. Not only would that help people

---

[23]Contract voidable for fraudulent misrepresentation, 6 Couch on Ins. § 82:21 ("where a representation is intentionally false, and calculated to mislead the insurer into issuing the policy, and is material, the policy may be avoided . . .").

Cite this article as *Cold Spring Harb Perspect Med* doi: 10.1101/cshperspect.a036665

make more informed choices, it would also engage the public in a larger discussion of the proper reach of genetic nondiscrimination laws.

GINA's shortcomings reflect, to a large degree, the lack of political will to go further (Ray 2010; Zhang 2017). Greater public awareness could change the political discourse and perhaps motivate efforts to expand protections against discriminatory uses of genetic and genomic information. As it becomes more difficult to control access to genetic information, privacy protections may prove too weak to prevent genetic discrimination. Instead, we should tighten restrictions on uses of genetic information, which requires a serious debate about the social value of access to things like life, disability, and long-term care insurance. How and where we should draw those lines is too large an issue for this review, but genetic counselors should be part of this conversation.

Legal Duty to Warn At-Risk Relatives?

Insurers are not the only third parties interested in a patient's genetic information; biological relatives also have a personal interest in this information. Because genetic information can have significant value to family members, the consensus among professional organizations and ethicists is that genetic counselors should inform patients about the clinical relevance of their genetic information to family members and encourage patients to disclose such information to their relatives (Rothstein 2018b). The thornier question is whether genetic counselors owe a legal duty to those relatives and, if so, what the nature of that duty is, particularly when the patient chooses not to share clinically actionable information with relatives.

Only three legal cases have addressed this issue. Although they involve physicians, their analysis applies to health-care providers, including genetic counselors. The first case, *Pate v. Threlkel*, involved a mother, Marianne New, who was treated for medullary thyroid carcinoma in 1987. New's physicians failed to inform her that the condition was autosomal dominant, posing a risk of cancer to her children. When her daughter, Heidi Pate, was later diagnosed with

the same condition, she sued the physicians, claiming they violated their duty to warn her about the heritable risk. Pate argued that had she been warned, she would have been tested and received preventive treatment.

In its 1995 ruling, the Florida Supreme Court held that physicians owe a duty of care to "identified third parties" in the "zone of foreseeable risk," like New's daughter, if the standard of care requires physicians to warn *patients* about the heritability of the condition. The court reasoned that any such standard of care exists precisely for the benefit of individuals like a patient's children. Nevertheless, the court pointedly narrowed the scope of this obligation, declaring any duty of care to a patient's close relatives "to warn of a genetically transferable disease … will be satisfied by warning the *patient*."[24] The court found that requiring the physician to "seek out and warn" the patient's family "would often be difficult or impractical and would place too heavy a burden upon the physician" and would likely conflict with statutory obligations to protect patient confidentiality.[25]

A few years later, in *Safer v. Estate of Pack*, a New Jersey appellate court faced a similar issue. In the mid-1950s, Dr. Pack treated Robert Batkin for retroperitoneal cancer. Unfortunately, Batkin succumbed to the cancer when his daughter, Donna, was 10 years old. More than 30 years after her father was treated for cancer, newly married Donna Safer was diagnosed with metastatic colorectal cancer. Arguing her cancer resulted from untreated multiple polyposis, which her father had, Safer sued Dr. Pack's estate for failing to warn her father about the hereditary nature of the disease.

Assuming the standard of care required Dr. Pack to warn Safer's father of the "genetic threat" to his children, the New Jersey appellate court had no difficulty concluding a duty was owed to "members of the [patient's] immediate family" facing "an avertable risk from genetic causes." The court did not, however, follow *Pate* in lim-

---

[24]*Pate v. Threlkel* (661 So. 2d 278, 282) (Fla. 1995) (emphasis added).
[25]*Id.*

iting the scope of the duty to informing the *patient* of the familial risk. Instead, it declared it "may be necessary, at some stage, to resolve a conflict between the physician's broader duty to warn and his fidelity to an expressed preference of the patient that nothing be said to family members about the details of the disease."[26]

Although the *Safer* court did not resolve whether "there are or ought to be any limits on physician–patient confidentiality," the New Jersey legislature did so 5 years later by enacting a genetic privacy statute. The statute prohibits the disclosure of identifiable genetic information without the individual's written consent,[27] with some exceptions, including to "furnish genetic information relating to a decedent for medical diagnosis of" the decedent's genetically related relatives.[28] This statute effectively overturned *Safer* (Rothstein 2018b), at least with respect to living patients who have not consented to such disclosures.

The third case, *Molloy v. Meier*, involved the failure to identify the fragile X mutation in Kimberly Molloy's developmentally delayed daughter and to counsel Molloy about the risk of passing on the mutation to future children. Molloy remarried and had a son with similar disabilities. He was found to carry the fragile X mutation, which ultimately led to identification of the mutation in Molloy's daughter. The court held that Molloy was owed a duty of care even though she was not a patient because "a physician's duty regarding genetic testing and diagnosis extends beyond the patient to biological parents who foreseeably may be harmed by a breach of that duty."[29] Unlike *Pate* and *Safer*, however, *Molloy* did not implicate confidentiality issues because Molloy was entitled to her minor child's medical information. In contrast, the parties alleging the physician had a duty to warn them in *Pate* and *Safer* had no underlying right to the adult patients' information.

The upshot of this limited case law is that genetic counselors, as health-care providers, owe a duty of care to identifiable, closely related familial members, if a patient's medical diagnosis or genetic information reveals an "avertable" or foreseeable risk to the relatives. The trickier question concerns the scope of that duty. Would a genetic counselor fulfill it by informing the *client* about the medical implications to family members, as *Pate* decided? Or must a genetic counselor, as *Safer* suggested, sometimes warn the *relative* directly, creating tension between the counselor's duty of confidentiality to the patient and duty of care to the relative?

In some jurisdictions, state law offers guidance. New Jersey's privacy law seems to allow a breach of confidentiality only after the patient has died. Florida's genetic privacy statute, however, consistent with *Pate*, prohibits disclosure of an individual's genetic analysis without the individuals' consent, with no exception for decedents.[30] Whether other states set such limits depends on whether they have genetic privacy statutes and their scope.

The HIPAA Privacy Rule is also relevant in prohibiting disclosure of protected health information, including genetic information, without written authorization if such disclosure is not for treatment, payment, or health-care operations. These protections are not absolute, however. The Rule contains 12 exceptions in which an individual's personal health information can be disclosed without the individual's authorization.[31] One exception allows unauthorized disclosures "necessary to prevent or lessen a serious and imminent threat to the health or safety of a person or the public."[32] The regulatory language describes this exception to "avert a serious threat to health or safety" as consistent with the "'duty to warn' third persons" articulated in *Tarasoff v. Regents of the University of California*.[33] In that case, a therapist's patient revealed his intention to kill a woman whom he ultimately mur-

---

[26]*Safer v. Pack* (677 A.2d 1188) (N.J. Super. Ct. App. Div. 1996).

[27]N.J. STAT. ANN. § 10:5-47 (2001).

[28]N.J. STAT. ANN. § 10:5-47 (a)(6) (2001).

[29]*Molloy v. Meier* (679 N.W.2d 711) (Minn. 2004).

[30]FLA. STAT. ANN. § 760.40.

[31]45 C.F.R. § 164.512 (2016).

[32]45 C.F.R. § 164.512(j) (2016).

[33]65 Fed. Reg. 82462, 82538 (Dec. 28, 2000).

dered. The California Supreme Court held that the therapist owed a duty of care to the woman, leaving open the possibility that such a duty might have required warning her directly.[34]

Although differences exist between the third-party risks in *Tarasoff* and in genetics (Suter 1993; King 2000), unauthorized disclosures might be permissible under this Privacy Rule exception if a genetic counselor believes in good faith that (1) the client's genetic information reveals a "serious and imminent threat" to the relative's health, (2) disclosure of this information would be "necessary to prevent or lessen the threat," (3) the relative would be "reasonably able to prevent or lessen the threat," and (4) such disclosure would be consistent with "applicable law and standards of ethical context."[35]

In some genetic counseling scenarios, the second and third elements would apply. If a client's medical information indicates clinically actionable risks to a family member, by definition, the relative could prevent or lessen the risk, meeting the third element. Additionally, if the client declines to share such information with the relative and the relative cannot otherwise learn about the risk, unauthorized disclosures might be necessary to avert the risk, meeting the second element.

In most genetic counseling cases, however, the first and last elements are hard to satisfy. Even when a client's genetic information suggests a high risk to relatives of a serious disease, the risk is rarely *imminent* in the way a homicidal patient's intent to murder might be. Even a 50% risk of inheriting a genetic variant for a serious condition presents less than a 50% risk (unless the variant is fully penetrant) of developing the disease at some undefined point in the future, which does not seem imminent. In certain specialties, genetic counselors may have clients with a heritable condition like arrhythmogenic right ventricular cardiomyopathy, which can pose temporally imminent risks like sudden cardiac death. In the vast majority of genetic counseling situations, however, predic-

tive genetic information does not present an *imminent* risk, failing the first element of the Privacy Rule exception.

Second, even if genetic counselors find themselves in the "rare circumstances" in which the risk seems imminent,[36] HIPAA's regulatory language makes clear it does not "preempt any state law" prohibiting disclosure of protected health information.[37] Thus, in jurisdictions with privacy statutes like Florida's, HIPAA would not override state prohibitions against unauthorized disclosure. Of course, many jurisdictions do not have legal restrictions to limit this exception.

The Privacy Rule exception also requires disclosures to be "consistent with … ethical standards of conduct." Professional groups (American Society of Clinical Oncology 2003; American Medical Association 2015), including genetic counselors, emphasize the duty of confidentiality. The National Society of Genetic Counselors' Code of Ethics stresses the importance of maintaining the "privacy and security of their client's confidential information and individually identifiable health information, unless released by the client or disclosure is required by law" (National Society of Genetic Counselors). This code of ethics describes no situation in which a genetic counselor must or may disclose information to an at-risk biological relative.

The American Society of Human Genetics (ASHG), however, considers the possibility that genetic counselors *may*, though they are not required to, disclose information to relatives. But it limits permissible disclosures to very narrow circumstances:

> … where attempts to encourage disclosure on the part of the patient have failed; where the harm is highly likely to occur and is serious and foreseeable; where the at-risk relative(s) is identifiable; and where either the disease is preventable/treatable or medically accepted standards indicate that early monitoring will reduce the genetic risk (American Society of Human Genetics 1998).

---

[34]551 P.2d 334 (Cal. 1976).
[35]45 C.F.R. § 164.512(j)(1)(i) (2016).

[36]65 Fed. Reg. 82462-01, 82703 (Dec. 28, 2000).
[37]65 Fed. Reg. 82462-01, 82704 (Dec. 28 2000).

Although narrow, this exception is slightly broader than HIPAA's exception because it does not require an "imminent" risk.

These different professional statements create uncertainty about whether breaching client confidentiality to warn a relative to "avert a serious threat to health" would be consistent with ethical standards. Such disclosure seems inconsistent with the Genetic Counselors' Code of Ethics, which prohibits breaches of confidentiality not *required* by law. However, if the risk of harm is "highly likely" and "preventable/treatable," it seems consistent with ASHG's policy statement.

Finally, although the original HIPAA Privacy Rule does not explicitly address disclosures of genetic information to family members, the Office for Civil Rights of the Department of Health and Human Services, which enforces the rule, addresses scenarios in which patients' family members want "to identify their own genetic health risks." It states that a health-care provider may share genetic information about the patient "with providers treating family members of the individual …, provided the individual has not requested and the health-care provider has not agreed to a restriction on such disclosure."[38] If upheld by courts, this interpretation seems to allow genetic counselors to disclose genetic information concerning a client to family members' *providers* if the patient has not requested, and the genetic counselor has not granted, a request, to restrict disclosure (Rothstein 2018b).

This mix of case, statutory, and regulatory law is clearer in some respects than others. It suggests genetic counselors have a legal duty of care to a client's relatives, if the client's genetic information reveals a "foreseeable risk" to closely related relatives.[39] The limited law suggests that telling the *client* about the risk satisfies the duty. In other words, genetic counselors do not seem to have a legal obligation to warn relatives or their physicians directly and would likely not

risk liability for failing to warn, if they explain the relatives' genetics risks to the patient.

It is less clear, however, whether disclosures to at-risk relatives are legally permissible. If the HIPAA Privacy Rule exception to "avert serious threats to health" applies in the genetics context, it does so rarely: when the health risk is serious and imminent, when disclosure is *necessary*, and when disclosure could avert the risk. Alternatively, under the 2013 interpretations of the Privacy Rule, genetic counselors may sometimes be permitted to share health information about a patient to a relative's health-care *provider* for treatment purposes. Both exceptions only *permit* rather than *require* disclosures, and neither has been tested in court.

In the rare instance when a client chooses not to warn relatives, genetic counselors face an ethical dilemma. On the one hand, disclosure to relatives could help them prevent serious medical harms. The necessity of disclosure to relatives or their patients, however, may depend on the state of technology and access to genetic testing. For example, if whole-genome sequencing becomes more routine, relatives with access to such technology could learn about actionable genetic risks without a genetic counselor having to breach patient confidentiality. Sometimes, however, disclosure might be the only way to avert harm.

On the other hand, clients may be trying to protect relatives. For example, a mother reportedly forbade her married daughters to tell her unmarried daughter about the mother's *BRCA* mutation for fear it could make the unmarried daughter "less marriageable" (Lewin 2000). Breaches of confidentiality could cause various harms, such as embarrassment, impaired family relationships (Rothstein 2018b), stigmatization, or discrimination. Finally, they could lead to distrust of genetic counselors, deterring genetic testing, genetic counseling, or participation in genetics research, the very concerns that motivated GINA and related laws.

For these reasons it is ethically appropriate not to *require* genetic counselors to do more than encourage clients to warn relatives about genetic risks (Rothstein 2018b). Nevertheless, an absolute rule forbidding disclosure to rela-

---

[38]78 Fed. Reg. 5668 (Jan. 25, 2013).

[39]The case law does not make clear the outer reaches of genetic relatedness for such an obligation to apply.

tives or their physicians raises ethical concerns. Imagine a genetic counseling client with arrhythmogenic right ventricular cardiomyopathy. In trying to cultivate a tough image as a police officer and avoid seeming weak, he refuses to disclose his condition to relatives. His marathon-runner sister would be at great risk of sudden cardiac arrest if she inherited the genetic variant.[40] Should the client's strong interest in confidentiality trump her strong(er) interest in living? Though the genetic counselor likely has no legal obligation to do so, should she be permitted to warn the sister if she cannot persuade her client to do it himself?

The HIPAA interpretation that *permits* a provider to inform a relative's *physician* about genetic risks for treatment purposes might offer the best alternative in these rare scenarios in which serious and imminent risks can only be prevented by disclosure. The genetic counselor could inform the relative's physician, who could recommend genetic tests as part of "routine care" without disclosing information about the genetic counseling client. Although this might prevent some harms the patient seeks to avoid, it puts the provider in the position of keeping secrets, raising other ethical concerns. Sometimes telling the relative, if possible, might be the best way to prevent harm.

If courts agree with the interpretations of HIPAA allowing, in exceedingly rare instances, disclosures to relatives or their physicians to avert serious harms, genetic counselors would have discretion to balance the potential harms of disclosure and nondisclosure. Given that these exceptions depend in part on ethical principles, the genetic counseling community has a role to play in considering whether disclosures are ever ethically permissible in the genetic counseling context and, if so, under what narrow circumstances. Until such cases are litigated, however, the permissibility of disclosures remains uncertain.

---

[40]This scenario was shared with me by a cardiac genetic counselor.

CONCLUSION

As we have seen, many legal issues in genetic counseling are unresolved or depend on the jurisdiction. As is often the case, the law remains a few steps behind emerging technologies. Unfortunately, this makes it difficult to predict with certainty how courts will rule when confronted with some of these genetic counseling dilemmas. Because norms within the genetic counseling community influence the law's approach to some of these issues, genetic counselors must remain engaged in the discussions and debates concerning the complex issues genetics and genomics pose.

REFERENCES

American Medical Association. 2015. Code of Medical Ethics § 2:131 Disclosure of Familial Risk in Genetic Testing, 2014–2015.

American Society of Clinical Oncology. 2003. American Society of Clinical Oncology Policy Statement Update: Genetic testing for cancer susceptibility. *J Clin Oncol* **21:** 2397–2406. doi:10.1200/JCO.2003.03.189

American Society of Human Genetics. 1998. Professional disclosure of familial genetic information Social Issues Subcommittee on Familial Disclosure, Professional Disclosure of familial Genetic Information. *Am J Hum Genet* **62:** 474–483. doi:10.1086/301707

Everett B. 2018. Republicans give up on Obamacare repeal. *Politico* https://www.politico.com/story/2018/02/01/obamacare-repeal-republican-status-381470

Goodnough A. 2019. Appeals court seems skeptical about constitutionality of Obamacare mandate. *New York Times* https://www.nytimes.com/2019/07/09/health/obamacare-appeals-court.html

Hall M. 1999. Legal rules and industry norms: the impact of laws restricting health insurers' use of genetic information. *Jurimetrics* **40:** 97.

Hazel J, Sogbogan C. 2018. Who knew what and when? A survey of the privacy policies preferred by U.S. direct-to-consumer genetic testing companies. *Cornell J Law Public Policy* **28:** 35.

Holger D. 2018. DNA testing for ancestry is more detailed for white people. Here's why, and how it's changing. *PCWorld* https://www.pcworld.com/article/3323366/dna-testing-for-ancestry-white-people.html

Keith K. 2018. Trump administration declines to defend the ACA. *Health Affairs Blog* https://www.healthaffairs.org/do/10.1377/hblog20180608.355585/full/

King M. 2000. Physician duty to warn a patient's offspring of hereditary genetic defects: balancing the patient's right to confidentiality against the family member's right to know—can or should Tarasoff apply. *Quinnipiac Health Law J* **4:** 1.

Lewin T. 2000. Boom in genetic testing raises questions on sharing tests. *New York Times* A1.

National Society of Genetic Counselors. NSGC Code of Ethics sec. 22, pt. 7. https://www.nsgc.org/p/cm/ld/fid=12#section2.

Pear R. 2019. Trump officials broaden attack on health law, arguing courts should reject all of it. *New York Times* https://www.nytimes.com/2019/03/25/us/politics/obamacare-unconstitutional-trump-aca.html?action=click&module=RelatedCoverage&pgtype=Article&region=Footer

Ray T. 2010. After GINA, where do life insurance firms stand on using genomic information for coverage decisions. *GenomeWeb* https://www.genomeweb.com/dxpgx/after-gina-where-do-life-insurance-firms-stand-using-genomic-information-coverag#.XA_8ZmhKg2w

Roberts J. 2010. Preempting discrimination: lessons from the Genetic Information Nondiscrimination Act. *Vanderbilt Law Rev* **63:** 457–462 & fn 76.

Roberts J. 2015. Protecting privacy to prevent discrimination. *William Mary Law Rev* **56:** 2128.

Rothstein M. 2009. GINA's beauty is only skin deep. *Gene Watch* **22:** 9.

Rothstein M. 2018a. GINA at ten and the future of genetic nondiscrimination law. *Hastings Center Report* **48:** 6.

Rothstein MA. 2018b. Reconsidering the duty to warn genetically at-risk relatives. *Genet Med* **20:** 285–290. doi:10.1038/gim.2017.257

Sarata A, DeBergh J, Staman J. 2011. The Genetic Information Nondiscrimination Act of 2008 and the Patient Protection and Affordable Care Act of 2010: overview and legal analysis of potential interactions. *Congressional Research Service* **R41314:** 3–4.

Suter S. 1993. Whose genes are these anyway?: familial conflicts over access to genetic information. *Mich Law Rev* **91:** 1854. doi:10.2307/1289655

Suter S. 2001. The allure and peril of genetics exceptionalism: do we need special genetics legislation? *Wash Univ Law Q* **79:** 669–748.

Suter S. 2004. Disentangling privacy from property: toward a deeper understanding of genetic privacy. *George Washington Law Rev* **72:** 737–814.

Suter S. 2019. GINA at 10 years: the battle over 'genetic information' continues in court. *J Law Biosci* **5:** 495–526. doi:10.1093/jlb/lsz002

Wieczner J. 2013. Your company wants to make you healthy. *Wall Street Journal* https://www.wsj.com/articles/SB10001424127887323933304578360252284151378

Zhang S. 2017. The loopholes in the law prohibiting genetic discrimination, *Atlantic* https://www.theatlantic.com/health/archive/2017/03/genetic-discrimination-law-gina/519216/

Birds of a Feather? Genetic Counseling, Genetic Testing, and Humanism

Robert Resta

Swedish Cancer Institute, Swedish Medical Center, Seattle, Washington 98104, USA

Correspondence: robert.resta@swedish.org

Humanism is a philosophy that emphasizes rational, scientific, and empiric analysis of the world we live in to improve the physical, social, and psychological life of humanity. Although individual genetic counselors may or may not identify as humanists, genetic counseling and genetic testing are primarily humanistic endeavors because they are situated in the context of humanistic medicine in the westernized world. Humanistic goals are also implicit and explicit in the profession and practice of genetic counselors. This review examines the relationship between humanism and genetic counseling, highlighting situations in which the two may be discordant, and suggests ways that genetic counselors can reconcile these discordances.

In this review I address three questions:

1. What role has humanism played in shaping the development of the genetic counseling profession, the practice of genetic counseling, and the utilization of genetic testing?

2. What aspects of genetic counseling practice and genetic testing are or are not consistent with humanistic philosophy?

3. Does it matter if genetic counseling and genetic testing are or are not consistent with humanism and how can inconsistencies be reconciled?

To develop the arguments in this paper, I will first define humanism and genetic counseling.

WHAT IS HUMANISM?

Humanistic philosophy is very much a product of Western European thought and ethics. It had its beginnings in the Renaissance, with roots in classical Greece and Rome. Broadly speaking, classical humanism was based on the idea that individual human and social betterment could be achieved through the study of the liberal arts and not just through adherence to religious beliefs.

In its modern sense, humanism blossomed in the late seventeenth, the eighteenth, and early nineteenth centuries during the Enlightenment (also called the Age of Reason) and the beginnings of scientific/philosophical inquiry (what we now call science was, until 1834, often called philosophy). Think Isaac Newton, Antonie van Leeuwenhoek, Voltaire, wife and husband Marie

Lavoissier and Antoine Lavoisier, Mary Somerville, Jean-Jacques Rosseau, Immanuel Kant, James Watt, James Hutton, Benjamin Franklin, sister and brother Caroline and William Herschel, Anne Conway, and Mary Wollstonecraft, to name a few. This period also witnessed women making inroads into being recognized in male-dominated fields such as science/philosophy, mathematics, and literature, albeit often constrained by social and economic limitations. Genetic counselors—who, around the world, are overwhelmingly female—are the current benefactors of the efforts of these Enlightenment women to integrate themselves into the scientific enterprise and to lay the groundwork for feminist philosophy (the word "scientist" first appeared in print in William Whewell's 1834 review of Mary Sommerville's book *On the Connexion of the Physical Sciences.* the best-selling book for 25 years for the venerable British publisher John Murray until Darwin's *Origin of Species*).

Humanism became a more formalized movement in the late nineteenth and twentieth centuries, and there are now many local, national, and international humanist organizations. Twenty-first century humanism emphasizes rational, scientific, and empiric analysis of the world we live in to improve the physical, social, and psychological life of humanity. According to the American Humanist Association, humanism "is a progressive philosophy of life that, without theism or other supernatural beliefs, affirms our ability and responsibility to lead ethical lives of personal fulfillment that aspire to the greater good" (https://americanhumanist.org/what-is-humanism/definition-of-humanism/, accessed 1/15/2019). It rejects explanations of the universe based on religious dogma, scripture, or divine revelation.

This nontheistic view has resulted in tensions and conflicts between and within the scientific and religious communities, as the two worldviews are often seen as being in opposition to each other. This perceived polarization between scientific and religious communities is unfortunate. This misperception is based on a limited understanding of the complexity of the human mind, which is perfectly capable of harboring two apparently conflicting ideas without producing cognitive dissonance. It is entirely possible for biologists, geologists, and physicists to be devoutly religious just as it is equally true that many people with strong spiritual and religious beliefs accept the validity of global warming, evolution, plate tectonics, and the Big Bang theory (Martin 2010; Ecklund et al. 2016; https://www.interfaithpowerandlight.org/religious-statements-on-climate-change/, accessed 1/17/2019). Despite the supposedly conflicting views of humanism and religion, humanism has been a critical component of the modern scientific and medical enterprise, regardless of the religious beliefs of individual scientists. Although some health-care providers and patients may pray to God or some deity to assist with cure and recovery from disease, medical therapy is firmly grounded in reason, logic, scientific investigation, and empiric observation rather than relying primarily on divine intervention.

WHAT IS GENETIC COUNSELING?

Genetic counseling as a formal term to describe a clinical practice was introduced in 1947 by Sheldon Reed, who famously and vaguely defined it as "a kind of genetic social work." Genetic counseling as a profession emerged in the United States with the establishment of a graduate-level genetic counseling program at Sarah Lawrence College in Bronxville, New York in 1968 (the first students matriculated in fall of 1969) (Stern 2012). Similar programs sprouted up over the next 50 years in the United States and globally; the Transnational Alliance of Genetic Counseling lists more than 80 training programs from 18 countries on six continents (http://tagc.med.sc.edu/education.asp, accessed 1/15/2019).

The practice and scope of genetic counseling has evolved in the half-century since the founding of the Sarah Lawrence program (Resta 2006). The most widely accepted and cited definition of genetic counseling is the one crafted and endorsed by the National Society of Genetic Counselors (2006):

> Genetic counseling is the process of helping people understand and adapt to the medical, psycho-

logical and familial implications of genetic contributions to disease. This process integrates the following:

- Interpretation of family and medical histories to assess the chance of disease occurrence or recurrence.
- Education about inheritance, testing, management, prevention, resources, and research.
- Counseling to promote informed choices and adaptation to the risk or condition.

The practice and practitioners of genetic counseling vary around the world but this definition probably captures the spirit of what most genetic counselors are trying to achieve, as judged by the large body of international publications about the goals, effectiveness, and conduct of genetic counseling (Resta 2019). Genetic counselors everywhere strive to be compassionate people who work toward providing highly competent, ethical, and compassionate care to their patients in an effort to improve their medical and psychological well-being.

WHAT ROLE HAS HUMANISM PLAYED IN THE DEVELOPMENT OF THE PROFESSION AND PRACTICE OF GENETIC COUNSELING?

Biological and medical sciences are inherently humanistic in their practice and ethos (Sinsheimer 1977), and genetic counseling is embedded within this larger sociocultural framework. In other words, genetic counseling is humanistic inasmuch as the practice of medicine in which it takes place as well as the scientific research that provides its basis are both largely humanistic endeavors.

Several important figures relevant to the history of genetic counseling were celebrated with awards by the American Humanist Association: Linus Pauling, who identified the molecular basis of sickle cell anemia; Herman Muller, who was awarded a Nobel Prize for his work on radiation-induced genetic mutations; and the psychologist Carl Rogers. On the other side of the coin, some geneticists have maintained deeply held religious beliefs, such as Theodosius Dobzhansky, R.A.F. Fisher, Francis Collins, and, of course, the monk Gregor Mendel. The religious views of these geneticists demonstrate how the

human mind can readily accommodate the coexistence of religious and humanist worldviews.

But it was the work of Carl Rogers that has the most direct relevance to the practice and humanistic ethos of genetic counseling. Rogers was awarded the American Humanist Association's "Humanist of The Year" Award in 1964, just 4 years before the establishment of the Sarah Lawrence genetic counseling training program. Rogers was a leading figure in the development of a humanistic model of psychology that emphasized the individual's inherent potential to grow and develop as an emotionally healthy, mature, and self-aware person. Humanistic psychology, like genetic counseling, places great value on patient autonomy: "… humanistic psychology fully acknowledges individual autonomy and the pursuit of self-actualization in the lives of all individuals" (http://oxfordmedicine.com/view/10.1093/med/9780199571390.001.0001/med-9780199571390-chapter-007, accessed 1/17/2019).

Rogers incorporated humanistic psychology into clinical practice with his counseling technique of client-centered therapy, which is based on the philosophy that the therapist should be nonjudgmental and nondirective and that patients should take an active role in their own therapy. Rogers had no direct connection with the genetic counseling profession and early on distanced himself from the term "nondirectiveness" because he felt it was misinterpreted. He maintained that the term "client-centered therapy" better reflected his philosophy than "nondirectiveness" did. In Rogers' view, clients should "self-actualize" utilizing their own inherent skills, aided by the therapist, rather than achieving a goal dictated by the therapist's view of human nature (Evans 1978). Rogers' client-centered therapy formed the psychological underpinnings of the genetic counseling model used first at the Sarah Lawrence Program, and for several decades was the predominant counseling model in genetic counseling practice (Veach et al. 2007; Stern 2012).

Nondirectiveness, which, as interpreted by genetic counselors, is somewhat different than how Rogers used it, also became a central ethical value of genetic counseling practice for many years (Kessler 1997). But this conflates nondirectiveness as an "ethical value" with nondirectiveness as a "counseling technique." Rogers would agree that it would be unethical for genetic counselors to impose their personal values on their clients. But the Rogersian approach would be to try to help clients strengthen and utilize their inherent psychological skills to direct clients to decisions that are consistent with clients' best interests, as defined by clients themselves. In Rogers' view, directiveness toward a specific outcome is desirable, as long as the therapist and client work together to help the client figure out what goal the client wanted and how the client could best work to achieve it.

Humanistic philosophy is still pervasive in the field of genetic counseling. Humanism is explicitly stated as a key value of several genetic counseling training programs, as illustrated by the following examples:

> The Ohio State University Genetic Counseling Program will provide graduate students … with a culture embodied in its values of altruism, compassion, diversity, education, ethics, honesty, **humanism**, integrity, lifelong learning, personal and professional growth… —https://medicine.osu.edu/residents/masters_programs/genetic_counseling/program/pages/index.aspx, accessed 1/16/2019

> Join the … Joan H. Marks Graduate Program in Human Genetics at Sarah Lawrence College. Integrating education, healthcare, and **humanism**, …. —https://www.sarahlawrence.edu/genetic-counseling/, accessed 1/16/2019

> The Columbia University Genetic Counseling Graduate Program curriculum combines basic science and clinical medicine with **humanism** and professionalism. —https://www.ps.columbia.edu/education/academic-programs/program-genetic-counseling, accessed 1/16/2019

> The [Case Western] genetic counseling program fosters the development of professional skills to provide students with the tools to become knowledgeable, competent and caring genetic counselors and emphasize personal growth including:
> • **Humanism**, compassion, integrity, and respect for others; based on the characteristics of an empathetic genetic counselor —http://genetics.case.edu/files/Genetic_Counseling_Student_Handbook_2018.pdf, accessed 1/16/2019

Even a recent job posting for a genetic counselor mentions humanism (https://www.linkedin.com/jobs/view/genetics-counselor-at-florida-atlantic-university-1036727676, accessed 1/16/2019):

> The mission of … a community-based medical school is to advance the health and well-being of our community by training future generations of **humanistic** clinicians and scientists and translating discovery to patient-centered care.

The guiding ethical principles of the U.S. genetic counseling community are found in the Code of Ethics of the National Society of Genetic Counseling (NSGC) (https://www.nsgc.org/p/cm/ld/fid=12, accessed 1/27/2019). Very similar ethical codes have been promulgated by the Human Genetics Society of Australasia, the Association of Genetic Nurses and Genetic Counsellors in the United Kingdom, the Canadian Association of Genetic Counsellors, and the European Board of Medical Genetics. None of these codes specifically state—or negate—a humanistic philosophy. Indeed, the original NSGC Code of Ethics, formulated in 1992, was developed with a very conscious attempt to be modeled on feminist ethics rather than humanism (J Benkendorf, pers. comm.). Nonetheless, the language of the NSGC Code of Ethics is concordant with humanistic principles—"Enable their clients to make informed decisions, free of coercion…," "Provide genetic counseling services to their clients regardless of their clients' abilities, age, culture, religion, ethnicity, language, sexual orientation, and gender identity," "These values are drawn from the ethical principles of autonomy, beneficence, nonmaleficence, and justice… ." This vocabulary draws on the principles of humanism and is quite similar to the language used by humanist organizations, as illustrated by these examples:

> Humanism is a democratic and ethical life stance that affirms that human beings have the right and responsibility to give meaning and shape to their own lives. Humanism stands for the building of a more humane society through an ethics based on human and other natural values in a spirit of reason and free inquiry through human capabilities. —The International Humanist and Ethical

Union Minimum Statement on Humanism (https://iheu.org/about/humanism/, accessed 1/1/2019)

Regardless of race, ethnicity, economic status, ability, sexual orientation, gender identity, religious beliefs or nonbelief, or citizenship, all individuals have universal human rights that must be respected and protected. —https://americanhumanist.org/key-issues/social-justice/, accessed 1/18/2019

The [American Humanist Association] also supports every woman's unequivocal moral and legal right to autonomy over her own body and reproductive choices, meaning we work to ensure women's access to family planning, contraception, birth control, emergency contraception, and healthcare services. —https://americanhumanist.org/key-issues/social-justice/, accessed 1/16/2019

Although no studies have been conducted to determine if genetic counselors typically identify as humanists, at least in the United States they are less likely than the general population to believe in a supreme deity, attend religious services, pray, or believe in an afterlife (Cragun et al. 2009). This is compatible with the demographic profile of religious beliefs of scientists (http://www.people-press.org/2009/07/09/public-praises-science-scientists-fault-public-media/, accessed 1/17/2019). On the other hand, physicians generally tend to be more religious than scientists (Curlin et al. 2005), so being a medical care provider is not necessarily an indicator of religiosity. But professional medical societies, medical institutions, and thought leaders often emphasize the importance of humanistic medicine (e.g., see https://www.abp.org/professionalism-guide/chapter-8/humanism, accessed 1/17/19).

To be clear, I am not suggesting that the practice of genetic counseling was developed to intentionally promote humanism or that most genetic counselors are card-carrying humanists. Rather, the point is that humanism or philosophies compatible with humanism are woven into the fabric of the genetic counseling tradition in the Western world. This influence is reinforced by the scientific training and the personal beliefs of genetic counselors themselves.

WHAT ASPECTS OF GENETIC COUNSELING PRACTICE ARE OR ARE NOT CONCORDANT WITH HUMANISTIC PHILOSOPHY?

Although the underlying philosophy and ethos of genetic counseling is compatible with humanistic tradition, whether this is also true of the day-to-day practice of genetic counseling and genetic testing is a more complicated question.

Some areas of genetic counseling practice are compatible with humanism. Genetic counseling for germline genetic testing for hereditary cancer, such as the *BRCA1/2* genes, aims to identify women at high risk of developing breast cancers to help them decide if they wish to pursue genetic testing with the aim of offering appropriate screening and risk-reducing strategies to reduce the chances of patients being diagnosed with, or treated for, cancer. Genetic counseling for hereditary cardiomyopathies can guide patients through the process of deciding whether to pursue genetic testing so that appropriate measures can be taken to reduce the risk of sudden cardiac death. Exome and genome sequencing of children or adults with rare genetic conditions can help establish a diagnosis and a prognosis and aid in the treatment and management of some conditions. The intention of counseling and testing in these settings is to help patients make autonomous decisions with the goal of reducing or preventing future suffering.

Genetic counseling is not only about genetic testing. Genetic counselors also work to enrich patients' sense of autonomy; help them adapt to living with or being at risk for a genetic condition; communicate critical, complex, and at times devastating clinical information; and provide comfort and empathy in difficult moments. These tasks all fit comfortably within a humanistic framework.

Other aspects of genetic counseling and testing are more complicated when it comes to being humanistic in that they do not necessarily "affirm[s] our ability and responsibility to lead ethical lives of personal fulfillment that aspire to the greater good" or "stand ... for the building of a more humane society" (https://iheu.org/about/humanism/, accessed 1/1/2019).

Depending on one's viewpoint and interpretation, some aspects of genetic counseling can be seen as either supportive of or contrary to humanistic values and principles.

Prenatal screening for Down syndrome and other chromosomal abnormalities has long been criticized by some people with disabilities, their families, and their supporters as a form of extreme bias based on ability (Parens and Asch 2003). They view prenatal screening as an attempt to rid the world of people with disabilities and judges them to be a drain on families and society, lives not worth living. From these critics' perspective, prenatal screening can hardly be thought of as building a more humane society and essentially amounts to ability-based prejudice.

Supporters of prenatal screening counter that the reality is that not all families are financially or functionally capable of caring for a child with multiple health, developmental, and behavioral issues. Prenatal screening, they maintain, is not intended to eliminate people with disabilities but rather to allow individuals and families to make choices that are most appropriate for their unique situations (Markens 2013). This would be compatible with humanistic values such as autonomy, reproductive freedom, and the right to shape one's own life.

The ethical waters are further muddied with the offer of prenatal testing for all pregnancies, not just for so-called high-risk women such as those 35 and older or who have certain ultrasound findings. Is universal offering of prenatal screening further proof of prejudice against people with disabilities or is it a way of enhancing the autonomy of all women and not just a minority of women who are somewhat arbitrarily labeled "high risk" (after all, a 40-yr-old pregnant woman has only an ~2% probability of having a liveborn baby with a chromosomal disorder)?

The argument that prenatal testing does not aim to eliminate people with disabilities is partially undercut by the history of prenatal testing and genetic counseling. Until the 1990s, almost all studies that attempted to measure the success of genetic counseling did so by examining its impact on reproductive choices and/or the incidence of babies born with disabling conditions (Resta 2019). On the other hand, the genetic counseling profession today does not typically evaluate its effectiveness by these measures and is, somewhat paradoxically, a strong advocate for people with disabilities and individual patient choices about whether or not to continue a pregnancy to term. Serving in these—at times—opposing roles of screeners and advocates can result in discordance between different humanistic values.

The complexity of the issues surrounding prenatal screening also arises in the context of offering "expanded carrier screening"—testing individuals of all ancestries for carrier status of sometimes hundreds of genetic conditions. The stated justification for carrier screening is to help couples make more informed reproductive choices. However, the vast majority of these conditions are very rare, there is a wide range of severity among and within the conditions, and determining the pathogenicity of rare variants can be extremely difficult (Kraft et al. 2019). In addition, the impetus for the development of expanded carrier screening—and noninvasive prenatal testing (NIPT)—came primarily from commercial laboratories rather than from a widespread demand from the clinical or public sectors (Holton et al. 2017; Swanson and Goldberg 2018). Is such screening a means of enhancing and informing reproductive autonomy, further evidence of prejudice against people with disabilities, or a natural outgrowth of the business model of medical testing? Well, arguably, expanded carrier screening is all three.

Conflicts with the values of humanism can also arise with exome/genome sequencing (ES/GS) of the healthy population, as distinct from such testing of the population of patients who have or who are at higher risk for suspected genetic conditions. The explicit humanistic goal of ES/GS is to reduce the risk of morbidity and mortality for genetic disorders such as cancer, heart disease, and cardiovascular disease by identifying people who carry clinically significant genetic variants but who have none of the standard risk factors such as clinical or family history.

One concern about ES/GS, when applied to the healthy population, is that interpretation of

Cite this article as *Cold Spring Harb Perspect Med* doi: 10.1101/cshperspect.a036673

the clinical significance of genomic variants, especially in healthy populations, is extraordinarily difficult and complicated. There is a significant risk of both false-positive and false-negative results (Wright et al. 2019). Patients may be told that they are at high risk of developing a disease or, conversely, that they are not at high risk for a disease when in fact they carry a highly penetrant pathogenic variant. Although this is also true for at-risk populations, there the problem is somewhat attenuated in that setting by having family history or other risk factors to assist with risk assessment. Furthermore, for many pathogenic variants detected on ES/GS there are very few interventions with adequate data to assess their utility in reducing disease morbidity and mortality. National and transnational large-scale research initiatives such as the United States's *All Of Us* (https://allofus.nih .gov), Europe's *1+ Millions Genomes* (https://ec .europa.eu/digital-single-market/en/news/eu-cou ntries-will-cooperate-linking-genomic-databases- across-borders), and Asia's *Genome Asia 100K* (http://www.genomeasia100k.com) will eventually provide data that will help with the interpretation and health-care application of genomic data, though it will likely be many years before the results of these studies will be useful in clinically meaningful ways.

Thus, although the goals of ES/GS in a low-risk population may be humanistic, the net effect for now of uncertain results and results of unproven clinical benefit may work at cross-purposes to informed autonomy and beneficence. This is particularly important in light of the growing number of commercial enterprises that are now offering ES/GS analysis directly to consumers.

Another type of genetic testing that may create conflicts with humanistic values is DNA-based ancestry testing. Companies promote such analysis as a means of better understanding people's ancestral origins and hence to improve their understanding of themselves as human beings. Nominally, this is very much in keeping with humanistic ideals of using science to help people understand humanity's place in the world and their ancestral relationships with other populations. Questions about the accuracy

of the ancestral categories aside (Huml et al. 2019), such testing can serve to treat our personhood as the sum of our DNA. The message that largely nonfunctional bits of DNA somehow dictate your fashion preferences, behaviors, and personality is communicated through advertising for these products that often depict people changing their clothing to match their stereotype of what a particular ethnic group might wear (and conflating ancestry with ethnicity) or speculating on why they might have particular personality traits or idiosyncratic preferences based on DNA analysis ("I never knew I was part Italian until I had that DNA testing. That explains why I like pasta so much."). DNA testing can reinforce racial stereotypes and exacerbate us versus them social dynamics—"My DNA says I am one of Them, not one of You." (Wailoo et al. 2012). It creates the potential to reduce humanism to DNAism and downplay the many developmental, environmental, and stochastic factors that shape our humanity. This is not just unfounded speculation—DNA-based ancestry testing has been used to justify racist agendas (Zhang 2016).

DOES IT MATTER IF GENETIC COUNSELING AND GENETIC TESTING ARE OR ARE NOT COMPATIBLE WITH HUMANISM?

As noted above, genetic counselors are situated within the larger local, regional, and national delivery of health care, a largely humanistic endeavor. In addition, humanism is implicitly and explicitly pervasive throughout the history and practice of genetic counseling. Therefore, one would expect genetic counselors' professional practice, ethos, and goals to be broadly similar to the ethos and goals of humanism, even if—as individuals—genetic counselors do not consider themselves humanists.

Genetic counselors should be prepared to justify any policies and activities that appear to be contrary to humanistic values. Furthermore, if genetic counselors are committed to the humanistic values expressed in the NSGC's Code of Ethics, then they are obligated to provide services that are reflective of this document and, just as importantly, criticizing and working to change applications of genetic testing and information that are contradictory to those values.

Aligning Genetic Counseling with Humanistic Goals: Suggestions and Barriers

There are several areas of genetic counseling—prenatal screening, expanded carrier testing, incorporating innovative genetic testing into clinical practice, awareness of the potential for abuse of genetic data—in which genetic counselors can work to better align their clinical services with humanistic goals (Madeo et al. 2011).

As discussed above, some people with disabilities, their families, and their advocates have raised concerns about the stigmatizing effects of prenatal testing. In addition, people with disabilities do not derive any direct medical or social benefit from prenatal screening programs. The humanistic benefit of prenatal testing is not readily apparent for this group of people.

Despite decades of prenatal testing and thousands of studies investigating its accuracy, sensitivity, program implementation, effectiveness, outcomes, etc., it is telling that there are almost no studies that even address the question of whether there are benefits to prenatal testing such as aneuploidy screening beyond the option of pregnancy termination. Does NIPT help produce better medical outcomes if a child with Down syndrome and a heart defect or duodenal atresia has a planned birth at a tertiary care center? Do families adapt better if a diagnosis is made prenatally? Does it improve children's ability to live up to their developmental potential? Not only do we not know the answers to these questions, researchers have not even systematically attempted to learn the answers.

It is not uncommon for prospective parents to say that they undergo prenatal tests so that they, their doctors, and their families "can be better prepared for the birth of a child with Down syndrome" (Bowman-Smart et al. 2019). Prenatal screening can take an emotional and psychological toll on women (Harris et al. 2012), and when it comes to diagnostic testing, there is a small but real risk of fetal loss. The benefits of testing should be worth its risks. Genetic counselors would be better able to counsel

pregnant women if they were able to provide data that could help inform such decisions. If research can show that, in addition to allowing parents to make difficult and situated reproductive decisions, prenatal testing also improves the lives of people with disabilities, then there would be a stronger ethical argument that prenatal testing is compatible with humanistic values for all people, regardless of their physical or developmental potential.

The increasing number of legal limitations that are being put on abortion in the United States and worldwide (Conti et al. 2016) could actually help address this research deficiency. If abortion becomes less widely available or is no longer a legal option in the future, then a prime justification for prenatal testing—abortion of affected fetuses—would no longer be an option for many women (although economically and socially advantaged women would still likely be able to maintain some access to abortion [Lowy 2018]). This could motivate the obstetrical and genetics communities to conduct more research on the potential medical, adaptational, and psychological benefits of prenatal diagnosis.

To complicate matters, prenatal screening has expanded into the care of all pregnant women. Hence more of the frontline education, counseling, and guidance on prenatal and carrier screening will be provided by obstetricians, midwives, and family practitioners. Ethical ramifications and fine points can fade to the background when a medical service becomes routinized—it's just one more test to run along with antibody titers, fasting glucose levels, blood typing, and the myriad other screening tests that now make up a goodly part of the medicalized pregnancy experience. Lengthy ethical discussions are not typically part of pregnant women's appointments with their providers. Administrators do not reward practitioners for this and there are no insurance reimbursement codes for Ethical Discussion with Pregnant Woman. Thus, genetic counselors need to partner with their colleagues from other specialties to ensure that genetic services and testing are provided in a manner consistent with aims of humanism.

Commercial laboratories can be important sources of innovation in genetic testing and play

Cite this article as *Cold Spring Harb Perspect Med* doi: 10.1101/cshperspect.a036673

a critical role in the introduction of new, better, and more cost-effective testing technologies (Evans and Vermeesch 2016). However, laboratories should not be the primary arbiters of which genetic tests are introduced into clinical practice. The voices of clinicians, patients, and those with no direct financial benefit from testing need to play dominant roles in the introduction and evaluation of new testing technologies. And governments need to provide the funding to conduct the necessary research that assesses the medical, economic, ethical, and social impact of genetic testing. Commercial entities should profit from their products but, to be consistent with humanistic values, those who derive the greatest financial benefit from a test should not be deciding which tests are best for patients. Genetic counselors and their colleagues from related specialties need to partner with laboratories, patients, social scientists, and ethicists to develop a more balanced approach to introducing innovative testing and policies into patient care.

This goal of cooperative test development may be difficult to achieve as care providers, clinics, and laboratories all have a vested interest in expanding the number and availability of genetic tests because they generate income. For medical centers and providers, genetic tests can generate more screening and other interventions, and for labs, more tests can mean greater profits. The profit motive, although often unacknowledged, can be a powerful incentive to incorporate more testing into clinical practice when the risks and benefits have not been clearly established (Kushnick 2015). Such financial conflicts of interest can produce blind spots such that relevant parties are psychologically incapable of recognizing when their recommendations are at least partially driven by financial gain (Sezer et al. 2015).

Harm from genetic testing and information can arise in ways beyond financial concerns. There is a long history of genetic data collected for ostensibly benign or helpful goals to be repurposed to ethically questionable ends. For example, during the nineteenth century, many so-called lunatic asylums in Europe and the United States collected data on the causes of insanity for all individuals admitted to these institutions.

In the majority of cases, heredity was assigned as the primary cause, often based on the family report or supposition on the part of the clinician (Porter 2018). These data, as well as data collected from schools and the military on intelligence tests, were eventually used by supporters of eugenics as justification for eugenic programs such as nonvoluntary sterilization, racist policies such as antimiscegenation laws, limits on educational financing, and immigration restriction (Reilly 1992; Gottesman and Bertelsen 1996; Ferri and Connor 2005; Stern 2005; Porter 2018; http://www.eugenicsarchive.org/html/eugenics/essay7 text.html). In much the same way, data from DNA-based ancestry testing, direct-to-consumer genetic testing, and even from clinical testing laboratories who sell or share their data for research purposes could be accessed and utilized by special interest groups and governmental organizations to inform a political agenda or a repressive policy such as immigration restriction (yet again) (Lynch 2012; Abel 2018), racial stigmatization that was reinforced when sickle cell carrier screening was introduced in the United States in the 1970s (Naik and Haywood 2015), or even more extreme actions such as deportation, internment, or social isolation, as has happened to many minority people at many times and in many places around the world, even in progressive democracies (Robinson 2010; Golash-Boza 2016). One does not need to turn to Nazi Germany to find an extreme example of the social abuse of genetics.

Genetic counselors can play a greater role in maintaining transparent guidelines that limit appropriate and safeguarded access to genetic data in several ways:

- Working with legislators to develop and enact privacy laws and racially unbiased immigration policies.

- Utilizing and working for genetic testing laboratories that do not share or sell patient data or least have data sharing policies that are spelled out clearly so patients can decide if they wish to use a particular laboratory.

- Greater awareness of the history of genetics so that they can better understand how even the best-intentioned research and databases can be subverted to nefarious or ethically questionable purposes.

The concern for data privacy becomes even greater as genomic testing expands beyond use in specialized clinics to broader applications such as newborn screening and testing of healthy adults by a greater number of laboratories and with less government oversight. The intentional or unintentional release of information from genetic databases, or hacking of these databases, provides very real opportunities for abuse or for beneficent use with unanticipated consequences that are not compatible with humanism.

It is by no means easy to create just and powerful laws that try to protect against the abuse of genetic information. For example, in the United States, the Genetic Information Nondiscrimination Act (GINA) of 2008 largely bans the use of genetic information of healthy people in determining their eligibility for health insurance. This legislation appears to be consistent with humanism. However, GINA privileges genetic information over other kinds of health-related information. Why should genetic disorders deserve special protection over communicable diseases or multifactorial conditions such as psychiatric disorders, diabetes, hypertension, and multiple sclerosis? For all of the good that GINA has achieved, at the same it violates the principle of distributive justice—the equitable distribution of goods and benefits across society. The lofty goals of humanism are straightforward to state but it can be difficult to actually apply them.

CONCLUDING REMARKS

Genetic counselors view themselves as highly ethical people who strive to deliver clinical care that is fair, equitable, and just. This strong ethical commitment can sometimes blind genetic counselors to some of the justifiable criticisms leveled against them, such as conflicts of interest or bias against people with disabilities. It is very difficult for ethical people to acknowledge that their behavior and choices may be viewed by others as unethical or biased. Such blind spots are inherent in human nature. Rather than denying the

validity of such criticisms, genetic counselors must learn to be able to accept criticism in positive ways and use it to improve the care provided to the patient community. This is necessary for genetic counseling to continue to be compatible with humanistic goals in the rapidly expanding and evolving field of genomic medicine.

This is easier said than done, but not impossible. Professional supervision, balanced and honest assessment of criticisms by nongenetic counselors, thoughtful and collegial discussion in conferences and scholarly articles, and open-minded meetings with critics in which differences can be discussed and solutions negotiated are some of the ways that the genetic counseling community can address criticism.

The future of genetic counseling and genomic medicine will likely be a mix of hope, uncertainty, change, painful failures, and profound achievements. Embryo editing with technologies such as CRISPR is just one example of the clinical and ethical challenges facing not only genetic counselors but also the entire planet. There is a justifiable concern for abuse, but also reason to be hopeful. Genetic counseling can continue to be a humanistic endeavor if genetic counselors do not lose sight of the goal of working to improve the lives of all of their patient and stay open to criticism when their practices deviate from this aim.

REFERENCES

Abel S. 2018. What DNA can't tell: problems with using genetic tests to determine the nationality of migrants. *Anthropol Today* **34:** 3–6. doi:10.1111/1467-8322.12470

Bowman-Smart H, Savulescu J, Mand C, Gyngell C, Pertile MD, Lewis S, Delatycki MB. 2019. 'Is it better not to know certain things?': views of women who have undergone non-invasive prenatal testing on its possible future applications. *J Med Ethics* **45:** 231–238. doi:10.1136/medethics-2018-105167

Conti JA, Brant AR, Shumaker HD, Reeves MF. 2016. Update on abortion policy. *Curr Opin Obstet Gynecol* **28:** 517–521.

Cragun RT, Woltanski AR, Myers MF, Cragun DL. 2009. Genetic counselors' religiosity & spirituality: are genetic counselors different from the general population? *J Genet Couns* **18:** 551–566. doi:10.1007/s10897-009-9241-0

Curlin FA, Lantos JD, Roach CJ, Sellergren SA, Chin MH. 2005. Religious characteristics of U.S. physicians: a national survey. *J Gen Intern Med* **20:** 629–634. doi:10.1111/j.1525-1497.2005.0119.x

Cite this article as *Cold Spring Harb Perspect Med* doi: 10.1101/cshperspect.a036673

Ecklund EH, Johnson DR, Scheitle CP, Matthews KRW, Lewis SW. 2016. Religion among scientists in international context: a new study of scientists in eight regions. *Socius* **2**: 1–9. doi:10.1177/2378023116664353

Evans R. 1978. *Carl Rogers: the man and his ideas*. E.P. Dutton, New York.

Evans MI, Vermeesch JR. 2016. Current controversies in prenatal diagnosis 3: industry drives innovation in research and clinical application of genetic prenatal diagnosis and screening. *Prenat Diagn* **36**: 1172–1177. doi:10.1002/pd.4967

Ferri BA, Connor DJ. 2005. Tools of exclusion: race, disability, and (re)segregated education. *Teach Coll Rec* **107**: 453–474. doi:10.1111/j.1467-9620.2005.00483.x

Golash-Boza T. 2016. The parallels between mass incarceration and mass deportation: an intersectional analysis of state repression. *J World-Syst Res* **22**: 484–509. doi:10.5195/JWSR.2016.616

Gottesman II, Bertelsen A. 1996. Legacy of German psychiatric genetics: hindsight is always 20/20. *Am J Med Genet* **67**: 317–322. doi:10.1002/(SICI)1096-8628(19960726)67:4<317::AID-AJMG1>3.0.CO;2-J

Harris JM, Franck L, Michie S. 2012. Assessing the psychological effects of prenatal screening tests for maternal and foetal conditions: a systematic review. *J Reprod Infant Psychol* **30**: 222–246. doi:10.1080/02646838.2012.710834

Holton AE, Canary HE, Wong B. 2017. Business and breakthrough: framing (expanded) genetic carrier screening for the public. *Health Commun* **32**: 1051–1058. doi:10.1080/10410236.2016.1196515

Huml AM, Sullivan C, Figueroa M, Scott K, Sehgal AR. 2019. Consistency of direct to consumer genetic testing results among identical twins. *Am J Med*. doi:10.1016/j.amjmed.2019.04.052

Kessler S. 1997. Psychological aspects of genetic counseling. XI. Nondirectiveness revisited. *Am J Med Genet* **72**: 164–171. doi:10.1002/(SICI)10968628(19971017)72:2<164::AID-AJMG8>3.0.CO;2-V

Kraft SA, Duenas D, Wilfond BS, Goddard KAB. 2019. The evolving landscape of expanded carrier screening: challenges and opportunities. *Genet Med* **21**: 790–797. doi:10.1038/s41436-018-0273-4

Kushnick HA. 2015. Medicine and the market. *AMA J Ethics* **17**: 727–728. doi:10.1001/journalofethics.2015.17.8.fred1-1508

Lowy I. 2018. *Tangled diagnoses: prenatal testing, women, and risk*. University of Chicago Press, Chicago.

Lynch, J. 2012. From fingerprints to DNA: biometric data collection in U.S. immigrant communities and beyond. Available at https://ssrn.com/abstract=2134481 or http://dx.doi.org/10.2139/ssrn.2134481

Madeo AC, Biesecker BB, Brasington C, Erby LH, Peters KF. 2011. The relationship between the genetic counseling profession and the disability community: a commentary. *Am J Med Genet A* **155**: 1777–1785. doi:10.1002/ajmg.a.34054

Markens S. 2013. "Is this something you want?": genetic counselors' accounts of their role in prenatal decision making. *Sociol Forum* **28**: 431–451. doi:10.1111/socf.12032

Martin JW. 2010. Compatibility of major U.S. Christian denominations with evolution. *Evol: Educ Outreach* **3**: 420–431. doi:10.1007/s12052-010-0221-5

Naik RP, Haywood C Jr. 2015. Sickle cell trait diagnosis: clinical and social implications. *Hematology* **2015**: 160–167. doi:10.1182/asheducation-2015.1.160

National Society of Genetic Counselors' Definition Task Force, Resta R, Biesecker BB, Bennett RL, Blum S, Hahn SE, Strecker MN, Williams JL. 2006. A new definition of genetic counseling: National Society of Genetic Counselors' Task Force report. *J Genet Couns* **15**: 77–83.

Parens E, Asch A. 2003. Disability rights critique of prenatal genetic testing: reflections and recommendations. *Ment Retard Dev Disabil Res Rev* **9**: 40–47. doi:10.1002/mrdd.10056

Porter TM. 2018. *Genetics in the madhouse—the unknown history of human heredity*. Princeton University Press, Princeton, NJ.

Reilly PR. 1992. *The surgical solution: a history of involuntary sterilization in the United States*. Johns Hopkins University Press, Baltimore.

Resta RG. 2006. Defining and re-defining the scope and goals of genetic counseling. *Am J Med Genet C* **142C**: 269–275. doi:10.1002/ajmg.c.30093

Resta RG. 2019. What have we been trying to do and have we been any good at it? A history of measuring the success of genetic counseling. *Eur J Med Genet* **62**: 300–307. doi:10.1016/j.ejmg.2018.11.003

Robinson G. 2010. A tragedy of democracy: Japanese confinement in North America. *J Transnatl Am Stud* https://cloudfront.escholarship.org/dist/prd/content/qt0qn09858/qt0qn09858.pdf?t=l04ho6&v=lg

Sezer O, Gino F, Bazerman MH. 2015. Ethical blind spots: explaining unintentional unethical behavior. *Curr Opin Psychol* **6**: 77–81. doi:10.1016/j.copsyc.2015.03.030

Sinsheimer RL. 1977. Humanism and science. *Leonardo* **10**: 59–62. doi:10.2307/1573635

Stern AM. 2005. Sterilized in the name of public health: race, immigration, and reproductive control in modern California. *Am J Public Health* **95**: 1128–1138. doi:10.2105/AJPH.2004.041608

Stern AM. 2012. *Telling genes—the story of genetic counseling in America*. Johns Hopkins University Press, Baltimore.

Swanson A, Goldberg JD. 2018. Industry perspectives on prenatal genetic testing. *Semin Perinatol* **42**: 314–317. doi:10.1053/j.semperi.2018.07.021

Veach P, Bartels DM, LeRoy BS. 2007. Coming full circle: a reciprocal-engagement model of genetic counseling practice. *J Genet Couns* **16**: 713–728. doi:10.1007/s10897-007-9113-4

Wailoo K, Nelson A, Lee C. 2012. *Genetics and the unsettled past: the collision of DNA, race, and history*. Rutgers University Press, New Brunswick, NJ.

Wright CF, West BF, Tuke M, Jones SE, Patel K, Laver TW, Beaumont RN, Tyrrell J, Wood AR, Frayling TM, et al. 2019. Assessing the pathogenicity, penetrance, and expressivity of putative disease-causing variants in a population setting. *Am J Hum Genet* **104**: 275–286. doi:10.1016/j.ajhg.2018.12.006

Zhang S. 2016. Will the alt-right promote a new kind of racist genetics? *The Atlantic*. December 29, 2016. https://www.theatlantic.com/science/archive/2016/12/genetics-race-ancestry-tests/510962/

Regulating Preimplantation Genetic Testing across the World: A Comparison of International Policy and Ethical Perspectives

Margaret E.C. Ginoza[1] and Rosario Isasi[2]

[1]University of Miami Miller School of Medicine, Miami, Florida 33136, USA

[2]Dr. John T. Macdonald Foundation Department of Human Genetics, University of Miami Miller School of Medicine, Miami, Florida 33136, USA

Correspondence: risasi@miami.edu

Preimplantation genetic testing (PGT) is a reproductive technology that, in the course of in vitro fertilization (IVF), allows prospective parents to select their future offspring based on genetic characteristics. PGT could be seen as an exercise of reproductive liberty, thus potentially raising significant socioethical and legal controversy. In this review, we examine—from a comparative perspective—variations in policy approaches to the regulation of PGT. We draw on a sample of 19 countries (Australia, Austria, Belgium, Brazil, Canada, China, France, Germany, India, Israel, Italy, Japan, Mexico, Netherlands, Singapore, South Korea, Switzerland, United Kingdom, and the United States) to provide a global landscape of the spectrum of policy and legislative approaches (e.g., restrictive to permissive, public vs. private models). We also explore central socioethical and policy issues and contentious applications, including permissibility criteria (e.g., medical necessity), nonmedical sex selection, and reproductive tourism. Finally, we further outline genetic counseling requirements across policy approaches.

Preimplantation genetic testing (PGT) is an assisted reproductive technology (ART) that, in the course of in vitro fertilization (IVF), enables the practitioner to select embryos by genotype prior to transfer to the womb (Knoppers et al. 2006; French Republic rev. 2019). Although IVF was originally developed to assist couples who were infertile or unable to conceive naturally, advances in PGT allow fertile individuals or couples to undergo IVF in order to conceive a child who does not have a genetic condition for which they would otherwise be at risk (e.g., because one prospective parent is affected or a carrier). PGT can also be used to conduct human leukocyte antigen (HLA) matching, in order to select for a child that is an HLA match for an existing sibling, a concept commonly referred to as a "savior child" (Kakourou et al. 2017). An HLA-matched child can then serve as a hematopoietic stem cell donor for a sibling affected by a disease such as leukemia. However, in theory, PGT can be used to select for any genetically determined characteristic desired by the parents, including sex and certain physical characteristics.

In 1967, scientists used fluorescence microscopy to identify the sex of living rabbit blastocysts, setting the path for the development of the technology that would later become PGT (Edwards and Gardner 1967). Clinical application of PGT would only follow in 1990, in the case of two couples who were at risk for adrenoleukodystrophy and X-linked intellectual disability (Handyside et al. 1990). In these cases, U.K. physicians used single-cell biopsies from embryos followed by Y chromosome–specific DNA amplification to select for female embryos, which would be unaffected by these X-linked conditions. Since then, techniques such as polymerase chain reaction (PCR) and fluorescent in situ hybridization (FISH) have been used to identify embryos carrying an expanding range of genetic conditions with increasing accuracy (Harper and Sengupta 2012; Van der Aa et al. 2013). Advances in sequencing and interpretation methods have further increased the possibilities provided by PGT, including using rapid embryo genotyping in combination with parental genome sequencing to predict the whole genome of an embryo, a step that would allow us to identify most Mendelian disorders, as well as some well-defined complex diseases (Kumar et al. 2015). Alongside these technical advances, there has been an "ever-increasing" number of PGT cycles performed, for a constantly increasing number of indications (Harper et al. 2012) as observed by the European Society of Human Reproduction and Embryology (ESHRE).

The possibilities provided by PGT have subjected the technology to significant legal, ethical, and social controversy since it was first developed. The "specter of eugenics" looms large, as the world grapples with the legacy of past eugenic practices (e.g., United States, Canada, and Germany) and the implications of a technology that enables selection of offspring according to preference for specific characteristics (Bayefsky 2016) without a medical indication. Similarly present in the PGT discussion, as with other forms of ART, are concerns regarding interference with the natural processes of reproduction.

Although PGT allows individuals or couples to avoid the issue of choosing to terminate a pregnancy that is posed by postimplantation prenatal diagnostic techniques, issues of moral status and personhood of the embryo still play a significant role (Baertschi 2008). Additionally, the tension between individual reproductive autonomy and the interest of the state in regulating reproduction runs throughout the discussion.

Because of these socioethical concerns, the use of PGT is the subject of many laws and international policies. In this review, we will analyze the regulation of PGT across a diverse sample of jurisdictions and policy approaches, covering a continuum from restrictive to permissive policy models. Based on this continuum, we selected as an illustrative sample the following 19 countries: Australia, Austria, Belgium, Brazil, Canada, China, France, Germany, India, Israel, Italy, Japan, Mexico, Netherlands, Singapore, South Korea, Switzerland, United Kingdom, and the United States. For each country, we identified the normative and legislative documents that govern the use of PGT. In addition, we identified international regulations set forth by regional regulatory bodies, such as the Council of Europe, as they complement the national policy framework in multiple countries in our sample. In the first section, we examine these norms in terms of their regulatory approaches (public or private ordering) and degree of permissibility (restrictiveness on the use of PGT). We also discuss the implications of recent changes in regulations. In the second section, we identify and discuss several central issues present in these regulations (or consequences thereof), including criteria for permissibility, nonmedical sex selection, genetic counseling requirements, and reproductive tourism.

THE REGULATORY LANDSCAPE

In this section, we analyze the regulation of PGT in 19 countries, according to their regulatory approaches and degree of permissiveness. The laws and regulations referenced in this section can be found in the supplementary material (see Supplemental Table S1 online).

Cite this article as *Cold Spring Harb Perspect Med* doi: 10.1101/cshperspect.a036681

Regulatory Approaches

A country's approach to regulating PGT can be categorized in several ways. Countries may approach regulation through public ordering (top-down, state-led), private ordering (bottom-up self-regulatory approach), or a mixture of the two models (Table 1; Knoppers and Isasi 2004). In a public ordering approach, the use of PGT is governed by statute or legislation. These statues may vary widely in their criteria for acceptable use, from establishing blanket prohibitions to restricting the application of PGT to limited indications. Conversely, a private ordering approach instead relies on guidelines or self-regulation. As with public ordering, these approaches range in degree of permissiveness. Additionally, regulations governing PGT fall on a spectrum frequently characterized by the binary distinction between "soft" and "hard" laws or policies, based on the degree to which a policy is legally binding or enforceable. Although recognizing the practical benefits of such categorization, it is important to note that "soft" and "hard" laws often act as mutually reinforcing or complementary policy instruments, thus rendering any binary classification as arbitrary.

A "soft" policy is not legally binding and thus is observed only through voluntary compliance by physicians and other practitioners. This includes guidelines or position statements from professional organizations such as the American Society for Reproductive Medicine (ASRM), which suggest best practices but have limited coercive powers as they often do not carry sanctions or entail consequences for failure to comply. However, professional guidelines can positively contribute to complementing a policy framework as they often detail clinical practice, determining what constitutes the "standard of care." Another benefit of this type of "soft" policy is that guidelines are easier to update than legislative mandates and thereby are faster to respond to scientific and social developments (Simpson et al. 2006). In contrast, a "hard" law or policy carries binding, legally enforceable obligations, which may be enforced by a coercive body such as a court or licensing authority and which carry sanctions as a consequence for failure to comply. Depending on a given framework, such policies impose penalties ranging from criminal sanctions to pecuniary fines and to professional or licensing suspensions or even debarment. One example of such a policy is India's *Pre-conception and Prenatal Diagnostic Techniques (Prohibition of Sex Selection) Act*, which outlines consequences, for both patient and practitioner, for the use of PGT to determine the sex of an embryo (Republic of India 1994 rev. 2003; Tribune News Service 2017; Kumar and Sinha 2018). A first offense by a practitioner "shall be punishable with imprisonment for a term which may extend to three years and with fine which may extend to ten thousand rupees," while the patient who sought the use of PGT may receive "imprisonment for a term which may extend to three years and with fine which may extend to fifty thousand rupees." (The reasons for this policy regarding sex selection, as well as its limitations, are discussed in a later section.) "Hard" policies may be issued by both national legislative bodies as well as by professional organizations, if granted the ability to place sanctions on members. One such example is Brazil's Federal Medical Council (CFM), which has the constitutionally granted power of applying sanctions for violations to the Code of Medical Ethics (Federal Medical Council 2010). Although CFM does not have the power to pass legislation regarding PGT (no such legislation exists in Brazil), it does

Table 1. Regulatory frameworks by country

Legislative	Guidelines	No identifiable regulation
Austria	Australia	Mexico
Belgium	Brazil	
Canada	China (mix model)	
France	Israel	
Germany	Japan	
India	Singapore	
Italy	United States	
Netherlands		
South Korea		
Switzerland		
United Kingdom		

have the ability to adopt resolutions regarding ethical conduct of physicians, as it did for PGT with, for instance, the *Resolution CFM No. 2.013/2013*, which enforced sanctions on those physicians who were found in violation of the ethical code (Federal Medical Council 2013).

Countries with government-sponsored health care (such as France, Italy, and the United Kingdom) adopt a public ordering approach to regulate if and how PGT is covered by government funding, though this approach is not exclusive to countries with such programs (Bayefsky 2016). National and professional organizations may still weigh in on the practice of PGT through policy statements and guidelines, which may serve to fill in the gaps left by vague or outdated legislation. For example, although Canadian federal law regulates the use of ART, including PGT, on a broad level, organizations such as the Society of Obstetricians and Gynaecologists of Canada (SOGC) provide more detailed guidelines governing the medical practice of PGT, thus defining the contours of what constitute the standard of care at a given time (Canada 2004 rev. 2012; Dahdouh et al. 2015). In this way, public and private ordering as well as "soft" and "hard" policy work together to form the regulatory landscape in which PGT is practiced.

In some countries, regulatory approaches to PGT differ based on their political structure, such as between those at the state and federal level. Although Australia has only guidelines at the national level, at least three states have enacted legislation regulating PGT. In the United States, regulation of PGT would similarly fall to individual states; however, no state currently has laws regulating the use of PGT. This split jurisdiction can sometimes result in conflict, such as when Canada passed a 2004 federal law regulating assisted human reproduction (AHR), which included provisions relating to PGT. Much of this law was struck down in 2010, for overstepping provincial authority (Supreme Court of Canada 2010; Snow 2018). Although the ruling significantly reduced federal regulatory power, certain prohibitions were left in place, including prohibitions relevant to PGT. The province of Québec, meanwhile, introduced several laws and guidelines of its own to regulate PGT (Québec 2010).

Degree of Permissiveness

Within each regulatory approach, control of PGT may range from complete prohibition to the placement of moderate to strict restrictions. Across the globe, no country has adopted policy conferring absolute or unfettered access, yet policy vacuums, legal loopholes, and/or ineffective oversight have resulted in some jurisdictions adopting a de facto laissez faire approach.

Permissive approaches toward use of technologies such as PGT may arise from differing underlying values. A permissive approach may be adopted "with the belief that it is beneficial for humanity" or as a result of a consumerist approach to regulation that values self-enforcement (Isasi et al. 2016). A mix of both approaches appears to be at play in the regulation of PGT. Among the countries examined, the most permissive approaches to PGT can be found in the United States and Mexico. In both countries, which each follow a private ordering model, PGT is actively practiced and commercially available, including for nonmedical sex selection (Dondorp et al. 2013). In the United States, multiple professional societies have published policy statements or guidelines offering recommendations regarding the practice of PGT. The Society for Assisted Reproductive Technology (SART) and the ASRM describe PGT as "a major scientific advance" for parents at risk of passing a "heritable and debilitating genetic disease" to their children, indicating a belief in the beneficial nature of the technology (SART, ASRM 2008). These guidelines are generally supportive of the use of PGT, although varying in opinion on such issues as sex selection and medical necessity. However, lacking enforcement provisions, the degree to which the recommendations set forth in these guidelines are followed is up to the discretion of the provider, in line with the consumerist approach. In Mexico, very little regulation exists related to PGT. It has been stated that in Mexico "there are no rules," because the federal *General Health Law* does not specifically regulate any form of assisted reproduction (Mexico 1984 rev. 2018; Palacios-González and Medina-Arellano 2017). Whereas the *Regulations of the General Health Law on Health*

Cite this article as *Cold Spring Harb Perspect Med* doi: 10.1101/cshperspect.a036681

Research do address research related to assisted reproduction, these regulations do not relate to routine clinical practice of PGT (Mexico 1986 rev. 2014).

It is worth noting that although Japan takes a similar private ordering approach, PGT has not been as quickly or widely adopted there as in the United States or Mexico. PGT was practiced actively in the United States by 1994 and Mexico by at least 2001 (Handyside 1994; Jones and Cohen 2001). Although clinical PGT research was approved in Japan in 1998, as late as 2004, PGT had never been implemented (Sato et al. 2015). This is in large part because of the reluctance of the Japan Society of Obstetrics and Gynecology (JSOG), which requires members abiding by its guidelines to seek approval for PGT from its ethics committee, and which feared approving of discrimination toward persons with genetic conditions (Munné and Cohen 2004). Despite the lack of legal restrictions, these professional guidelines seem to carry more moral weight than those of professional organizations in the United States, effectively prohibiting the use of PGT through reluctance to approve individual cases. However, approval and subsequent use of the technique is now steadily increasing, demonstrating a shift in acceptance of PGT from research to common practice (Sato et al. 2015).

The remaining 16 countries in our sample place some degree of restriction on the use of PGT. These restrictions may take the form of substantive requirements, regulating which patients and conditions qualify for PGT, or procedural safeguards, regulating the process by which PGT is provided (Knoppers and Isasi 2004). Procedural requirements often serve to protect patient's rights, including requirements for informed consent from couples or an individual undergoing PGT or legislative mandates for the use of genetic counseling (as found in several countries including Germany, Switzerland, and France). Procedural requirements may also concern the civil status of the patient and oversight for the use of PGT, such as the requirement that PGT be performed only in specifically licensed centers. Substantive requirements, meanwhile, address the way in which PGT may be used, within the procedural framework. Substantive requirements might include the requirement for "medical necessity" as a component of eligibility for PGT, and more restrictive countries further limit the use of PGT to only such inheritable conditions deemed suitably "serious" or "severe." The prohibition of nonmedical sex selection (Table 2) is another common substantive requirement. These restrictions, which will be addressed at length later in this review, serve to ensure safe medical practices and to limit the practice of PGT to only those situations deemed ethically and morally sound by that country's cultural and regulatory standards.

Table 2. Regulation of nonmedical sex selection by country

Nonmedical Sex Selection			
Legislatively prohibited	Prohibited by guidelines	Allowed (limited circumstances)	Allowed/not regulated
Australia	China	Israel	Mexico
Austria	Brazil		United States
Belgium	Japan		
Canada	Singapore		
France			
Germany			
India			
Italy			
Netherlands			
South Korea			
Switzerland			
United Kingdom			

A Changing Landscape

No countries identified in this analysis have prohibitive regulations on PGT currently in force. This reflects a recent transformation, as over the past decade changes in national regulation have demonstrated an increasing acceptance of PGT. As previously described, PGT has seen increasing uptake in countries such as Japan, which had previously been reluctant to endorse the practice. Germany, Switzerland, and Austria had previously been singled out as having among the most prohibitive stances toward PGT (Knoppers et al. 2006; IFFS Surveillance 07 2007; Jones and Cohen 2007). Each of these countries has since implemented legislation explicitly permitting PGT, albeit under heavy restriction. This shift toward permissiveness reflects changing cultural attitudes, broader adoption of PGT in clinical practice, and judicial interest in the right to reproductive freedom.

National and international courts have played a role in the shift toward permissiveness. The legislative shift in Germany was predicated by a 2010 German Federal Supreme Court decision acquitting a Berlin physician of violating the prohibitive 1990 *Embryo Protection Act* (EPA), finding that there are some cases in which PGT may be permissible (Federal Republic of Germany 1990 rev. 2011; Turner 2011). Prior to this ruling, the EPA had commonly been interpreted to prohibit PGT because of its strict prohibition of the production and use of a human embryo for any purpose other than inducing a pregnancy (or preservation for future use). The 2010 ruling may reflect a shift away from the legislative and cultural view of PGT as a tool for selecting *against* unwanted embryos, toward a view of PGT as a means to select *for* healthy embryos, with the end goal of inducing a "healthy" pregnancy (Bock von Wülfingen 2016). Following this ruling, the German Parliament voted to approve a 2011 amendment to the EPA explicitly addressing PGT, allowing the use of PGT in select cases, including risk of "serious hereditary disease."

In Italy, a series of court rulings challenged the 2004 *Rules on Medically Assisted Procreation*, which, under the strict interpretation initially followed, restricted access to ART and outright prohibited the use of PGT (Italian Republic 2004; Turillazzi et al. 2015). The law restricted the use of medically assisted reproduction to cases in which no other method exists to "remove the causes of sterility or infertility" and limited the production of embryos to only the number "absolutely necessary," capped at three. In addition, the law prohibited experimentation on embryos, including "to predetermine genetic characteristics" except for "diagnostic and therapeutic purposes which are exclusively associated with it for the protection of health and development of the embryo itself." Although Italian courts legitimated the use of PGT to prevent serious genetic disease, access to ART (including PGT) remained restricted to infertile heterosexual couples (Turillazzi et al. 2015). In 2012, the European Court of Human rights ruled in favor of a fertile Italian couple seeking access to PGT, finding the Italian law to be in violation of the "right to respect for his private and family life" provided by the Convention for the Protection of Human Rights and Fundamental Freedoms (Council of Europe 1950; *Costa and Pavan v. Italy*, ECHR 2012).

At the permissive end of the spectrum, particularly in the United States, there appears to be a shift of focus from *if* to *how* PGT should be used. Most guidelines released by U.S. professional organizations in recent years begin with the assumption that PGT is an accepted and increasingly useful technique in reproductive medicine (ASRM 2017). These newer guidelines, rather than addressing the acceptability of the broad practice of PGT, focus on specific ethically contentious scenarios such as the transfer of embryos with genetic conditions or testing for adult-onset conditions.

REGULATION OF GENETIC COUNSELING AND KEY ISSUES IN THE PRACTICE OF PGT

In this section, we discuss several specific issues within PGT regulation, including the role of genetic counseling in PGT, as well as key issues such as nonmedical sex selection, medical necessity, and reproductive tourism.

Cite this article as *Cold Spring Harb Perspect Med* doi: 10.1101/cshperspect.a036681

Regulation of Genetic Counseling

Genetic counseling can play a key role in supporting potential parents in decisions related to PGT. Several countries have taken steps to require specific counseling regarding genetic conditions. This counseling may be provided either by a professional genetic counselor (in countries where such a profession exists) or by the physician providing treatment (including in countries that do not have an independent genetic counseling profession). These regulations typically specify what information must be provided during the process of obtaining PGT, although the exact persons or methods for providing that information are often unspecified. Germany, for example, does not have a genetic counseling profession, but requires that a woman seeking PGT be informed and advised "on the medical, psychological and social consequences of the genetic examination of embryonic cells" prior to providing consent for the procedure (Federal Republic of Germany 1990 rev. 2011). In Switzerland, the *Reproductive Medicine Act* requires undergoing counseling before any ART technique can be used, followed by a period of reflection before treatment. In addition, the act requires that "comprehensive genetic counseling must be provided to couples if assisted reproductive techniques" such as PGT "are used to avoid the transmission of a serious, incurable disease" (Swiss Confederation 1998, revised 2017).

Specific information to be provided at these counseling sessions is outlined within the law or respective guideline, including the risks and benefits of testing, as well as resources and information regarding specific conditions. In the United Kingdom, the Human Fertilization and Embryo Authority (HFEA) code of practice requires centers offering PGT to ensure that patients are provided with access to genetic counselors before and after treatment (HFEA 2019). However, in reference to prenatal testing, French law requires that a pregnant woman be referred to a "multidisciplinary center of prenatal diagnosis" if there is a "definite risk" of a condition, in order to receive further information (French Republic rev. 2019).

Although requirements for genetic counseling ensure that patients have access to accurate information regarding PGT and the conditions being tested for, some requirements may have the effect (intended or not) of discouraging patients from using PGT or pursuing certain actions (such as abortion) upon receiving the results. Although the United States does not require genetic counseling for all uses of PGT, federal law and statutes in a number of states require the provision of specific information following a prenatal diagnosis of Down syndrome or "other prenatally ... diagnosed conditions" (United States 2008). Advocates argue that these laws ensure patients receive accurate information regarding the lives of persons with Down syndrome (or other conditions), whereas critics of these laws suggest that legislatively regulating the information provided to patients in this manner violates the ethical norm of "strict neutrality" in genetic counseling (Caplan 2015). In either case, it is generally understood that the goal of this legislation is to influence the patient's decision-making following prenatal testing, with the goal to reduce the number of patients choosing to terminate their pregnancy after receiving a prenatal diagnosis (Reilly 2009). In this case, PGT may be seen as a better alternative to prenatal screening (PNS), in which testing is conducted during pregnancy, as it avoids the need to make a choice to have an abortion if a genetic condition is identified. However, PGT still raises the question of whether selecting against conditions such as Down syndrome is discriminatory toward persons with disabilities, who live fulfilling lives in spite of their medical conditions. This discussion is playing out beyond the United States, particularly in light of recent controversy surrounding the widespread adoption of prenatal screening and low Down syndrome birth rates in Iceland. Critics argue that the Icelandic approach to prenatal testing has led to the abortion of nearly all pregnancies identified with Down syndrome, based on discriminatory attitudes toward persons with Down syndrome (Quinones and Lajka 2017). The Icelandic government maintains that the controversy is misleading, and that although it is the "responsibility of health-care personnel to explain clearly and objectively what options a woman has," the choice of whether to conduct prenatal testing or abort a pregnancy is fully left to the mother (Ministry of Health 2017).

Medical Necessity

One of the major substantive requirements placed on the application of PGT is the limitation of its use to only situations of medical necessity, prohibiting the use of PGT for personal or social reasons. This may take the form of a requirement that the fetus be at "substantial risk" of "severe genetic disease" (Knoppers and Isasi 2004) or the criteria might remain undefined. Additional substantive requirements identified in our selected countries included prioritization of the welfare of the (future) child and treatability of the condition. Specifically, countries including a treatability requirement mandate that there be no treatment available at the time of diagnosis. Countries utilizing each of these criteria are outlined in Table 3. The most common criteria among the countries examined is a requirement that the condition being tested for is "serious" or "severe," as reflected in some form by regulations or guidelines in Australia, Austria, Canada, France, Germany, Japan, Netherlands, Singapore, and Switzerland.

Because most countries do not define which specific conditions meet the severity criteria, some room is left for debate. However, several countries have outlined more detailed criteria or specific conditions, demonstrating a range of restrictiveness. For example, Austria defines the severity of a condition by its outcome, allowing for PGT only in conditions in which

the child becomes so ill during pregnancy or after birth that it 1. can only be kept alive by the constant use of modern medical technology or the constant use of other medical or nursing aids which severely impair his or her life, or 2. has severe brain damage or 3. in the long run will suffer from not effectively treatable severe pain and moreover, the cause of this disease cannot be treated. (Republic of Austria 1992 rev. 2018)

The United Kingdom and South Korea were the only countries in our sample that identify specific medical diagnoses for which PGT may be used. In South Korea, the 2005 *Bioethics and Safety Act* states that PGT may be used "only for diagnosing muscular dystrophy or any other hereditary disease specified by presidential decree" (Republic of Korea 2005). To date, 153 additional conditions have been approved by subsequent presidential decrees (Kim 2015). In the United Kingdom, conditions must be approved by the HFEA, the statutory body tasked with overseeing fertility treatment and research in the United Kingdom. At the time of this writing, more than 400 conditions had been approved, with at least 20 new conditions awaiting approval (HFEA n.d.).

There is no significant change apparent in the application of these criteria over time, with the exception of countries that previously prohibited the use of PGT. All countries in which we have identified a shift in the permissibility of PGT from prohibited to permitted now utilize the "serious or severe" criteria, along with other restrictive criteria, in line with continued restrictive attitudes toward PGT. The approach taken in South Korea and the United Kingdom of approving conditions on a case-by-case basis

Table 3. Criteria for allowing PGT

Criteria	Countries/States
Welfare of the child/embryo	Southern Australia (AU), Victoria (AU), Italy, Switzerland
Genetic defect (no severity criteria)	Victoria (AU), Brazil,[a] United States[a]
Significant risk	Western Australia (AU), Singapore,[a] Switzerland, United Kingdom
"Serious" or "severe"	Australia,[a] Southern Australia (AU), Western Australia (AU), Austria, Quebec (CA),[a] France, Germany, Japan,[a] Netherlands, Singapore,[a] Switzerland
Treatability	Austria, France, Germany, Netherlands, Switzerland
Specific conditions	South Korea, United Kingdom

[a]Suggested by guidelines.

Cite this article as *Cold Spring Harb Perspect Med* doi: 10.1101/cshperspect.a036681

requires a dynamic regulatory approach, increasing the scope of use each time a new condition is approved; however, the process itself does not appear to have changed significantly since being implemented. In countries with less specific criteria, however, shifts in interpretation of the severity required may not be reflected in the regulations themselves, as the interpretation of severity falls instead to the physicians and institutions providing PGT and within the private confines of the patient–doctor relationship.

In addition to severity, several countries discuss age of onset as a condition for approval; however, the age specified varies widely. Testing for conditions with onset beyond early childhood (particularly in the case of risk factors for adult-onset conditions with incomplete penetrance) faces criticism because of the excessive time, financial, and emotional cost for limited immediate benefit or effect on pregnancy outcome and raises questions regarding a (future) child's "right to an open future" (Kopelman 2007; Noble et al. 2008). In establishing a threshold for age of onset, countries must weigh these concerns against the individual reproductive liberties of the prospective parent(s), and the desire to prevent severe suffering even later in life. For instance, Germany prohibits testing for conditions with onset after age 18, whereas Switzerland permits testing for conditions with onset up to age 50. In the United States, guidelines from the ASRM suggest that testing for serious, untreatable adult-onset conditions is justified, and that testing for less penetrant or serious conditions may still be allowable.

Regulations in Australia, Belgium, and the United Kingdom include preimplantation HLA matching as an allowable condition for PGT. Although not specified by statute, the technique is also allowed or practiced in a number of other countries, including the United States, France, Brazil, Singapore, and Canada (Speechley and Nisker 2010; Bayefsky 2016). HLA matching is used in order to select for a child that will be able to serve as a hematopoietic stem cell donor for an existing sibling, often referred to as a "savior child" (Kakourou et al. 2017). The concept of savior siblings has stirred ethical debate, as the interests of the donor child, including potential risks of harm from PGT, the donation process, or emotional burden on the family, must be weighed against the medical benefit to the existing sibling. Although the success rate of HLA matching through PGT is limited, the procedure has gained increasing acceptance and more widespread use in recent years (Kakourou et al. 2018).

Nonmedical Sex Selection

One of the most heavily regulated and contentious applications of PGT is sex selection for nonmedical purposes. In nonmedical sex selection, preimplantation screening techniques can be used purposely to select embryos of a specific sex for personal or social reasons, rather than to avoid the selection of an embryo that may carry a sex-linked disease (Milliez 2007). However, the sex of an embryo may also be incidentally identified in the process of diagnosing another, medically indicated condition, at which point the provider (or the prospective parent[s]) must choose whether to utilize this information when selecting embryos to transfer (ASRM 2018).

The controversy over nonmedical sex selection weighs the potential for sex discrimination, particularly in light of historical sex discrimination practices in many countries, against the reproductive liberty of the individual (Milliez 2007). The vast majority of countries included in our sample weigh the former more heavily, adopting some sort of prohibition against or condemnation of nonmedical sex selection, as outlined in Table 2. In countries such as India, broad regulations place a blanket prohibition on prenatal sex determination by any means. In others, a prohibition on sex selection is included in legislation specifically focused on PGT. In spite of these prohibitions, sex selection by various methods appears to be widely practiced in several countries, including China and India, reflecting the limitations of even "hard" regulations that are not well supported by "soft" policies or social support (George 2006; Milliez 2007).

Three countries included in our sample allow some degree of nonmedical sex selection.

Israel, for instance, allows sex selection for "extremely exceptional, rare, and special cases," including for reasons of family balancing or harm to the "mental well-being" of parents or child (State of Israel Ministry of Health n.d.). The United States and Mexico are deemed to be the most permissive, having no legislation and few guidelines addressing sex selection. In each of these countries, sex selection is actively and increasingly practiced (Baruch et al. 2008; Whittaker 2011; Capelouto et al. 2018). However, professional organizations in the United States differ in their stances on this issue. The American College of Obstetricians (ACOG) has opposed the practice, whereas ASRM acknowledges the controversy surrounding nonmedical sex selection, yet stops short of prohibiting or encouraging the practice (ACOG 2007; ASRM 2015).

Conflicting Regulation, Reproductive Tourism, and International Norms

Because of the differences in regulation of PGT and other reproductive technologies across countries, cross-border travel for the purposes of seeking access to ART (or "reproductive tourism") is a significant topic of controversy and socioethical and legal concern (Spar 2005; Whittaker 2011). Through reproductive tourism, patients from countries with more restrictive regulations travel to countries with more permissive (often private ordering) ones to receive fertility treatments, including PGT, which they would not otherwise be eligible for in their home country. In this manner, they seek to avoid restrictions on access to such reproductive services in general, but also to controversial ones (e.g., nonmedical sex selection), or to obtain lower-cost services. The United States and Mexico, the most permissive countries in our sample, are commonly cited as destinations for reproductive tourism (Blyth 2010). Reproductive tourism pits individual reproductive autonomy and open international commerce against the attempts of each country to regulate the use of PGT by their own citizens in a way that reflects prevailing societal values.

Reproductive tourism made headlines around the world with the birth of the first child born following mitochondrial replacement therapy (MRT), which occurred in Mexico in 2016 (Zhang et al. 2017). In this case, MRT was performed by a U.S.-based researcher for Jordanian parents, sparking global controversy (Chan et al. 2017). Although there are significant differences in the technologies and ethical concerns involved in MRT and PGT, this case shone a spotlight on the way permissive scientific and medical regulations can encourage cross-border travel for access to reproductive services.

Some attempts have been made to harmonize or bring these differing regulations in line with each other, particularly in Europe. As previously described, the European Court of Human Rights has ruled that access to PGT (at least in cases of medical necessity) is protected under the Convention for the Protection of Human Rights and Fundamental Freedoms, to which the 47 member states of the Council of Europe are held (Council of Europe n.d.; ECHR 2012). The Council of Europe Convention on Human Rights and Biomedicine (referred to as the Oviedo Convention) outlines more restrictions, limiting all predictive genetic testing to use "only for health purposes or for scientific research linked to health purposes, and subject to appropriate genetic counseling" and prohibiting nonmedical sex selection (Council of Europe 1997). As of 2017, the Oviedo Convention had been ratified by 29 European countries, including those from our sample.

CONCLUDING REMARKS

PGT is subject to a wide array of regulatory approaches, which vary significantly by country. Although international regulations exist, they apply only to countries that choose to ratify and enforce them. Similarly, the strength of self-regulatory policy is predicated on voluntary compliance. In the course of our analysis, we observed a recent trend toward more permissive regulation of PGT, in both private and publicly ordered countries as well as at both ends of the permissibility spectrum. This shift may be driven by advances in technology, changes in cul-

Cite this article as *Cold Spring Harb Perspect Med* doi: 10.1101/cshperspect.a036681

tural attitudes, or court-driven alterations to legal practice. Medical necessity remains, to varying degrees, a common criterion for eligibility in more restrictive countries, a term that goes generally ill-defined and therefore necessarily subject to local interpretation. PGT for nonmedical purposes is actively practiced in countries with permissive regulation, yet does not necessarily follow a market or an unfettered laissez faire model. Although genetic counseling is widely accepted and encouraged by professional guidelines, only a few countries require it by legislative mandate. Because genetic counseling does not exist as a profession in all of these countries, the responsibility for providing counseling often falls to the physicians providing PGT. These conflicting regulatory approaches are certainly a driver of cross-border travel for reproductive purposes.

During the last decade, the regulation of PGT has followed mostly a linear path, with incremental changes driven by scientific advances as well as greater societal uptake. No longer considered an experimental practice, stakeholders across the world have challenged the status quo, recognizing that both technological and policy developments do not take place in a moral vacuum. They have done so under the flag of reproductive freedom, equitable access, and welfare considerations, among other ethical concerns. Despite the passage of time, the ubiquitous presence of eugenic fears constantly casts a shadow over technologies that open the door for selecting the genetic characteristics of a human offspring. For this reason, we predict that despite greater societal acceptance, widening the permissibility criteria for PGT will remain controversial and a matter of continued social and policy debate. Our hope is that such debates and ensuing changes in the regulatory landscape will always be inclusive so as to consider the voices of those who will ultimately be affected.

ACKNOWLEDGMENTS

Partially supported (R.I.) by the Chinese Academy of Science President's International Fellowship Initiative (PIFI). The opinions expressed above are those of the authors alone.

REFERENCES

Assisted Human Reproduction Act (2010). 3 SCR 457, 2010 SCC61 (CanLII). Supreme of Canada. https://www.canlii.org/en/ca/scc/doc/2010/2010scc61/2010scc61.html

Assisted Human Reproduction Act (S.C. 2004, c.2). rev. 2012, Canada

Baertschi B. 2008. The question of the embryo's moral status. Bioethica Forum 1: 76–80. https://www.unige.ch/medecine/ieh2/files/9714/3472/9164/b48-Statut_embryon.pdf

Baruch S, Kaufman D, Hudson KL. 2008. Genetic testing of embryos: practices and perspectives of US in vitro fertilization clinics. Fertil Steril 89: 1053–1058. doi:10.1016/j.fertnstert.2007.05.048

Bayefsky MJ. 2016. Comparative preimplantation genetic diagnosis policy in Europe and the USA and its implications for reproductive tourism. Reprod Biomed Soc Online 3: 41–47. doi:10.1016/j.rbms.2017.01.001

Blyth E. 2010. Fertility patients' experiences of cross-border reproductive care. Fertil Steril 94: e11–e15. doi:10.1016/j.fertnstert.2010.01.046

Bock von Wülfingen B. 2016. Contested change: how Germany came to allow PGD. Reprod Biomed Soc Online 3: 60–67. doi:10.1016/j.rbms.2016.11.002

Capelouto SM, Archer SR, Morris JR, Kawwass JF, Hipp HS. 2018. Sex selection for nonmedical indications: a survey of current pre-implantation genetic screening practices among U.S. ART clinics. J Assist Reprod Genet 35: 409–416. doi:10.1007/s10815-017-1076-2

Caplan AL. 2015. Chloe's Law: A powerful legislative movement challenging a core ethical norm of genetic testing. PLoS Biol 13: e1002219. doi:10.1371/journal.pbio.1002219

Chan S, Palacios-González C, De Jesús Medina Arellano M. 2017. Mitochondrial replacement techniques, scientific tourism, and the global politics of science. Hastings Cent Rep 47: 7–9. doi:10.1002/hast.763

Committee on Ethics American College of Obstetricians Gynecologists. 2007. ACOG Committee Opinion No. 360: sex selection. Obstet Gynecol 109: 475–478. doi:10.1097/00006250-200702000-00063

Council of Europe. n.d. The European Convention on Human Rights: impact in 47 countries. https://www.coe.int/en/web/human-rights-convention/impact-in-47-countries

Council of Europe. 1950. Convention for the Protection of Human Rights and Fundamental Freedoms. https://www.coe.int/en/web/conventions/full-list/-/conventions/treaty/005

Council of Europe. 1997. Convention for the Protection of Human Rights and Dignity of the Human Being with regard to the Application of Biology and Medicine: Convention on Human Rights and Biomedicine. https://www.coe.int/en/web/conventions/full-list/-/conventions/treaty/164

Dahdouh EM, Balayla J, Audibert F, Genetics C, Wilson RD, Audibert F, Brock JA, Campagnolo C, Carroll J, Chong K, et al. 2015. Technical update: preimplantation genetic diagnosis and screening. J Obstet Gynaecol Can 37: 451–463. doi:10.1016/S1701-2163(15)30261-9

Dondorp W, De Wert G, Pennings G, Shenfield F, Devroey P, Tarlatzis B, Barri P, Diedrich K. 2013. ESHRE Task Force on Ethics and Law 20: sex selection for non-medical reasons. *Hum Reprod* **28:** 1448–1454. doi:10.1093/humrep/det109

Edwards RG, Gardner RL. 1967. Sexing of live rabbit blastocysts. *Nature* **214:** 576–577. doi:10.1038/214576a0

Ethics Committee of the American Society for Reproductive Medicine. 2015. Use of reproductive technology for sex selection for nonmedical reasons. *Fertil Steril* **103:** 1418–1422. doi:10.1016/j.fertnstert.2015.03.035

Ethics Committee of the American Society for Reproductive Medicine. 2017. Transferring embryos with genetic anomalies detected in preimplantation testing: an Ethics Committee Opinion. *Fertil Steril* **107:** 1130–1135. doi:10.1016/j.fertnstert.2017.02.121

Ethics Committee of the American Society for Reproductive Medicine. 2018. Disclosure of sex when incidentally revealed as part of preimplantation genetic testing (PGT): an Ethics Committee opinion. *Fertil Steril* **110:** 625–627. doi:10.1016/j.fertnstert.2018.06.019

Federal Medical Council. 2010. The institution. *Conselho Federal de Medicina.* http://portal.cfm.org.br/index.php?option=com_content&view=article&id=20671&Itemid=23m

Federal Medical Council. 2013. RESOLUTION CFM No. 2.013/2013. *Federal Offical Gazette* 119. http://www.portalmedico.org.br/resolucoes/cfm/2013/2013_2013.pdf (Portuguese); http://www.huntington.com.br/wp-content/uploads/2013/04/Resolution-CFM-N-2013-2013.pdf (English)

Federal Republic of Germany. 1990 rev. 2011. Embryo Protection Act (BGBl. I P. 2746). http://www.gesetze-im-internet.de/eschg/BJNR027460990.html

French Republic. rev. 2019. Public Health Code—Article L2131-1. https://www.legifrance.gouv.fr/affichCodeArticle.do;jsessionid=68080FD62939971356A17F4BB1E1A328.tplgfr33s_1?idArticle=LEGIARTI000031972271&cidTexte=LEGITEXT000006072665&categorieLien=id&dateTexte=

George SM. 2006. Millions of missing girls: from fetal sexing to high technology sex selection in India. *Prenat Diagn* **26:** 604–609. doi:10.1002/pd.1475

Handyside AH. 1994. Preimplantation diagnosis of genetic diseases: a new technique in assisted reproduction. *Am J Hum Genet* **54:** 117–119. https://www.ncbi.nlm.nih.gov/pmc/articles/PMC1918056/?page=1

Handyside AH, Kontogianni EH, Hardy K, Winston RM. 1990. Pregnancies from biopsied human preimplantation embryos sexed by Y-specific DNA amplification. *Nature* **344:** 768–770. doi:10.1038/344768a0

Harper JC, Sengupta SB. 2012. Preimplantation genetic diagnosis: state of the art 2011. *Hum Genet* **131:** 175–186. doi:10.1007/s00439-011-1056-z

Harper JC, Wilton L, Traeger-Synodinos J, Goossens V, Moutou C, SenGupta SB, Pehlivan Budak T, Renwick P, De Rycke M, Geraedts JP, et al. 2012. The ESHRE PGD Consortium: 10 years of data collection. *Hum Reprod Update* **18:** 234–247. doi:10.1093/humupd/dmr052

Human Fertilization and Embryo Authority. n.d. Approved PGD and PTT conditions. https://www.hfea.gov.uk/treatments/embryo-testing-and-treatments-for-disease/approved-pgd-and-ptt-conditions/

Human Fertilization and Embryo Authority. 2019. Code of Practice. London. https://www.hfea.gov.uk/about-us/news-and-press-releases/2019-news-and-press-releases/new-version-of-the-code-of-practice-has-been-launched/

IFFS Surveillance 07. 2007. Chapter 14: preimplantation genetic diagnosis. *Fertil Steril* **87:** S47–S49. doi:10.1016/j.fertnstert.2007.01.080

Isasi R, Kleiderman E, Knoppers BM. 2016. Genetic technology regulation. Editing policy to fit the genome? *Science* **351:** 337–339. doi:10.1126/science.aad6778

Italian Republic. 2004. Rules on medically assisted procreation (Law 40/2004). https://www.normattiva.it/uri-res/N2Ls?urn:nir:stato:legge:2004;40

Jones HW Jr, Cohen J. 2001. IFFS surveillance 01. *Fertil Steril* **76:** S5–S36. doi:10.1016/S0015-0282(01)02931-4

Jones HW Jr, Cohen J. 2007. IFFS surveillance 07. *Fertil Steril* **87:** S1–S67. doi:10.1016/j.fertnstert.2006.11.079

Kakourou G, Vrettou C, Moutafi M, Traeger-Synodinos J. 2017. Pre-implantation HLA matching: the production of a saviour child. *Best Pract Res Clin Obstet Gynaecol* **44:** 76–89. doi:10.1016/j.bpobgyn.2017.05.008

Kakourou G, Kahraman S, Ekmekci GC, Tac HA, Kourlaba G, Kourkouni E, Sanz AC, Martin J, Malmgren H, Giménez C, et al. 2018. The clinical utility of PGD with HLA matching: a collaborative multi-centre ESHRE study. *Hum Reprod* **33:** 520–530. doi:10.1093/humrep/dex384

Kim NK. 2015. Legislation on genetic diagnosis: comparison of South Korea and Germany—with focus on the application and communication structure. *Dev Reprod* **19:** 111–118. doi:10.12717/DR.2015.19.2.111

Knoppers BM, Isasi RM. 2004. Regulatory approaches to reproductive genetic testing. *Hum Reprod* **19:** 2695–2701. doi:10.1093/humrep/deh505

Knoppers BM, Bordet S, Isasi RM. 2006. Preimplantation genetic diagnosis: an overview of socio-ethical and legal considerations. *Annu Rev Genomics Hum Genet* **7:** 201–221. doi:10.1146/annurev.genom.7.080505.115753

Kopelman LM. 2007. Using the Best Interests Standard to decide whether to test children for untreatable, late-onset genetic diseases. *J Med Philos* **32:** 375–394. doi:10.1080/03605310701515252

Kumar S, Sinha N. 2018. *Preventing more "missing girls": a review of policies to tackle son preference.* World Bank Policy Research Working Papers. Poverty & Equity Global Practice Group, Washington, DC.

Kumar A, Ryan A, Kitzman JO, Wemmer N, Snyder MW, Sigurjonsson S, Lee C, Banjevic M, Zarutskie PW, Lewis AP, et al. 2015. Whole genome prediction for preimplantation genetic diagnosis. *Genome Med* **7:** 35. doi:10.1186/s13073-015-0160-4

Mexico. 1984 rev. 2018. General Health Law. http://www.diputados.gob.mx/LeyesBiblio/ref/lgs.htm

Mexico. 1986 rev. 2014. Regulations of the General Health Law on Health Research. http://www.salud.gob.mx/unidades/cdi/nom/compi/rlgsmis.html

Milliez J. 2007. Sex Selection for non-medical purposes. *Reprod Biomed Online* **14:** 114–117. doi:10.1016/S1472-6483(10)60742-0

Ministry of Health. 2017. Facts about Down syndrome and pre-natal screening in Iceland. https://www.government

.is/news/article/2017/12/11/Facts-about-Down-syndrome-and-pre-natal-screening-in-Iceland/

Munné S, Cohen J. 2004. The status of preimplantation genetic diagnosis in Japan: a criticism. *Reprod Biomed Online* **9**: 258–259. doi:10.1016/S1472-6483(10)62138-4

Noble R, Bahadur G, Iqbal M, Sanyal A. 2008. Pandora's box: ethics of PGD for inherited risk of late-onset disorders. *Reprod Biomed Online* **17**: 55–60. doi:10.1016/S1472-6483(10)60332-X

Palacios-González C, Medina-Arellano MJ. 2017. Mitochondrial replacement techniques and Mexico's rule of law: on the legality of the first maternal spindle transfer case. *J Law Biosci* **4**: 50–69. doi:10.1093/jlb/lsw065

Practice Committee of Society for Assisted Reproductive Technology, Practice Committee of American Society for Reproductive Medicine. 2008. Preimplantation genetic testing: a Practice Committee opinion. *Fertil Steril* **90**: S136–S143. https://www.ncbi.nlm.nih.gov/pubmed/19007612

Québec. 2010. Act Respecting Clinical and Research Activities Relating to Assisted Procreation Publications Quebec. http://legisquebec.gouv.qc.ca/fr/ShowDoc/cs/A-5.01

Quinones J, Lajka A. 2017. *What kind of society do you want to live in?: inside the country where Down syndrome is disappearing.* CBSN: On Assignment. https://www.cbsnews.com/news/down-syndrome-iceland/

Reilly PR. 2009. Commentary: the federal 'prenatally and postnatally diagnosed conditions awareness act'. *Prenat Diagn* **29**: 829–832. doi:10.1002/pd.2304

Republic of Austria. 1992 rev. 2018. Law on Reproductive Medicine. https://www.ris.bka.gv.at/GeltendeFassung.wxe?Abfrage=Bundesnormen&Gesetzesnummer=10003046

Republic of India. 1994 rev. 2003. Pre-conception and Prenatal Diagnostic Techniques (Prohibition of Sex Selection) Act. http://legislative.gov.in/actsofparliamentfromtheyear/pre-conception-and-pre-natal-diagnostic-techniques-prohibition-sex

Republic of Korea. 2005. Bioethics and Safety Act. https://elaw.klri.re.kr/eng_mobile/viewer.do?hseq=33442&type=part&key=36

Sato K, Sueoka K, Iino K, Senba H, Suzuki M, Mizuguchi Y, Izumi Y, Sato S, Nakabayashi A, Tanaka M. 2015. Current status of preimplantation genetic diagnosis in Japan. *Bioinformation* **11**: 254–260. doi:10.6026/97320630011254

Simpson JL, Rebar RW, Carson SA. 2006. Professional self-regulation for preimplantation genetic diagnosis: experience of the American Society for Reproductive Medicine

and other professional societies. *Fertil Steril* **85**: 1653–1660. doi:10.1016/j.fertnstert.2006.02.072

Snow D. 2018. *Assisted reproduction policy in Canada: framing, federalism, and failure.* University of Toronto Press, Toronto.

Spar D. 2005. Reproductive tourism and the regulatory map. *N Engl J Med* **352**: 531–533. doi:10.1056/NEJMp048295

Speechley KN, Nisker J. 2010. Preimplantation genetic diagnosis in Canada: a survey of Canadian IVF units. *J Obstet Gynaecol Can* **32**: 341–347. doi:10.1016/S1701-2163(16)34479-6

State of Israel Ministry of Health. n.d. Infant gender selection —the National Commission on Infant Gender Selection through Pre-implantation Genetic Diagnosis. https://www.health.gov.il/English/Services/Citizen_Services/Pages/gender.aspx

Swiss Confederation. 1998, revised 2017. 810.11 Federal Act on Medically Assisted Reproduction. https://www.admin.ch/opc/en/classified-compilation/20001938/index.html

Tribune News Service. 2017. Violation of PNDT act: ultrasound centre's registration suspended. *The Tribune.* https://www.admin.ch/opc/en/classified-compilation/20001938/index.html

Turillazzi E, Frati P, Busardò FP, Gulino M, Fineschi V. 2015. The European Court legitimates access of Italian couples to assisted reproductive techniques and to pre-implantation genetic diagnosis. *Med Sci Law* **55**: 194–200. doi:10.1177/0025802414532245

Turner M. 2011. Germany to allow preimplantation diagnosis. *Nature newsblog.* Springer Nature. http://blogs.nature.com/news/2011/07/germany_to_allow_preimplantati.html

United States. 2008. Prenatally and Postnatally Diagnosed Conditions Awareness Act. https://www.congress.gov/bill/110th-congress/senate-bill/1810/text

Van der Aa N, Zamani Esteki M, Vermeesch JR, Voet T. 2013. Preimplantation genetic diagnosis guided by single-cell genomics. *Genome Med* **5**: 71. doi:10.1186/gm475

Whittaker AM. 2011. Reproduction opportunists in the new global sex trade: PGD and non-medical sex selection. *Reprod Biomed Online* **23**: 609–617. doi:10.1016/j.rbmo.2011.06.017

Zhang J, Liu H, Luo S, Lu Z, Chávez-Badiola A, Liu Z, Yang M, Merhi Z, Silber SJ, Munné S, et al. 2017. Live birth derived from oocyte spindle transfer to prevent mitochondrial disease. *Reprod Biomed Online* **34**: 361–368. doi:10.1016/j.rbmo.2017.01.013

Evolving Roles of Genetic Counselors in the Clinical Laboratory

Megan T. Cho[1] and Carrie Guy[2]

[1]National Human Genome Research Institute, Bethesda, Maryland 20894, USA

[2]Quest Diagnostics, Secaucus, New Jersey 07094, USA

Correspondence: megan.cho@nih.gov

Genetic counselors (GCs) possess several core competencies that provide direct benefit in the clinical laboratory setting. Communication with clients about complex information such as test methodology or results and the skills of facilitation and translation of complex information were recognized as important skills early in the establishment of GCs in laboratories. The clinical expertise of GCs serves as the background and experience from which they facilitate complex laboratory cases. Early roles for GCs in the laboratory also included result reporting, case management, and test development. The scope of roles has broadened to include management, business development, education, telemedicine, research, and variant interpretation. With increasing value being placed on genetic counseling skills both in and outside of a clinical laboratory, the roles and positions of GCs will likely continue to expand.

An overarching role of genetic counselors (GCs) is to function as a guide and advocate for their patients, which includes connecting patients with other members of the health-care team. As hospitals and university laboratories began expanding their biochemical, cytogenetic, and molecular test menus, individuals with clinical and molecular expertise were sought out as liaisons between the laboratories and patients, and in this capacity, GCs were a natural fit. Most of these first laboratory positions in the 1980s were filled by GCs who maintained some of their clinical responsibilities while adding on part-time work in their institution's laboratory (Zetzsche et al. 2014). Interestingly, because in these instances the clinical and laboratory roles were both housed under a single institution (or its affiliates), there was little concern at that time about any conflict of interest (COI) for the GCs.

Later, a similar boom occurred in the menu of tests offered by commercial laboratories (those not affiliated with a hospital or university), and those laboratories also hired GCs to partner with their clients. As single-gene disorder testing became prevalent in the United States market, GCs were employed by laboratories to educate physicians about the uses and benefits of molecular testing. Many non-genetics providers interested in ordering genetic tests did not have clinical GCs in their practice, did not have enough time for the nuances of ordering and interpreting this new testing, and did not have much, if any, training in the clinical applications of molecular genetic testing (Kotzer et al. 2014). In addition to guidance about appropriate test ordering, these pioneering laboratory GCs provided risk revision information, reviewed implications and limitations of test results, and made referrals for the patient to a local genetics provider. Additionally, laboratories and insurance companies became aware of GCs' value in improving test utilization, which resulted in cost savings. With the emergence of laboratory genetic counseling as a new subspecialty, physicians and patients were granted increased access to diagnostic testing and genetic counseling services, and laboratory policies and procedures were influenced by the expertise of these early pioneers (Amos and Gold 1998).

Noting the expansion within the field, an ad hoc committee appointed by The National Society of Genetic Counselors (NSGC) in 1985 examined how GCs were broadening their areas of practice (Heimler 1997). Genetic counselors in laboratories or other nondirect patient care positions were labeled as "nontraditional," which took on some negative connotations within the field. One main reason for discontent centered around a notion of these GCs straying too far from their clinical foundation by not providing direct patient care. Additionally, the movement of some GCs to fill these laboratory roles naturally led to reduced numbers of GCs providing direct patient care, resulting in a challenge to fill their previous clinical positions in order to maintain patient access to care. This trend of an increasing number of "nontraditional" GCs continued, referring to an increasingly diverse assortment of roles and responsibilities. Although that term still is used on occasion, language has shifted toward "nondirect patient care" in an effort to recognize the important contributions of laboratory GCs and that their roles are not outside of a tradition. Here we describe the expansion and diversification of the role of the GC in the laboratory.

INITIAL LABORATORY GENETIC COUNSELOR ROLES

Client Communication

One of the earliest roles of GCs in the clinical laboratory involved communication with laboratory clients. In a 2012 survey of GCs, client communication was indicated as the primary role for GCs in a clinical laboratory setting (Christian et al. 2012). The laboratory GC acted as a primary resource to clients, answering routine questions about test requirements, addressing concerns about sample conditions, and providing a wide variety of genetic testing information. Communication with clients about complex information such as test methodology or results, confrontation management around result delay, assay failures, or laboratory error; rapport building; and support of clients through difficult situations are a few examples of the application of genetic counseling communication skills in the laboratory setting (Goodenberger et al. 2015, 2017). Strong communication skills are ever important in the role of a liaison, facilitating communication between the laboratory, the ordering provider, and the patient (Scacheri et al. 2008; McWalter et al. 2018).

Result Reporting

Disclosure of results through written and verbal communication is another key role of laboratory GCs. In fact, the 2012 survey of laboratory GCs lists report writing as the second most common role (Christian et al. 2012). Genetic counselors' experience in a wide variety of clinical settings helps develop expertise in the review of test results, analysis of family history, or other medical records that may be needed for the gathering of phenotypic data and clarification of clinical indication for testing (Goodenberger et al. 2017). Thus, the laboratory GC's role in result reporting may include obtaining information from the ordering provider to assist in the interpretation of results, generation or editing of reporting comments for both clinical accuracy and clear communication, and verbal communication of the results to the provider to address questions or highlight recommendations (Swanson et al. 2014). Genetic counselors also provide risk interpretation and aid in the development of follow-up testing recommendations, providing guidance for ordering providers who may be less experienced or have less familiarity with genetic testing.

Cite this article as *Cold Spring Harb Perspect Med* doi: 10.1101/cshperspect.a036574

Beyond patient-specific reporting, laboratory GCs are often involved in the development of reporting language for new tests and the implementation of appropriate laboratory policy around testing and reporting based on their clinical experience of communicating results to patients and clients (Swanson et al. 2014). For example, they may provide clinical insight in the development of new result reporting protocols and technical structure that impacts when, in what format, and how results are transmitted to clients and patients. They also develop supplemental educational materials that are provided as part of the genetic testing report that facilitate the clear communication of results to patients by providers.

Case Management

The varied expertise of GCs also provides them the background and experience from which they are able to manage complex cases. Case management may include communication with ordering providers to assist in selection of the most clinically appropriate test, discussion of laboratory policy and requirements, and coordination of testing logistics. This clinical expertise is demonstrated in the growing laboratory specialization of test utilization management. In this process, select test orders are examined to determine clinical appropriateness through the review of medical records or communication with the ordering provider. Laboratory GCs frequently fill this role as they have the background to review medical records, family history, and records of previously performed genetic tests to help determine the most clinically appropriate test (Kotzer et al. 2014). They may also contribute to the development of organizational policy that describes best ordering practices. The number of GCs involved in test utilization management has increased as the scope of genetic testing options has expanded. Ordering providers and health-care systems have increasingly recognized the value of a systematic review of genetic test orders to reduce duplicate and incorrect test orders. Several studies have demonstrated the value that laboratory GCs provide in the review of genetic testing orders and the reduction of cost

based on the correction of incorrect ordering practices (Kotzer et al. 2014; Miller et al. 2014). Thus, test utilization management provides great value to laboratories in quality management, the reduction of liability, and increase of trust in the laboratory assay by both clinicians and patients.

Test Development

Genetic counselor training in facilitation and translation of complex information influences internal laboratory processes as well as external client relations. Relevant roles may include working with interdisciplinary teams in test development, test launch coordination, and research and development (R&D). In this setting, facilitation and translation of information as well as logistical considerations and scientific coordination are often required for an interdisciplinary team to function efficiently in a test launch. Genetic counselors involved in test development help determine which genes and variants should be included in new tests and collaborate with R&D to decide which tests or new technologies are the best investment for the laboratory. They may also design research protocols or participate in the coordination of research protocols designed to provide scientific evidence of test performance. Genetic counselors' understanding of health-care delivery and deep knowledge of genetics provide valuable insight into test development and enhancement (Swanson et al. 2014). In addition to focusing on test development at established genetic testing laboratories, GCs are increasingly involved in start-up genetic testing companies. In this unique setting, GCs may take on multiple responsibilities in a fast-paced and less structured work environment (Rabideau et al. 2016). In a start-up company, the GC may play a role in developing not only the actual assay and educational materials but also the information technology utilized to manage the product.

Policies and Ethical Considerations

Genetic counselors are trained to recognize and manage ethical dilemmas, and this skill was quickly seen to be of value for the ethical practice of genetic testing in a clinical laboratory (Balcom et al. 2016). With internal ethical challenges emerging from interpretation or return of results, GCs contribute a patient-centered focus in the laboratory. Acting as a liaison between the laboratory staff and the ordering provider, GCs have helped to maintain focus on the patient in difficult discussions. Genetic testing has a long history of ethical dilemmas. Because genetics is familial in nature, testing has implications that extend beyond the individual being tested (Balcom et al. 2016). Training in the management of ethical conflicts starts in genetic counseling programs and is guided in practice by the NSGC Code of Ethics (https://www.nsgc.org/p/cm/ld/fid=12). Although prior studies have demonstrated several areas of interspecialty concordance between GCs in the laboratory setting and the clinical setting, the laboratory setting also provides a unique set of ethical considerations related to indirect patient care (Groepper et al. 2015). Potential ethical challenges in a clinical laboratory include patient privacy, appropriateness of testing, incidental findings, and the duty to reanalyze (Balcom et al. 2016). They may also include issues surrounding variant classification and reclassification (Wain 2018). A GC's role in the clinical laboratory may extend beyond the management of an immediate ethical dilemma into mentoring and educating others within the laboratory setting. For example, in the discussion of test development, they may provide important feedback on the appropriateness of test development in sensitive areas or the manner in which tests are researched, marketed, or reported. The role of GCs may also extend to the creation of guidelines or protocols to offset or address ethically problematic situations.

FURTHER EXPANSION AND FUTURE ROLES

Business Development

Although a follow-up survey of laboratory GCs confirmed the findings by Christian et al. that the most common role remains case coordination and result reporting, the survey also found an expansion into sales and marketing, and management responsibilities (Waltman et al. 2016). The area of sales and marketing support in the industry was an early area of expansion for GCs filling positions of medical affairs and medical science liaisons. Genetic counselors in these positions provide clinical and scientific expertise to support the work of regional or specialty sales representatives. This support may include attending in-person client visits and providing educational presentations to clinicians. They may provide a perspective on the clinical application of tests that gives specialty sales representatives insight into how clients may use the tests offered and which patients may benefit from certain specialized testing. In 2016, Waltman et al. found that sales and marketing accounted for 9.8% of laboratory GC time, of which the most common activity was editing marketing materials, followed by staffing of booths at professional conferences where scientific and clinical experience is drawn upon to answer client questions about the performance or clinical application of tests offered. Marketing support may range from providing clinical review of patient and provider facing materials for medical accuracy to having an active role in the design and development of genetic testing educational materials.

The role of the product manager is an example of further expansion of GCs into the business side of the industry. In this position, GCs provide insight and direction in the development and enhancement of genetic tests offered by their employer. They may advise on emerging technology and provide insight into the current market based on clinical trends or discussions in the provider community. As more GCs move into product director positions, opportunities for promotion into senior or executive director positions may develop. Skills learned while focused on genetic testing products may be directly adapted to nongenetic testing, potentially leading to increasing numbers of GCs demonstrating the core competency of "Understand how to adapt genetic counseling skills for varied service delivery models" in the clinical laboratory business setting (Counseling 2015).

Once the value of GCs' skills in one clinical specialty has been recognized, movement into new space has followed. Cardiology, ophthal-

Cite this article as Cold Spring Harb Perspect Med doi: 10.1101/cshperspect.a036574

mology, infertility, and endocrinology are all specialties that have seen an increased presence of GCs in direct patient care. The increased presence of GCs in these clinical specialties suggests that GCs will likely play an increased role in laboratory business development in these specialty areas. As genetic testing becomes more common in family medicine and general medicine, GCs may come to be involved in the development of genetic tests and business models in primary care. Perhaps it is the application of genetic testing into the routine medical setting in which GCs will lead business development and bridge the translational gap between specialty genetic testing and integration into routine care.

Additional expansion of GCs into emerging biotechnical companies has included movement into the biotechnical and pharmaceutical industry (Field et al. 2016). This emerging field of work for GCs has built upon similar strengths demonstrated in the clinical laboratory. Those in medical affairs and medical science liaison positions cited "education" as the ACGC competency most utilized in their current role, followed by "professional development and practice" (Field et al. 2016). Clinical practice has witnessed a decline in the availability of medical geneticists, with fellowship training programs struggling to fill positions, and GCs have responded to this clinical need by expanding their traditional practice model. Clinical laboratories have historically hired medical geneticists as medical directors to help guide clinical utility of developed assays. Mirroring the increased independence of GCs in clinical positions, increasingly independent roles for GCs in the laboratory may follow a similar pattern to help fill this gap.

Leadership and Management

In 2016, Waltman and colleagues found that management accounted for 5% of laboratory GC time. The role of leadership and management reinforced the earlier finding by Zetzsche et al. (2014) that laboratory GCs described spending more time in management roles. Such roles may include supervision of other lab-

oratory GCs or staff that support genetic services, such as genetic assistants or those providing client services. The increasing numbers of GCs stepping into informal and formal leadership roles has resulted in the development of resources to support these roles. Such resources include the establishment of the Leadership and Management Special Interest Group within NSGC and active discussion and formal presentations on the career ladder and professional development. Although leadership positions exist in many practice settings, the laboratory industry has provided substantial opportunity for the development and expansion of such roles. Genetic counselors are skilled in facilitating difficult discussions and guiding decision-making. The background of respect for autonomy and skill in empowering others gives GCs tools to assist a colleague or client in making difficult decisions and working toward the identification of a decision that allows them to move forward in a direction with which they are comfortable. Such decisions may include the decision to launch a particular test or educational material or when and how to communicate laboratory errors. Genetic counselors are increasingly recognized as employees that can engage in difficult conversations and demonstrate leadership in arriving at solutions.

Education

In a clinical laboratory setting, education occurs both in-person and virtually and is both internally directed at company employees and externally directed at clients. Providing internal education about clinical ramifications of testing may provide context, greater appreciation, and the generation of ideas for improved efficiency or advances in technology. Educational outreach within the laboratory includes the training of interns, students, or fellows and the development of educational literature or lectures. Additionally, GCs in laboratories may participate in educating colleagues within their own company including scientists, sales representatives, and other lab personnel. This education may be via in-person or virtual lectures or by contributing to the development of internal training materials. Larger laboratories often have continuing education opportunities for their board-certified staff that also double as learning opportunities for other staff who might be investigating additional degree programs or are interested in learning more about fields of their colleagues. Genetic counselors are involved with inviting guest speakers, such as parent advocates, or by speaking themselves on clinical implications and considerations of the laboratory's work. Sales representatives, medical science liaisons, and other similar employees may or may not have a background in genetics, yet they are the individuals representing the laboratory and answering questions in-person with ordering clinicians. Therefore, it is very important that they have at least a basic accurate understanding of the test products and the processes of the laboratory as they relate to the client. Genetic counselors often participate in educating these externally facing laboratory personnel so that they understand the most pertinent information and are able to communicate that effectively to clients.

As for the training of genetic counseling graduate students, initial laboratory exposure for these students focused mostly on learning about technologies and techniques through observation and hands-on laboratory experience (e.g., pipetting), often with a non-GC supervisor, because the laboratory tasks were not ones that GCs typically performed. As the presence and roles of GCs in academic and commercial laboratories increased, student exposure broadened. In 2013, a new edition of the ACGC Standards required that GC students be instructed in and observe genetic laboratory activities. With the expansion of laboratory activities and roles of GCs, student rotations must evolve as well. Many of the GCs entering laboratory positions have previous experience in the supervision of students in direct patient care and welcome the opportunity to work with them in a new capacity. Although time constraints can be an issue for any GC supervisor, commercial laboratory management may have more stringent guidelines about employee time allocated to education, whereas those requirements may be more flexible in academic environments. One way to address limited time for education is to have a student work with many GCs and other laboratory scientists, spending only a short amount of time with each. Another advantage of that structure is that the student is exposed to a wide variety of roles and viewpoints within the laboratory. As with any learning experience, applied exercises and practical case-based examples are beneficial. Laboratory GCs are encouraged to have students take on projects of varying sizes that will actually be used in future operations—for example, researching appropriate genes to be included in a new test offering or drafting language for a results report. An optimal rotation might be a collaboration between a clinical site and commercial clinical laboratory in which a student participates by following and guiding the patient and their sample through the entire counseling and testing process. Another opportunity for GC students is to be involved in the business and education roles of laboratory GCs, such as the education of internal scientists or working with external providers and clients alongside medical science liaisons or sales representatives.

In early 2019, ACGC distributed a draft of the proposed update to the previous Standards. The revisions under consideration include changes to fieldwork training requirements and a new definition of what qualifies as a participatory encounter, or "case," for students to add to the logbook that must be submitted in partial fulfillment of requirements for ABGC board certification. Specifically, under the proposed changes, the maximum number of telecounseling cases (telephone, video, group sessions) will be unlimited, whereas previously only a few such cases would be eligible. This change reflects an increasing pool of data suggesting a relative equivalence between in-person and telecounseling across a variety of measures (Buchanan et al. 2016; Athens et al. 2017; McCuaig et al. 2018). Another modification allows for a health-care provider to be classified as a recipient of genetic counseling services—or, in other words, a "client" for logbook cases purposes—so long as the discussion is about an individual patient. Collectively, these proposed changes acknowledge and place value on roles commonly engaged in by laboratory GCs such as providing counseling to ordering providers. If approved, they would facilitate more comprehensive student engagement and rotation opportunities with laboratory GCs. Student training could expand into participating with direct client and patient interactions, enriching the students' experiences and allowing laboratory GCs to engage students in more aspects of their job.

Alternative Service Delivery Models and Conflicts of Interest

Studies of laboratory GC roles found that an estimated 42% in 2012 and 24% in 2016 provided face-to-face clinical counseling (Christian et al. 2012; Waltman et al. 2016). That downward trend parallels the shift away from a primary academic or hospital work setting and toward commercial laboratories, with 28% of participants noting a commercial laboratory as their work setting in 2012 and 57% in 2016. When access to genetic counseling started becoming an issue, one solution provided by commercial laboratories was to employ GCs to perform traditional laboratory roles as well as direct patient care at the clinical centers ordering the testing/screening. Early positions were typically in cancer for *BRCA1/2* testing only or in preconception or prenatal settings, for maternal serum screening and carrier screening. These GCs did not need to live geographically close to the laboratory itself, but rather were contracted to work at regional centers without their own genetic counseling services. These services are still being provided in certain areas, suggesting a system that works well for the laboratories, provides patients access to services they might not otherwise have, and reduces the cost or burden to the medical institution.

Although most direct patient care is still provided in-person (96%), 59% of GCs providing direct patient care report that they counsel patients over the phone and 19% through video conferencing. Out of those using video conferencing, only 3% ($n = 8$) used that model all of the time, and 62.5% of those individuals were prenatal GCs (Counselors 2018). Out of the 754 full-time GCs using phone counseling, only 5% ($n = 36$) used that model all of the time, and 47% of those were primarily providing cancer services. In terms of GCs who identify their primary work setting as in a diagnostic laboratory, 3.1% exclusively provide direct patient care, a number that rises to ~26% when you add in individuals who report a mix of direct and nondirect patient care (Counselors 2018).

One new technology making its way into health care and the field of genetics is the chatbot (Garg et al. 2018; Pereira and Díaz 2019). Chatbots are trained via machine learning, and consumers can ask questions and receive answers without directly interacting with a human. The application of this technology in the laboratory setting allows clients and patients to receive tailored responses to their questions or concerns beyond what a FAQs page can provide. Some issues may be fully resolved through the chatbot encounter, whereas others will be forwarded after being triaged as needing intervention from a GC or other laboratory member. Such an application would make better use of genetic counseling resources while increasing consumer access during hours of the day or night when staff are not available to answer questions. Some laboratories already have chat features and many may move toward these technologies in response to payor plan requirements and an increasingly competitive genetic testing market. Another system that could increase access is an application based on-demand system with live chat or video features. This type of system could also be integrated into the laboratories' existing interface along with sample tracking and other traditionally portal-based tools. With evolving definitions and parameters of clinical utility and clinical testing, integrating technology to increase genetic counseling access will be crucial (Rashkin et al. 2019).

In some instances, insurance companies have created added pressure for genetic counseling services. In 2013, Cigna (Lee 2013) began requiring that patients undergo genetic counseling prior to genetic testing. These policies were instituted to reduce costs associated with incorrect test ordering and recognized the value of GCs' expertise in appropriate test selection. A few companies followed suit, although the trend has not continued, and some companies later removed those requirements. With lack of GC services sometimes being an impediment to ordering testing, some commercial laboratories began offering remote genetic counseling sessions directly to patients who were referred by their ordering clinicians and otherwise may not have had access to a GC (Stoll et al. 2018). The session was offered not as stand-alone counseling, but in conjunction with testing that was being ordered, and typically took the form of posttesting consultations. In this practice setting, the GC may have a conversation with the ordering clinician similar to a more traditional customer service role, but the patient was the direct recipient of a personalized posttest counseling session provided by a certified GC. Complex situations such as variants of uncertain significance and testing of other family members could be assessed and managed by the GC in that encounter as well. Some have suggested that the COI of a laboratory GC providing direct patient care cannot be mitigated, and laboratories should only be offering genetic counseling as a temporary solution until the workforce increases and more nonlaboratory GCs are accessible (Stoll et al. 2017). Pretest counseling by a laboratory GC raises obvious concern that the GC may be persuasive, even if unintentionally and without conscious awareness, in steering the patient toward their laboratory employer for testing. A GC can be careful to minimize bias and disclose who he or she works for to the patient, but the COI still may not be apparent to the patient.

Currently, disclosure of COI is not mandatory for GCs. Although COIs are required to be declared in peer-reviewed journals and in scientific conference presentations, those disclosures do nothing to inform the patient of a GC's financial ties. Some have suggested that a COI may be mitigated by notifying the patient multiple times about the laboratory GC's source of employment (Iacoboni et al. 2018) or perhaps through a system whereby GCs would register their COI in a national database. A database in which GCs' COIs could be deposited, similar to that which exists for physicians, may be an acceptable route to address the main issue of transparency for pa-

tients. After declaring their COIs, physicians may continue to receive payment or other benefits from pharmaceutical companies for prescribing their products. A COI is not typically mentioned directly to patients by physicians; rather, the national database registry is deemed an acceptable minimum step to achieve transparency. Generally, GCs in direct patient care do not receive payments or benefits dependent on the number of tests ordered, whether employed by a laboratory or not. Yet some interpretations of COI indicate that even a perceived financial conflict can be as or more damaging than an actual COI (https://thednaexchange.com/2016/04/25/appearances-are-important/). If true, then a publicly searchable database of COI listings would offer no additional benefit beyond a policy of verbally notifying patients during appointments except for anyone who decides to search it in advance and potentially switch providers. Would the required use of a COI database actually prime the field to move toward an environment more similar to the one physicians exist in, whereby GCs would be in a position to seek or accept financial support from laboratories or pharmaceutical companies and not mention their COI verbally to patients, because of the illusion of transparency provided by database registration?

With laboratory GCs providing direct patient care, the emphasis on quality becomes increasingly important to maintain. Oftentimes when a patient is referred to genetic counseling through a laboratory, the referring provider has already recommended a specific genetic test(s). The consulting GC at the laboratory must then balance the expectations of their employer, the patient, and the referring provider. Not only must the patient be made aware up-front that the GC is employed by the laboratory which would be performing any testing ordered, the GC must ensure that decisions around testing are focused on what is most appropriate for and desired by the patient. The GC should then communicate to the referring provider why a test was performed, if relevant. Perhaps some of this conflict is mitigated by ensuring that the laboratory provides transparent guidance around its expectations for GCs to emphasize quality even if (and especially when) cost of testing decreases with increased quality. Other options include blinding the GC to the cost of testing and having clear protocols based on clinical guidelines. Some preliminary data suggests that laboratory GCs who are in direct patient care roles do alter or cancel genetic testing orders to ensure the right test for the right patient is ordered (Iacobani et al. 2018), often reducing revenue to the laboratory. This evidence suggests that GCs are aiming to be transparent to their employers and to their patients as they abide by ethical codes and prioritize high-quality, patient-centered care.

Research

The connection between academic and commercial laboratory companies is a long-standing one, with one study estimating that >85% of consultants in a sample of 19 for-profit clinical genetics laboratories have dual affiliations in academic centers (Natowicz and Ard 1997). In a 2012 study, 42% of 43 GCs working mainly in laboratory settings performed research/study coordination, amounting to 2.7% of their time (ranging from 0% to 40%) (Christian et al. 2012). Later, a survey of 121 GCs working primarily in laboratory settings found that 2.5% (n = 3) listed research/study coordinator as their primary role (Waltman et al. 2016). In that study, the average overall time spent on research and publications by all surveyed laboratory GCs was 6.9%, with the main tasks being collecting and analyzing data. Fifty-seven percent of the respondents worked in a commercial laboratory setting. Of those GCs, the most common research tasks were collecting data, analyzing data, and writing publications (each task had 15%–20% of respondents who performed them frequently). Tasks that were done frequently by <5% of commercial laboratory GCs included recruiting or consenting patients, writing IRB protocols, and submitting to databases.

With the increase in clinical exome and genome sequencing, laboratories have become the primary source of valuable genomic data from both affected and unaffected individuals (Boycott et al. 2013). To move the science forward,

even in instances where no direct revenue is being generated, some laboratories have identified a need to actively engage in research with collaborators. As such, some GCs began to have significant or even full research responsibilities within commercial laboratories. The main tasks in this role typically include evaluating internal data for suitability of inclusion in research proposals, communicating with ordering clinicians to ensure provider and patient consent, navigating ethical concerns and research expectations across collaborators, and advising internal laboratory colleagues on research topics and best practices (McWalter et al. 2018). At scientific conferences or through peer-reviewed publications, laboratory GCs present high-quality genetic or genomic laboratory data that enriches the field of translational genomics by opening doors to unique or larger data sets not otherwise accessible. As more scientific and laboratory companies merge and pursue involvement with therapeutics, GCs may be involved with the genomics side of companies that also have pharmaceutical ties. Because pharmaceutical companies face barriers in presenting at scientific conferences because of the COI, a GC or other health-care provider employed by a company with pharmaceutical ties may face barriers or even be banned from presenting data at certain conferences regardless of whether their work is related to pharmaceuticals. This type of ban-by-association limits the field's access to valuable data and reduces motivation for collaborative research with laboratory GCs.

Having research GCs in commercial laboratories has been extremely useful in facilitating rare disease discovery, which depends greatly on large data sets and communication across institutions. Various Web-based platforms exist for this purpose, and often a research laboratory GC is the point person for a laboratory and facilitates these discussions and collaborations with clinicians, researchers, and patients. GeneMatcher, a tool launched by an academic consortia in the Fall of 2013 to connect researchers, clinicians, laboratory scientists, and patients to discuss rare variants in "candidate genes," aims to elucidate new genetic disorders (Sobreira et al. 2015). A study by research GCs at a commercial laboratory over eight retrospective months and four prospective months looked at the outcomes of 642 requests from external GeneMatcher users (requesters), which were triaged by the laboratory GCs (Fisher et al. 2018). Approximately one-quarter of those discussions led to agreements to collaborate and resulted in 13 peer-reviewed journal publications. An additional 25%–31% of requests had enough potential for the laboratory GC to facilitate connections between the external requesters and the ordering clinicians, to continue comparisons. These types of discussions contribute to the scientific literature on rare disease, allow other laboratories to diagnose conditions they otherwise might not, and allow families to receive diagnoses and build a supportive community and research networks.

Variant Interpretation

Variant interpretation has become a key area of contribution for GCs (Swanson et al. 2014). The ability to synthesize laboratory and clinical information increases the accuracy of result interpretation. In laboratories, GCs are often sought out for their expertise on various aspects of the company's products, because other members of the laboratory may have little or no clinical experience. With their clinical counterparts also in need of variant interpretation skills, laboratory GCs have been an important source of peer education on the topic (Reuter et al. 2018). Common responsibilities requiring clinical expertise include reviewing and comparing the literature to a patient's phenotype, determining the relevance of particular genes or variants for inclusion on a gene panel test, and results reinterpretation or determining test appropriateness in light of new medical or family information. Approximately 42% of laboratory GCs in one study report that investigating the clinical significance of variants is an important part of their job (Waltman et al. 2016).

With the introduction of clinical exome and genome sequencing, the need for variant interpretation has increased greatly. Genetic counselors were involved with the creation of laboratories' internal variant interpretation protocols and are now key contributors in refining the American College of Medical Genetics and Genomics (ACMG) variant classification guidelines. Laboratories factor the ACMG guidelines into their own detailed protocols for variant interpretation but may have more specific criteria and guidelines for which only certain laboratory personnel are qualified to assess and apply. In addition, laboratory GCs may be responsible for outlining the rationale to report variants aside from clinical factors, including applying population database frequencies and incorporating known mechanism(s) of disease for the disorder. Currently there are GCs who function as clinical analysts for exome or genome sequencing, supplementing geneticists who have primarily filled that role (McWalter et al. 2018). Although workflows may differ between laboratories, a GC who analyzes genomic data typically applies filters, factors in clinical presentation, and considers ACMG or other proprietary guidelines, with the aim of identifying a variant(s) that may explain symptoms or have health implications. Laboratory GCs are also involved with making submissions to or responding to inquiries for more information about their entries in ClinVar, a resource for variant interpretation that would not exist in its current form without laboratory data sharing. Genetic counselors not only interpret variants but also process and interpret all data from the point at which it is generated, such as creating filtering pipelines, etc. Genetic counselors are involved at various points in the process of issuing reanalysis results from previously completed exome/genome reports in which the initial result needs to be updated based on new data about the variant, gene, or patient. Although some of the steps in that process are automated, depending on the laboratory, others require clinical correlation and interpretation and contact with the ordering provider.

Although variant interpretation is cited as a collaborative effort, there remains a divide between clinical and laboratory groups. From both the clinical and laboratory perspectives, there can be concerns around privacy, accuracy of interpretation, ownership, and resources, causing perspectives and goals to not always be fully aligned between the groups. Clinical groups sometimes provide little or no phenotypic information about the patient when sending in samples for genetic testing because of a concern that the laboratory may sell or publish the patient's data without their knowledge. The clinical group may be working on research or wanting to publish the data themselves, so they might not want others to have access to all the information. That concern is mounting because of increasing venture capital investments in some laboratories in which the data may be viewed and treated as a profitable commodity. Upon receiving a results report, the clinical team may have concerns that the laboratory's interpretation was incorrect, although the laboratory may have made a different determination had they access to the same medical and family history information as the clinicians. Sometimes a solution is to create a small in-house laboratory so that all the data are within one institution, although it may not be fully equipped with the breadth and depth of scientific expertise nor the sizeable genomic database of larger laboratories that have performed the testing for many years.

Should variant interpretation fall under a clinical GC's purview? If so, in what circumstances or to what degree? In settings such as cardiology or cancer in which genomic results may influence surgical and other life-altering decisions, GCs providing direct patient care commonly perform their own full variant interpretation. Perhaps in the future, collaborative data sharing, improved understanding of variants, and use of only high-quality laboratories will reduce the need for those efforts to be repeated by clinicians. Regardless of new roles, GCs and other medical and scientific team members all need to collaborate, not only across institutions, but also with data scientists and other health-care structures aiming to integrate genomic testing more seamlessly into medical care. It is possible that variant consultation centers will be created to serve as hubs for collaboration and best practices around variant interpretation. As it expands, the field of genomics will move away from a focus on classifying individual variants in favor of a more holistic approach that factors in polygenic data including environmental factors, sets of genomic variants, epigenetic variables, etc. In this multifactorial approach, collaboration and consultation centers will become even more necessary to power these connections and improve patient health.

CONCLUSION

The application of genetic counseling skills in the laboratory setting has been increasingly recognized and valued. Areas of expansion within the laboratory include laboratory test utilization, sales and marketing, research, bioinformatics, variant interpretation, product management, and direct patient care through telemedicine. Genetic counselors are contributing to medical scientific knowledge through research collaborations and clinical trials, phenotypic database development, and variant classification guidelines. As GCs' skills become even more valuable alongside new technologies and precision medicine, clinical laboratories will continue to be a place for GCs to play a significant role in defining and creating high-quality patient care, with many more exciting avenues on the horizon.

ACKNOWLEDGMENTS

This research was supported in part by the Intramural Research Program of the National Institutes of Health (NIH), National Human Genome Research Institute (NHGRI).

REFERENCES

Amos J, Gold B. 1998. Testing environment for single-gene disorders in U.S. reference laboratories. *Hum Mutat* **12:** 293–300. doi:10.1002/(SICI)1098-1004(1998)12:5<293:: AID-HUMU1>3.0.CO;2-F

Athens BA, Caldwell SL, Umstead KL, Connors PD, Brenna E, Biesecker BB. 2017. A systematic review of randomized controlled trials to assess outcomes of genetic counseling. *J Genet Couns* **26:** 902–933. doi:10.1007/s10897-017-0082-y

Balcom JR, Kotzer KE, Waltman LA, Kemppainen JL, Thomas BC. 2016. The genetic counselor's role in managing ethical dilemmas arising in the laboratory setting. *J Genet Couns* **25:** 838–854. doi:10.1007/s10897-016-9957-6

Boycott KM, Vanstone MR, Bulman DE, MacKenzie AE. 2013. Rare-disease genetics in the era of next-generation sequencing: discovery to translation. *Nat Rev Genet* **14:** 681–691. doi:10.1038/nrg3555

Buchanan AH, Rahm AK, Williams JL. 2016. Alternate service delivery models in cancer genetic counseling: a mini-review. *Front Oncol* **6**: 1–6. doi:10.3389/fonc.2016.00120

Christian S, Lilley M, Hume S, Scott P, Somerville M. 2012. Defining the role of laboratory genetic counselor. *J Genet Couns* **21**: 605–611. doi:10.1007/s10897-011-9419-0

Counseling ACfG. 2015. Practice-based competencies for genetic counselors. https://www.gceducation.org/wp-content/uploads/2019/02/ACGC-Core-Competencies-Brochure_15_Web.pdf

Field T, Brewster SJ, Towne M, Campion MW. 2016. Emerging genetic counselor roles within the biotechnology and pharmaceutical industries: as industry interest grows in rare genetic disorders, how are genetic counselors joining the discussion? *J Genet Couns* **25**: 708–719. doi:10.1007/s10897-016-9946-9

Fisher L, Holmes H, Forster C, Yang S, Torti E, McWalter K, Cho M. 2018. Use of GeneMatcher platform to facilitate connections between laboratories, clinicians, and researchers interested in candidate genes. In *National Society of Genetic Counselors Annual Conference*, Atlanta, GA. https://www.nsgc.org/p/cm/ld/fid=157

Garg S, Williams NL, Ip A, Dicker AP. 2018. Clinical integration of digital solutions in health care: an overview of the current landscape of digital technologies in cancer care. *JCO Clin Cancer Inform* **2**: 1–9. doi:10.1200/CCI.17.00159

Goodenberger M, Thomas B, Wain KE. 2015. The utilization of counseling skills by the laboratory genetic counselor. *J Genet Couns* **24**: 6–17. doi:10.1007/s10897-014-9749-9

Goodenberger M, Thomas B, Kruisselbrink T. 2017. *Practical genetic counseling for the laboratory*. Oxford University, New York.

Groepper D, McCarthy Veach P, LeRoy BS, Bower M. 2015. Ethical and professional challenges encountered by laboratory genetic counselors. *J Genet Couns* **24**: 580–596. doi:10.1007/s10897-014-9787-3

Heimler A. 1997. An oral history of the National Society of Genetic Counselors. *J Genet Couns* **6**: 315–336. doi:10.1023/A:1025680306348

Iacoboni D, Lynch K, Esplin ED, Nussbaum RL. 2018. Conflicts of interest in genetic counseling: addressing and delivering. *Genet Med* **20**: 1094–1095. doi:10.1038/gim.2017.234

Kotzer KE, Riley JD, Conta JH, Anderson CM, Schahl KA, Goodenberger ML. 2014. Genetic testing utilization and the role of the laboratory genetic counselor. *Clin Chim Acta* **427**: 193–195. doi:10.1016/j.cca.2013.09.033

Lee J. 2013. *Late news: Cigna requires genetic counseling.* Retrieved 3/5/2019, from https://www.modernhealthcare.com/article/20130727/MAGAZINE/307279986/late-news-cigna-requires-genetic-counseling.

McCuaig JM, Armel SR, Care M, Volenik A, Kim RH, Metcalfe KA. 2018. Next-generation service delivery: a scoping review of patient outcomes associated with alternative models of genetic counseling and genetic testing for hereditary cancer. *Cancers* **10**: E435. doi:10.3390/cancers10110435

McWalter K, Cho MT, Hart T, Nusbaum R, Sebold C, Knapke S, Klein R, Friedman B, Willaert R, Singleton A, et al. 2018. Genetic counseling in industry settings: opportunities in the era of precision health. *Am J Med Genet C Semin Med Genet* **178C**: 46–53. doi:10.1002/ajmg.c.31606

Miller CE, Krautscheid P, Baldwin EE, Tvrdik T, Openshaw AS, Hart K, LaGrave D. 2014. Genetic counselor review of genetic test orders in a reference laboratory reduces unnecessary testing. *Am J Med Genet C Semin Med Genet* **164A**: 1094–1101. doi:10.1002/ajmg.a.36453

National Society of Genetic Counselors. 2018. *National Society of Genetic Counselors Professional Status Survey.* Retrieved 5/1/2019, from https://www.nsgc.org/p/cm/ld/fid=68

Natowicz MR, Ard C. 1997. The commercialization of clinical genetics: an analysis of interrelations between academic centers and for-profit clinical genetics diagnostics companies. *J Genet Couns* **6**: 337–355. doi:10.1023/A:1025632423186

Pereira J, Díaz O. 2019. Using health chatbots for behavior change: a mapping study. *J Med Syst* **43**: 135. doi:10.1007/s10916-019-1237-1

Rabideau MM, Wong K, Gordon ES, Ryan L. 2016. Genetic counselors in startup companies: redefining the genetic counselor role. *J Genet Couns* **25**: 649–657. doi:10.1007/s10897-015-9923-8

Rashkin MD, Bowes J, Dunaway DJ, Dhaliwal J, Loomis E, Riffle S, Washington NL, Ziegler C, Lu J, Levin E. 2019. Genetic counseling, 2030: an on-demand service tailored to the needs of price conscious, genetically literate, and busy world. *J Genet Couns* **28**: 456–465. doi:10.1002/jgc4.1123

Reuter C, Grove ME, Orland K, Spoonamore K, Caleshu C. 2018. Clinical cardiovascular genetic counselors take a leading role in team-based variant classification. *J Genet Couns* **27**: 751–760. doi:10.1007/s10897-017-0175-7

Scacheri C, Redman JB, Pike-Buchanan L, Steenblock K. 2008. Molecular testing: improving patient care through partnering with laboratory genetic counselors. *Genet Med* **10**: 337–342. doi:10.1097/GIM.0b013e31817283a5

Sobreira N, Schiettecatte F, Valle D, Hamosh A. 2015. GeneMatcher: a matching tool for connecting investigators with an interest in the same gene. *Hum Mutat* **36**: 928–930. doi:10.1002/humu.22844

Stoll KA, Mackison A, Allyse MA, Michie M. 2017. Conflicts of interest in genetic counseling: acknowledging and accepting. *Genet Med* **19**: 864–866. doi:10.1038/gim.2016.216

Stoll K, Kubendran S, Cohen SA. 2018. The past, present and future of service delivery in genetic counseling: keeping up in the era of precision medicine. *Am J Med Genet* **178C**: 24–37. doi:10.1002/ajmg.c.31602

Swanson A, Ramos E, Snyder H. 2014. Next generation sequencing is the impetus for the next generation of laboratory-based genetic counselors. *J Genet Couns* **23**: 647–654. doi:10.1007/s10897-013-9684-1

Wain K. 2018. A commentary on opportunities for the genetic counseling profession through genomic variant interpretation: reflections from an ex-lab rat. *J Genet Couns* **27**: 747–750. doi:10.1007/s10897-018-0247-3

Waltman L, Runke C, Balcom J, Riley JD, Lilley M, Christian S, Zetzsche L, Goodenberger ML. 2016. Further defining the role of the laboratory genetic counselor. *J Genet Couns* **25**: 786–798. doi:10.1007/s10897-015-9927-4

Zetzsche LH, Kotzer KE, Wain KE. 2014. Looking back and moving forward: an historical perspective from laboratory genetic counselors. *J Genet Couns* **23**: 363–370. doi:10.1007/s10897-013-9670-7

Cite this article as *Cold Spring Harb Perspect Med* doi: 10.1101/cshperspect.a036574

Informed Consent in the Genomics Era

Shannon Rego,[1] Megan E. Grove,[2] Mildred K. Cho,[3,4] and Kelly E. Ormond[4,5]

[1]Institute for Human Genetics, University of California San Francisco, San Francisco, California 94143, USA

[2]Stanford Medicine Clinical Genomics Program, Stanford, California 94305, USA

[3]Division of Medical Genetics, Stanford University Department of Pediatrics, Stanford, California 94305, USA

[4]Stanford Center for Biomedical Ethics, Stanford, California 94305, USA

[5]Department of Genetics, Stanford University School of Medicine, Stanford, California 94305, USA

Correspondence: kormond@stanford.edu

Informed consent, the process of gathering autonomous authorization for a medical intervention or medical research participation, is a fundamental component of medical practice. Medical informed consent assumes decision-making capacity, voluntariness, comprehension, and adequate information. The increasing use of genetic testing, particularly genomic sequencing, in clinical and research settings has presented many new challenges for clinicians and researchers when obtaining informed consent. Many of these challenges revolve around the need for patient comprehension of sufficient information. Genomic sequencing is complex—all of the possible results are too numerous to explain, and many of the risks and benefits remain unknown. Thus, historical standards of consent are difficult to apply. Alternative models of consent have been proposed to increase patient understanding, and several have empirically demonstrated effectiveness. However, there is still a striking lack of consensus in the genetics community about what constitutes informed consent in the context of genomic sequencing. Multiple approaches are needed to address this challenge, including consensus building around standards, targeted use of genetic counselors in nongenetics clinics in which genomic testing is ordered, and the development and testing of alternative models for obtaining informed consent.

As use of genomic testing continues to increase, genetic counselors play a valuable role in genetics clinics and in other clinical and research settings. A genetic counselor's skill set includes the ability to facilitate the consent process for genomic testing, and this can be particularly useful in settings in which other team members may be less familiar with genetics and the complexities of informed consent for genomic sequencing. The increasing use of genomic sequencing by specialists outside of genetics also presents an opportunity for genetic counselors to help ensure that patients providing informed consent understand the information relevant to making an informed decision about testing and have the opportunity to get answers to their questions. Here we describe the history and evolution of informed consent in general medicine, in genetic counseling, and most recently in the genomics era; proposed approaches for adapting the informed consent process in a genomics context; and future directions, including the important role for genetic counselors in obtaining informed consent for genomic testing.

HISTORY OF INFORMED CONSENT

Informed consent is the process by which clinicians and biomedical researchers gather *autonomous* authorization for a medical intervention or research participation (Appelbaum 2007; Beauchamp 2011). The concept of informed consent evolved from the idea that patients and research participants have the right to make autonomous decisions about their bodies. Medical informed consent, whether clinical or research focused, assumes decision-making capacity of the person consenting, voluntariness, and comprehension of adequate information (Bunnik et al. 2013a).

Obtaining informed consent is a process. In a genetic counseling setting, this process generally entails a conversation in which the clinician or researcher conveys information about the medical intervention or research study, including the associated risks and benefits, and the patient conveys information about their health-related goals, expectations, priorities, preferences, and concerns. The clinician may assess patient understanding of the information conveyed by asking questions or utilizing a teach-back method. Based on this feedback, the clinician may try different teaching approaches to tailor the information and discussion to the patient's needs. The patient has the opportunity to ask questions, and the person obtaining consent facilitates a discussion with the goal of determining if the proposed intervention or study aligns with the patient's goals and values. Ideally the dialogue takes place over time—a patient may feel differently about the intervention or study at different times, and ongoing communication between the clinician and the patient is key to ensuring responsiveness to the patient's viewpoint as his or her circumstances change (Bernat and Peterson 2006). The process of informed consent is typically documented through both parties signing a form.

Although the concept of informed consent seems as though it must be nearly as old as medicine, in reality it is remarkably new. In fact, for most of the history of medicine, medical ethicists were more concerned with how best to conceal potentially stress-inducing information from patients to reduce harm and anxiety. Even the Hippocratic oath encouraged physicians to keep their patients in the dark as a means of protecting them from undue stress (Beauchamp 2011). It was not until 1947 that the Nuremberg code, created in response to medical experiments conducted in Nazi concentration camps, established the need for voluntary consent and an informed assessment of risks and benefits. It does not, however, specify what information should be conveyed to potential research subjects (Moreno et al. 2017).

Informed consent is not just an ethical and moral concept but also a legal concept developed in the courts, starting with a series of cases beginning in the 1950s. In 1957 the term "informed consent" first appeared in the case of *Salgo v. Leland Stanford Jr University Board of Trustees* (Salgo 1957), in which a patient underwent a novel medical procedure without his physician warning him of the potential risks, including the paralysis that ultimately afflicted him. The ruling by the court focused on the duty of clinicians to disclose risks associated with any recommended procedure but also emphasized the need for patient understanding, reinforcing the idea of patients as autonomous individuals who should be empowered with information to make decisions in their own best interest.

Over the next 15 years, courts refined the concept of informed consent by establishing standards and defining exceptions for when informed consent is not required (Nelson-Marten and Rich 1999; Beauchamp 2011). A series of court cases in the 1970s, in particular *Canterbury v. Spence*, laid the foundations for a patient-oriented standard for disclosure (Spence 2008; Beauchamp 2011). In this 1972 case, a patient undergoing surgery because of a herniated disc was paralyzed. He sued his physician, claiming that he had not been warned of the risks involved in the surgery. The physician argued the risk was small and disclosing it could have provoked unnecessary anxiety. The Court of Appeals of the District of Columbia ruled against the physician, stating that he had had an obligation to disclose the risk of paralysis to the patient and described the following types of information as necessary to achieve informed consent: the condition being treated, the nature of the intervention, the likely results, alternative forms of treatment, and serious risks or complications that could result from the intervention (Murray 2012). This ruling dramatically shifted the standard practice of disclosure from the previous physician-centric approach, in which the appropriate amount of information required for informed consent was determined by what other physicians would have done in similar circumstances, to a patient-centric approach, in which the appropriate amount of information is defined as the amount of information a "reasonable patient" would need to make an informed choice about whether to proceed with a given intervention (Beauchamp 2011). Court cases that followed *Canterbury v. Spence* added further disclosure requirements to this standard, but the precise details of the legal requirements for achieving informed consent vary by state (Spector-Bagdady et al. 2018).

As the standards for informed consent have continued to evolve in the courts and in practice, genetic testing technology has evolved alongside it. In the 20-plus years since genetic testing became commonplace in clinic and research settings, physicians, researchers, and ethicists have attempted to apply the standards of informed consent used in medicine more generally and found some of them to be particularly challenging in a genetic context. For example, informed consent requires decision-making capacity, but decision-making capacity may be uncertain in families with suspected genetic conditions that include intellectual disability, autism, or neurodegeneration. The bar for appropriate decision-making capacity may vary depending on the level of risk involved in the intervention or genetic test—riskier interventions generally require a higher level of decision-making capacity and comprehension. Voluntariness may be compromised by family members for whom the testing can also have implications (resulting in

subtle, or not-so-subtle, coercion within the family), or clinicians who convey strongly that testing is in the patient's best interest. Comprehension is difficult to achieve and assess when the topic is complicated, as with genetic tests. Finally, the question of what counts as sufficient information has been the topic of numerous court cases and publications and still remains very subjective and dependent upon the patient's needs. Although some of these challenges also arise in consent processes in other areas of medicine, they arise much more frequently in genetics, in part because genetics, by definition, involves the whole family, and because patients in this setting are more likely to have developmental disabilities that may compromise understanding.

INFORMED CONSENT IN GENETICS AND GENOMICS

Informed Consent in the Pregenomics Era

Prior to next-generation sequencing, genetic testing was an iterative process in which a clinician created a differential diagnosis, began by testing the one or two genes most likely to yield a diagnosis, and then tested other genes if the prior test was negative. The traditional genetic counseling approach to informed consent in this context had an educational focus, providing information about the testing process; potential benefits, risks, and limitations of testing; and education about the natural history and potential treatment(s) of the condition(s) for which testing was being performed (Ormond 2013). Although this approach is similar to that used for other medical tests or interventions, the details of the consent conversation differed, particularly with regard to the risks. The most significant risks associated with clinical genetic testing are psychological and social—for example, the risk of anxiety or stress related to presymptomatic knowledge that one may develop a genetic condition for which there may not be any treatments or preventive measures, and the potential risk of stigma or discrimination due to such a genetic test result. The familial nature of genetic information also presents unique risks. Individuals within a family may have experience living with a genetic condition and may approach an informed consent process with strong feelings about what they do and do not want from testing and what they may do in response. There can be differences in opinion among family members about testing, creating stress or conflict. In research genetic testing, additional risks occur at a group or community level and can include the potential for stigmatization. For example, in 2003 members of the Havasupai tribe discovered that DNA samples collected for type II diabetes research had been used to study schizophrenia, migration, and inbreeding without their knowledge or consent (Garrison and Cho 2013). Community engagement and consent are critical to mitigate these types of problems.

The psychoeducational approach to consent for genetic testing emerged at a time when there were limited data about the potential psychological and social harms of genetic testing and even more limited interventions available for most genetic conditions than exist today. Extensive information was shared in hopes of providing patients with as comprehensive an understanding of the risks as possible. This was possible because it was typically only one condition being tested for (such as Huntington's disease), and the penetrance of the first conditions for which clinical testing was available was typically quite high (as these were the conditions for which it was easiest to establish gene–disease relationships), which removes much of the need to assess the patient's tolerance for uncertainty as part of the consent discussion. Studies suggest that this traditional model was not particularly effective at equipping patients with an understanding of information the authors considered necessary for providing informed consent (Joffe et al. 2001; Lee et al. 2011). Over time, a more traditional psychoeducational model of informed consent for genetic testing evolved to be an interactional and patient-centered approach, mirroring the evolution toward shared decision-making models in other parts of medicine (Kunneman and Montori 2017).

Despite the differences in the nature of the risks of genetic testing as compared to many other types of medical procedures, approaches to genetic testing consent remained primarily information-laden, and this was readily modified to accommodate the single-gene (or small-panel) genetic testing that was available until 2011. However, it has become increasingly clear that this consent approach cannot be scaled to genomic sequencing, and clinicians and researchers face a number of challenges in their efforts to achieve informed consent in the context of genomic sequencing (Ormond et al. 2010; Berg et al. 2011; Bester et al. 2016).

Genomic Sequencing Has Changed Models of Consent

Next-generation sequencing made it possible to obtain a great deal of genetic information quickly and cost-effectively, leading to widespread adoption of panel-based genetic testing and increasing use of exome and genome sequencing in both clinical and research settings. The availability of genomic data created new challenges around informed consent as we have traditionally understood it (Patch and Middleton 2018), including how and whether it is even possible to satisfy the previously accepted standards for informed consent for tests with such expansive implications (Williams et al. 2014; Samuel et al. 2017). Many of the specific challenges of achieving informed consent in the context of genomic sequencing relate to defining the requirement for adequate information and comprehension (see Table 1; Bunnik et al. 2013a). The question of how much information is "adequate" is deeply subjective and poorly defined, and the appropriate amount of information required for a patient to make an informed decision as per the predominant patient-centered standard presents challenges for clinicians. Some patients and families desire significantly more information than others to feel prepared to decide about genomic testing. For example, patients with more lived experience with a condition may have different information needs than a patient with no such familiarity. Clinicians often struggle to determine how much detail to provide about genomic sequencing and how to appropriately simplify the information to

Table 1. Major and unique concepts in informed consent for genomic sequencing

Concepts that differ between genetic and genomic consent	
Concept	*Difference between genetic versus genomic consent*
Scope of test	Scope of test is much larger in genomic testing (e.g., includes all genes rather than just one or a few genes)
Limitations of test	Sequencing-based tests like genomic cannot identify all mutation types
	Higher chance of uncertain results with genomic sequencing
Potential risks	Genomic data is identifiable in ways genetic data is not, increasing the privacy risks
	Insurance risks due to secondary or incidental findings
	Genomic testing can cause stress or anxiety due to:
	• Learning about unexpected family relationships (e.g., nonpaternity or consanguinity)
	• Learning about unexpected health risk (e.g., due to secondary or incidental findings)

Concepts that are similar between genetic and genomic consent

Description of test process/procedures
Expected benefits
Costs or payments
Voluntary nature of the test
Alternatives to the test
Confidentiality/privacy
How/when/to whom results will be returned
Future use of data/samples

make it understandable (Ormond et al. 2010; Klima et al. 2014). Too much detail or complexity can negatively impact comprehension, lead to a loss of attention, and/or create unnecessary anxiety. Too little can lead to decisions that are inconsistent with the patient's values or goals. Achieving balance represents the primary challenge when obtaining informed consent for genomic testing and is another reason for the importance of approaching consent as a dialogue between clinician and patient, as the clinician must elicit information about the patient's values, objectives, preferences, and concerns in order to determine what information that patient might need in order to make an informed decision. The patient must also be given an opportunity to ask questions about the information conveyed and to cover any additional information he or she may need to incorporate into the decision-making process.

Challenges of Obtaining Informed Consent for Genomic Sequencing

Genomic sequencing is complex, and it can be difficult for a clinician or researcher to know how deep to dive into that complexity when

having a consent conversation with a patient. Currently, genomic testing is most often used when the patient's differential diagnosis is too broad to make single-gene or panel testing practical. The potential range of results is limited only by the number of conditions associated with specific genes (now more than 4000 and rising) (OMIM). This means the possible results are too numerous and diverse to discuss using the traditional consent approach. Rather, clinicians should describe the types of results to patients in broader strokes—for example, describing the possibility that the patient could learn he or she has a genetic difference that provides information about prognosis, including whether treatments or preventative measures are currently available. It can be challenging for patients to understand that, unlike many medical tests they may have had in the past in which the result is binary and immediately actionable, their doctor cannot tell them in pretest what will happen next if the genomic sequencing result is positive, because it depends on what that result is. Patients may also struggle to understand the possibility that the result may change over time (e.g., if a variant of uncertain significance is reclassified as pathogenic several years later) or the possibility that the test will yield unexpected findings that are unrelated to the primary reason for testing (often called secondary or incidental findings). The issue of future re-analysis of genomic data also raises questions about the extent of clinicians' and researchers' duties to conduct such analyses and their obligations to discuss the possibility as part of the initial informed consent, which have recently been addressed in policy statements by the American College of Medical Genetics and the American Society of Human Genetics (Bombard et al. 2019; David et al. 2019).

Describing the potential benefits and risks of genomic sequencing presents similar challenges. For patients, one of the most significant benefits is the possibility of receiving a diagnosis (Dillon et al. 2018; Dragojlovic et al. 2018) or, in some cases, identifying multiple diagnoses that are at play (Posey et al. 2017). This shortens their "diagnostic odyssey" and can potentially lead to new management recommendations or treatments, provide information about recurrence risk, shed light on prognosis, and provide a support network of other families dealing with the same diagnosis (Lambertson et al. 2015). It is faster than a more iterative approach to genetic testing (Petrikin et al. 2015) and increasingly cost-effective (Farnaes et al. 2018; Stark et al. 2019). As for traditional genetic testing, the main risks associated with genomic sequencing are psychological, social, and familial, including learning about unexpected biological relationships (consanguinity, misattributed parentage) when multiple family members contribute samples for interpretation.

As genomic testing has gained traction in the media and popular science, patients tend to overestimate its capabilities (Roberts et al. 2018; Wynn et al. 2018), necessitating a discussion about the patient's expectations for the testing and the likelihood of receiving diagnostic results in order to provide a realistic projection of the clinical utility of genomic sequencing (Bernhardt et al. 2015). Additional logistical concerns that participants may want to consider at the time of consent include time limitations, provider expertise, and the rapidly changing potential scope of the test (McGuire and Beskow

2010; Hallowell et al. 2015; Burke and Clarke 2016).

Incidental and Secondary Findings Present Specific Challenges and Opportunities

One of the most consequential differences between single-gene or panel testing compared to genomic sequencing is the potential to receive results that are not related to the primary indication for testing. Incidental findings refer to results with potential health significance that are identified in the process of searching for a variant that explains the patient's phenotype but are not sought out purposefully. Secondary findings are also unrelated to the reasons for testing, but these are actively sought out by the testing lab, usually because the identification of such results would allow for prevention or early identification of a serious but preventable or treatable medical condition (Green et al. 2013). Practices have varied widely among laboratories regarding the options and specifics around the return of incidental or secondary findings (O'Daniel et al. 2017), and in an effort to provide guidance to laboratories, the American College of Medical Genetics and Genomics (ACMG) first issued recommendations for the return of secondary findings for patients undergoing genomic sequencing in 2013 (Green et al. 2013). These guidelines proposed a list of 56 genes in which secondary findings be returned to patients undergoing genomic sequencing. Subsequent revisions have included minor changes to the list of genes, as well as a major revision to allow patients to opt out of receiving secondary findings (ACMG Board of Directors 2015). Independent work groups have also assessed the medical actionability of the genes included on the secondary findings list (Hunter et al. 2016). The most recent update by the ACMG includes a list of 59 genes, most of which have been implicated in autosomal-dominant hereditary cancer or cardiovascular conditions for which there are well-established guidelines for early detection of disease or prevention (Kalia et al. 2017).

Most laboratories performing clinical genomic sequencing now offer patients the option to receive secondary/incidental findings, at a min-

Cite this article as *Cold Spring Harb Perspect Med* doi: 10.1101/cshperspect.a036582

imum pathogenic variants in the 59 genes endorsed by ACMG for return. Many labs offer additional options beyond the 59 genes, and there remains a great deal of variability among laboratory policies for reporting secondary findings and how those options are presented (Ackerman and Koenig 2018).

Secondary and incidental findings present both opportunities and challenges that are largely unique to genomic sequencing. Patients who consent to receive secondary findings could potentially learn that they have a significant risk of developing a life-threatening condition for which medical interventions are available. For example, if a female patient receives a pathogenic *BRCA1* variant secondary finding, she can mitigate the high risk of developing cancer by undergoing earlier and more frequent screening. Additionally, her family can be offered cascade testing, which is especially important as emerging data suggests that perhaps 50% of these families do not have a family history that would meet guidelines for clinical testing (Manickam et al. 2018).

Similar to the risks associated with primary genomic findings, those associated with secondary findings are largely psychological and social, including the short-term distress patients may face when learning about significant unexpected risks for serious genetic conditions (Sapp et al. 2018; Hart et al. 2019), discordance among family members about whether to test for secondary findings, insurance implications, and false-negative or false-positive results (Appelbaum et al. 2014; Ormond et al. 2019a).

Obtaining Informed Consent for Secondary Findings

Recent studies suggest patients undergoing genomic sequencing struggle with comprehension (Hellwig et al. 2018; Roberts et al. 2019). Information overload may contribute to patients' challenges with comprehension, and that problem can be exacerbated when the consent discussion includes secondary findings in addition to the primary purpose of testing (Turbitt et al. 2018). In such cases patients sometimes struggle to understand the difference between the two,

confounding the risks and benefits and in some cases misunderstanding the main purpose of the test. Discussing secondary findings may add time to the informed consent process. This creates logistical challenges and can also lead to information overload, which can make it difficult to achieve the level of comprehension necessary to consent to testing. Family discordance about who wishes to learn secondary findings can further complicate decision-making. For example, a secondary finding in a child is most often assumed to be present in one of the parents; if one parent elects to receive secondary findings and the other does not, one parent's status may reveal the other's by default. Practically, the parents must therefore be in agreement about the decision. Even outside the immediate family, it is possible that secondary findings could cause disruption to the family—for example, some family members may resent being presented with the knowledge of a genetic condition in the family (Hallowell et al. 2013). The risk for insurance discrimination is a particularly complicated part of any consent conversation regarding secondary findings, because the individual is being identified with a risk that was previously unknown, as compared to primary findings, in which the patient by definition has a preexisting medical condition that led to testing. Laws such as the Genetic Information Nondiscrimination Act [GINA (2008)] protect such patients from being denied health insurance because of a genetic difference, but GINA does not apply to life insurance, long-term care insurance, or disability insurance. These risks are complicated and largely hypothetical, making them difficult to explain, which in turn may inflate their significance in the minds of patients.

EXPLORING ALTERNATIVE MODELS OF CONSENT

Recognizing that people have struggled with using traditional models of consent to accommodate the new challenges presented by genomic sequencing, many have proposed alternatives. One such model is called staged (also known as tiered or layered) consent (Bunnik et al. 2013b; Appelbaum et al. 2014). This is a variation on the concept of generic consent first coined by Elias and Annas (1994), which proposes that patients should receive general pretest information that covers the possible test outcomes and their significance broadly and have the opportunity to ask relevant individual questions and personalize the content that they need. Staged consent and disclosure would occur as new findings arise, in part to alleviate the challenge of time pressure, to avoid overwhelming the patient with information, and to separate two decisions at the time of consent: whether or not to have the test for the diagnostic reasons and whether or not to receive secondary findings. Staged consent and disclosure processes also have the potential benefit that family members can independently decide whether to learn information related to the primary test indication and whether to learn about secondary findings. Yu et al. (2013) have demonstrated that patients and families are interested and engaged in a staged consent and disclosure process when offered it through an online portal. In fact, online platforms for genomic sequencing consent and return of results have continued to evolve as a cost-effective way of promoting participant understanding and engagement (Tabor et al. 2017), and increasing patients' comprehension and recall of what they learned in the consent process (Angiolillo 2004; Appelbaum et al. 2014). However, a staged process may be more logistically complicated and time-consuming for clinicians, who would be required to schedule multiple clinic visits or telemedicine sessions. This would add cost and create risk that patients who are lost to follow-up would not receive important information about their health. Although the staged approach could go a long way toward mitigating the challenges of achieving informed consent in the context of genomic sequencing, it may not be practical in many clinics or for underserved patient populations with limited access to the internet or with language or literacy barriers.

Increasingly, researchers and clinicians are turning to new conceptual approaches and technology in search of ways to streamline information exchange for both clinicians and patients. For example, the Consent and Disclosure Recommendations Workgroup of the National Institutes of Health (NIH)-funded Clinical Genome Resource has developed a conceptual framework and tools that genetics experts can apply to suggest a "communication starting place" for ordering clinicians. The recommended communication approach can then be tailored based on patient characteristics and using clinical judgment (Ormond et al. 2019b). A common application of technology to genetic testing has been in the creation of decision aids to facilitate increased patient participation in health-care decision-making by defining the decisions that must be made, providing information about the interventions or tests being offered, and clarifying the patient's personal values (https://decisionaid.ohri.ca/). Even before genomic sequencing was widely available in clinical and research settings, decision aids created in the context of more traditional genetic testing were shown to improve patient understanding (Kuppermann et al. 2014; Ekstract et al. 2017). More recently, researchers and clinicians have created online decision tools to help facilitate the consent and disclosure process for genomic sequencing, and results show promise that patients find these tools helpful (Birch et al. 2016; Bombard et al. 2018; Shickh et al. 2018). A major advantage to using technology as part of the consent and disclosure process is that it can reduce the need for clinician time, which decreases costs. Further, potential participants can review information or seek further information using available links to enhance their understanding of the goals of the research and the potential harms and benefits. It can also alleviate some of the challenges associated with the current shortage of genetics clinicians and provide valuable assistance to nongenetics health-care providers who will increasingly encounter genomic testing in their clinics. Previous studies of primary care physicians and nongenetics specialists have demonstrated these providers are enthusiastic about the potential benefits of genomic sequencing but many lack confidence in their knowledge and abilities surrounding genomics (Nippert et al. 2011; Carroll et al. 2016). There remains much to learn about the efficacy of decision aids for consenting to undergo genomic sequencing with regard to their impact

Cite this article as *Cold Spring Harb Perspect Med* doi: 10.1101/cshperspect.a036582

on patient knowledge, satisfaction, and posttest outcomes such as psychological well-being. Decision aids can also present practical challenges in clinic, particularly if they were created in a language or cultural context that is foreign to the patient. Additionally, overreliance on decision aids without access to providers to answer questions may allow individual patient misunderstandings about testing to go unchecked.

As a large part of the burden of the consent process for genomic sequencing relates to secondary findings, some models have been proposed to address these challenges specifically. In a research setting, one such proposal is a governance model. This model addresses the challenges associated with consent for secondary findings by suggesting the creation of a governing group in which stakeholders (including research participants) can provide input about key questions, including whether research participants should receive unexpected findings, what types of unexpected results, and how these should be returned. This approach takes some of the burden off of the consent process because participants agree to be governed by the decisions of others (Koenig 2014). This approach is being used in large genomics studies such as the Clinical Sequencing Evidence-generating Research (CSER) consortium (Amendola et al. 2018). Another approach to secondary findings in research settings involves outsourcing such results to a third party. In this approach, a research participant undergoing sequencing would not receive any incidental or secondary findings but would be provided with their data in order to pursue secondary findings with another genetic specialist or company. Therefore, consent for secondary findings and the determination of which types of results to receive would no longer be part of the consent process for the primary genomic research (Appelbaum et al. 2014). Many clinical labs offer patients the opportunity to request their raw data, as do some genomic research studies, and although most clinical labs also offer, at a minimum, the secondary findings options recommended by the ACMG, access to raw data allows patients to go much further in their search for potentially medically relevant data (Kalia et al. 2017; Ackerman

and Koenig 2018; Fowler et al. 2018). Although this is helpful in that it reduces the burden on the initial consent process and preserves patient autonomy, patient access to raw genomic data also has significant risks in that its analytic and clinical validity can be variable and there is no guarantee outside genomics services will be accurate in their interpretations or have the expertise to convey the significance of findings appropriately —limitations that patients should comprehend before obtaining their raw data for further interpretation (Appelbaum et al. 2014). Finally, a model that has been proposed to address challenges with consent for secondary findings is that of "binning" secondary findings into categories to simplify the options presented to the patient/ participant (Berg et al. 2011). With the binning model, patients have the option of saying yes or no to receiving categories of secondary findings stratified, for example, by level of risk and/or actionability. This model is already widely used in both clinical and research settings—for example, when patients are given the option to receive actionable secondary findings from the ACMG's list of 59 genes described above (Shahmirzadi et al. 2014; Ackerman and Koenig 2018).

While the alternative models described above have improved the consent process for genomic sequencing, none have completely addressed the challenges posed by this complicated test. There is still much more work to be done by researchers and clinicians, in partnership with our patients, to better promote informed consent and decision-making in the context of genomics.

FUTURE DIRECTIONS AND THE ROLE FOR GENETIC COUNSELORS

Rethinking Standards for Genomic Consent— a Need for a More Individualized, Flexible Approach

It is in the nature of genomic sequencing to be adapted and personalized, and appropriate standards for informed consent should recognize this feature of modern testing. Going forward, it is essential for clinicians and researchers to take a more individualized and flexible approach to informed consent when it comes to genomic sequencing. The widely used standard for the appropriate amount of information to disclose to patients in order to obtain informed consent is based on what a "reasonable person" would want to know, but experienced genetics clinicians know that patients differ in their information preferences about genomic testing (Regier et al. 2015). A more practical standard may be one that is personalized based on the individual or family and their information preferences. Genetic counselors are specialists in tailoring consent discussions based on the individual patient in front of them. This tailoring can include determining the appropriate amount of information to provide a patient to facilitate informed, values-based decision-making without sacrificing comprehension and also by adapting the complexity of the information based on the patient's level of understanding. Decision-making capacity, another important requirement for valid consent, is subjective and difficult to define; thus, its definition may also need to be flexible depending on the level of risk and potential benefit involved in the context of genomic sequencing. For some families, the level of risk involved in genomic sequencing is higher than others. A higher bar for decision-making capacity may make sense in situations with more significant risks and/or lower potential for direct benefits. The challenges clinicians face in obtaining informed consent for genomic sequencing arise primarily from the use of outdated standards and models for consent, and so it will be important for genetic counselors to continue to be involved in the process of adapting such standards and models of consent for genomic sequencing.

Scaling Up for the Genomics Era

To meet the increasing demand for genomic sequencing, scalable approaches to informed consent are needed. The presence of genetic counselors in nongenetics specialty clinics and research settings can help, but as genomic testing becomes more commonplace, other approaches will be needed to meet the demand. The goals of a more individualized and a more scalable approach to informed consent may at first seem contradictory, but they need not be. Technological solutions, such as interactive video platforms and applications, hold great promise when it comes to meeting this challenge, as they are both scalable and can also allow for patients to customize their own experience based on their preferences and values. Educating nongenetics health-care providers about genomic sequencing and consent can also help facilitate the scaling up of genomic sequencing. These approaches may also help free up time for genetic counselors to spend in settings where specialized knowledge of genetics is more necessary—in the disclosure of genomic testing results to patients. Genetic counselors can further this goal by devoting time to educating nongenetics providers about genomic sequencing and consent, as well as by participating in the development of technology-based educational tools and decision aids to facilitate the integration of patient values into decision-making (Lewis et al. 2016; Adam et al. 2018; Reumkens et al. 2019). Genetic counselors can also use their skills and knowledge on a broader scale by participating in larger policy discussions about informed consent in the genomics era.

CONCLUSIONS

The widespread availability of genomic sequencing in clinical and research settings has stretched traditional approaches to informed consent. Many new models of consent have been proposed to meet the goals of informed consent in the context of genomic testing, including staged or tiered models of consent, technological approaches, and others. Going forward, genetic counselors will be integral in developing more individualized, flexible, and scalable approaches to obtaining informed consent for genomic testing.

REFERENCES

Ackerman S, Koenig B. 2018. Understanding variations in secondary findings reporting practices across U.S. genome sequencing laboratories. *AJOB Empir Bioeth* **9**: 48–57. doi:10.1080/23294515.2017.1405095

ACMG Board of Directors. 2015. ACMG policy statement: updated recommendations regarding analysis and reporting of secondary findings in clinical genome-scale sequencing. *Genet Med* **17:** 68–69. doi:10.1038/gim.2014.151

Adam S, Birch PH, Coe RR, Bansback N, Jones AL, Connolly MB, Demos MK, Toyota EB, Farrer MJ, Friedman JM. 2018. Assessing an interactive online tool to support parents' genomic testing decisions. *J Genet Couns* **28:** 10–17. doi:10.1007/s10897-018-028-1

Amendola LM, Berg JS, Horowitz CR, Angelo F, Bensen JT, Biesecker BB, Biesecker LG, Cooper GM, East K, Filipski K, et al. 2018. The Clinical Sequencing Evidence-Generating Research Consortium: integrating genomic sequencing in diverse and medically underserved populations. *Am J Hum Genet* **103:** 319–327. doi:10.1016/j.ajhg.2018.08.007

Angiolillo AL, Simon C, Kodish E, Lange B, Noll RB, Ruccione K, Matloub Y. 2004. Staged informed consent for a randomized clinical trial in childhood leukemia: impact on the consent process. *Pediatr Blood Cancer* **42:** 433–437. doi:10.1002/pbc.20010

Appelbaum PS. 2007. Assessment of patients' competence to consent to treatment. *N Engl J Med* **357:** 1834–1840. doi:10.1056/NEJMcp074045

Appelbaum PS, Parens E, Waldman CR, Klitzman R, Fyer A, Martinez J, Price WN II, Chung WK. 2014. Models of consent to return of incidental findings in genomic research. *Hastings Cent Rep* **44:** 22–32. doi:10.1002/hast.328

Beauchamp TL. 2011. Informed consent: its history, meaning, and present challenges. *Camb Q Healthc Ethics* **20:** 515–523. doi:10.1017/S0963180111000259

Berg JS, Khoury MJ, Evans JP. 2011. Deploying whole genome sequencing in clinical practice and public health: meeting the challenge one bin at a time. *Genet Med* **13:** 499–504. doi:10.1097/GIM.0b013e318220aaba

Bernat JL, Peterson LM. 2006. Patient-centered informed consent in surgical practice. *Arch Surg* **141:** 86–92. doi:10.1001/archsurg.141.1.86

Bernhardt BA, Roche MI, Perry DL, Scollon SR, Tomlinson AN, Skinner D. 2015. Experiences with obtaining informed consent for genomic sequencing. *Am J Med Genet A* **167A:** 2635–2646. doi:10.1002/ajmg.a.37256

Bester J, Cole CM, Kodish E. 2016. The limits of informed consent for an overwhelmed patient: clinicians' role in protecting patients and preventing overwhelm. *AMA J Ethics* **18:** 869–886. doi:10.1001/journalofethics.2016.18.9.peer2-1609

Birch P, Adam S, Bansback N, Coe RR, Hicklin J, Lehman A, Li KC, Friedman JM. 2016. DECIDE: a decision support tool to facilitate parents' choices regarding genome-wide sequencing. *J Genet Couns* **25:** 1298–1308. doi:10.1007/s10897-016-9971-8

Bombard Y, Clausen M, Mighton C, Carlsson L, Casalino S, Glogowski E, Schrader K, Evans M, Scheer A, Baxter N, et al. 2018. The Genomics ADvISER: development and usability testing of a decision aid for the selection of incidental sequencing results. *Eur J Hum Genet* **26:** 984–995. doi:10.1038/s41431-018-0144-0

Bombard Y, Brothers KB, Fitzgerald-Butt S, Garrison NA, Jamal L, James CA, Jarvik GP, McCormick JB, Nelson TN,

Ormond KE, et al. 2019. The responsibility to recontact research participants after reinterpretation of genetic and genomic research results. *Am J Hum Genet* **104:** 578–595. doi:10.1016/j.ajhg.2019.02.025

Bunnik EM, de Jong A, Nijsingh N, de Wert GMWR. 2013a. The new genetics and informed consent: differentiating choice to preserve autonomy. *Bioethics* **27:** 348–355. doi:10.1111/bioe.12030

Bunnik EM, Janssens ACJW, Schermer MHN. 2013b. A tiered-layered-staged model for informed consent in personal genome testing. *Eur J Hum Genet* **21:** 596–601. doi:10.1038/ejhg.2012.237

Burke K, Clarke A. 2016. The challenge of consent in clinical genome-wide testing. *Arch Dis Child* **101:** 1048–1052. doi:10.1136/archdischild-2013-304109

Carroll J, Makuwaza T, Manca D, Sopcak N, Permaul JA, O'Brien MA, Heisey RE, Eisenhauer EA, Easley JK, Krzyzanowska MK, et al. 2016. Primary care providers' experiences with and perceptions of personalized genomic medicine. *Can Fam Physician* **62:** e626–e635.

David KL, Best RG, Brenman LM, Bush L, Deignan JL, Flannery D, Hoffman JD, Holm I, Miller DT, O'Leary J, et al. 2019. Patient re-contact after revision of genomic test results: points to consider—a statement of the American College of Medical Genetics and Genomics (ACMG). *Genet Med* **21:** 769–771. doi:10.1038/s41436-018-0391-z

Dillon OJ, Lunke S, Stark Z, Yeung A, Thorne N, Melbourne Genomics Health Alliance, Gaff C, White SM, Tan TY. 2018. Exome sequencing has higher diagnostic yield compared to simulated disease-specific panels in children with suspected monogenic disorders. *Eur J Hum Genet* **26:** 644–651. doi:10.1038/s41431-018-0099-1

Dragojlovic N, Elliott AM, Adam S, van Karnebeek C, Lehman A, Mwenifumbo JC, Nelson TN, du Souich C, Friedman JM, Lynd LD. 2018. The cost and diagnostic yield of exome sequencing for children with suspected genetic disorders: a benchmarking study. *Genet Med* **20:** 1013–1021. doi:10.1038/gim.2017.226

Ekstract M, Holtzman GI, Kim KY, Willis SM, Zallen DT. 2017. Evaluation of a web-based decision aid for people considering the APOE genetic test for Alzheimer risk. *Genet Med* **19:** 676–682. doi:10.1038/gim.2016.170

Elias S, Annas GJ. 1994. Generic consent for genetic screening. *N Engl J Med* **330:** 1611–1613. doi:10.1056/NEJM199406023302213

Farnaes L, Hildreth A, Sweeney NM, Clark MM, Chowdhury S, Nahas S, Cakici JA, Benson W, Kaplan RH, Kronick R, et al. 2018. Rapid whole-genome sequencing decreases infant morbidity and cost of hospitalization. *NPJ Genom Med* **3:** 10. doi:10.1038/s41525-018-0049-4

Fowler SA, Saunders CJ, Hoffman MA. 2018. Variation among consent forms for clinical whole exome sequencing. *J Genet Couns* **27:** 104–114. doi:10.1007/s10897-017-0127-2

Garrison NA, Cho MK. 2013. Awareness and acceptable practices: IRB and researcher reflections on the Havasupai lawsuit. *AJOB Prim Res* **4:** 55–63. doi:10.1080/21507716.2013.770104

Genetic Information Nondiscrimination Act of 2008 (Pub.L. 110–233, 122 Stat. 881).

Green RC, Berg JS, Grody WW, Kalia SS, Korf BR, Martin CL, McGuire AL, Nussbaum RL, O'Daniel JM, Ormond

KE, et al. 2013. ACMG recommendations for reporting of incidental findings in clinical exome and genome sequencing. *Genet Med* **15**: 565–574. doi:10.1038/gim.2013.73

Hallowell N, Alsop K, Gleeson M, Crook A, Plunkett L, Bowtell D, Mitchell G; Australian Ovarian Cancer Study Group, Young MA. 2013. The responses of research participants and their next of kin to receiving feedback of genetic test results following participation in the Australian Ovarian Cancer Study. *Genet Med* **15**: 458–465. doi:10.1038/gim.2012.154

Hallowell N, Hall A, Alberg C, Zimmern R. 2015. Revealing the results of whole-genome sequencing and whole-exome sequencing in research and clinical investigations: some ethical issues. *J Med Ethics* **41**: 317–321. doi:10.1136/medethics-2013-101996

Hart MR, Biesecker BB, Blout CL, Christensen KD, Amendola LM, Bergstrom KL, Biswas S, Bowling KM, Brothers KB, Conlin LK, et al. 2019. Secondary findings from clinical genomic sequencing: prevalence, patient perspectives, family history assessment, and health-care costs from a multisite study. *Genet Med* **21**: 1100–1110. doi:10.1038/s41436-018-0308-x

Hellwig LD, Biesecker BB, Lewis KL, Biesecker LG, James CA, Klein WMP. 2018. Ability of patients to distinguish among cardiac genomic variant subclassifications. *Circ Genom Precis Med* **11**: e001975. doi:10.1161/CIRCGEN.117.001975

Hunter JE, Irving SA, Biesecker LG, Buchanan A, Jensen B, Lee K, Martin CL, Milko L, Muessig K, Niehaus AD, et al. 2016. A standardized, evidence-based protocol to assess clinical actionability of genetic disorders associated with genomic variation. *Genet Med* **18**: 1258–1268. doi:10.1038/gim.2016.40

Joffe S, Cook EF, Cleary PD, Clark JW, Weeks JC. 2001. Quality of informed consent in cancer clinical trials: a cross-sectional survey. *Lancet* **358**: 1772–1777. doi:10.1016/S0140-6736(01)06805-2

Kalia SS, Adelman K, Bale SJ, Chung WK, Eng C, Evans JP, Herman GE, Hufnagel SB, Klein TE, Korf BR, et al. 2017. Recommendations for reporting of secondary findings in clinical exome and genome sequencing, 2016 update (ACMG SF v2.0): a policy statement of the American College of Medical Genetics and Genomics. *Genet Med* **19**: 249–255. doi:10.1038/gim.2016.190

Klima J, Fitzgerald-Butt SM, Kelleher KJ, Chisolm DJ, Comstock RD, Ferketich AK, McBride KL. 2014. Understanding of informed consent by parents of children enrolled in a genetic biobank. *Genet Med* **16**: 141–148. doi:10.1038/gim.2013.86

Koenig BA. 2014. Have we asked too much of consent? *Hastings Cent Rep* **44**: 33–34. doi:10.1002/hast.329

Kunneman M, Montori VM. 2017. When patient-centred care is worth doing well: informed consent or shared decision-making. *BMJ Qual Saf* **26**: 522–524. doi:10.1136/bmjqs-2016-005969

Kuppermann M, Pena S, Bishop JT, Nakagawa S, Gregorich SE, Sit A, Vargas J, Caughey AB, Sykes S, Pierce L, et al. 2014. Effect of enhanced information, values clarification, and removal of financial barriers on use of prenatal genetic testing. *J Am Med Assoc* **312**: 1210. doi:10.1001/jama.2014.11479

Lambertson KF, Damiani SA, Might M, Shelton R, Terry SF. 2015. Participant-driven matchmaking in the genomic era. *Hum Mutat* **36**: 965–973. doi:10.1002/humu.22852

Lee R, Lampert S, Wilder L, Sowell AL. 2011. Subjects agree to participate in environmental health studies without fully comprehending the associated risk. *Int J Environ Res Public Health* **8**: 830–841. doi:10.3390/ijerph8030830

Lewis MA, Paquin RS, Roche MI, Furberg RD, Rini C, Berg JS, Powell CM, Bailey DB Jr. 2016. Supporting parental decisions about genomic sequencing for newborn screening: the NC NEXUS Decision Aid. *Pediatrics* **137**: S16–S23. doi:10.1542/peds.2015-3731E

Manickam K, Buchanan AH, Schwartz MLB, Hallquist ML, Williams J, Rahm AK, Rocha H, Savatt JM, Evans AE, Butry LM, et al. 2018. Exome sequencing-based screening for *BRCA1/2* expected pathogenic variants among adult biobank participants. *JAMA Netw Open* **1**: e182140. doi:10.1001/jamanetworkopen.2018.2140

McGuire AL, Beskow LM. 2010. Informed consent in genomics and genetic research. *Annu Rev Genomics Hum Genet* **11**: 361–381. doi:10.1146/annurev-genom-082509-141711

Moreno JD, Schmidt U, Joffe S. 2017. The Nuremberg Code 70 years later. *J Am Med Assoc* **318**: 795–796. doi:10.1001/jama.2017.10265

Murray B. 2012. Informed consent: what must a physician disclose to a patient? *Virtual Mentor* **14**: 563–566.

Nelson-Marten P, Rich BA. 1999. A historical perspective of informed consent in clinical practice and research. *Semin Oncol Nurs* **15**: 81–88. doi:10.1016/S0749-2081(99)80065-5

Nippert I, Harris HJ, Julian-Reynier C, Kristoffersson U, ten Kate LP, Anionwu E, Benjamin C, Challen K, Schmidtke J, Nippert RP, et al. 2011. Confidence of primary care physicians in their ability to carry out basic medical genetic tasks—a European survey in five countries—Part 1. *J Community Genet* **2**: 1–11. doi:10.1007/s12687-010-0030-0

O'Daniel JM, McLaughlin HM, Amendola LM, Bale SJ, Berg JS, Bick D, Bowling KM, Chao EC, Chung WK, Conlin LK, et al. 2017. A survey of current practices for genomic sequencing test interpretation and reporting processes in US laboratories. *Genet Med* **19**: 575–582. doi:10.1038/gim.2016.152

OMIM Entry Statistics. www.omim.org/statistics/entry (accessed December 21, 2018a).

Ormond KE. 2013. From genetic counseling to "genomic counseling". *Mol Genet Genomic Med* **1**: 189–193. doi:10.1002/mgg3.45.

Ormond KE, Wheeler MT, Hudgins L, Klein TE, Butte AJ, Altman RB, Ashley EA, Greely HT. 2010. Challenges in the clinical application of whole-genome sequencing. *Lancet* **375**: 1749–1751. doi:10.1016/S0140-6736(10)60599-5

Ormond KE, O'Daniel JM, Kalia SS. 2019a. Secondary findings: how did we get here, and where are we going? *J Genet Couns* **28**: 326–333. doi:10.1002/jgc4.1098

Ormond KE, Hallquist MLG, Buchanan AH, Dondanville D, Cho MK, Smith M, Roche M, Brothers KB, Coughlin CR II, Hercher L, et al. 2019b. Developing a conceptual, reproducible, rubric-based approach to consent and result disclosure for genetic testing by clinicians with minimal genetics background. *Genet Med* **21:** 727–735. doi:10 .1038/s41436-018-0093-6

Patch C, Middleton A. 2018. Genetic counselling in the era of genomic medicine. *Br Med Bull* **126:** 27–36. doi:10.1093/ bmb/ldy008

Petrikin JE, Willig LK, Smith LD, Kingsmore SF. 2015. Rapid whole genome sequencing and precision neonatology. *Semin Perinatol* **39:** 623–631. doi:10.1053/j.semperi .2015.09.009

Posey JE, Harel T, Liu P, Rosenfeld JA, James RA, Coban Akdemir ZH, Walkiewicz M, Bi W, Xiao R, Ding Y, et al. 2017. Resolution of disease phenotypes resulting from multilocus genomic variation. *N Engl J Med* **376:** 21–31. doi:10.1056/NEJMoa1516767

Regier DA, Peacock SJ, Pataky R, van der Hoek K, Jarvik GP, Hoch J, Veenstra D. 2015. Societal preferences for the return of incidental findings from clinical genomic sequencing: a discrete-choice experiment. *CMAJ* **187:** E190–E197. doi:10.1503/cmaj.140697

Reumkens K, Tummers MHE, Gietel-Habets JJG, van Kuijk SMJ, Aalfs CM, van Asperen CJ, Ausems MGEM, Collée M, Dommering CJ, Kets CM, et al. 2019. The development of an online decision aid to support persons having a genetic predisposition to cancer and their partners during reproductive decision-making: a usability and pilot study. *Fam Cancer* **18:** 137–146. doi:10.1007/s10689-018-0092-4

Roberts JS, Robinson JO, Diamond PM, Bharadwaj A, Christensen KD, Lee KB, Green RC, McGuire AL, MedSeq Project team. 2018. Patient understanding of, satisfaction with, and perceived utility of whole-genome sequencing: findings from the MedSeq Project. *Genet Med* **20:** 1069–1076. doi:10.1038/gim.2017.223

Roberts JS, Gornick MC, Le LQ, Bartnik NJ, Zikmund-Fisher BJ, Chinnaiyan AM; MI-ONCOSEQ Study team. 2019. Next-generation sequencing in precision oncology: patient understanding and expectations. *Cancer Med* **8:** 227–237. doi:10.1002/cam4.1947

Salgo v. 1957. Leland Stanford Jr. Univ Bd Trustees 154: 2d.

Samuel GN, Dheensa S, Farsides B, Fenwick A, Lucassen A. 2017. Healthcare professionals' and patients' perspectives on consent to clinical genetic testing: moving towards a more relational approach. *BMC Med Ethics* **18:** 47. doi:10 .1186/s12910-017-0207-8

Sapp JC, Johnston JJ, Driscoll K, Heidlebaugh AR, Sagardia AM, Dogbe DN, Umstead KL, Turbitt E, Alevizos I, Baron J. et al. 2018. Evaluation of recipients of positive and negative secondary findings evaluations in a hybrid CLIA-research sequencing pilot. *Am J Hum Genet* **103:** 358–366. doi:10.1016/j.ajhg.2018.07.018

Shahmirzadi L, Chao EC, Palmaer E, Parra MC, Tang S, Farwell Gonzalez KD. 2014. Patient decisions for disclosure of secondary findings among the first 200 individuals undergoing clinical diagnostic exome sequencing. *Genet Med* **16:** 395–399. doi:10.1038/gim.2013.153

Shickh S, Clausen M, Mighton C, Casalino S, Joshi E, Glogowski E, Schrader KA, Scheer A, Elser C, Panchal S, et al. 2018. Evaluation of a decision aid for incidental genomic results, the Genomics ADvISER: protocol for a mixed methods randomised controlled trial. *BMJ Open* **8:** e021876. doi:10.1136/bmjopen-2018-021876

Spector-Bagdady K, Prince AER, Yu JH, Appelbaum PS. 2018. Analysis of state laws on informed consent for clinical genetic testing in the era of genomic sequencing. *Am J Med Genet C Semin Med Genet* **178:** 81–88. doi:10.1002/ ajmg.c.31608

Spence C. Cir FDC. 2008. Canterbury v. Spence. *464 F.2d 772 (D.C. Cir. 1972)* [11] 772: 1–31.

Stark Z, Schofield D, Martyn M, Rynehart L, Shrestha R, Alam K, Lunke S, Tan TY, Gaff C, White S. 2019. Does genomic sequencing early in the diagnostic trajectory make a difference? A follow-up study of clinical outcomes and cost-effectiveness. *Genet Med* **21:** 173–180. doi:10 .1038/s41436-018-0006-8

Tabor HK, Jamal SM, Yu JH, Crouch JM, Shankar AG, Dent KM, Anderson N, Miller DA, Futral BT, Bamshad MJ. 2017. My46: a Web-based tool for self-guided management of genomic test results in research and clinical settings. *Genet Med* **19:** 467–475. doi:10.1038/gim.2016.133

Turbitt E, Chrysostomou PP, Peay HL, Heidlebaugh AR, Nelson LM, Biesecker BB. 2018. A randomized controlled study of a consent intervention for participating in an NIH Genome Sequencing Study. *Eur J Hum Genet* **26:** 622–630. doi:10.1038/s41431-018-0105-7

Williams J, Faucett W, Smith-Packard B, Wagner M, Williams MS. 2014. An assessment of time involved in pretest case review and counseling for a whole genome sequencing clinical research program. *J Genet Couns* **23:** 516–521. doi:10.1007/s10897-014-9697-4

Wynn J, Lewis K, Amendola LM, Bernhardt BA, Biswas S, Joshi M, McMullen C, Scollon S. 2018. Clinical providers' experiences with returning results from genomic sequencing: an interview study. *BMC Med Genomics* **11:** 45. doi:10.1186/s12920-018-0360-z

Yu JH, Jamal SM, Tabor HK, Bamshad MJ. 2013. Self-guided management of exome and whole-genome sequencing results: changing the results return model. *Genet Med* **15:** 684–690. doi:10.1038/gim.2013.35

Genetic Counseling, Personalized Medicine, and Precision Health

Erica Ramos

Director, Clinical and Product Development, Geisinger National Precision Health, Geisinger Health System, North Bethesda, Maryland 20852, USA

Correspondence: eramos@geisinger.edu

Millions of individuals in the United States will have their exomes and genomes sequenced over the next 5 years as the use of genomic sequencing technologies in clinical care grows and as initiatives in personalized medicine and precision health move forward. As a result, we will see a shift away from the patient population of early adopters who pursued direct-to-consumer (DTC) testing and paid thousands of dollars to get their genomes sequenced and toward a different and more diverse set of test takers. Early data suggest that these individuals will have different motivations for pursuing genomic sequencing and will be less knowledgeable about and less confident of the benefits of genetic testing. To serve this growing population, genetic counselors must understand our future patients as well as the changing landscape of genomic testing, DTC offerings, and population sequencing initiatives.

The practice of genetic counseling is facing several significant changes as the promise of personalized medicine is realized. Anytime the status quo shifts, particularly when changes appear to be rapid and all-encompassing, anxiety, concern, consternation, and a penetrating sense of uncertainty can flood in. However, history shows that in the face of change, the field of genetic counseling will be ready—not only to adapt—but to lead. Genetic counseling has seen multiple shifts in the areas in which we practice, in our employers, in the health-care and political landscapes, and in our patient populations. However, the core components of genetic counseling have remained relatively stable, and this stability has enabled us to move nimbly into these new areas of practice. As the promise of personalized medicine is realized, we will leverage that same core to shift once again as we provide services to the increasingly broader swath of people who require our expertise, guidance, and care.

PERSONALIZED, PRECISION, OR INDIVIDUALIZED MEDICINE?

Although the term "personalized medicine" has become part of the lexicon, there remains a lack of clarity on what this term means and how it differs from related terms, including "precision medicine," "individualized medicine," and others. One common complaint from physicians about the term "personalized medicine" is that it can be perceived as downplaying the extent to which they have always practiced medicine in a way that took the personal characteristics of their patients into consideration. And although this is true, the vision of "personalized medicine" as it is portrayed today implies a deeper level of customization, including diagnostic, screening, treatment, and management approaches based on genomics and other factors, and a systematized integration of this customization into care.

One early proposal for how to describe this realignment of medical care came from Dr. Lee Hood, who coined the term "P4 medicine"—predictive, personalized, preventive, and participatory. He felt that a single term could not fully represent the scope of how medicine would evolve given advancements in genomics, monitoring devices, and our ability to assess risk (Hood 2008). Although this approach attempted to capture the complexity of a future state in which these characteristics converged, it was not widely adopted. Another suggestion came from the National Research Council's Committee on a Framework for Developing a New Taxonomy of Disease, in a 2011 report called *Toward Precision Medicine: Building a Knowledge Network for Biomedical Research and a New Taxonomy of Disease* (www.nap.edu/catalog.php?record_id=13284):

> Personalized medicine refers to the tailoring of medical treatment to the individual characteristics of each patient. It does not literally mean the creation of drugs or medical devices that are unique to a patient, but rather the ability to classify individuals into subpopulations that differ in their susceptibility to a particular disease or their response to a specific treatment. Preventive or therapeutic interventions can then be concentrated on those who will benefit, sparing expense and side effects for those who will not (PCAST 2008). This term is now widely used, including in advertisements for commercial products, and it is sometimes misinterpreted as implying that unique treatments can be designed for each individual. For this reason, the Committee thinks that the term "precision medicine" is preferable to "personalized medicine" to convey the meaning intended in this report.

Despite the Committee's advocacy, the use of both terms has persisted. Additionally, with the

rapid growth of genomics-guided therapies in oncology, the reverse has happened: "Precision medicine" is often used to suggest cancer treatments that are specific to somatic mutations that drive cancer growth. In a blog post addressing this issue, Dr. Muin Khoury, Director of the Office of Public Health Genomics for the Centers for Disease Control and Prevention (CDC), questions, "Does this change in words represent just semantics or is it an important conceptual shift in the scientific understanding of health and disease and its application to treatment and prevention?" (https://blogs.cdc.gov/genomics/2016/04/21/shift/). An important extension to his question is whether this terminology will provide useful distinctions for the general population or mask their true concerns, and if that question should guide how these terms are used with genetic counseling patients in practice.

A more recent addition to the lexicon describing this new initiative in medicine is "precision health," which some suggest is a rebranding of personalized medicine in the face of underwhelming success. Although personalized medicine always included a strong aspect of prevention, precision health is presented as having a specific focus on the identification of high-risk individuals and provision of targeted risk mitigation to prevent disease symptoms (Juengst and McGown 2018). This concept extends to precision public health as well, as "there is an emerging list of genomic applications that merit a targeted public health approach to find people with selected genetic conditions (e.g., hereditary breast/ovarian and colorectal cancers and familial hypercholesterolemia)" (Khoury et al. 2016).

SEQUENCING EVERYONE: POPULATION GENOMIC SEQUENCING INITIATIVES

Regardless of the terminology, genetics and genomics remains central to most applications of personalized medicine. Around the world, major publicly funded research initiatives are underway to assess potential precision public health and personalized medicine benefits, among other goals, via population-scale genomic sequencing. Australia, France, India, Singapore, South Korea, and the United Kingdom have all announced genomics initiatives that will sequence tens of thousands to millions of individuals in their countries (https://www.genomeweb.com/sequencing/illumina-says-population-sequencing-dtc-will-continue-grow-long-term).

The United Kingdom's National Health Service (NHS) has completed a significant pilot program in population genomics and upon its completion, announced that they would soon undertake the largest population-scale genomics screening initiative proposed to date. In 2012, Genomics England, a company funded by the Department of Health & Social Care, announced that they would sequence 100,000 genomes from NHS patients (https://www.genomicsengland.co.uk/about-genomics-england/the-100000-genomes-project/). This effort was exclusively focused on individuals who had an undiagnosed rare disease that appeared to be genetic and those with cancer, and the project completed the 100,000th genome in December 2018. In October 2018, Health and Social Care Secretary Matt Hancock announced the expansion of the 100,000 Genomes Project to one million genomes and the intent to sequence 5 million genomes in 5 years, the most ambitious target thus far. Key to the latter announcement was a controversial "genomic volunteering" proposal that would allow individuals to purchase genome sequencing and receive health-related results even if they did not have any medical concerns, with the data going to support a greater research effort. Critics of this proposal highlighted that there were insufficient data on how to use a genome in a "healthy" person and decried the unrolling of a two-tiered health system (https://www.theguardian.com/society/2019/jan/26/nhs-to-sell-dna-tests-to-healthy-people-in-push-to-find-new-treatments). In July 2019, an official government consultation called "Advancing our health: prevention in the 2020s—consultation document" was published confirming the goal of sequencing 5 million people by 2024, including healthy individuals, but stating that it would be free to all NHS patients (https://www.gov.uk/government/consultations/advancing-our-health-prevention-in-the-2020s/advancing-our-health-prevention-in-the-2020s-consultation-document#contents).

In the United States, the Million Veterans Program (MVP) began enrolling participants in 2011. As of July 2019, the MVP boasted 750,000 enrolled veterans, including the largest existing cohort of minority participants in the United States (Rubin 2019). In addition to collecting participant questionnaires and electronic health record data from the Veterans Administration, the MVP has performed a combination of genotyping and exome and genome sequencing on participant samples. Research targets include schizophrenia, bipolar disorder, posttraumatic stress disorder, cardiac disease, metabolic disease, renal disease, and substance abuse (Gaziano et al. 2016).

In 2015, during his State of the Union address, President Barack Obama announced his intention to launch the Precision Medicine Initiative—a population-wide research effort, led by researchers, clinicians and patients, to advance personalized medicine (https://www.whitehouse.gov/the-press-office/2015/01/20/remarks-president-state-union-address-january-20-2015). The National Institutes of Health–funded research initiative, the *All of Us* Research Program, launched in 2016 with the intent of enrolling a diverse group of participants, with an emphasis on including individuals from groups who have been underrepresented to date in biomedical research. As of May 2019, 143,000 participants were enrolled as direct volunteers or through health-care partner organizations. Of all participants, 53% are ethnic or racial minorities, which if maintained through the proposed enrollment goal of one million individuals, would surpass the minority cohort size of the MVP (https://www.joinallofus.org/en/in-the-news/nih-says-its-1-million-person-health-study-good-start).

In addition to these government-funded research efforts, several health systems have launched population genomic screening efforts focused on their patient populations. These programs each target up to 500,000 participants, use a variety of genetic testing methodologies, and have been enabled to a significant degree by public-private partnerships (Table 1).

The Geisinger Health System MyCode Community Health Initiative (MyCode) was the first study to start sequencing patients in

Table 1. Selected public-private population genomic screening initiatives

Project	Collaborators	Target population	Year initiated	Assay	Results returned
Geisinger MyCode[a]	Regeneron Pharmaceuticals	Geisinger patients	2014	Exome	60+ genes for actionable disorders
Renown Healthy Nevada[b]	23andMe	Nevada residents	2016	Array	Ancestry + health
	Helix		2018	Exome	Ancestry + traits
Sanford Imagenetics[c]	Illumina	Sanford patients	2018	Array	59 genes for actionable disorders, pharmacogenomics
NorthShore DNA-10K[d]	Color	NorthShore patients	2019	Genome	Cancer, cardiac, pharmacogenomics
Mayo Clinic[e]	Regeneron Pharmaceuticals	Mayo Clinic patients	2019[g]	Exome	Planning to research
Intermountain HerediGene: Population Study[f]	deCODE Genetics/ Amgen	Utah and Idaho residents	2019[g]	Genome	Planned

[a]www.geisinger.org/precision-health/mycode
[b]healthynv.org/
[c]imagenetics.sanfordhealth.org/sanford-chip/
[d]northshore.org/personalized-medicine/our-services/color-genetics-test/
[e]genomeweb.com/genetic-research/regeneron-mayo-ink-pact-sequence-genotype-100k-patient-samples
[f]intermountainhealthcare.org/heredigene
[g]Not yet enrolling patients.

2014 and remains the largest private cohort to date with almost 250,000 consented participants and 145,000 individuals from whom they have obtained exome sequencing data (D Ledbetter pers. comm.). In 2015, MyCode began to consent patients for return of results for clinically actionable findings following extensive discussions and assessments by the clinical and bioethics teams, community advisory groups, and scientific leaders, as well as patient-participant focus groups (Faucett and Davis 2016). The MyCode Initiative has enabled the study of the return of results protocols and processes in the context of population screening (Schwartz et al. 2018), as well as patient-centric initiatives such as patient-facing genomic (Goehringer et al. 2018) and pharmacogenomic (Jones et al. 2018) laboratory reports. MyCode has also generated data on the prevalence of familial hypercholesterolemia (Abul-Husn et al. 2016), hereditary breast and ovarian cancer (Manickam et al. 2018), and arrhythmogenic right ventricular cardiomyopathy (Haggerty et al. 2017) in an unselected health system research population. Genetic counseling is fundamental to the MyCode Initiative with genetic counselors leading many of these research efforts, and is inclusive of the primary care physicians and specialists who will follow these patients over time.

Although the other studies cited in Table 1 launched more recently and are just starting to generate data, Renown Health has also published findings regarding the frequencies of rare variants across the exome (Cirulli et al. 2019) and the detection of individuals at risk for hereditary cancers and familial hypercholesterolemia (Grzymski et al. 2019), demonstrating the potential for the rapid generation of insights in very large cohorts.

The expanding scope of population screening is not new. One specific example is cystic fibrosis (CF) carrier screening, although the progression has been similar for many other genes and conditions (Fig. 1). Over the years, practice progressed from testing affected individuals only, to familial variant testing, to population-specific testing, to broad-based testing. As the tested population expands, new insights into the molecular underpinnings of the disease are elucidated.

Cite this article as *Cold Spring Harb Perspect Med* doi: 10.1101/cshperspect.a036699

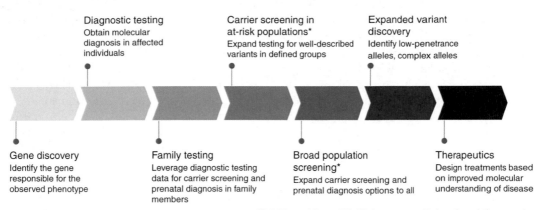

Figure 1. Example of genetic testing trajectory.

For example, in 2004, the American College of Medical Genetics (ACMG) CF Carrier Screening Working Group issued an update to their 2001 CF carrier screening guidelines based on the evidence that had been gathered by screening more than 400,000 individuals. The Working Group recommended changes to testing with regards to two of the original 25 variants. First, it was determined that individuals with CF who carried the I148T mutation had the same second mutation (3199del6), which was subsequently determined to cause the disease. The Working Group recommended that the I148T mutation be removed from the panel. Second, testing for the R117H variant was found to be more complicated than originally understood, as it had to be found in combination with the 5T variant. Although the Working Group recommended keeping R117H on the panel, they recommended screening for the 5T variant as a reflex only when R117H was identified (Watson et al. 2004).

As next-generation sequencing technologies evolved, pan-ethnic expanded carrier screening (ECS) was introduced. In contrast to historical carrier screening efforts which typically limited testing to specific, predefined variants, ECS sequenced all exons of the gene. Continuing the example of CF, Beauchamp et al. (2018) compared the diagnostic yield of ECS with three ACMG CF screening panels across multiple ethnic backgrounds. The addition of ECS increased

disease risk detection by >80% in individuals of East, South, and Southeast Asian descent and by >40% in individuals of Hispanic and African or African–American descent. Although the performance of the ACMG panels was already high in individuals of European descent, ECS increased disease risk detection by ~20% in individuals of Northern European and Mixed/ Other Caucasian descent.

When considering an international patient population, these findings suggest that ECS would improve detection rates for most, as individuals from European background are in the minority worldwide. The American College of Obstetricians and Gynecologists and the European Society of Human Genetics have both stated that ECS or ethnic-specific or pan-ethnic carrier screening are acceptable offerings for individuals of reproductive age, assuming the tests are performed with high standards of clinical validity and the content demonstrates clinical utility and severity. However, these organizations have done so while advocating for ECS to be offered in the context of engagement with clinicians, patients, and communities; access to quality genetic counseling services and support; and ongoing outcomes assessments and informed consent (Henneman et al. 2016; American College of Obstetricians and Gynecologists. 2017). Concerns have also been raised about the psychological well-being of patients, including emotional impact, psychological harm, and misunderstanding of a negative result. Limited data from couples undergoing ECS does not suggest harm in individuals with negative results (Matar et al. 2019) and demonstrates utility in changes to reproductive decision-making (Ghiossi et al. 2018). Further, ethical issues have been raised about the elimination of individuals with certain genetic disorders and equitable access because of ability/willingness to pay (Clarke et al. 2018).

Similar concerns have been put forward about population genomic screening. Yet, multiple cohort studies spanning a range of other testing scenarios and indications including ostensibly healthy individuals who underwent direct-to-consumer (DTC) testing (Bloss et al. 2013; Francke et al. 2013), *ApoE* testing in individuals at-risk for Alzheimer's disease (Green et al. 2009), clinical exome/genome sequencing (Robinson et al. 2019), and molecular screening in patients with cancer (Best et al. 2019). Although these findings may not hold in all groups who undergo population genomic screening, these studies suggest that people do not widely experience long-lasting harm as a result of genetic testing. Rather, motivations for testing and behaviors when receiving test results, such as limited sharing of results with family members, may be more impactful to assess as population genomic screening grows.

AN EXPANDING AND EVOLVING GENETIC TESTING LANDSCAPE

In addition to the data being generated by various population-level genomic sequencing initiatives, a substantial wave of increased genetic testing showing sustained growth is that of at-home genetic testing, also commonly referred to as DTC genetic testing, direct access testing, or consumer-directed/consumer-initiated testing (Ramos and Weissman 2018). The first significant versions of at-home testing for genetics, as we know it today, came in 2008 with the broad launch of 23andMe's $399 genetic test (Hamilton 2008; http://ti.me/1fhRH7z). Tests from several other companies entered the market, provoking immediate concern from regulatory authorities. In 2010, the U.S. Government Accountability Office (GAO) commissioned an in-

vestigation into several DTC companies and found multiple instances of conflicting results between the companies and other discrepancies (Kutz 2010). Following the Kutz report, the Food and Drug Administration (FDA) sent warning letters to several DTC companies instructing them to stop marketing their tests for health and medical indications. In 2013, following several years of intermittent communication and discussion of regulatory scrutiny, 23andMe was forced to stop providing health information until they developed an FDA-authorized assay and reports. They are currently the only DTC company that can offer FDA-authorized genetic testing for health indications without a physician order. However, although consumers can access these 23andMe genetic tests for health-related conditions directly, the FDA specifically states that "results obtained from the tests should not be used for diagnosis or to inform treatment decisions" (FDA News Release 2017). As a result, consumers who receive a positive result and wish to integrate the result into their medical care must confirm the presence of the variant through a clinical laboratory.

Several other laboratories have taken a path that facilitates consumer-directed testing by using third-party physicians to order the test. These labs are offering a variety of tests for adult and pediatric patients (Ramos and Weissman 2018; Weissman et al. 2019). In many of these cases, posttest genetic counseling is provided for individuals with positive results by genetic counselors used by the labs themselves or third-party groups. However, they are rarely fully connected back to the health-care system and the patient is often made to serve as the expert while communicating results back to their treating clinicians.

PUBLIC AWARENESS, PERCEPTIONS, AND PRIORITIES

Despite excitement in the research, medical, and technology communities, we have yet to see if personalized medicine and genomics will bridge the adoption gap to become routine clinical care. Public Perspectives on Personalized Medicine: A Survey of U.S. Public Opinion, a 2018 poll conducted for the Personalized Medicine Coalition and GenomeWeb (http://www.personalizedmedicinecoalition.org/Userfiles/PMC-Corporate/file/Public_Perspectives_on_PM1.pdf), found that 66% of respondents had not heard of either "personalized medicine" or "precision medicine." Although almost all were positive or neutral to the concept of personalized medicine when it was presented, many respondents immediately jumped to practical concerns:

> When shown a list of known concerns, most said they have major concerns that a test might not be covered by their insurer (62%), they might not be able to afford personalized medicine (59%), or that test results could be used to deny coverage for a treatment (52%) or affect long-term care or life insurance policies (51%). Only 10% of Americans indicated that they are aware of the Genetic Information Nondiscrimination Act, which prohibits employers or health insurers from asking for genetic test results when making employment or coverage decisions.

This study indicates that the intricacies of the terminology are not important to our patients and, more importantly, suggests that they may attribute certain characteristics to any incarnation of "personalized medicine," including that it is risky and expensive. If these perceptions remain unaddressed, fewer people may benefit from these medical advances.

A second 2018 poll, conducted by The Associated Press-NORC Center for Public Affairs Research, focused more on genetic testing specifically. They found that only 17% of respondents had taken a genetic test themselves and the vast majority of those (65%) reported that they had undergone testing to learn more about their heritage or family history, which suggests that many of these individuals may have pursued DTC ancestry testing rather than a clinical genetic test. Only 14% of those who had taken a genetic test reported having amniocentesis or genetic counseling during pregnancy, which may be more indicative of clinical testing. This represents just 27 out of the 1109 respondents, or 2.4%, of all those who were polled. Historically, patients primarily learned of genetic testing by experts such as genetic counselors, and clinical genetic testing was predominantly of-

fered for diagnostic purposes and within the health-care system. Although most individuals who present for genetic counseling will not have had any genetic testing, a larger swath will have had some experience with genetics—good or bad—without any expert interpretation, guidance, or support. Additionally, they are more likely to have had tests that may contain information about health-related conditions but are not of the clinical quality needed to guide medical care.

As one of the key proposed advantages of personalized medicine is disease prevention, the willingness to learn about risk is critical. Seventy-one percent of individuals under age 40 reported that they would want to know if they "carried the gene" for an incurable disease. This dropped to 53% for individuals age 40 and up.

Finally, this poll suggests that there is a need for initiatives to convey accurate information about the utility and accuracy of genetic testing. When asked how reliable genetic tests were at diagnosing certain conditions, 7% said not very or not at all reliable. Confidence was lower when asked how reliable genetic testing was at predicting whether a person might develop certain diseases or conditions (11% said not very or not at all reliable) (Genetic Testing: Ancestry Interest, But Privacy Concerns. Chicago: AP-NORC. https://reports.norc.org/issue_brief/genetic-testing-ancestry-interest-but-privacy-concerns/). As those who are unaffected but concerned about disease risk and prevention or who learn about their risk via consumer-directed tests or research efforts begin to present for genetic counseling services, these preconceptions will need to be addressed.

A key takeaway from these two studies, as well as the population genomic screening studies conducted to date, is that genomic testing is not commonplace nor has it penetrated into the zeitgeist of most communities. This is particularly true in underrepresented communities as, with the exception of the MVP, most testing has been performed in individuals of European descent. For instance, despite the rapid growth in the at-home genetic testing market, 23andMe's research database of more than five million people has about 845,000 who were identified as

African–American, East Asian, Middle Eastern, South Asian, or "other" and 820,000 who were identified as Latino (https://research.23andme .com/research-innovation-collaborations/). However, with the announcements of international screening programs in non-European countries and the efforts of the All of Us Research Program to focus on underrepresented minorities, that is starting to shift.

This limitation is mirrored in the demographics of the field of genetic counseling and other professions. In the 2019 Professional Status Survey from the National Society of Genetic Counselors, 90% of respondents identified as White (https://www.nsgc.org/page/whoaregene ticcounselors). This lack of diversity plagues most health-care specialties, and the genetic counseling profession will need to identify and address the barriers to entry for individuals from ethnic minority and other excluded backgrounds. In addition to training more professionals to provide genetic counseling who represent the patient populations, we must also ensure that the entire workforce is provided the tools for cultural competence. As Dr. Rhea W. Boyd stated in *The Lancet*, "To be clear, nonwhite patients should not require a race-matched provider to receive a standard of care that is equitable and dignified" (Boyd 2019). Although cultural competence is core to the training and continuing education of genetic counselors, to optimally serve a more diverse population and ensure that all people have the opportunity to benefit from advances in personalized medicine and precision health, our cultural competence must continue to grow (McGinniss et al. 2018).

GENETIC COUNSELING: THE KEYSTONE

Millions of individuals worldwide will be sequenced in the next few years, and it is clear that genetic counselors will need to play a significant role in shepherding these personalized medicine and precision health efforts forward, from research and health system initiatives to population sequencing initiatives to DTC genetic testing. There are limited studies looking at the impact of genetic counseling in efforts of this

scale, both in terms of the cohort size and the quantity of genomic information. Those available paint a positive picture about the importance of genetic counseling, but also highlight the need to adapt counseling models and service delivery to fit the needs of different patient populations. Additionally, and not surprisingly, the motivations of these patients can differ from those seen for diagnostic testing.

Sweet et al. (2017) report on individuals who were part of a Coriell Personalized Medicine Collaborative study. Each participant received a report assessing their risks for 19 potentially actionable complex diseases and seven pharmacogenomic drug reports. Participants could request genetic counseling appointments at any time and the content of the counseling appointment was driven by patient questions and concerns. Of the 51 individuals who were interviewed about their experience, the majority reported a positive impression of genetic counseling and a desire to access genetic counseling in the future if additional results were returned. Key comments reflected the ability of the counselor to synthesize information for them: One participant said she joined the study because of the option to see a genetic counselor and noted it was important to "have somebody with expertise to really give me the insight on what it all means and how it all fits together was what I was looking for" (female, age 54). Another said that her counselor "gave me a tremendous amount of interpretation. It was just very helpful to me. He was able to say well look, this is what this means" (female, age 68).

However, the participant feedback also suggests a need to consider alternatives to traditional service delivery models and flexibility in the content and length of genetic counseling services, which should be considered strongly by counselors providing services for personalized medicine patients.

Schmidt et al. (2019) present data from 70 patients seen over the course of 2 years for genetic counseling following exome or genome sequencing. Similar to Sweet et al. the counselors followed a Reciprocal Engagement Model for their counseling, allowing the client to set the priorities of the session throughout. During the sessions, patients were determined to be

Cite this article as *Cold Spring Harb Perspect Med* doi: 10.1101/cshperspect.a036699

one of three patient types based on their motivations for testing: (1) the "healthy and curious" who were not expecting any findings based on their histories but interested in any potential "red flags"; (2) those who were interested in a specific finding or risk, either because of family history or a pathogenic variant; and (3) the "undiagnosed and searching" who were suffering from unexplained symptoms and seeking validation and "proof" that they were ill. Interestingly, the motivations expressed by the first two patient types were also reflected in another study of 543 early adopters pursuing elective genome sequencing, but the "undiagnosed and searching" type was not (Zoltick et al. 2019). This third group may represent a shift from the very early adopter population, and one that may especially benefit from genetic counseling, as the authors note that these patients expressed frustration, despair, defeat, desperation, and hopelessness with their physicians and the health-care system.

CONCLUSION

As the genetic counselor workforce grows and our roles expand, there are professional skills and areas of focus that will become increasingly important. Training programs are evolving to incorporate bioinformatics and variant interpretation and to view long-standing hallmarks of genetic counseling such as ethics, education, counseling, and cultural competence through a genomic lens (Hooker et al. 2014). Although the scope of the changes in the genomics landscape appear vast, genetic counselors have the training, professional resources, and continuing education to adapt and thrive in this evolving environment.

REFERENCES

Abul-Husn NS, Manickam K, Jones LK, Wright EA, Hartzel DN, Gonzaga-Jauregui C, O'Dushlaine C, Leader JB, Lester Kirchner H, Lindbuchler DM, et al. 2016. Genetic identification of familial hypercholesterolemia within a single U.S. health care system. *Science* 354: pii:aaf7000.

American College of Obstetricians and Gynecologists. 2017. Carrier screening in the age of genomic medicine. Committee Opinion No. 690. *Obstet Gynecol* 129: e35–e40.

Beauchamp KA, Muzzey D, Wong KK, Hogan GJ, Karimi K, Candille SI, Mehta N, Mar-Heyming R, Kaseniit KE, Kang HP, et al. 2018. Systematic design and comparison of expanded carrier screening panels. *Genet Med* 20: 55–63. doi:10.1038/gim.2017.69

Best MC, Bartley N, Jacobs C, Juraskova I, Goldstein D, Newson AJ, Savard J, Meiser B, Ballinger M, Napier C, et al. 2019. Patient perspectives on molecular tumor profiling: "why wouldn't you?". *BMC Cancer* 19: 753. doi:10.1186/s12885-019-5920-x

Bloss CS, Wineinger NE, Darst BF, Schork NJ, Topol EJ. 2013. Impact of direct-to-consumer genomic testing at long term follow-up. *J Med Genet* 50: 393–400. doi:10.1136/jmedgenet-2012-101207

Boyd RW. 2019. The case for desegregation. *Lancet* 393: 2484–2485. doi:10.1016/S0140-6736(19)31353-4

Cirulli ET, White S, Read RW, Elhanan G, Metcalf WJ, Schlauch KA, Grzymski JJ, Lu J, Washington NL. 2019. Genome-wide rare variant analysis for thousands of phenotypes in 54,000 exomes. bioRxiv doi:10.1101/692368

Clarke EV, Schneider JL, Lynch F, Kauffman TL, Leo MC, Rosales AG, Dickerson JF, Shuster E, Wilfond BS, Goddard KAB. 2018. Assessment of willingness to pay for expanded carrier screening among women and couples undergoing preconception carrier screening. *PLoS One* 13: e0200139. doi:10.1371/journal.pone.0200139

Faucett WA, Davis FD. 2016. How Geisinger made the case for an institutional duty to return genomic results to biobank participants. *Appl Transl Genom* 8: 33–35. doi:10.1016/j.atg.2016.01.003

FDA News Release. 2017. Press Announcements—FDA allows marketing of first direct-to-consumer tests that provide genetic risk information for certain conditions. Retrieved from https://www.fda.gov/NewsEvents/Newsroom/PressAnnouncements/ucm551185.htm

Francke U, Dijamco C, Kiefer AK, Eriksson N, Moiseff B, Tung JY, Mountain JL. 2013. Dealing with the unexpected: consumer responses to direct-access *BRCA* mutation testing. *Peer J* 1: e8. doi:10.7717/peerj.8

Gaziano JM, Concato J, Brophy M, Fiore L, Pyarajan S, Breeling J, Whitbourne S, Deen J, Shannon C, Humphries D, et al. 2016. Million veteran program: a mega-biobank to study genetic influences on health and disease. *J Clin Epidemiol* 70: 214–223. doi:10.1016/j.jclinepi.2015.09.016

Ghiossi CE, Goldberg JD, Haque IS, Lazarin GA, Wong KK. 2018. Clinical utility of expanded carrier screening: reproductive behaviors of at-risk couples. *J Genet Couns* 27: 616–625. doi:10.1007/s10897-017-0160-1

Goehringer JM, Bonhag MA, Jones LK, Schmidlen T, Schwartz M, Rahm AK, Williams JL, Williams MS. 2018. Generation and implementation of a patient-centered and patient-facing genomic test report in the EHR. *EGEMs (Wash DC)* 6: 14. doi:10.5334/egems.256.

Green RC, Roberts JS, Cupples LA, Relkin NR, Whitehouse PJ, Brown T, Eckert SL, Butson M, Sadovnick AD, Quaid KA, et al. 2009. Disclosure of *APOE* genotype for risk of Alzheimer's disease. *N Engl J Med* 361: 245–254. doi:10.1056/NEJMoa0809578

Grzymski JJ, Elhanan G, Smith E, Rowan C, Slotnick N, Dabe S, Schlauch K, Read R, Metcalf WJ, Lipp B et al. 2019. Population health genetic screening for tier 1 inherited diseases in Northern Nevada: 90% of at-risk carriers are missed. bioRxiv doi:10.1101/650549.

Haggerty CM, James CA, Calkins H, Tichnell C, Leader JB, Hartzel DN, Nevius CD, Pendergrass SA, Person TN, Schwartz M, et al. 2017. Electronic health record phenotype in subjects with genetic variants associated with arrhythmogenic right ventricular cardiomyopathy: a study of 30,716 subjects with exome sequencing. *Genet Med* **19:** 1245–1252. doi:10.1038/gim.2017.40

Hamilton A. 2008. INVENTION OF THE YEAR 1. The retail DNA test. *Time Magazine.* Retrieved from http://content.time.com/time/specials/packages/article/0,28804,1852747_1854493,00.html

Henneman L, Borry P, Chokoshvili D, Cornel MC, van El CG, Forzano F, Hall A, Howard HC, Janssens S, Kayserili H, et al. 2016. Responsible implementation of expanded carrier screening. *Eur J Hum Genet* **24:** e1–e12. doi:10.1038/ejhg.2015.271

Hood L. 2008. A personal journey of discovery: developing technology and changing biology. *Annu Rev Anal Chem (Palo Alto Calif)* **1:** 1–43. doi:10.1146/annurev.anchem.1.031207.113113

Hooker GW, Ormond KE, Sweet K, Biesecker BB. 2014. Teaching genomic counseling: preparing the genetic counseling workforce for the genomic era. *J Genet Counsel* **23:** 445–451. doi:10.1007/s10897-014-9689-4

Jones LK, Rahm AK, Gionfriddo MR, Williams JL, Fan AL, Pulk RA, Wright EA, Williams MS. 2018. Developing pharmacogenomic reports: insights from patients and clinicians. *Clin Transl Sci* **11:** 289–295. doi:10.1111/cts.12534

Juengst ET, McGown ML. 2018. Why does the shift from "personalized medicine" to "precision health" and "wellness genomics" matter? *AMA J Ethics* **20:** E881–E890. doi:10.1001/amajethics.2018.881

Khoury ML, Iademarco MF, Riley WT. 2016. Precision public health for the era of precision medicine. *Am J Prev Med* **50:** 398–401. doi:10.1016/j.amepre.2015.08.031

Kraft SA, Schneider JL, Leo MC, Kauffman TL, Davis JV, Porter KM, McMullen CK, Wilfond BS, Goddard KAB. 2018. Patient actions and reactions after receiving negative results from expanded carrier screening. *Clin Genet* **93:** 962–971. doi:10.1111/cge.13206

Kutz G. 2010. *Direct-to-consumer genetic tests misleading test results are further complicated by deceptive marketing and other questionable practices.* United States Government Accountability Office. Retrieved from https://doi.org/Serial No. 111-138

Manickam K, Buchanan AH, Schwartz MLB, Hallquist MLG, Williams JL, Rahm AK, Rocha H, Savatt JM, Evans AE, Butry LM, et al. 2018. Exome sequencing-based screening for *BRCA1/2* expected pathogenic variants among adult biobank participants. *JAMA Netw Open* **1:** e182140. doi:10.1001/jamanetworkopen.2018.2140

Matar A, Hansson MG, Höglund AT. 2019. "A perfect society"—Swedish policymakers' ethical and social views on preconception expanded carrier screening. *J Community Genet* **10:** 267–280. doi:10.1007/s12687-018-0389-x

McGinniss MA, Tahmassi AG, Ramos E. 2018. Towards cultural competence in the genomic age: a review of current health care provider educational trainings and interventions. *Curr Genet Med Rep* **6:** 187–198. doi:10.1007/s40142-018-0150-0

PCAST (President's Council of Advisors on Science and Technology). 2008. *Priorities for personalized medicine.* President's Council of Advisors on Science and Technology, September 2008. Available at http://www.whitehouse.gov/files/documents/ostp/PCAST/pcast_report

Ramos E, Weissman SM. 2018. The dawn of consumer-directed testing. *Am J Med Genet C Semin Med Genet* **178:** 89–97.

Robinson JO, Wynn J, Biesecker B, Biesecker LG, Berhardt B, Brothers KB, Chung WK, Christensen KD, Green RC, McGuire AL, et al. 2019. Psychological outcomes related to exome and genome sequencing result disclosure: a meta-analysis of seven Clinical Sequencing Exploratory Research (CSER) Consortium studies. *Genet Med.* doi:10.1038/s41436-019-0565-3

Rubin R. 2019. Veterans genomic cohort nearing 1 million. *J Am Med Assoc* **321:** 2395.

Schmidt JL, Maas R, Altmeyer SR. 2019. Genetic counseling for consumer-driven whole exome and whole genome sequencing: a commentary on early experiences. *J Genet Couns* **28:** 449–455.

Schwartz MLB, McCormick CZ, Lazzeri AL, Lindbuchler DM, Hallquist MLG, Manickam K, Buchanan AH, Rahm AK, Giovanni MA, Frisbie L, et al. 2018. A model for genome-first care: returning secondary genomic findings to participants and their healthcare providers in a large research cohort. *Am J Hum Genet* **103:** 328–337. doi:10.1016/j.ajhg.2018.07.009

Sweet K, Hovick S, Sturm AC, Schmidlen T, Gordon E, Bernhardt B, Wawak L, Wernke K, McElroy J, Scheinfeldt L, et al. 2017. Counselees' Perspectives of Genomic Counseling Following Online Receipt of Multiple Actionable Complex Disease and Pharmacogenomic Results: a qualitative research study. *J Genet Couns* **26:** 738–751. doi:10.1007/s10897-016-0044-9

Watson MS, Cutting GR, Desnick RJ, Driscoll DA, Klinger K, Mennuti M, Palomaki GE, Popovich BW, Pratt VM, Rohlfs EM, et al. 2004. Cystic fibrosis population carrier screening: 2004 revision of American College of Medical Genetics mutation panel. *Genet Med* **6:** 387–391. doi:10.1097/01.GIM.0000139506.11694.7C

Weissman SM, Kirkpatrick B, Ramos E. 2019. At-home genetic testing in pediatrics. *Curr Opin Pediatr* doi:10.1097/MOP.0000000000000824

Zoltick ES, Linderman MD, McGinniss MA, Ramos E, Ball MP, Church GM, Leonard DGB, Pereira S, McGuire AL, Caskey CT, et al. 2019. Predispositional genome sequencing in healthy adults: design, participant characteristics, and early outcomes of the PeopleSeq Consortium. *Genome Med* **11:** 10. doi:10.1186/s13073-019-0619-9

Cite this article as *Cold Spring Harb Perspect Med* doi: 10.1101/cshperspect.a036699

Index

A

ACA. *See* Affordable Care Act
AD. *See* Alzheimer's disease
ADA. *See* Americans with Disabilities Act
Adrenal carcinoma, genetic testing, 75
Affordable Care Act (ACA), 227
ALS. *See* Amyotrophic lateral sclerosis
Alzheimer's disease (AD)
    APOE testing, 49–50
    direct-to-consumer testing, 49–50
    gene mutations, 48
    genetic counseling, 48–49, 53
    polygenic risk scores, 50
    predictive testing, 48
Americans with Disabilities Act (ADA), 228
Amyotrophic lateral sclerosis (ALS), genetic testing
        and counseling, 51
Angelman syndrome, 63
Anxiety. *See* Psychiatric genetic counseling
APOE, 49–50, 294
APP. See Alzheimer's disease
Arrhythmogenic right ventricular cardiomyopathy
        (ARVC), 145, 148
ARVC. *See* Arrhythmogenic right ventricular
        cardiomyopathy
Assisted reproductive technologies. *See* Gamete donation;
        Intracytoplasmic sperm injection; In vitro
        fertilization; Preimplantation genetic testing
Attention deficit-hyperactivity disorder.
        See Neurodevelopmental disorders
Autism. *See* Neurodevelopmental disorders

B

BATHE assessment, 162–164
Bipolar disorder. *See* Psychiatric genetic counseling
BRAF, 99
BRCA1, 78, 83, 97, 101–102, 106
BRCA2, 78, 83, 101–102, 106
Breast cancer, genetic testing, 77–78, 83

C

CADD Score. *See* Combined Annotation
        Dependent Depletion Score
Cancer
    cell-free DNA, 105–106
    germline variant identification through somatic
        analysis, 106–110
Cancer. *See also specific cancers*

genetic counseling
    historical perspective, 83
    overview, 73
    pretest counseling, 84–86
    recent developments, 83–84
genetic testing
    biosample selection, 86
    direct-to-consumer testing, 78
    germline testing, 87
    guidelines, 76, 78–82
    indications, 74–76
    interpretation, 87–88
    laboratory selection, 86
    polygenic risk scores, 87
    test selection, 86–87
pediatric considerations in testing and
        counseling, 88–89
risk assessment tools, 77
tumor gene sequencing
    germline implications
        allelic fraction, 102
        discordant tumor types, 103
        filtration, 101
        founder variants, 102–103
        Lynch syndrome case example, 104–105
        overview, 101
        transcript use, 101–102
        tumor heterogeneity, 102
        tumor mutational burden, 103
        variant classification, 103–104
    techniques, 100–101
Cardiovascular disease. *See also* Sudden cardiac death
    genetic counseling
        BATHE assessment, 162–164
        coping promotion, 161–162
        diagnosis assimilation, 142–143
    genetic testing
        decision-making
            literature, 158–159
            overview, 156, 158
        indications, 156–157
        interpretation, 160–161
    psychological and social issues
        implantable cardioverter defibrillator
            decision-making, 147
            outcomes, 147–148
        research findings, 143–146, 159–160
        sports restrictions, 148–149
        sudden cardiac death
            grief, 149–151

Cardiovascular disease. (*Continued*)
  risks, 146–147
 syndromes, 141–142
CDH1, 88, 109
Cell-free DNA (cfDNA)
 prenatal screening
  aneuploidy, 27–29
  monogenetic disorders, 29–30
  overview, 15
  prospects, 34–36
  recommendations, 33
 tumors, 105–106
Cerebral palsy. *See* Neurodevelopmental disorders
cfDNA. *See* Cell-free DNA
CHARGE syndrome, 170
CHIP. *See* Clonal hematopoiesis of indeterminate
  potential
CHMP2B, 51
Chromosomal microarray analysis (CMA), 16, 31,
  61, 63, 170, 188
Clonal hematopoiesis of indeterminate potential
  (CHIP), 105
CMA. *See* Chromosomal microarray analysis
Combined Annotation Dependent Depletion (CADD)
  Score, 173
Counsyl Complete, 7
Cystic fibrosis
 carrier screening, 292–293
 genetic counseling, 187, 194
 newborn screening, 190
 prenatal diagnosis, 32
 treatment, 192

D

DCM. *See* Dilated cardiomyopathy
Decision-making, prenatal genetic testing
 challenges, 14–15
 educational video, 11–12
 historical perspective, 14
 nondirective counseling, 12
 patient coaching with decision aid, 20–21
 shared decision-making model
  decision aids, 17–21
  overview, 13–14
 uncertain information and informed decisions, 16
Depression. *See* Psychiatric genetic counseling
DGC. *See* Diffuse gastric cancer
Diagnostic workup
 confirmation and results disclosure, 191–192
 test selection and interpretation for pediatrics,
  188–189
 uncertainty, 189
Diffuse gastric cancer (DGC), genetic testing, 75, 108–109
Dilated cardiomyopathy (DCM), 143, 145, 158
Direct-to-consumer (DTC) testing
 Alzheimer's disease, 49–50
 cancer, 78
 legal issues, 230

 minors, 221
 personalized medicine, 294
Down syndrome, preimplantation genetic
  testing, 202–204, 255
DTC testing. *See* Direct-to-consumer testing
Duty to warn, at-risk relatives, 231–235

E

Eating disorders. *See* Psychiatric genetic counseling
ECS. *See* Expanded carrier screening
EGFR, 103
Elective genetic testing
 minors
  direct-to-consumer testing, 221
  exceptions to deferment, 217–218
  exome/genome sequencing, 218–220
  prospects, 222
  rationale for discouragement, 214–215
 position statements, 213–214
 prenatal testing
  exome/genome sequencing, 220–221
  overview, 216–217
 right not to know, 215–216
 value assessment, 218
EPCAM, 98–99
ES. *See* Exome sequencing
Eugenics, 14, 250
Everolimus, 69
Exome sequencing (ES), 33, 35–36, 170, 218–221
Expanded carrier screening (ECS), 30, 293

F

False-negative rate, cell-free DNA screening, 28–29
False-positive rate, cell-free DNA screening, 28
Family-centered counseling, pediatric common
  diseases, 187–188
Fetal cell, identification in maternal blood, 36
Fetal fraction (FF), cell-free DNA screening, 28–29
FF. *See* Fetal fraction
FISH. *See* Fluorescence in situ hybridization
FLCN, 97
Fluorescence in situ hybridization (FISH), 31, 203
FMR1, 61
Fragile X syndrome, 63
Frontotemporal dementia (FTD), genetic testing
  and counseling, 51–52
FTD. *See* Frontotemporal dementia

G

Gamete donation
 donors
  screening, 208–209
  types, 209–210
 genetic issues, 210
 overview, 207–208
Gastric cancer. *See* Diffuse gastric cancer

GBA, 52–53
GDD. *See* Global developmental delay
Genetic counseling, definition, 238–239
Genetic Information Assistant (GIA), 7
Genetic Information Nondiscrimination Act
       (GINA), 226–231, 234, 246
Genetic testing, genetic counseling relationship, 5
GIA. *See* Genetic Information Assistant
GINA. *See* Genetic Information Nondiscrimination Act
Global developmental delay (GDD).
       See Neurodevelopmental disorders

H

HCM. *See* Hypertrophic cardiomyopathy
HD. *See* Huntington's disease
Health Insurance Portability and Privacy Act
       (HIPAA), 229, 232–235
Hereditary nonpolyposis colorectal cancer syndrome.
       See Lynch syndrome
HIPAA. *See* Health Insurance Portability and Privacy Act
Humanism
   genetic counseling
      aligning with humanistic goals, 244–246
      nonhumanistic aspects, 241–243
      role, 239–241
   overview, 237–238
Huntington's disease (HD)
   gene discovery, 43–44
   genetic counseling for predictive testing
      prospects, 47–48
      protocol, 44–45
      session characteristics, 46–47
   predictive testing
      outcomes, 45–46
      protocol, 44–45
Hypertrophic cardiomyopathy (HCM), 143, 145,
       148–149, 158

I

ICD. *See* Implantable cardioverter defibrillator
ICSI. *See* Intracytoplasmic sperm injection
Implantable cardioverter defibrillator (ICD)
   decision-making, 147
   outcomes, 147–148
In vitro fertilization (IVF). *See also* Preimplantation
       genetic testing
   historical perspective, 199
   principles, 200–201
Informed consent
   alternative models, 281–283
   challenges in genomics era, 279–280
   genetic counseling role, 283–284
   genomic sequencing impact on models, 278–279
   historical perspective, 276–278
   incidental and secondary findings impact, 280–281
   prospects, 283–284
   secondary findings, 281

Intellectual disability. *See* Neurodevelopmental disorders
Intracytoplasmic sperm injection (ICSI), 200
IVF. *See* In vitro fertilization

J

Jar model, psychiatric disorders, 120–122

K

Kabuki syndrome, 69, 170
KRAS, 99

L

Laboratory genetic counselor
   conflicts of interest, 270–271
   overview, 263–264
   prospective roles
      business development, 266–267
      education of clients and staff, 267–269
      leadership and management, 267
      research, 271–272
      variant interpretation, 272–273
   roles
      case management, 265
      communication with clients, 264
      result reporting, 264–265
      test development, 265–266
   work settings, 269–270
Learning disorders. *See* Neurodevelopmental disorders
Legal issues, genetic testing, 225–235
LMNA, 156
Long QT syndrome (LQTS), 143, 146–147, 149, 160
LQTS. *See* Long QT syndrome
LRKK2, 52–53
Lynch syndrome
   clonal hematopoiesis of indeterminate potential, 105
   immunohistochemistry, 98–99
   microsatellite instability, 98
   mismatch repair screening, 99
   tumor gene sequencing
      germline implications
         allelic fraction, 102
         case example, 104–105
         discordant tumor types, 103
         filtration, 101
         founder variants, 102–103
         overview, 101
         transcript use, 101–102
         tumor heterogeneity, 102
         tumor mutational burden, 103
         variant classification, 103–104
      techniques, 100–101

M

MAPT, 51–52
Massively parallel sequencing (MPS), 35

MECP2, 62
Million Veterans Program (MVP), 291
MLH1, 98–99, 104
Molloy v. Meier, 232
MPS. *See* Massively parallel sequencing
MSH2, 98–99, 104
MSH6, 98–99, 104
MVP. *See* Million Veterans Program
MyCode, 291–292

N

NBS. *See* Newborn screening
NDD. *See* Neurodevelopmental disorders
Neurodevelopmental disorders (NDD)
    definitions, 60
    etiology, 60–61
    genetic counseling
        decision-making about testing, 65–66
        delivery of diagnosis, 67
        hope fostering, 68
        meaning-making, 68–69
        parental attitudes toward diagnosis, 63–64
        recurrence risk, 66–67
        relationship between genetics and disorders, 64–65
        social support opportunities, 69
        stigma, shame, and guilt, 69
        uncertainty management, 67–68
    genetic testing, 61–63
    treatment prospects, 69–70
Newborn screening (NBS), referrals, 189–191
Next-generation sequencing (NGS), 31–32, 62–63, 98, 100
NGS. *See* Next-generation sequencing
NOTCH1, 110

O

OPDG. *See* Ottawa Personal Decision Guide
Ottawa Personal Decision Guide (OPDG), 17–19
Ovarian cancer, genetic testing, 75, 77–78, 83

P

Pancreatic cancer, genetic testing, 75–77
Parkinson's disease (PD), genetic testing and
                counseling, 52–53
Pate v. Threlkel, 231–232
PD. *See* Parkinson's disease
Personalized medicine
    genetic counseling as keystone, 296–297
    overview, 289–290
    precision health, 290
    public perspective, 294–296
PGRN, 51
PGT. *See* Preimplantation genetic testing
Pheochromocytoma, genetic testing, 76
PMS2, 98–99, 104
POLD, 105–106
POLE, 105–106

Polygenic risk score (PRS)
    Alzheimer's disease, 50
    cancer, 87
Polygenic risk score, psychiatric disorders, 132–133, 136
Population genomic sequencing, initiatives, 290–294
Practice models, genetic counseling, 3–4
Precision Medicine Initiative, 291
Preimplantation genetic testing (PGT)
    aneuploidy, 202–204
    common pediatric diseases, 193
    historical perspective, 250
    HLA matching, 206–207, 249
    mitochondrial disease, 207, 258
    monogenetic disease, 205
    nondisclosing monogenetic testing, 205–206
    principles, 201–202
    regulation
        genetic counseling regulation, 255–256, 259
        medical necessity requirement, 256–257
        permissiveness degree, 252–253
        policy approach, 251–252
        reproductive tourism, 258
        sex selection, 257–258
        trends, 254
    sex selection, 250
    structural chromosome rearrangement, 204–205
    variants of uncertain significance, 207
Prenatal genetic testing. *See also specific diseases*
    counseling session components, 26–27
    decision-making. *See* Decision-making,
                prenatal genetic testing
    diagnostic tests
        aneuploidy, 31
        microdeletions/microduplications, 31–32
        monogenetic disorders, 32
        recommendations, 32–33
        techniques, 31
    patient perspectives, 36
    prenatal treatment options, 36–37
    prospects, 34–36
    screening tests
        aneuploidy, 27
        cell-free DNA, 27–30
        microdeletions, 29
        monogenetic disorders, 29–30
        multiple marker screening, 27
        recommendations, 32
        reproductive carrier screening, 30–31
PRS. *See* Polygenic risk score
PSEN1. *See* Alzheimer's disease
PSEN2. *See* Alzheimer's disease
Psychiatric genetic counseling
    appointment and follow-up, 118
    case load, 118
    diagnosis confirmation, 117–118
    historical perspective, 116–117
    outcome evaluation, 118, 125–126
    overview, 115–116

preappointment procedures, 117
process and content
    etiology, 120–122
    familial recurrence chances, 123–124
    initial contracting, 120
    overview, 118–119
    recovery, 122–123
    session closing, 124
    symptom impact on session, 124–125
prospects, 126
rationale, 116
risk assessment
    family history
        data collection, 133–134
        provision of risks, 136
        recurrence risks, 123–124, 134–136
    overview, 129–130
    polygenic risk score, 132–133
    prospects, 136–137
schizophrenia
    comorbidity with other disorders, 131
    genetics
        copy number variants, 131
        functional categories of genes, 131
        rare variants, 131
        variants with small effect sizes, 130–131
    nongenetic risk factors, 132
team, 117
training, 118
PTEN, 62

R

Rare disease
    definition, 169–170
    diagnostic odyssey, 177–179
    genetic counseling, 175
    genetic testing
        incidental and secondary findings, 175–176
        sequencing
            exome versus genome sequencing, 171

indications, 171–172
        limitations, 171
        neonates, 172
    techniques, 170
    variant classification, 172–175
neonatal intensive care unit genetic testing and
        counseling, 172, 179
Reciprocal engagement model (REM), tenets for genetic
        counseling, 2–3
RECQL1, 110
REM. See Reciprocal engagement model
RET, 83
Retinoblastoma, genetic testing, 76
REVEAL study, 49

S

Safer v. Estate of Pack, 231–232
Scalability, genetic counseling resources, 5–8
SCD. See Sudden cardiac death
Schizophrenia. See Psychiatric genetic counseling
SDM model. See Shared decision-making model
Service delivery models, genetic counseling, 5–7
Shared decision-making (SDM) model
    decision aids, 17–21
    overview, 13–14
Siblings, carrier testing, 192–193
SIDS. See Sudden infant death syndrome
SNCA, 52
SOD1, 51
Sperm donor. See Gamete donation
Sudden cardiac death (SCD)
    grief, 149–151
    psychological impact of young death, 144
    risks, 146–147
Sudden infant death syndrome (SIDS), 143

T

Tarasoff v. Regents of the University of California, 232–233
TP53, 88, 101